Fungicides: Risks and Management in Crop Production

Fungicides: Risks and Management in Crop Production

Editor: Chris Frost

www.callistoreference.com

Callisto Reference,
118-35 Queens Blvd., Suite 400,
Forest Hills, NY 11375, USA

Visit us on the World Wide Web at:
www.callistoreference.com

ISBN: 978-1-63239-905-2 (Hardback)

Cataloging-in-Publication Data

Fungicides : risks and management in crop production / edited by Chris Frost.
      p. cm.
Includes bibliographical references and index.
ISBN 978-1-63239-905-2
1. Fungicides. 2. Crop science. 3. Crops. 4. Agriculture. I. Frost, Chris.
SB951.3 .F86 2018
632.952--dc23

Table of Contents

Preface

Diseases are a common occurrence in plants, which results in a significant impact on their yield. Managing these diseases is an essential component in the production of crops. Fungicides are chemicals that are used to kill fungi. They are mainly required in order to stop the damage that can be caused by fungi resulting in the loss of yield in agriculture. Most of the topics introduced in this book cover new applications of fungicides. While understanding the long-term perspectives of the topics, the book makes an effort in highlighting their impact as a modern tool for the growth of the discipline. Those in search of information to further their knowledge in this field will be greatly assisted by this book.

After months of intensive research and writing, this book is the end result of all who devoted their time and efforts in the initiation and progress of this book. It will surely be a source of reference in enhancing the required knowledge of the new developments in the area. During the course of developing this book, certain measures such as accuracy, authenticity and research focused analytical studies were given preference in order to produce a comprehensive book in the area of study.

This book would not have been possible without the efforts of the authors and the publisher. I extend my sincere thanks to them. Secondly, I express my gratitude to my family and well-wishers. And most importantly, I thank my students for constantly expressing their willingness and curiosity in enhancing their knowledge in the field, which encourages me to take up further research projects for the advancement of the area.

Editor

Inhibition of Fungal Plant Pathogens by Synergistic Action of Chito-Oligosaccharides and Commercially Available Fungicides

Md. Hafizur Rahman, Latifur Rahman Shovan, Linda Gordon Hjeljord, Berit Bjugan Aam, Vincent G. H. Eijsink, Morten Sørlie, Arne Tronsmo*

Department of Chemistry, Biotechnology and Food Science, Norwegian University of Life Sciences (NMBU), Ås, Norway

Abstract

Chitosan is a linear heteropolymer consisting of β 1,4-linked N-acetyl-D-glucosamine (GlcNAc) and D-glucosamine (GlcN). We have compared the antifungal activity of chitosan with DP_n (average degree of polymerization) 206 and F_A (fraction of acetylation) 0.15 and of enzymatically produced chito-oligosaccharides (CHOS) of different DP_n alone and in combination with commercially available synthetic fungicides, against *Botrytis cinerea*, the causative agent of gray mold in numerous fruit and vegetable crops. CHOS with DP_n in the range of 15–40 had the greatest anti-fungal activity. The combination of CHOS and low dosages of synthetic fungicides showed synergistic effects on antifungal activity in both *in vitro* and *in vivo* assays. Our study shows that CHOS enhance the activity of commercially available fungicides. Thus, addition of CHOS, available as a nontoxic byproduct of the shellfish industry, may reduce the amounts of fungicides that are needed to control plant diseases.

Editor: Joy Sturtevant, Louisiana State University, United States of America

Funding: The work is funded by Norwegian University of Life Sciences (UMB), P.O. 5003, N-1432 Ås, Norway. The funders had no role in study design, data collection and analysis, decision to publish, or preparation of the manuscript.

Competing Interests: The authors have declared that no competing interests exist.

* E-mail: arne.tronsmo@nmbu.no

Introduction

Botrytis cinerea Pers.: Fr. (anamorph of *Botryotinia fuckeliana*) causes gray mold in over 200 plant species worldwide, which results in great damage to agricultural crops. For example, in Bangladesh, gray mold has caused near complete yield losses of chickpea [1] and in Norway the pathogen causes 30–60% yield reductions in strawberry production [2]. Other economically important plant pathogenic fungi include *Mucor piriformis* Fischer, causing postharvest rots on strawberries as well as on several other fruit crops [3–4], and *Alternaria brassicicola* (Schw.) Wiltshire, causing black spot on crucifers [5]. The control of these plant pathogens relies heavily on synthetic fungicides. Excessive use of synthetic fungicides has caused environmental pollution and development of fungicide resistance in plant pathogens [6]. Thus, there is a need to reduce the use of synthetic fungicides by increasing their efficacy or by finding alternatives.

Chitin, a linear biopolymer consisting of β 1,4-linked N-acetyl-D-glucosamine (GlcNAc) residues, is insoluble in water, aqueous acidic solutions and most organic solvents due to strong intra- and inter-chain hydrogen bonds [7]. The fraction of acetylation (F_A) of chitin is usually above 0.90 [8], meaning that there are very few D-glucosamine (GlcN) units present. Chitosan, which is obtained by partial deacetylation of chitin, is a heteropolymer consisting of GlcNAc and GlcN residues. Chitosan with an F_A of around 0.65 or lower is soluble in aqueous acid solutions [7–9], Both chitin and chitosan can be hydrolyzed into chito-oligosaccharides (CHOS) by synthetic or enzymatic methods. CHOS are known to have several

beneficial biological effects and may be used as fungicides, bactericides, bone-strengthener in osteoporosis, vector for gene delivery, hemostatic agent in wound-dressings, antimicrobial agents, and as inducer of plant defense responses against pathogens [10–12].

Hydrolysis of chitosan into CHOS can be done chemically or by glycosyl hydrolases (GH) classified as chitinases or chitosanases [13]. Chitinases are found in the GH families 18 and 19. Besides chitin, these enzymes also hydrolyze chitosans to varying extents, depending on the F_A [14–15]. Chitosanases are found in GH families 5, 7, 8, 46, 75 and 80 (see www.cazy.org for more details on the classification). Of these, the GH46, GH75 and GH80 families only contain chitosanases and the GH46 enzymes are probably the best studied. The key difference between chitinases and chitosanases is that only chitosanases can cleave GlcN-GlcN bonds and only chitinases can cleave GlcNAc- GlcNAc bonds. Apart from this clear difference the enzymes have varying and to some extent overlapping cleavage specificities that have been analyzed in several studies (the term "cleavage specificity" alludes to the specific sequences of GlcNAc and GlcN sugars that are being cleaved) [10].

Recently, we showed that CHOS fractions of DP_n 40 and DP_n 23 obtained from enzymatic hydrolysis of a chitosan ($F_A = 0.15$; $DP_n = 206$) by a family 46 chitosanase [16] significantly inhibited germination of isolates of *B. cinerea* and *M. piriformis* [17]. In the present study, we have investigated the antifungal effects that can be obtained by combining such CHOS with commercially available synthetic fungicides. To test anti-fungal effects, we have primarily studied inhibition of *B. cinerea in vitro* and *in vivo*, but effects on other fungal pathogens have also been addressed. Our results reveal remarkable synergistic effects of combining CHOS with synthetic fungicides, thus opening up new avenues towards the use of these oligosaccharides in environmentally benign plant protection strategies.

Materials and Methods

Fungal Cultures

B. cinerea (isolate BC 101), *A. brassicicola* (isolate A 328), and *M. piriformis* (isolate M119J) were obtained from the culture collection at the Norwegian University of Life Sciences (NMBU). For the *in vitro* and *in vivo* bioassays, conidia were collected from cultures grown on potato dextrose agar (PDA) (Difco Laboratories, Detroit, MI) under regular laboratory light for 2 weeks at $23 \pm 1°C$. Concentrations of conidia in aqueous suspensions were determined by haemocytometer count at $400 \times$ magnification and adjusted to the required concentration with sterile water.

Synthetic Fungicides

Five fungicides were tested: (1) Teldor® WG 50 (Bayer Crop Science Pty Ltd., Germany); active ingredient: 500 g kg^{-1} fenhexamid; chemical group: hydroxyanilide. (2) Switch® 62.5 WG (Syngenta Crop Protection Pty. Ltd., Switzerland); active ingredients: 375 g kg^{-1} cyprodinil and 250 g kg^{-1} fludioxonil; chemical groups: anilinopyramidine and phenylpyrrole respectively. (3) Amistar® (Syngenta Crop Protection Pty. Ltd.); active ingredient: 500 g kg^{-1} azoxystrobin; chemical group: strobilurin. (4) Signum® WG (BASF, Germany); active ingredients: 26.7% w/w boscalid and 6.7% w/w pyraclostrobin; chemical groups: pyridinecarboximide and methoxy-carbamate, respectively. (5) Delan® (BASF, Germany); active ingredient: 700 g kg^{-1} dithianon; chemical group: quinone.

Enzymatic Production of CHOS

Chitosan (KitoNor, F_A 0.15, DP$_n$ 206) was obtained from Norwegian Chitosan, Gardermoen, Norway. This chitosan was used for all experiments in this work. CHOS were produced by enzymatic hydrolysis of chitosan. Chitosanase ScCsn46A was produced as described by Heggset and coworkers [16]; briefly, the chitosanase, originally from *Streptomyces coelicolor* (UniProt accession code q9rj88), was purified from the culture supernatant of a recombinant *Escherichia coli* BL21Star (DE3) strain, following the published protocol, without removal of the (His)$_6$-tag after purification. The enzyme was dialyzed against 20 mM Tris-HCl, pH 8.0, and stored at 4°C. Chitinase ChiA from *Serratia marcescens* was produced according to Brurberg and coworkers [18].

Chitosan (10 mg mL^{-1}) in 0.04 M NaAc, 0.1 M NaCl, 1% HCl was incubated at 37°C and 225 rpm until the chitosan was dissolved (approximately 15 min). The pH was then adjusted to 5.5 with 0.5 M NaOH.

ScCsn46A [16] or ChiA [18] (0.5 μg mg^{-1} chitosan) were added to the chitosan solution and the mixture was incubated for various lengths of time at 37°C and 225 rpm. The enzymatic reaction was stopped by decreasing the pH to 2.5 with 0.5 M HCl, followed by immersing the tube in boiling water for at least 10 minutes to permanently inactivate the enzymes. CHOS samples were dialyzed against distilled water for 48 hours (water was changed every 12 hours) using a cellulose membrane (Float-A-Lyzer® MWCO 500 Da from Spectrum Labs, USA) to remove buffer salts from the sample. Dialyzed samples were sterile filtered through Filtropur S 0.2 μm sterile filters (Sarstedt, Germany), lyophilized and stored at 4°C [10]. ChiA was used to produce CHOS with predominantly GlcNAc on the reducing ends and ScCsn46A was used to produce CHOS with predominantly GlcN on the reducing ends (ChiA has an absolute preference for cleaving after GlcNAc [19]; ScCSn46 has a strong but not absolute preference for cleaving after a GlcN, and will essentially only cleave after GlcN under the conditions used here) [16]. It is important to note that the degree of degradation of chitosan cannot be monitored online (^{1}H-NMR needs to be used; see below). This complicates reproducible production of CHOS batches with identical DP$_n$ and explains why CHOS batches used in this study show slight variations in DP$_n$.

1H-NMR Analysis °f CHOS

Lyophilized CHOS (10 mg) were dissolved in deuterium oxide (D$_2$O) (0.5 mL) and the pH was adjusted to 4.2 with sodium deuteroxide (NaOD) prior to lyophilization. The lyophilized CHOS was redissolved in D$_2$O and lyophilized again to secure that all the H$_2$O had been removed. Finally the lyophilized CHOS were dissolved in D$_2$O (700 μL) and ^{1}H-NMR analysis was performed on a 300 MHz Varian Gemini instrument (Varian, USA) at 85°C. The DP$_n$ was calculated by the equation (Dα+Dβ+D+Aα+Aβ+A)/(Dα+Dβ+Aα+Aβ), where Dα, Dβ, Aα and Aβ are the integrals of the reducing end signals of the α and β anomers of the deacetylated (D, GlcN) and acetylated (A, GlcNAc) units respectively, D is the integral of the signals from GlcN in internal positions and non-reducing end positions, and A is the integral of the signals from GlcNAc in internal and non-reducing end positions [14].

Separation of CHOS by Size Exclusion Chromatography (SEC)

A CHOS sample (100 mg) generated by enzymatic hydrolysis of chitosan (DP$_n$ 206) with ScCsn46A was applied to three Superdex™ 30 columns (XK columns from GE Healthcare) coupled in series with an overall dimension of 2.6×180 cm. The flow rate of the mobile phase (0.15 M NH$_4$Ac, pH 4.5) was maintained at 0.8 ml min^{-1} [14]. A refractive index detector (Gilson model 133, UK) was used to monitor the relative amounts of the CHOS fractions.

Effect of CHOS on Germination of B. cinerea, A. brassicicola and M. piriformis

Activity against *B. cinerea* was assessed using minimal salt medium (MSM) pH 5.2, with the following final concentrations: 2.5 mM NH$_4$NO$_3$; 0.28 mM CaCl$_2$·2H$_2$O; 0.16 mM MgSO$_4$·7H$_2$O; 0.002 mM MnSO$_4$·4H$_2$O; 0.002 mM ZnSO$_4$·7H$_2$O; 1 mM KH$_2$PO$_4$; 0.06 mM FeC$_6$H$_5$O$_7$·5H$_2$O and 55.5 mM glucose. Experiments were set up by adding 100 μL of CHOS or chitosan dissolved in 2×MSM to a 100 μL conidial suspension (2×10^4 conidia mL^{-1} in water), in wells of a flat-bottom 96-well microtiter plate (Nunc™, Roskilde, Denmark). There were three replicate wells for each treatment. The microtiter plates were incubated at $23 \pm 1°C$ for 24 hours. Germination was visually estimated at 400×magnification using an invert microscope (Fluovert FU, Ernst Leitz Wetzlar GmbH, Wetzlar, Germany). The conidia were scored as germinated when the germ tube length was at least as long as the diameter of the conidium.

The germination inhibition percentage was calculated by the following equation:

$$\text{Germination inhibition}(\%) = \{(a-b)/a\} \times 100$$

Where, a = number of germinated conidia in the control (conidia in MSM) b = germinated conidia in the treatment (conidia and chitosan/CHOS and/or fungicides in MSM).

The pH of the conidia suspension in the microtiter wells with and without CHOS was between 5.2 and 5.3 at the start of the experiment, and remained about the same 24 hours after inoculation.

Activity against *M. piriformis*, and *A. brassicicola* was tested in the same manner. Germinated *M. piriformis* M199J conidia showed abnormal swelling with amoeba-like structures and one or more protrusions. These conidia were counted as germinated if the length of at least one of the protrusions was at least as long as the diameter of the swollen conidia 12 hours after inoculation. Conidia of *A. brassicicola* were counted as germinated when the length of the germ tube was half of the conidia length.

Synergism Between Fungicides and Chitosan or CHOS in Inhibiting *B. cinerea* and *M. piriformis in vitro*

Germination experiments were set up as described above, meaning that 100 µL of the to-be-tested samples were added to 100 µL of a conidia suspension in MSM. The samples were: a) control treatment (only MSM), b) chitosan or CHOS in MSM, c) chitosan or CHOS combined with synthetic fungicides (Teldor, Switch, Amistar or Signum) in MSM, and d) individual synthetic fungicides in MSM.

The interaction between synthetic fungicides and chitosan or CHOS was determined using Abbott's equation for synergy calculation [20]. The synergistic effect was calculated by determining the ratio between the observed efficacy E_{obs} (% inhibition) and the expected efficacy (E_{exp}): $E_{exp} = a+b - (ab/100)$. Here a = % germination inhibition by synthetic fungicides (Teldor, Switch, Amistar or Signum) alone, b = % germination inhibition by chitosan or CHOS alone. An E_{obs}/E_{exp} ratio equal to 1 indicates additivity, ratios >1 indicate synergy, and ratios <1 indicate an antagonistic interaction [20].

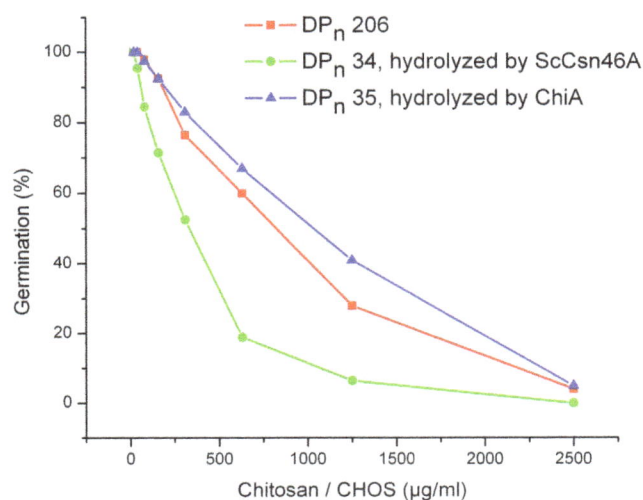

Figure 1. Effect of the reducing end sugars (GlcN vs GlcNAc) on the ability of CHOS to inhibit germination of *B. cinerea* (measured 24 hours after inoculation). Squares: chitosan, DP_n 206, 85 % D at the reducing end; circles: CHOS generated by ScCsn46A, DP_n 33.5, >90% GlcN at the reducing end; triangles: CHOS generated by ChiA, DP_n 34.6, about 35% GlcNAc at the reducing end. Data points represent the mean of three replicate wells.

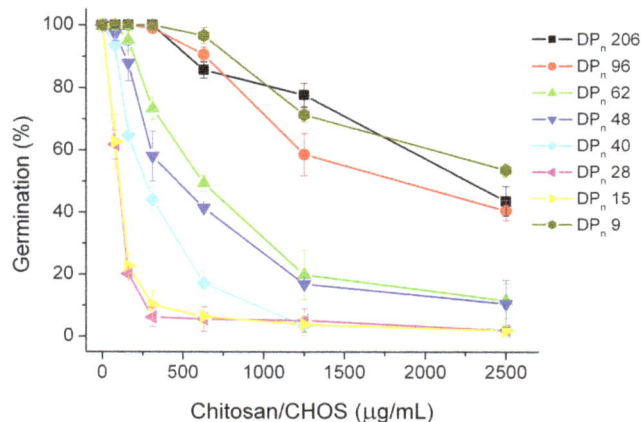

Figure 2. Effect of chitosan (DP_n 206) or CHOS obtained by hydrolysis of chitosan with ScCsn46A on germination of *Botrytis cinerea* (measured 24 hours after inoculation). The data points are the mean of three experiments with standard deviation.

In Vivo Bioassay: Inhibition of Infection of Strawberry Flowers and Chickpea Leaves By *B. cinerea*

Synergism between synthetic fungicides and chitosan or CHOS in inhibiting flower infection by *B. cinerea* was tested on newly opened strawberry (*Fragaria × ananassa*) flowers (cv. Corona). Strawberry plants were grown in a greenhouse with controlled temperature (18°C day; 12°C night), light (16 hours, light intensity: 150 µmols m^{-2} sec^{-1}) and relative humidity (65%). Newly opened flowers were cut off with a 1½-2 cm stem and placed in empty pipette tip racks placed in plastic containers filled with 1–2 cm water. After mixing the conidia suspension (final concentration $1×10^6$ conidia mL^{-1}) with each test solution, 10 µL drops of the mixtures were placed at the base of three petals on each flower using an automatic pipette (Finnpipette 4027, Thermo Labsystems, Finland). There were six replicates of three flowers (i.e., nine inoculation points per replicate) for each treatment. The

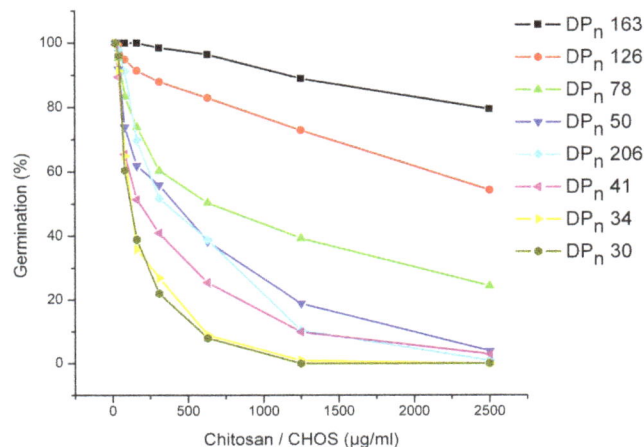

Figure 3. Dose-response relationships for the inhibitory effect of chitosan DP_n 206 and various CHOS fractions on germination of *Botrytis cinerea* (measured 24 hours after inoculation). CHOS DP_n 34 was produced by hydrolysis of chitosan (DP_n 206) with ScCsn46A. CHOS DP_n 34 was separated by size exclusion chromatography to fraction with DP_n 30, 41, 50, 78, 126 and 163.

sets of inoculated flowers were randomized and placed in containers covered with aluminium foil and incubated at $23\pm1°C$. The relative humidity around the flowers was 90–95%, as measured using a thermo-hygrometer (Lambrecht, Germany). The experiments were repeated once. The infection incidence was determined visually: necrotic regions on the abaxial surface of the flowers (under the inoculation points) were registered daily for eight days and recorded as percent infected inoculation sites. The area under the disease progress curve (AUDPC) was calculated on the basis of the accumulated percent infection by the following equation:

$$AUDPC = \sum_i (D_i - D_{i-1}) \times \{S_{i-1} + 0.5(S_i - S_{i-1})\}$$

where i = number of assessment, D_i = day of the ith assessment and S_i = percent infected inoculation points at the ith assessment.

The protection index was calculated using the AUDPC values in the following formula [21]:

$$100 \times (AUDPC_{control} - AUDPC_{treatment})/AUDPC_{control}$$

where $AUDPC_{control}$ is derived from infection in flowers inoculated with *B. cinerea* conidia alone and $AUDPC_{treatment}$ is derived from infection in flowers treated with synthetic fungicides and/or CHOS premixed with *B. cinerea* conidia.

Similar tests were performed using detached chickpea (*Cicer arientinum* L.) leaves. Chickpea were grown in the green house at $22\pm3°C$ under twelve hours light. Three compound chickpea leaves were used for each treatment and each chickpea leaf had one inoculation point on six of its leaflets. There were three replicates of each treatment. The chickpea leaves were inoculated with 10 μL drops of a 2×10^6 mL^{-1} suspension of *B. cinerea* conidia in water, supplemented with sterile water (control) or solutions of the to-be-tested compounds in sterile water. The infection was recorded when a brown (necrotic) spot appeared under the inoculation point, and the cumulative disease development was recorded daily up to eight days after inoculation.

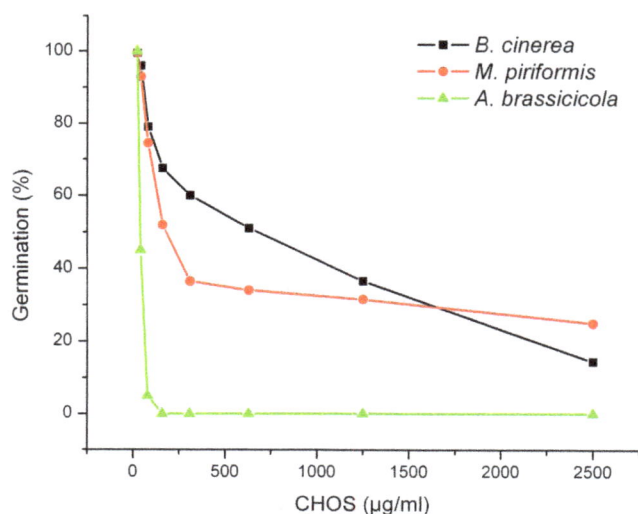

Figure 4. Dose-response relationships for the inhibitory effect of CHOS (DP$_n$ 37) on germination of *Botrytis cinerea*, *Alternaria brassicicola* and *Mucor piriformis* (measured 24 hours after inoculation).

Table 1. Inhibition of germination of *Botrytis cinerea* by chitosan (DP$_n$ 206) and synthetic fungicides, alone and in combination.

Treatment (μg mL^{-1})	Germination inhibition (% ± SD)[a]	E$_{obs}$/E$_{exp}$[b]
Chitosan 80	10.4±2.5	-
Teldor 60	1.3±0.6	-
Teldor 15	1.3±0.6	-
Chitosan 80 + Teldor 60	64.2±8.9	4.8
Chitosan 80 + Teldor 15	35.6±5.2	3.0
		-
Switch 25	74.8±6.0	-
Switch 5	35.3±4.2	-
Chitosan 80 + Switch 25	91.6±7.8	1.2
Chitosan 80 + Switch 5	56.2±7.9	1.3
		-
Amistar 100	20.0±4.5	-
Amistar 10	3.6±3.6	-
Chitosan 80 + Amistar 100	31.7±1.7	1.0
Chitosan 80 + Amistar 10	14.8±3.8	1.1
		-
Signum 10	17.7±1.5	-
Signum 2	3.8±3.1	-
Chitosan 80 + Signum 10	24.7±4.2	1.1
Chitosan 80 + Signum 2	17.8±5.5	1.3

Germination was recorded 24 hours after inoculation.
[a]All data are the mean of three experiments ± standard deviation
[b]An E$_{obs}$/E$_{exp}$ ratio of 1 indicates additivity; ratios >1 indicate synergy.

Sporulation of *B. cinerea* on the chickpea leaves was recorded at the end of the experiment. To do so all leaves from each treatment were soaked in sterile water (10 mL) for 20 min at 25°C and vortexed several times. Subsequently, the conidia concentration in the water was determined by counting in a hemacytometer.

Field trial: Inhibition of Infection of Apple Fruits by *Venturia ineuqualis*

Apple trees (*Malus domestica* Broch) of the cultivar Aakerø in the apple orchard at the Norwegian University of Life Sciences, Ås, Norway were used. The experiment was conducted in 2013 and there were three replicates of each treatment and three trees in each replicate. The trees were sprayed to runoff once in the flowering period (28Th of May) and three times in the fruiting season (24th of June, 7th of July and 17th of August). At harvest (3rd of September) the fraction of apples with infection of apple scab (*Venturia inaequalis*) was recorded.

Data Analysis

In the microtiter plate assay, the percentages of germination inhibition of pathogens by chitosan and CHOS were transformed by arcsine transformation and tested by one way ANOVA (only non transformed data are presented). In the strawberry flower assay, the AUDPC was calculated based on cumulative daily infection from one to eight days, and tested by one way ANOVA. When appropriate, means were separated by Tukey's Honestly Significant Difference method. All statistical analysis was done

Table 2. Effect of chitooligosaccharides (CHOS DP_n 23) and synthetic fungicides on germination inhibition of *Botrytis cinerea*.

Treatment (µg mL^{-1})	Germination inhibition (% ± SD)[a]	E_{obs}/E_{exp}[b]
CHOS 5	4.8±3.7	-
Teldor 150	4.4±4.2	-
Teldor 15	0.6±0.9	-
CHOS 5 + Teldor 150	21.0±5.1	2.3
CHOS 5 + Teldor 15	21.7±5.8	4.0
Switch 25	81.7±4.9	-
Switch 5	18.3±11.7	-
CHOS 5 + Switch 25	94.2±5.7	1.1
CHOS 5 + Switch 5	96.2±2.7	4.3
Amistar 100	4.6±1.7	-
Amistar 10	1.6±0.7	-
CHOS 5 + Amistar 100	95.6±3.4	10.4
CHOS 5 + Amistar 10	96.4±3.6	15.3
Signum 10	1.6±0.8	-
Signum 2	1.7±1.7	-
CHOS 5 + Signum 10	93.2±7.1	16.5
CHOS 5 + Signum 2	89.0±7.0	15.8

Germination was recorded 24 hours after inoculation.
[a]All data are the mean of three experiments ± standard deviation
[b]An E_{obs}/E_{exp} ratio of 1 indicates additivity; ratios >1 indicate synergy.

Table 3. Inhibition of disease development in strawberry flowers inoculated with a mixture of *Botrytis cinerea* conidia and chitosan (DP_n 206) and/or synthetic fungicides.

Treatment (µg mL^{-1})	AUDPC (± SD)[a]	Protection index (% ± SD)[a,b]
Control[c]	5.0±0.2	-
Chitosan 400	3.8±0.2	24±3
Teldor 1500[d]	1.5±0.3	70±6
Teldor 15	3.8±0.5	23±7
Chitosan 400 + Teldor 15	2.4±0.2	53±11
Switch 500[d]	± 0.2	80±5
Switch 5	3.2±1.0	36±10
Chitosan 400 + Switch 5	2.1±0.3	58±4
Amistar 1000[d]	2.0±0.3	60±5
Amistar 10	3.5±0.1	31±2
Chitosan 400 + Amistar 10	2.0±0.3	60±4
Signum 1000[d]	1.3±0.3	74±5
Signum 10	3.7±0.2	26±7
Chitosan 400 + Signum 10	2.5±0.1	50±4

Disease development was scored as development of visual necrotic regions under the inoculation point up to eight days after inoculation and is quantified as the area under the disease progress curve (AUDPC). The protection index was calculated on the basis of the AUDPC values.
[a]All data are the mean of two experiments ± standard deviation with 6×3 flowers in each treatment.
[b]The AUDPC was used to calculate the protection index.
[c]Conidia in sterile water.
[d]Recommended dose.

using Microsoft Office Excel 2007 and Minitab 16 (MINITAB, USA).

Results

Enzymatic Production of CHOS

CHOS were produced by degrading chitosan (DP_n of 206 and F_A of 0.15) with either ChiA or ScCsn46A, as described above. By varying the incubation time CHOS fractions with DP_n values between 96 and 9 could be obtained. Note that the determination of DP_n needs to be done (by NMR) after the enzymatic reaction has been concluded, explaining why it is difficult to produce CHOS fractions with exactly the same DP_n. Since there are indications in the literature that the biological effects of CHOS depend not only on DP and F_A, but also on the pattern of acetylation [10] we initially tested the effect of the only controllable aspect of this pattern, namely the sugar at the reducing end of the CHOS. Hydrolysis by ChiA yields GlcNAc at the reducing end, whereas ScCsn46A almost exclusively yields GlcN.

Effect of the Reducing End Sugar on the Ability of CHOS To Inhibit Germination of *B. cinerea*

To test the effect of the reducing end sugars (GlcN vs GlcNAc) on the antifungal activity of CHOS, we tested the efficacy of chitosan (DP_n 206, 85% GlcN at the reducing ends), CHOS DP_n 33.5 prepared with ScCsn46A (>90% GlcN at the reducing ends),

and CHOS DP_n 34.6 prepared with ChiA (about 35% GlcNAc at the reducing end). Figure 1 shows that CHOS produced with ScCsn46A were more effective than CHOS produced with ChiA. Based on these observations all further studies were done with CHOS obtained from degradation of chitosan (DP_n 206; F_A 0.15) with ScCsn46A.

The Effect of the Degree of Polymerization on Inhibition of Germination of Fungal Conidia

Studies of the inhibitory effect of chitosan/CHOS with different DP_n (206 − 9) on *B. cinerea* germination showed that the most active fractions of CHOS had DP_n values around 28, but that also other CHOS samples with DP_n values in the range of 15 to 40 had good antifungal activities. All tested CHOS fractions (except DP_n 9) were more inhibitory than the chitosan (Fig. 2).

To investigate the antifungal effect of CHOS with a narrower range of chain lengths than could be obtained by hydrolyzing with ScCsn46A for various lengths of time, a CHOS mixture obtained by hydrolyzing chitosan (DP_n 206) with ScCsn46A to DP_n 34 was sub fractionated using size exclusion chromatography. The DP_n values of the resulting CHOS fractions were determined using NMR. Figure 3 shows that the fractions with DP_n in the range of 78 − 163 were less inhibitory to *B. cinerea* than the starting material chitosan, whereas the most inhibitory CHOS fractions were those with DP_n 30 and 34.

In another set of experiments, the effects of a CHOS fraction with DP_n 37 on germination of plant pathogenic fungi belonging

Table 4. Inhibition of disease development in strawberry flowers inoculated with a mixture of *Botrytis cinerea* conidia and chitooligosaccharides (CHOS DP$_n$ 23) and/or synthetic fungicides.

Treatment (µg mL^{-1})[a]	AUDPC (± SD)[b]	Protection index (% ± SD)[b,c]	E_{obs}/E_{exp}
Control[d]	4.7±0.2	-	-
CHOS 10	4.4±0.2	5±3	-
Teldor 150	2.8±0.5	39±11	-
Teldor 15	4.4±0.1	5±1	-
CHOS 10 + Teldor 150	0.6±0.2	87±5	2
CHOS 10 + Teldor 15	0.9±0.4	80±8	8
			-
Switch 25	4.3±0.1	9±3	-
Switch 5	4.5±0.2	3±1	-
CHOS 10 + Switch 25	0.6±0.4	87±4	6
CHOS 10 + Switch 5	0.4±0.4	92±8	12
			-
Amistar 100	4.5±0.2	3±1	-
Amistar 10	4.6±0.2	1±1	-
CHOS 10 + Amistar 100	0.9±0.3	79±8	10
CHOS 10 + Amistar 10	0.9±0.4	80±10	13
			-
Signum 10	4.4±0.1	4±2	-
Signum 2	4.6±0.1	2±1	-
CHOS 10 + Signum 10	0.7±0.3	85±7	10
CHOS 10 + Signum 2	0.6±0.4	86±8	12

Disease development and protection index were scored as in Table 3. The synergistic effect was calculated by determining the ratio between the observed efficacy E_{obs} (% inhibition) and the expected efficacy (E_{exp}) (see materials and methods). An E_{obs}/E_{exp} value of 1 indicates additivity, while E_{obs}/E_{exp}>1 indicates synergy.
[a]The recommended doses for the synthetic fungicides are 1500, 500, 1000 and 1000 µg mL^{-1} for Teldor, Switch, Amistar and Signum, respectively.
[b]All data are the mean of two experiments ± standard deviation, with 6×3 flowers in each treatment.
[c]The AUDPC was used to calculate the protection index.
[d]Conidia in sterile water.

to three different genera were tested. The three fungi showed quite different dose-response relationships (Fig. 4). While *B. cinerea* and *M. piriformis* showed decreasing germination over a broad concentration range of CHOS (20–2500 µg mL^{-1}), *A. brassicicola* was completely inhibited by 80 µg mL^{-1} CHOS. 50% germination inhibition of *A. brassicicola* was obtained at 40 µg mL^{-1}, whereas CHOS concentrations of 630 µg mL^{-1} and 160 µg mL^{-1} were needed to obtain 50% inhibition of *B. cinerea* and *M. piriformis*, respectively.

In Vitro Testing of the Effects of Combining Synthetic Fungicides with Chitosan or CHOS

Table 1 shows germination-inhibition data for *B. cinerea* treated with chitosan or a combination of chitosan and one of four synthetic fungicides, Teldor, Switch, Amistar or Signum. The data show minor effects when adding chitosan alone (10% inhibition, at 80 µg mL^{-1}), and reveal small synergistic effects for most of the combinations (Table 1). In the case of Teldor, however, quite strong synergistic effects were observed. For example, while application of 60 µg mL^{-1} Teldor alone gave only 1.3% inhibition, co-administration with 80 µg mL^{-1} chitosan yielded as much as 64% inhibition.

Strikingly, similar experiments on inhibition of germination of *B. cinerea* with CHOS (DP$_n$ 23) showed large synergistic effects for almost all combinations of CHOS (DP$_n$ 23) and the synthetic fungicides (Table 2). While CHOS alone (5 µg mL^{-1}) and the

synthetic fungicides alone, each applied at low concentrations, generally only slightly inhibited germination, in several cases more than 90% inhibition could be obtained by combining the two types of anti-fungal compounds. For example Amistar (10 µg mL^{-1}) and Signum (10 µg mL^{-1}) applied alone gave only 1.6% germination inhibition; upon addition of CHOS (5 µg mL^{-1}; yielding 4.8% inhibition when applied alone), germination inhibition increased to 96% and 93%, respectively (Table 2).

In Vivo Testing of the Effects of Combining Synthetic Fungicides with Chitosan or CHOS

In the strawberry flower assay, chitosan (400 µg mL^{-1}) gave approximately the same level of protection against *B. cinerea* as the synthetic fungicides applied at 1% of the recommended dose (Table 3). Clear synergistic effects were not observed. Interestingly though in one case (Amistar), the combination of the synthetic fungicide at 1% of the recommended dose and chitosan (400 µg mL^{-1}) yielded a level of protection that was similar to the protection level achieved by the recommended dose of fungicide (Table 3).

Like chitosan (DP$_n$ 206, 400 µg mL^{-1}), CHOS (DP$_n$ 23, at the low concentration of 10 µg mL^{-1}) hardly inhibited flower infection by *B. cinerea*, but combinations of CHOS with the synthetic fungicides revealed large synergistic effects and showed that effective inhibition of infection could be achieved with low concentrations of both CHOS and synthetic fungicides (Table 4).

Figure 5. Combined anti-fungal effects of CHOS and synthetic fungicides. The pictures illustrate the inhibitory effects of combinations of a synthetic fungicide (Teldor, Switch, Amistar or Signum, at 15, 5, 10 and 10 $\mu g\ mL^{-1}$, respectively) and CHOS (DP_n 23, 10 $\mu g\ mL^{-1}$) on disease caused by *Botrytis cinerea* applied to detached strawberry flowers, six days after inoculation. The flowers were considered 100% infected when all three inoculation points displayed necrotic signs. All treatments included 18 flowers, but only nine flowers are shown. Control flowers were inoculated with conidia in sterile water.

When co-administrated with 10 $\mu g\ mL^{-1}$ CHOS, the protection levels achieved with the synthetic fungicides at 1% of the recommended concentration were 80%, 92%, 80% and 85% for Teldor, Switch, Amistar and Signum, respectively.

In the control treatment (no anti-fungal compounds added) 100% of the strawberry flowers showed signs of infection 3 – 4 days after inoculation and a similar result was obtained when CHOS (DP_n 23, 10 $\mu g\ mL^{-1}$), Teldor (15 $\mu g\ mL^{-1}$), Switch (5 $\mu g\ mL^{-1}$), Amistar (10 $\mu g\ mL^{-1}$) or Signum (10 $\mu g\ mL^{-1}$) were

applied alone. However, when the inoculated flowers were treated with combinations of CHOS (DP_n 23) and synthetic fungicides (at the mentioned concentrations) no visible infection occurred before six days after inoculation (Fig. 5).

In a chickpea leaf bioassay, chitosan, CHOS (DP_n 30) and Switch were used alone and in combination against *B. cinerea* (Table 5). The combinations of chitosan (320 $\mu g\ mL^{-1}$) or CHOS DP_n 30 (320 $\mu g\ mL^{-1}$) and Switch (1% of the recommended dose) showed synergism, albeit less strongly than in the strawberry

Table 5. Effect of combinations of chitosan (DP$_n$ 206) or chito-oligosaccharides (CHOS DP$_n$30) and Switch on *Botrytis cinerea* infection of detached chickpea leaves.

Treatment (µg mL^{-1})	AUDPC (± SD)[a]	Protection index (% ± SD)[a,b]	E$_{obs}$/E$_{exp}$[c]
Control[d]	6.5	-	-
Chitosan 2500	4.4±0.4	33±7	-
Chitosan 320	6.1±0.2	5±2	-
CHOS 2500	2.8±0.4	58±7	-
CHOS 320	5.5±0.1	15±2	-
Switch 500[e]	0.1±0.1	98±1	-
Switch 10	1.3±0.5	80±7	-
Switch 5	3.5±0.4	46±6	-
Chitosan 320 + Switch 10	1.4±0.1	79±2	1
Chitosan 320 + Switch 5	1.3±0.2	80±4	2
CHOS 320 + Switch 10	0.3±0.1	96±1	1
CHOS 320 + Switch 5	0.7±0.3	90±5	2

Disease development was scored daily up to eight days after inoculation.
[a]All data are the mean of three replicates (each replicate contained three compound leaves with 6 inoculated leaflets) ± standard deviation.
[b]The AUDPC was used to calculate the protection index.
[c]E$_{obs}$/E$_{exp}$ 1 indicates additivity; E$_{obs}$/E$_{exp}$>1 indicates synergy.
[d]Conidia in sterile water.
[e]Recommended dose.

flower assay. The combination of CHOS (320 µg mL^{-1}) and Switch (10 µg mL^{-1}) was almost as protective (96%) as the recommended dose of Switch (500 µg mL^{-1}; 98% protection). CHOS consistently showed better effects than chitosan.

Similar studies with Signum (Fig. 6) showed no synergistic effects, but the effects of chitosan (320 µg mL^{-1}) or CHOS DP$_n$ 30 (320 µg mL^{-1}) and Signum (5 or 10 µg mL^{-1}) were additive, meaning that also in this case chitosan or CHOS may be used to reduce usage of the synthetic fungicide. For example, the combination of 320 µg mL^{-1} CHOS (DP$_n$ 30) and 10 µg mL^{-1} Signum (1% of recommended dose) yielded 98% inhibition. Again, CHOS consistently showed better effects than chitosan.

Sporulation of the plant pathogenic fungus on infected plant parts is an important source of secondary infections. Therefore,

experiments were performed to assess the effects of the combination of chitosan or CHOS (DP$_n$ 30) with Signum on the sporulation of *B. cinerea* on infected chickpea leaves. As shown in Table 6, combinations of chitosan or CHOS (DP$_n$ 30) and Signum reduced sporulation of *B. cinerea* more than each component alone. Of the tested conditions, the combination of CHOS (320 µg mL^{-1}) and Signum (10 µg mL^{-1}) was the most effective.

Field trial: Inhibition of Infection of Apple Fruits by *Venturia ineuqualis*

In a field trial we studied the effect of 0.1% (w/v) CHOS DP$_n$ 35 combined with Delan at recommended concentration (0.8% w/v)) or at 1/10 of the recommended concentration (0.08% w/v)

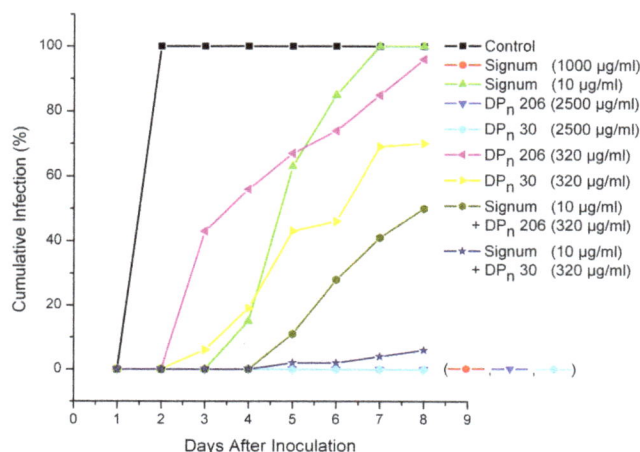

Figure 6. Effect of combinations of chitosan (DP$_n$ 206) or CHOS (DP$_n$ 30) and Signum on cumulative *Botrytis cinerea* infection of detached chickpea leaves.

Table 6. Effect of combinations of chitosan (DP$_n$ 206) or chito-oligosaccharides (CHOS; DP$_n$ 30) and Signum on sporulation of *Botrytis cinerea* on infected chickpea leaves.

Treatment (µg mL^{-1})	Conidia (10^3 mL^{-1})[b]
Control[a]	290±92
Signum 10	52±12
Chitosan 320	110±32
CHOS 320	57±17
Chitosan 320 + Signum 10	11±3
CHOS 320 + Signum 10	0.4±0.1

Spores were counted eight days after inoculation.
[a]Conidia in sterile water.
[b]The data are the mean of two experiments ± standard deviation. Each experiment had three replicates for each treatment and each replicate had three leaves with 6 inoculated leaflets.

Table 7. Effect of the combination of chito-oligosaccharides (CHOS; DP_n 35) and Delan on infection of apple by *Venturia inaequalis* in the field.

Treatment	% apples with apple scab
Untreated control	31.2 ± 9.7^a
Delan 0.8 g/L (800 $\mu g\ ml^{-1})^b$	20.9 ± 9.5
Delan 0.08 g/L (80 $\mu g\ ml^{-1}$)	27.5 ± 12.0
CHOS DP_n 30, 1.0 g/L (1000 $\mu g\ ml^{-1}$)	25.9 ± 13.3
Delan 80 $\mu g\ ml^{-1}$ + Chitosan DP_n 30, 1000 $\mu g\ ml^{-1}$	16.7 ± 5.2

[a]Standard deviation. The data are derived from one experiment (one season) with three replicates per treatment and three trees in each replicate.
[b]Recommended dose.

on development of scab in apples, which is due to infection by *Venturia inaequalis*.

The results in Table 7 show that the combination of CHOS and 1/10 of the recommended concentration of Delan was more effective in preventing scab development than the recommended concentration of Delan.

Discussion

It is well known from several studies that chitosan and CHOS have anti-microbial properties, and it is also known that the degree of acetylation of chitosan is an important factor affecting antifungal activity [22–23]. It has been proposed that the positive charge of the free amino groups of the glucosamine moieties in chitosan modulates interactions with the negatively charged cell surface, which under certain conditions may result in membrane destabilization and pore formation [22–23]. In the present study, we have focused on the effects of chain length, the particular role of the sugar moiety at the reducing end, and, first of all, on synergistic effects between chitosan or CHOS and synthetic fungicides.

To our knowledge there are no previous reports showing what is presented above, namely that the presence of GlcN at the reducing ends of CHOS is beneficial for antifungal activity. Interestingly, a common method to produce CHOS from chitin or chitosan is to treat the polymers with concentrated HCl in an acid catalyzed hydrolysis [24]. Due to the intrinsic chemistry of this reaction, hydrolysis after an acetylated sugar is favored 115 times more than hydrolysis after a deacetylated sugar [24]. Taking into account the beneficial effect of a deacetylated sugar at the reducing end on anti-fungal activity, chemical hydrolysis of chitosan could give less effective CHOS than hydrolysis using an enzyme such as ScCsn46A.

It has been suggested that CHOS are more inhibitory than polymeric chitosan due to better solubility in water [25–26]. The present results shows that the degree of polymerization (DP) is an important factor on the antifungal activity. Since all the chitosans and CHOS used in this study (with low F_A and at slightly acidic pH) are almost equally soluble at pH 5.3, it is unlikely that the antifungal activity of the chitosan and CHOS tested can be explained by a slight difference in water solubility.

Our *in vitro* assay showed that CHOS obtained using ScCsn46A were more inhibitory toward *B. cinerea* than the native chitosan (DP_n 206). CHOS in the DP_n range 15–40 were the most effective. The dosages were calculated by weight, rather than by moles, and thus the molar concentration of the smaller CHOS was higher than that of the longer CHOS. However, if one converts the data shown in Fig. 2 to molar dosages, the data still show a clear

optimal DP_n in the region 15 – 40 (Note that the inhibitory effect becomes strongly reduced at $DP_n < 15$). Interestingly, a previous study on the effect of CHOS on *Candida krusei* (the tested range was 5 to 27 kDa) [27] showed that antifungal activity was at is maximum for a 6 kDa CHOS fraction (DP_n around 40), whereas longer CHOS were less effective. The present results are in accordance with this observation.

It is of interest to note that the longer CHOS obtained after fractionating a CHOS sample with DP_n 34 by size exclusion chromatography were less inhibitory than the original chitosan (DP_n 206) (Figure 3). This indicates that the shorter CHOS molecules likely to be present in the chitosan DP_n 206, but not in the chromatographically purified DP_n 78, 126 and 163 fractions, are important for the antifungal activity.

The most important results of the present study is the demonstration of good effects of combining CHOS or chitosan with synthetic fungicides, which was observed in vitro and in vivo laboratory studies as well as in a field trial. In all cases, additivity was observed and in several cases the combinations were strongly synergistic in both *in vitro* and *in vivo* assays. The effects varied between the various fungi and plants tested, but the overall picture is that synergistic effects are common and that CHOS of the right DP_n tend to work better than chitosan, sometimes much better. The largest synergistic effects were observed with *B. cinerea*, in both the germination assay and the strawberry flower assay (Tables 2 & 4). For example, low concentrations of CHOS (DP_n 23) or Signum, which had almost no effect on *B. cinerea* germination when applied separately, achieved almost 90% reduction of germination when applied together.

The mechanisms for the synergism in inhibition of fungal growth are not known. Most likely, the synergism is due to the compounds' different modes of action. Teldor inhibits sterol biosynthesis, Switch inhibits protein synthesis and signal transduction, while Amistar and Signum inhibit respiration [28]. The mode of action of chitosan is not clearly understood [29–32] but previous studies suggest that electrostatic interactions between positively charged chitosan and the negatively charged cell surface may destabilize the cell wall and/or cell membrane, which ultimately increases the cell permeability and induces cell leakage [33–35]. The synergy could conceivably be the result of a general increase in stress when different cellular processes are attacked simultaneously. More specifically, increased cell wall permeability may have enabled Teldor (fenhexamid) to reach the conidial membrane earlier and thereby stop the germination at an earlier stage than if Teldor was applied alone. Increased cell membrane permeability [35] may enable Amistar and Signum to inhibit respiration or Switch to inhibit protein synthesis more easily than if the fungicides are applied alone. The reasons for the stronger synergism between CHOS and fungicides compared to chitosan and fungicides are not known, but this observation correlates with the observed clear optimum in chain length that was observed when applying CHOS alone (Fig. 2).

An issue not addressed in the present study but of major interest for future work concerns possible interactions between the CHOS and CHOS-binding proteins in the plant or the pathogenic fungus, in particular proteins containing LysM domains [36]. CHOS can stimulate plant immune responses by binding to specific receptor proteins, and such stimulation could contribute to the observed overall protective effects of CHOS and CHOS-fungicide mixtures. On the other hand plant pathogenic fungi may combat this response by secreting proteins that sequester CHOS [37], which could reduce protective effects. It is thus conceivable that variation in the protective effects described above to some extent is due to variation in the interactions between the CHOS and CHOS-binding proteins in plant or fungus. Notably, the *in vitro* data show

strong anti-fungal effects of CHOS-fungicide mixtures, which suggests that direct inhibition of fungal growth is a dominant contributor to the protective effects seen in the *in vivo* experiments.

In conclusion, our studies suggest that the use of CHOS of DP_n 15–40, with a deacetylated reducing end may reduce the need for synthetic fungicides by at least an order of magnitude. Thus, combinations of CHOS and synthetic fungicides should be considered for use in Integrated Pest Management (IPM) programs, where application of even small amounts of CHOS could reduce the need for synthetic fungicides considerably.

Author Contributions

Conceived and designed the experiments: LGH AT MHR MS. Performed the experiments: LGH MHR LRS AT. Analyzed the data: BBA MS VGHE MHR LGH AT. Contributed reagents/materials/analysis tools: BBA MS VGHE. Wrote the paper: MHR AT LGH VGHE.

References

1. Haware M (1998) Diseases of chickpea, In: Allen, D. and Lenné J, editors. The Pathology of Food and Pasture Legumes. CAB International, Wallingford. pp. 473–516.
2. Aarstad PA, Bjørlo B, Gundersen GI (2008) Bruk av plantevernmidler i jordbruket i 2008. Rapporter 2009/52. Statistisk sentralbyrå: Oslo–Kongsvinger.
3. Sholberg PL (1990) A new postharvest rot of peaches in Canada caused by *Mucor piriformis*. Can J Plant Pathol 12: 219–221.
4. Hjeljord LG, Stensvand A, Tronsmo A (2000) Effect of temperature and nutrient stress on the capacity of commercial *Trichoderma* products to control *Botrytis cinerea* and *Mucor piriformis* in Greenhouse Strawberries. Biological Control 19: 149–160.
5. Muto M, Takahashi H, Ishihara K, Yuasa H, Huang J (2005) Control of black leaf spot (*Alternaria brassicicola*) of crucifers by extracts of Black Nightshade (*Solanum nigrum*). Plant Pathol Bull 14: 25–34.
6. Brent KJ, Hollomon DW (2007) Fungicide resistance: the assessment of risk. In FRAC Monograph No. 2. second (reviced) edition. Fungicide Resistance Action Committee. 35p.
7. Hu X, Du Y, Tang Y, Wang Q, Feng T, et al. (2007) Solubility and property of chitin in NaOH or urea aqueous solution. Carbohyd Polym 70: 451–458.
8. Pillai C, Paul W, Sharma C (2009) Chitin and chitosan polymers: Chemistry, solubility and fiber formation. Prog Polym Sci 34: 641–678.
9. El-Ghaouth A, Arul J, Grenier J, Asselin A (1992) Antifungal activity of chitosan on two post harvest pathogens of strawberry fruits. Phytopathology 82: 398–402.
10. Aam BB, Heggset EB, Norberg AL, Sørlie M, Vårum KM, et al. (2010) Production of chitooligosaccharides and their potential applications in medicine. Mar Drugs 8: 1482–1517.
11. Allan CR, Hadwiger LA (1997) The fungicidal effect of chitosan on fungi of varying cell wall composition. Expt Mycol 3: 285–287.
12. Hamel LP, Beaudoin N (2010) Chitooligosaccharide sensing and downstream signaling: contrasted outcomes in pathogenic and beneficial plant–microbe interactions. Planta 232:787–806.
13. Cantarel BL, Coutinho PM, Rancurel C, Bernard T, Lombard V, et al. (2009) The Carbohydrate-Active EnZymes database (CAZy): an expert resource for Glycogenomics. Nucleic Acids Res 37: D233–238 (http://www.cazy.org/).
14. Sørbotten A, Horn SJ, Eijsink VGH, Vårum KM (2005) Degradation of chitosans with chitinase B from *Serratia marcescens* production of chito-oligosaccharides and insight into enzyme processivity. FEBS J 272: 538–549.
15. Heggset EB, Hoell IA, Kristoffersen M, Eijsink VGH, Vårum KM (2009) Degradation of chitosans with chitinase G from *Streptomyces coelicolor* A3(2): production of chito-oligosaccharides and insight into subsite specificities. Biomacromolecules 10: 892–899.
16. Heggset EB, Dybvik AI, Hoell IA, Norberg AL, Sørlie M, et al. (2010) Degradation of chitosans with a family 46 chitosanase from *Streptomyces coelicolor* A3(2). Biomacromolecules 11: 2487–2497.
17. Rahman MH (2013) Antifungal activity of chitosan/chito-oligosaccharides alone and in combination with chemical fungicides against fungal pathogens. PhD thesis. Norwegian University of Life Sciences. ISBN 978-82-575-1115-9. 12: 1–85.
18. Brurberg MB, Eijsink VGH, Nes IF (1994) Characterization of a chitinase gene (chiA) from *Serratia marcescens* BJL200 and one-step purification of the gene product. FEMS Microbiol Lett 124: 399–404.
19. Synstad B, Gaseidnes S, van Aalten DMF, Vriend G, Nielsen JE, et al. (2004) Mutational and computational analysis of the role of conserved residues in the active site of a family 18 chitinases. Eur. J. Biochem 271: 253–262.
20. Levy, Benderly M, Cohen Y, Gisi U, Bassand D (1986) The joint action of fungicides in mixtures: comparison of two methods for synergy calculation. EPPO Bulletin 16: 651–657.
21. Bardin M, Fargues J, Nicot P (2008) Compatibility between biopesticides used to control grey mould, powdery mildew and whitefly on tomato. Biol Control 46: 476–483.
22. Sudarshan NR, Hoover DG, Knorr D (1992) Antibacterial action of chitosan. Food Biotechnol 6: 257–272.
23. Choi B, Kim K, Yoo Y, Oh S, Choi J, et al. (2001) *In vitro* antimicrobial activity of chitooligosaccharide mixure against *Actinobacillus actinomycetemcomitans* and *Streptococcus mutans*. Int J Antimicrob Agent 18: 553–557.
24. Einbu A, Vårum K M (2008) Characterization of chitin and its hydrolysis to GlcNAc and GlcN. Biomacromolecules 9: 1870–1875.
25. Badawy EIM, Rabea EI (2011) A biopolymer chitosan and its derivatives as promising antimicrobial agents against plant pathogens and their applications in crop protection. Int J Carbohyd Chem 2011 29 pp. http://dx.doi.org/10.1155/2011/460381.
26. Rhoades J, Roller S (2000) Antimicrobial actions of degraded and native chitosan against spoilage organisms in laboratory media and foods. Appl Environ Microbiol 66: 80–86.
27. Gerasimenko DV, Avdienko ID, Bannikova GE, Zueva OY, Varlamov VP (2004) Antibacterial effects of water-soluble low-molecular-weight chitosans on different microorganisms. Appl Biochem Microbiol 40: 253–257.
28. Fishel FM, Dewdney MM (2012) Fungicide Resistance Action Committees (FRAC) Classification scheme of fungicidees according to mode of action. PI94. University of Florida. 7p.
29. Raafat D, von Bargen K, Haas A, Sahl H (2008) Insights into the mode of action of chitosan as an antibacterial compound. Appl Environ microbiol 74: 3764–3773.
30. Palma-Guerrero J, Jansson HB, Salinas J, Lopez-Llorca LV (2008) Effect of chitosan on hyphal growth and spore germination of plant pathogenic and biocontrol fungi. J Appl Microbiol 104: 541–553.
31. Palma-Guerrero J, Huang IC, Jansson HB, Salinas J, Lopez-Llorca LV, et al. (2009) Chitosan permeabilizes the plasma membrane and kills cells of *Neurospora crassa* in an energy dependent manner. Fungal Genet Biol 46: 585–594.
32. Palma-Guerrero J, Gómez-Vidal S, Tikhonov VE, Salinas J, Jansson HB, et al. (2010) Comparative analysis of extracellular proteins from *Pochonia chlamydosporia* grown with chitosan or chitin as main carbon and nitrogen sources. Enzyme Microb Tech 46: 568–574.
33. Hadwiger LA, Beckman JM (1980) Chitosan as a component of pea-*Fusarium solani* interactions. Plant Physio. 66: 205–211.
34. Vesentini D, Steward D, Singh AP, Ball R, Daniel G, et al. (2007) Chitosan-mediated changes in cell wall composition, morphology and ultrastructure in two wood-inhabiting fungi. Mycol Res 111: 875–890.
35. Xu J, Zhao X, Han X, Du Y (2007) Antifungal activity of oligochitosan against *Phytophthora capsici* and other plant pathogenic fungi *in vitro*. Pestic Biochem Physiol 87: 220–228.
36. Gust AA, Willmann R, Desaki Y, Grabherr HM, Nürnberger T (2012) Plant LysM proteins: modules mediating symbiosis and immunity. Trends Plant Sci 17: 495–502.
37. De Jonge R, van Esse HP, Kombrink A, Shinya T, Desaki Y, et al. (2010) Conserved fungal LysM effector Ecp6 prevents chitin-triggered immunity in plants. Science 329: 953–955.

2

Functional and Structural Comparison of Pyrrolnitrin- and Iprodione-Induced Modifications in the Class III Histidine-Kinase Bos1 of *Botrytis cinerea*

Sabine Fillinger[1]*, **Sakhr Ajouz**[2], **Philippe C. Nicot**[2], **Pierre Leroux**[1], **Marc Bardin**[2]*

1 INRA UR1290, BIOGER CPP, Thiverval-Grignon, France, **2** INRA, UR407, Plant Pathology Unit, Montfavet, France

Abstract

Dicarboximides and phenylpyrroles are commonly used fungicides against plant pathogenic ascomycetes. Although their effect on fungal osmosensing systems has been shown in many studies, their modes-of-action still remain unclear. Laboratory- or field-mutants of fungi resistant to either or both fungicide categories generally harbour point mutations in the sensor histidine kinase of the osmotic signal transduction cascade. In the present study we compared the mechanisms of resistance to the dicarboximide iprodione and to pyrrolnitrin, a structural analogue of phenylpyrrole fungicides, in *Botrytis cinerea*. Pyrrolnitrin-induced mutants and iprodione-induced mutants of *B. cinerea* were produced *in vitro*. For the pyrrolnitrin-induced mutants, a high level of resistance to pyrrolnitrin was associated with a high level of resistance to iprodione. For the iprodione-induced mutants, the high level of resistance to iprodione generated variable levels of resistance to pyrrolnitrin and phenylpyrroles. All selected mutants showed hypersensitivity to high osmolarity and regardless of their resistance levels to phenylpyrroles, they showed strongly reduced fitness parameters (sporulation, mycelial growth, aggressiveness on plants) compared to the parental phenotypes. Most of the mutants presented modifications in the osmosensing class III histidine kinase affecting the HAMP domains. Site directed mutagenesis of the *bos1* gene was applied to validate eight of the identified mutations. Structure modelling of the HAMP domains revealed that the replacements of hydrophobic residues within the HAMP domains generally affected their helical structure, probably abolishing signal transduction. Comparing mutant phenotypes to the HAMP structures, our study suggests that mutations perturbing helical structures of HAMP2-4 abolish signal-transduction leading to loss-of-function phenotype. The mutation of residues E529, M427, and T581, without consequences on HAMP structure, highlighted their involvement in signal transduction. E529 and M427 seem to be principally involved in osmotic signal transduction.

Editor: Zhengguang Zhang, Nanjing Agricultural University, China

Funding: This work was supported in part by the French National Research Agency (ANR-ECOSERRE project; Dr. Bardin and Dr. Nicot; http://www.agence-nationale-recherche.fr/en/project-based-funding-to-advance-french-research/) and by a grant from Institut National de la Recherche Agronomique (INRA - SPE DurLB project; Dr. Bardin, Dr. Ajouz, Dr. Nicot, Dr. Leroux, and Dr. Fillinger; http://www4.inra.fr/sante-plantes-environnement/). A grant for studies was provided by the Syrian government for Dr. Ajouz. The funders had no role in study design, data collection and analysis, decision to publish, or preparation of the manuscript.

Competing Interests: The authors have declared that no competing interests exist.

* E-mail: sabine.fillinger@versailles.inra.fr (SF); marc.bardin@avignon.inra.fr (MB)

Introduction

Gray mould, caused by the fungus *Botrytis cinerea* Pers.:Fr (teleomorph *Botryotinia fuckeliana* (de Bary) Whetzel), is a severe disease affecting a wide range of economically important crops [1]. Chemical control is the main approach to limit the incidence of this pathogen. However, the efficiency of fungicides is under threat, because isolates of *B. cinerea* resistant to fungicides have been found to evolve rapidly [2,3]. Biological control could be an alternative, or a complement, to chemical control because plant pathogens are considered to develop resistance to biocontrol agents less frequently than to fungicides [4]. Numerous biocontrol agents are effective against *B. cinerea* [5,6]. For some of them, production of antibiotics is one of the putative mechanisms of action [7].

Pyrrolnitrin [3-chloro-4-(3-chloro-2-nitrophenyl) pyrrole], first isolated from *Burkholderia pyrrocina* [8], is an antibiotic with a broad-spectral antifungal activity [7]. It was also found in several other bacteria used as biocontrol agents against various fungal plant pathogens [9] including *B. cinerea* [10,11,12]. Under laboratory conditions, *B. cinerea* mutants resistant to pyrrolnitrin have recently been reported, suggesting a possible loss of efficacy of pyrrolnitrin-producing biocontrol agents [13]. Resistance to synthetic phenyl-pyrrole fungicides (e.g., fenpiclonil, fludioxonil), structural analogues of pyrrolnitrin, has also been reported in laboratory-induced mutants [14,15,16,17]. The same studies revealed that mutants highly resistant to phenylpyrroles generally also displayed resistance to dicarboximide (e.g., iprodione) and aromatic hydrocarbon fungicides (e.g., dicloran). They also were found to be sensitive to osmotic stress [15,17]. However, under field conditions such phenotypes have not yet been observed for *B. cinerea*. To date, no specific resistance to phenylpyrroles is known for field isolates of *B. cinerea*.

Molecular studies have shown that an osmosensing histidine kinase (HK) mediates resistance to dicarboximides and phenyl-pyrroles in *B. cinerea* [18,19,20,21]. The same HK is also implicated in adaptation to adverse environmental conditions such as osmotic and oxidative stresses. Its essential role in the development and pathogenesis of *B. cinerea* [20,22] may explain why strains highly resistant to dicarboximides and phenylpyrroles are not found under field conditions. In other fungal species, the role of homologous HKs in resistance to dicarboximides and phenylpyrroles has also been demonstrated [30], but in contrast to *B. cinerea*, specific resistance to fludioxonil was also found in field isolates [23,24,25,26,27,28,29,30], giving rise to questions about the structure-function relationship of the Bos1 HK in *B. cinerea*.

Fungal osmosensing HKs belong principally to class III histidine kinases (according to the classification of Catlett and co-workers [31]) or, as in the case of *Saccharomyces cerevisiae*, which only has a unique HK,, to class VI HKs. Besides the typical HK domains the structural characteristics of the cytoplasmic class III HKs include five to six repeats of HAMP domains in their N-terminal half. HAMP domains are ubiquitous among eukaryotic and prokaryotic signal transduction (ST) proteins including histidine kinases, adenylate cyclases, methyl-accepting chemotaxis proteins and some phosphatases (reviewed in [32]). HAMP subunits contain two α helices, AS1 and AS2, bridged by a flexible connector of approximately 14 residues [33]. Each of the two helices is composed of a typical heptad repeat (a–g), with hydrophobic residues in positions a and d. Crystal structure and cysteine scanning studies of bacterial HAMPs revealed a typical four-helix bundle structure for HAMP dimers. Several putative mechanisms have been proposed for their role in signal transduction: piston movement, concerted rotation, or scissor-like movement between helices AS1 and AS2 or successive HAMPs [34,35,36,37,38]. In the case of eukaryotic proteins harbouring HAMP domains, in particular class III HKs, evidence of their role in ST has been brought to light by selecting osmo-sensitive and/or fungicide resistant mutants carrying point mutations in the HAMP domains [23,24,25,26,27,28,29,30]. More recently, *in vitro* mutagenesis of successive HAMP domains of the *Debaromyces hansenii* DhNik1 protein has been used [39]. Meena and colleagues [39] proposed the first functional model of a class III HK based on osmosensing in the heterologous system *S. cerevisiae*. According to their model of this five-HAMP-containing HK, HAMP4 cross-links to HAMP5 under high osmolarity, thereby inhibiting histidine kinase activity. Under normal osmolarity, HAMP4 is kept away from HAMP5 through its binding to HAMP1-3, thereby maintaining histidine kinase activity. In the case of the *B. cinerea* class III HK Bos1, six HAMP domains have been previously identified [19], leading us to think that the signal transduction mechanism could be different.

Dicarboximide and phenylpyrrole fungicidal signals are transduced through the Bos1 HK in addition to osmosensing [19,22]. In this study, from a functional point of view, we set out to investigate if these different fungicides induce mutations conferring different resistance phenotypes. Moreover, we asked what impact the resistance to fungicides has on fitness cost. From a molecular point of view, we examined the impact of the mutations on HAMP structure and its correlation with differential signal transduction.

Materials and Methods

Fungal isolates, culture conditions and mutant selection

Five single-spore isolates of *B. cinerea* (namely BC1, BC21, BC25, BC26 and H6) were selected from a collection maintained in the laboratory. BC1, BC25, BC26 and H6 were isolated from tomato between 1989 and 1991 and BC21 was isolated on strawberry in 1991. The choice was made on the basis of differences in patterns of resistance to 14 fungicides representing 11 chemical groups [12] and of differences in aggressiveness to tomato plants and apple fruits [13]. All isolates were sensitive to pyrrolnitrin. BC1 and BC26 were resistant (LR) to dicarboximides.

For DNA isolation, 10^7 spores were harvested and used to inoculate 100 ml liquid yeast-sugar-salt medium (YSS, 2 g L^{-1} of yeast extract, 10 g L^{-1} of glucose, 2 g L^{-1} of KH_2PO_4, 1.5 g L^{-1} of K_2HPO_4, 1 g L^{-1} of $(NH_4)_2SO_4$, 0.5 g L^{-1} of $MgSO_4$ $7H_2O$) and grown for 16 hours at 23°C with 150 rpm shaking.

Twenty successive conidial transfers were performed with the 5 isolates of *B. cinerea* on increasing doses of pyrrolnitrin and iprodione as described previously [13,40]. For each transfer, plates were incubated for 14 days at 21°C. As a control, twenty successive conidial transfers were performed on unsupplemented PDA medium. For each isolate, the whole experiment was carried out three times independently, aiming to provide three lineages of 20 transfers produced under selection pressure and three independent control lineages produced on PDA. To facilitate reading and avoid lengthy repetitions in the rest of this paper, control isolates produced on PDA medium and isolates produced in presence of pyrrolnitrin or iprodione will be labelled GnC, GnP and GnI respectively, where n indicates the transfer rank in the lineage. All isolates and mutants were maintained in stock cultures at −20°C in a 0.06 M phosphate buffer containing 20% (V/V) glycerol until they were used for phenotypic and genotypic characterizations. The biological and sequencing data are not available for mutants BC1G20P3, BC25G20P2, BC25G20P3, BC26G20P3 and H6G20I1 as they were lost before experiments could be achieved.

Antifungal assays

To determine the sensitivity to the antibiotic pyrrolnitrin, the mycelial growth was measured on PDA medium containing different concentrations of pyrrolnitrin as described previously [13]. We assessed the effect of iprodione, fenpiclonil and fludioxonil (technical grade quality, kindly provided respectively by BASF Agro, Germany and Syngenta Crop Protection AG, Switzerland) on spore germination and germ tube elongation as previously described [15]. The experiments were all repeated three times independently per lineage, each with three replicate plates. For each combination of strain/antifungal molecule, the concentration leading to 50% decrease in mycelial growth or germ-tube elongation (EC_{50}) was estimated by linear regression analysis of fungal development (as percentage of control values) [15]. To simplify the reading of the manuscript, we adopted the following nomenclature based on the established EC_{50} values. The ratio between the EC_{50} value of a tested strain and the mean EC_{50} value of fungicide-sensitive strains (BC21, BC25, and H6) were calculated for pyrrolnitrin, fludioxonil, fenpiclonil, and iprodione. For all compounds but iprodione, the strains were then considered as sensitive (S) if the ratio was between 0.5 and 2, slightly resistant (LR) between 2 and 20, moderately resistant (MR) between 20 and 100 and highly resistant (HR) for ratios over 100. In the case of iprodione, we adopted the classification LR (ratio between 2 and 10), MR (ratio between 10 and 20) and HR (ratio>20) from previous publications [15,18,19]. Resistance profiles to 10 other fungicides belonging to 9 different chemical families were established at discriminatory concentrations as already published [12].

Based on previous work, the resistance profiles of B05.10 transformants were established at the mycelial growth stage on discriminatory fungicide concentrations as follows. A transformant was considered as Iprodione LR if it was able to grow at 2.5 µg ml^{-1} but not at 25 µg ml^{-1}, and it was considered Iprodione HR if it grew at 25 µg ml^{-1}. It was considered as Phenylpyrrol LR if it grew on fludioxonil or fenpiclonil at 0.1 µg ml^{-1} but not at concentrations >1 µg ml^{-1} and Phenyl-pyrrol HR if it grew on fludioxonil or fenpiclonil at 5 µg ml^{-1}.

The discriminatory concentrations used for pyrrolnitrine LR and HR phenotypes were 0.05 µg ml^{-1} and 0.5 µg ml^{-1} respectively.

Osmotic stress assay

To determine *in vitro* sensitivity to osmotic stress, PDA or YSS plates were supplemented with 0.5 M or 1 M of either NaCl or sorbitol. The plates were then inoculated with 5-mm mycelial plugs taken from the periphery of a 4-day-old colony of the various tested strains. Cultures were incubated for 4 days at 21°C and growth was compared relative to that on PDA or YSS control plates as no growth, reduced or comparable growth. Strains showing comparable growth on 0.5 M osmolytes were considered as resistant. Those with reduced growth on 0.5 M and no growth on 1 M osmolytes were classified as osmosensitive. Two to three replicate plates were realized for each treatment.

In vitro and *in planta* estimation of fitness cost of the iprodione-induced mutants

Different fitness parameters were used to compare the strains. *In vitro* estimation of fitness was based on mycelial growth and spore production on PDA medium as described previously [13]. Statistical analyses were performed separately for each strain and each type of fitness parameter. In these analyses, we used the ANOVA module of Statistica software to test both a possible lineage effect and a number-of-transfer effect. When a significant effect was observed, the multiple comparison test of Student-Newman-Keuls was used to compare the means. To estimate fitness *in planta*, the aggressiveness of the various isolates was assessed on apple fruits cv. Golden Delicious (for all isolates) and on tomato cv. Monalbo plants (for strains BC1 and BC21), as previously described [13]. In addition, spore production was assessed on tomato plants for these two strains as follows. After 7 days of incubation in a growth chamber (21°C, relative humidity >90%, photoperiod of 14 hours), the stem segments carrying lesions were excised and placed in 5-mL water aliquots to collect spores. The concentration of the spore suspension was evaluated using a haematocytometer. Sporulation was then computed as the numbers of spores produced per mm of stem lesion. The experiments were all repeated two to three times independently per lineage, each with three replicate plants or fruits. To take into account the kinetics of disease development for each isolate, we computed the area under the disease progress curve (AUDPC) as already described [41]. Statistical analyses were performed separately for each isolate on the AUDPC values.

DNA manipulations

Genomic DNA was extracted from approximately 1 g of fresh fungal mycelium using a Sarcosyl-based protocol [42]. As most iprodione resistant mutants from *B. cinerea* resulted in modification of the class III histidine kinase Bos1 involved in osmosensing [19,22] the corresponding *bos1* gene from *B. cinerea* was sequenced. The primer pair bos1-F3 and bos1-R2 (Table S2), amplifying the fragment encoding the N-terminal half of the Bos1 histidine kinase between residues 192 and 741 harboring the HAMP domains, or the primer pair His1 and bos1-R5 (Table S2) amplifying the C-terminal half of the Bos1 protein (residues 593 to the end), were used. All primers used for PCR amplifications and DNA sequencing are listed in Table S2. The 3′ half of the gene (corresponding to the C-terminal half of the protein) was only sequenced in the absence of mutations in the 5′ half. PCR reactions were carried out with high-fidelity DNA polymerase (Phusion, Finnzymes) and were gel-purified prior to the sequencing reactions. The resulting sequences were quality analysed

(PHRED>20) and aligned to the reference sequences using the CodonCode Aligner software (CodonCode Corp., Dedham, MA). The *bos1* sequences of all isolates are available from Genbank under the accession numbers JX192607–JX192631.

bos1 site directed mutagenesis by homologous recombination in the B05.10 wild-type strain

2.5 kb fragments of mutated *bos1* alleles to be studied were amplified on genomic DNA of the sequenced strains listed in Table 1 using the primers bos1_promLP2 and bos1-R2 listed in Table S2 using proofreading Taq polymerase (Phusion, Finnzymes). The 5′ extremities of the primers are not located in the *bos1*-coding region, but in the promoter- and intron sequence respectively, in order to minimize the impact of mutations introduced during the recombination at the fragment's extremities. The PCR amplicons were gel-verified prior to purification with the Nucleo spin extraction kit (Macherey & Nagel, Düren, Germany). Protoplasts of the *B. cinerea* reference strain B05.10 [43] were prepared and transformed as described by Levis et al. [44] with 5–7 µg of each purified PCR product. Protoplasts were spread on non-selective, isotonic medium and incubated for 24 h at 23°C. The plates were then overlaid with YSS medium containing 3 µg ml^{-1} iprodione as selective agent and incubated at 23°C for an additional 48 h.

Transformant colonies were isolated twice on selective YSS medium containing either 3 µg ml^{-1} iprodione or 0.3 µg ml^{-1} fludioxonil prior to DNA extraction. The presence of the introduced mutation and the purity of the transformants were verified by sequencing the *bos1* fragment amplified with primers bos1-F1 and bos1-R1 (Table S2, located outside the fragment used for transformation). Heterokaryons were further purified on selective medium until purity of the introduced mutant allele was achieved.

HAMP structure modeling

All homology modeling analyses were performed on the SWISS-MODEL workspace [45]. Prior to modeling, we specified the positions of the HAMP domains of the Bos1 protein with an InterPro Domain Scan and HMM scan [46] on the protein sequence (BAB69486). The six identified HAMP domains including their connecting sequences were aligned using the ClustalW algorithm [47]. The coordinates from these newly defined HAMP domains (Figure 1) differ from those previously published [19]. In order to identify the best structural model template, a non-iterative blast search with the Bos1 peptide sequence covering residues 190 to 720 was performed against the SWISS-MODEL template library. For the fraction comprised between position 195 and 524, template scores >100 were found with the structure 3lnrA established on the aerotaxis receptor Aer2 of *Pseudomonas aeroginosa* [35] and for the fraction comprised between 548 and 678 with the structure 1qu7A, established on the chemotaxis receptor TarH of *Escherichia coli* [48], with p-values<10^{-19}. For reasons of homogeneity of our analysis, we also used the alignment of the C-terminal section (572–708) to the structure 3lnrA (score = 45; p-value = 2 10^{-9}).

HAMP domain sequences (wild-type or mutant peptide sequences) with the coordinates presented in Figure 1 were aligned to the 3lnrA peptide sequence (GI:295789465) using MUSCLE [49]. The alignments were submitted to the SWISS-MODEL server using the alignment modes. Rather than simply replacing the candidate amino acid in a model of wild-type HAMP, each sequence was modeled separately. All models were visualized using the molecular graphics program Chimera [50].

Table 1. Phenotypes and changes in Bos1 peptide sequence in pyrrolnitrine- and iprodione-induced mutants.

Isolate	Transfer generation[a]	Phenotype[b]				Bos1 peptide sequence	HAMP n°
		Osmotic stress	Pyrrolnitrin	Phenylpyrroles	Iprodione		
BC1	G0	R	S	S	LR	I365S	3
	G20C	R	S	S	LR	nd	
	G20P1	S	HR	HR	HR	I365S, G311R	3,2
	G20P2	S	HR	HR	HR	I365S, G311E	3,2
	G20I1	S	MR	MR	HR	I365S, E692K	3,6
	G20I2	S	HR	MR	HR	I365S, E692K	3,6
	G20I3	S	LR	MR	HR	I365S, V239F	1
BC21	G0	R	S	S	S	wt	
	G20C	R	S	S	S	nd	
	G20I1	S	HR	HR	HR	G278D	2
	G20I2	S	HR	LR	HR	G323C	2
	G20I3	S	nd	nd	HR	G323C	2
BC25	G0	R	S	S	S	wt	
	G20C	R	S	S	S	nd	
	G20P1	S	HR	HR	HR	G415D	3
	G20I1	S	HR	MR/HR	HR	wt	
	G20I2	S	MR	MR	HR	A493T	4
	G20I3	S	MR	HR	HR	A493T	4
BC26	G0	R	S	S	LR	I365S	1
	G20C	R	S	S	LR	nd	
	G20P1	S	HR	HR	HR	I365S, T581P	3,5
	G20P2	S	HR	HR	HR	I365S, T581P	3,5
	G20I1	S	LR	MR	HR	I365S, E529A	3,4
	G20I2	S	HR	nd	HR	I365S	3
	G20I3	S	MR	MR	HR	I365S, E529A	3,4
H6	G0	R	S	S	S	wt	
	G20C	R	S	S	S	nd	
	G20P1	S	HR	HR	HR	G81STOP, A157T	nonsense
	G20P2	S	HR	HR	HR	G81STOP, A157T	nonsense
	G20P3	S	HR	HR	HR	G81STOP, A157T	nonsense
	G20I2	S	MR	LR	HR	I365S, M427T	3
	G20I3	S	MR	LR	HR	I365S, M427T	3

[a]G0 is the wild-type parent isolate, G20C is the 20th transfer generation produced on PDA medium (control), G20P is the the 20th transfer generation produced on PDA supplemented with pyrrolnitrin and G20I is the 20th transfer generation produced on PDA supplemented with iprodione.
[b]S: sensitive, LR: low resistance, MR: moderate resistance, HR: high resistance, according to the resistance level classification explained in the Materials & Methods section and EC_{50} values indicated in Table S1.
nd: not determined.

Results

In vitro sensitivity profiles of pyrrolnitrin- and iprodione-induced mutants to pyrrolnitrin, fungicides and osmotic stress

The five *B. cinerea* parental strains used in this study were all susceptible to pyrrolnitrin and to the phenylpyrrole fungicides fludioxonil and fenpiclonil. Towards the dicarboximide iprodione they showed different behaviours, either sensitive (EC_{50} between 1 and 2.5 µg ml^{-1} in the case of BC21, BC25, H6) or displaying LR resistance (EC_{50} between 6 an 10 µg ml^{-1} for BC1 and BC26, Table S1). This last category corresponds to the previously described ImiR1 phenotype [15]. We then selected three iprodione-resistant mutant lines per parental strain by twenty successive conidial transfers on medium containing high concentrations of iprodione (steadily increasing from 5 to 200 mg ml^{-1}). After conidial isolation, the three mutants per parental strain – named G20I1 to G20I3 – were tested for their susceptibility to pyrrolnitrin, to iprodione, and to the phenylpyrroles fenpiclonil and fludioxonil in parallel with the analysis of pyrrolnitrin-induced mutants – named G20P1 to G20P3 – selected from the same parental strains (described in [13]). The results summarized in Table 1 (detailed in Table S1) show that the pyrrolnitrin-induced mutants (G20P) exhibited high resistance levels to pyrrolnitrin

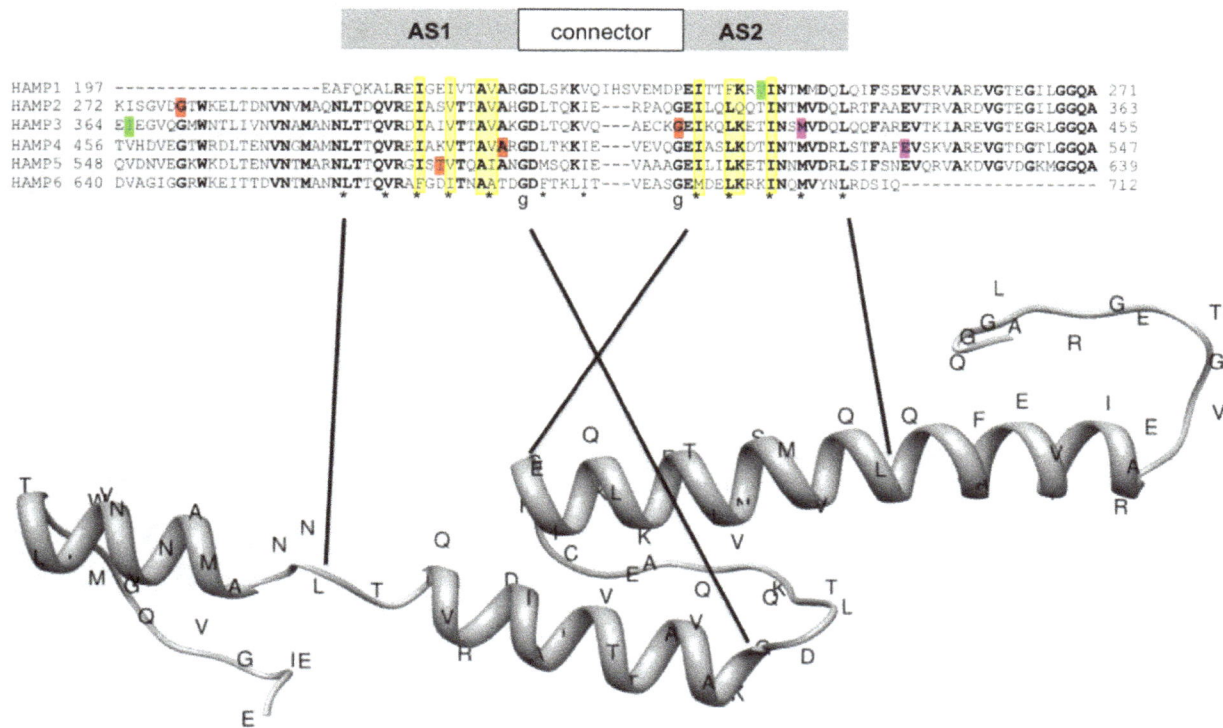

Figure 1. Sequence conservation and model structure of Bos1 HAMP domains. The amino-acid sequences of the six HAMP domains of the Bos1 protein were aligned with Clustal W. Amino acids identical over 80% are in bold. In the bottom panel "*" denotes hydrophobic core residues at critical heptad positions and hydrophobic core residues of the connector. "g" corresponds to the conserved glycine residues of the connector motif (according to [35] and [32]). Interacting residues derived from *in silico* structures are highlighted in yellow. Mutated residues are shaded with the following color code according to the phenotypes indicated in Table 3: red = HR to iprodione and phenylpyrroles, osmosensitivity; green = LR to iprodione; purple = MR to iprodione and osmosensitivity. The structure of the HAMP domain 3 was predicted on the SWISS-MODEL server using the alignment mode (for details see text). Portions of the HAMP sequences involved in the typical HAMP structures AS1, AS2 and the connector are indicated above the sequence.

(EC_{50}>4.5 µg ml^{-1}), but also to phenypyrroles (EC_{50}>8 µg ml^{-1}) and to iprodione (EC_{50}>25 µg ml^{-1}), whereas the iprodione-induced mutants (G20I), highly resistant to iprodione (EC_{50}>25 µg ml^{-1}), displayed variable resistance levels to pyrrolnitrin, in most cases much lower than the corresponding G20P mutants (EC_{50} reaching from 0.03 to over 0.5 µg ml^{-1}). Similarly, the resistance levels to the phenylpyrroles fenpiclonil and fludioxonil were extremely variable for the iprodione induced mutants (EC_{50} between 0.1 and over 10 µg ml^{-1}). In a few cases, resistance levels to pyrrolnitrin did not correlate with those observed on phenylpyrrole fungicides (Table 1, Table S1). For example, the mutant BC21G20I2, highly resistant to pyrrolnitrin, displayed only a low resistance level to the phenylpyrroles fludioxonil and fenpiclonil. The opposite was observed for BC1G20I3 with moderate resistance to phenylpyrroles and low resistance to pyrrolnitrin. The detailed pattern of resistance of the 5 parental strains (G0) to other fungicides is already known [12]. The selected mutants (G20C, G20P, G20I) displayed the same profiles as the parental strains on all these fungicides (data not shown).

Finally, all G20P and G20I mutants were highly sensitive to osmotic pressure resulting from 1 M sodium chloride and 1 M sorbitol compared to the wild-type parents G0 (data not shown).

Fitness of iprodione-induced mutants

We have previously observed that high level of resistance to pyrrolnitrin is correlated to a high fitness cost [13,51]. Since the iprodione-induced mutants displayed low levels of resistance to pyrrolnitrin, we investigated if they could also be associated with a fitness penalty.

Different parameters of fitness were studied for the parental strains G0, the control strains G20C and for all the lineages of the iprodione-induced mutants G20I. These fitness parameters included mycelial growth on PDA medium, spore production (assessed on PDA medium and on tomato plants for isolates derived from BC1 and BC21) and aggressiveness on apple fruits and on tomato plants (for isolates derived from BC1 and BC21). The average mycelial growth rate of the iprodione-induced mutants G20I was significantly reduced for each strain compared to G0 and G20C (56% reduction, P<0.0001 for strain BC1, 29% reduction, P=0.0005 for BC21, 42% reduction, P<0.00001 for BC25, 38% reduction, P<0.0001 for BC26 and 10% reduction, P=0.0001 for H6) (Table 2). Spore production of the iprodione-induced mutants G20I was significantly reduced on PDA medium (between 75 and 96%) for all lineages compared to G0 and G20C (P=0.002, 0.047, 0.027, 0.0007 and 0.002 for BC1, BC21, BC25, BC26 and H6, respectively) (Table 2). It was also greatly reduced on tomato plants (between 80 and 87%) for all mutant lines of BC1 and BC21 strains.

For each of the five strains tested, the iprodione-induced mutants G20I were significantly less aggressive on apple fruits than the parental strains G0 and the control G20C (Table 2). On tomato plants, a decrease in aggressiveness was also observed with the iprodione-induced mutants G20I of strains BC1 and BC21

Table 2. Comparison of fitness parameters between the iprodione-induced mutants and the parental strains.

Strain	Transfer generation[a]	Mycelium daily radial growth (mm/day)[b]	Sporulation ×10^6 spores/Petri plate[c]	Sporulation on tomato stem ×10^3 spores/mm lesion[d]	Aggressiveness (AUDPC) Apple fruit	tomato plant
BC1	G0	39.0 a[e]	144 a	234 a	178 a	125 a
	G20C	37.2 a	157 a	183 a	138 a	97 a
	G20I1	23.7 c	35 b	48 b	12 b	9 b
	G20I2	25.3 c	22 b	41 b	15 b	7 b
	G20I3	27.7 b	17 b	42 b	31 b	10 b
BC21	G0	35.3 a	57 a	207 a	165 a	66 a
	G20C	33.8 a	51 a	163 a	135 a	51 a
	G20I1	20.4 b	4 b	36 b	14 b	8 b
	G20I2	20.9 b	2 b	27 b	34 b	12 b
	G20I3	21.7 b	5 b	30 b	10 b	9 b
BC25	G0	28.0 a	94 a		120 a	
	G20C	27.5 a	98 a		102 a	
	G20I1	18.8 b	14 b		16 b	
	G20I2	15.9 c	3 b		24 b	
	G20I3	14.6 c	6 b		25 b	
BC26	G0	22.0 a	77 a		102 a	
	G20C	22.3 a	66 a		111 a	
	G20I1	15.3 c	24 b		11 b	
	G20I2	20.0 b	14 b		16 b	
	G20I3	15.0 c	16 b		6 b	
H6	G0	28.5 a	97 a		76 a	
	G20C	27.7 a	91 a		82 a	
	G20I2	18.9 b	6 b		8 b	
	G20I3	18.4 b	3 b		8 b	

[a]G0 is the wild-type parent isolate, G20C is the 20th transfer generation produced on PDA medium without iprodione (control) and G20I is the 20th transfer generation produced on PDA supplemented with iprodione.
[b]Daily radial growth rate between the 3rd and 4th day after inoculation.
[c]Spore produced on PDA medium 14 days after inoculation.
[d]Spore produced on tomato stem 7 days after inoculation.
[e]Data are means of two to three independent repetitions. For each isolate, means within a column followed by the same letter were not significantly different (ANOVA, $\alpha = 0.05$; Newman–Keuls test).

compared to G0 and G20C. Taken together, these results showed for all strains, a reduced *in vitro* growth rate, a severely reduced sporulation rate, and 5 to 10 fold reduction in lesion development. No significant lineage effect for the fitness parameter of any of the iprodione-induced isolate was observed ($P > 0.05$) and no correlation could be drawn between the levels of resistance to pyrrolnitrin and the degree of fitness modification.

Sequence analysis of the *bos1* gene in pyrrolnitrin- and iprodione-induced mutants

As underlined above, mutants selected on pyrrolnitrin or iprodione showed both comparable phenotypes on the basis of fungicide cross-resistance profiles, and clearly identical phenotypes in terms of sensitivity to hyperosmotic conditions. This suggests similar molecular mechanisms for resistance. Resistance to the dicarboximide iprodione and/or to the phenylpyrrole fludioxonil is conferred in many fungi by mutations of the osmosensing histidine kinase, which is orthologous to Bos1 in *B. cinerea* (reviewed in [52]). We therefore sequenced the *bos1* gene in the

G20P and G20I induced mutants and compared it to that of the parental G0 strains. We focused on the HAMP domains involved in signal transduction (reviewed in [32]). The amino acid changes observed are listed in Table 1.

The parental strains BC1G0 and BC26G0 that are resistant to low iprodione concentrations (ImiR1) harbor a modification of isoleucine at position 365 to serine (I365S). The same mutation was observed in all mutants derived from BC1 and BC26, but also in the iprodione resistant mutants derived from strain H6 (H6G20I2, H6G20I3). Other mutations in this N-terminal half of the Bos1 protein were selected in pyrrolnitrin- and iprodione-induced mutants displaying high resistance levels to iprodione (HR, Table 1). These mutations were more or less equally distributed over the six HAMP domains (Table 1). Only for two resistant mutants (BC25G20I1 and BC26G20I2) we did not detect any change in the Bos1 peptide sequence compared to the parental strain. All pyrrolnitrin-induced mutants derived from the H6 strain carry a nonsense mutation in the *bos1* gene at codon 81 leading to precocious stop of translation. It is interesting to note that in some cases, independently selected mutants from the same isolate displayed the same mutation (Table 1). However, identical mutations were selected only by identical treatments (pyrrolnitrin or iprodione, respectively).

bos1 site-directed mutagenesis

Since the above analysed mutants had been selected in 20 conidial transfers on steadily increasing iprodione or pyrrolnitrine concentrations, we could not exclude the inadvertent selection of mutations – even outside *bos1* – in addition to those identified by sequencing. We therefore chose to introduce the following 11 modifications into a *bos1* wild-type strain, the sequenced reference strain B05.10 [43]: I365S, I365S G331R, I365S E692K, I365S V239F, I365S T581P, I365S E529A, I365S M427T, G278D, G323C, G415D, A493T. We transformed the B05.10 wild-type strain with 2.5 kb fragments covering the 5′ half of each *bos1* allele obtained by PCR on the corresponding mutant strains (for details see Materials and Methods). After 24 h protoplast regeneration, transformants were selected on iprodione at 3 μg ml⁻¹. The absence of spontaneous mutations was validated by the negative control, without exogenous DNA. For each PCR fragment used we obtained from 30 to over 100 iprodione-resistant transformants. Fifteen transformants per allele were picked and isolated twice on selective medium. The insertion of the desired mutation at the native *bos1* locus was verified by sequencing after PCR amplification with the primers bos1-F1 and bos1-R1 (Table S2) located up- and downstream respectively of the PCR fragment used for transformation. Only homokaryotic transformants with validated *bos1* sequence - unmodified except for the studied mutation – were used for further analyses. We observed a high mutation rate within the transformed *bos1* fragment. We therefore could retain only few transformants; three mutations could not be validated using this approach (i.e., G323C, G311R, E692K).

The transformants with validated *bos1* sequence (n = 1 or 2 as indicated in Table 3) were tested for sensitivity to iprodione, to phenylpyrrole fungicides and to pyrrolnitrin. We also established their sensitivity to hyperosmolarity on NaCl and sorbitol. The observed phenotypes are listed in Table 3. Compared to the complex phenotypes observed for the G20P and G20I isolates described above and in Table 1, we can classify the *bos1* mutant phenotypes into three categories: 1/ those displaying high or moderate resistance levels to all tested fungicide categories associated with sensitivity to hyperosmolarity (i.e. G278D, G415D, I365S T581P, A493T); 2/ those displaying weak resistance to the dicarboximide iprodione, susceptibility to the phenylpyrroles including pyrrolnitrine and no osmosensitivity (i.e., I365S, I365S V239F), and 3/ mutants resistant to the dicarboximide, but not to phenylpyrroles and sensitive to hyperosmolarity (i.e. I365S E529A, I365S M427T). Comparing mutant phenotypes between Tables 1 and 3, the following mutations showed the expected phenotypes: I365S, G278D, G415D, A493T, I365S T581P. The *bos1* mutations I365S V239F, I365S E529A, I365S M427T introduced through site-directed mutagenesis lead to phenotypes diverging from those of the iprodione-induced mutants they have been identified in.

HAMP structure modelling

In order to get a clearer picture of the consequences on HAMP structures induced by the amino acid replacements, we conducted homology structure modelling on the Bos1 wild-type and mutant HAMP domains. Prior to performing the model analysis, we defined the HAMP domains in the Bos1 protein relative to the currently available domain databases rather than those published in 2002 [19] in order to adjust the coordinates of Bos1 HAMPs to

those with established crystal structures [38,51]. An HMM scan [53] revealed six HAMP domains with HMM scores >10 and c-values>0.01 (data not shown). We based our HAMP domain alignment presented in Figure 1 on the coordinates obtained in this search. The central core of the six repeats shows the typical feature of HAMP domains (reviewed in [32] and [54]): two α helices, AS1 and AS2 with heptad pattern of hydrophobic residues, as highlighted by the asterisks in Figure 1; the connector between AS1 and AS2 shows the conserved glycine residues (g) at each extremity in addition to the two hydrophobic residues indicated by "*" in Figure 1. This last feature is in agreement with the classification of divergent HAMPs proposed by Dunin-Horkawicz & Lupas [54]. The absence of the Pro6 residue conserved in canonical HAMPs also fits the classification as divergent HAMP.

Considering the different mutations selected and validated by mutagenesis, all HAMP domains were affected either in AS1 or AS2. The mutations related to high levels of cross-resistance to dicarboximides and phenylpyrroles affect HAMPs 2 to 5 (red highlights in Figure 1). The mutations leading to low resistance to iprodione without cross-resistance to pyrrolnitrin and phenylpyrroles (I365S, I365S V239, shaded in green in Figure 1) affect HAMPs 1 and 3, although the major phenotypic affect is probably due to I365S in the N-terminus of HAMP3. The third phenotypic category (moderate resistance to iprodione and osmosensitivity) represented by the mutations E529A, M427T, highlighted in pink in Figure 1, and found in association with I365S, affect helices AS2 of HAMP3 and HAMP4.

Homology-modeling analyses were performed on the SWISS-MODEL workspace [45] based on the structure 3lnrA of the poly-HAMP domains from the *Pseudomonas aeruginosa* receptor Aer2 [35]. All six HAMP domains have a structure similar to that presented in Figure 1, (see Figure S1). AS1 and AS2 can be easily distinguished and are separated by the connector encompassed between the two conserved glycine residues. The AS2s of all HAMPs have a perfect helical structure even C-terminal of the conserved fraction (gray box in Figure 1). In contrast, the helical structures N-terminal of AS1s are not perfectly conserved. The structures of HAMP2, 3 and 4 look highly similar, whereas those of HAMP1, 5 and 6 show differences especially in and N-terminal of AS1. In the case of HAMP1, no helix N-terminal of AS1 was predicted. In HAMP5 the helical part of AS1 seems shorter, and in HAMP6, the helical part of AS1 is shifted towards the N-terminus. The last four residues preceding the conserved glycine residue, namely AATD, do not seem to form a helix.

Impact of the Bos1 modifications on HAMP structures

We then compared the predicted structures of the mutated HAMP domains to the corresponding wild-type domains. Table 3 summarizes the differences observed between the mutant and the wild-type peptides. Briefly, all modifications replacing a hydrophobic, aliphatic residue by a polar or aromatic residue impacted the AS1 and/or AS2 helical structures, whereas the replacements of polar amino acids did not produce any change in the structure predictions (e.g., T581P, E529A). It became evident that all structural changes affected either the HAMP specific helixes AS1 helix or AS2. We therefore focused our analyses on these regions supposing that the interactions of both could be important for Bos1 function in signal transduction. When zooming the molecular models on the AS1 and AS2 helices proximal to the connector, one can identify for each HAMP, pairs of four to five amino acids facing each other, respectively between AS1 and AS2 (amino acids highlighted in yellow in Figure 1 and pointed out in Figures 2 and 3). The 2D representation of these helices (Figure 4)

Table 3. Modifications in Bos1 peptide sequence and associated phenotypic and structural changes.

Bos1 peptide sequence	Phenotype[a]				HAMP[b]	AS[c]	aa change[d]	structural changes[e]	model (HAMP n°)
	Osmotic stress	Pyrrolnitrin	Phenylpyrroles	Iprodione					
I365S (n = 2)	R	S	S	LR	3	out	NH to PU	Yes	HAMP3: start of AS1 and AS2 helical structures displaced. Interacting residues different from wt HAMP3.
I365S, V239F	R	S	S	LR	3, 1	2	ali to aro	Yes	HAMP1: middle helix of AS2 dist. to V239 lost, but most interactions maintained.
G278D	S	HR	HR	HR	2	1	NH to -	Yes	HAMP2: helix prox. to G311 lost, new helical region in AS1 incl. D278
G415D	S	HR	HR	HR	3	2	NH to -	Yes	HAMP3: severe impact on AS2 helixes
A493T	S	MR	MR/HR	HR	4	1	NH to PU	Yes	HAMP4: destruction of AS1's c-terminal helix
I365S, T581P	S	HR	HR	HR	3, 5	1	PU to NH	No	HAMP5: no visible changes
I365S, E529A (n = 2)	S	S/LR	S(Flu)/LR(Fen)[f]	MR	3, 4	2	PU to NH	No	HAMP4: no visible changes
I365S, M427T (n = 2)	S	S	S/LR	MR	3	2	NH to PU	Yes	HAMP3: similar to HAMP3[I365S]

[a]S: sensitive, LR: low resistance, MR: moderate resistance, HR: high resistance, according to the resistance level classification explained in Materials & Methods. nd: not determined.
[b]modified HAMP; numbering according to the peptide modifications cited in the first column.
[c]AS: amphiphilic helix in HAMP.
[d]Characteristics of amino acids: PU = polar uncharged, NH = non polar hydrophobic, + = positively charged, − = negatively charged, ali = aliphatic, aro = aromatic; na: not applicable.
[e]Structural changes as presented in Figure 2 and explained in the text.
[f]Flu = Fludioxonil; Fen = Fenpiclonil.

shows an identical arrangement and a strong sequence conservation of the putatively interacting hydrophobic residues in the four HAMP domains.

Figures 2 and 3 show the structural changes observed for the validated mutations. No structural changes were observed for HAMP5 with the selected mutants regardless of the model structure (3lnrA or 1qu7A) used (Table 3 and data not shown).

V239F. This mutation – observed in combination with I365S - replaces the aliphatic valine, located in the interacting region of AS2 of HAMP1, by the aromatic residue phenylalanine. The replacement abolishes the helical structure C-terminal of F239. Although the HAMP1 wt and mutant structures do not perfectly superimpose, most of the hydrophobic interactions seem to be maintained, such as those between amino acids I209 and I233, between A212, V213 and F236, K237. Only, the interacting amino acid I206 has been displaced from I233 (Figure 2A).

G278D. The replacement of glycine 278, located N-terminal of HAMP2 AS1, by the aspartic acid residue leads to a predicted perturbation of the AS1 helical structure distancing I301 and V308 from interaction (Figure 2B).

G415D. This mutation affects the conserved glycine between the connector and AS2 in HAMP3, by an aspartic acid residue instead of a cysteine. Its impact seems limited to HAMP3 AS2, perturbing the connector proximal helix and displacing I417 from interaction (Figure 3A).

I365S. Although the model of HAMP3 isoform I365 is not well supported (QMean Z score −4.18) by the alignment to the 3lnrA structure, one consequence of this amino acid replacement shown in Figure 3A is that both helical structures are shifted towards the N-terminus. The first helical structure is predicted between A381 and I393 instead of AS1 between L386 and K402. The second helix starts at Q407 in the middle of the connector sequence. Given the observed distance between both helices, the interacting residues are not the same as in wt HAMP3. Instead of I393, V396, A399 and V400 the interacting residues in the first helix seem to be M382, N385, L386, Q389, and I393. According to the model, they may interact with residues V409, E412, E416, and L420 of the second helix.

I365S M427T. The mutation of this hydrophobic residue located in AS2 by the polar threonine in HAMP3 harbouring already the I365S replacement, only produces a minor impact on AS2 compared to HAMP3^{I365S}, visible in the overlay of both predicted structures (Figure 3A, lower panel).

A493T. The replacement of the hydrophobic alanine in HAMP4's AS1 by the polar threonine abolishes the C-terminal helical structure of AS1, displacing the interacting residues A491 and V492 (Figure 3B).

T581P. No visible changes were observed for HAMP5 structure after the replacement of the polar threonine by the hydrophobic proline. However, the predictions for HAMP5 and HAMP6 based on the 3lnrA crystal structure are less well supported than those of the other HAMP domains (QMEAN Z-Score: −4.42, according to [55]). Consequently, we cannot exclude an impact of the mutation T581P on HAMP5 structure *per se*.

Discussion

Phenylpyrrole and dicarboximide fungicides affect the fungal osmotic signal transduction cascade. Although their precise mode of action remains unclear, Pillonel and Meyer [30] showed differences in protein kinase inhibition profiles between phenylpyrrole and dicarboximide fungicides. In many plant pathogenic fungi, isolates cross-resistant to both fungicides harbour mutations

in the osmo-sensing histidine kinase gene (reviewed in [52]). In this study we compared the selection exerted by the dicarboximide iprodione to that exerted by pyrrolnitrin, a natural and structural analogue of phenylpyrroles, on the function and structure of the histidine-kinase Bos1 of *B. cinerea*.

Given the mutants analyzed, different phenotypic categories were selected. Firstly, two parental strains (BC1 and BC26) displayed low resistance to the dicarboximide iprodione without cross-resistance to phenypyrroles or pyrrolnitrin as well as no sensitivity to hyperosmotic conditions. This phenotype matches the previously described ImiR1 phenotype observed among field populations of *B. cinerea* [18,19,56]. Secondly, the mutants selected on pyrrolnitrin showed high levels of cross-resistance to dicarboximides and phenylpyrroles including pyrrolnitrin, as well as osmosensitivity. This phenotype matches the ImiR4 phenotype selected only under laboratory conditions [15,16,56,57]. Moreover, it resembles the *bos1* loss-of-function phenotype observed after gene replacement [20,22], which is unable to transmit the signals derived from phenylpyrroles, dicarboximides and hyperosmotic stress. Mutants isolated on iprodione, with high resistance levels to this dicarboximide, presented low to moderate resistance to pyrrolnitrin and phenylpyrroles, as well as osmosensitivity. Mycelial growth of all selected mutants was clearly affected, no matter the resistance profile. This equally held true for conidiation and aggressiveness on plant tissue. These findings point to a fitness cost associated with the acquisition of fungicide resistance.

The results were comparable between the iprodione-induced mutants from this study and the pyrrolnitrin-induced mutants previously reported [13]. Reduced fitness has been suggested as a possible explanation for the absence of field isolates highly resistant to dicarboximides and phenylpyrroles [2]. There is a practical consequence. If pyrrolnitrin and iprodione cross-resistance occurred in the field, it would have a detrimental effect on the fitness in terms of viability and aggressiveness of the resistant mutants. In turn, the sensitive strains, would gain ground once the fungicides are no longer active. All together, this would ensure the lifespan of phenylpyrroles or of pyrrolnitrin-producing biological control agents.

Seeking to correlate the phenotypic changes to structural modifications in the signal transduction HK Bos1, we first set out to analyze the sequence of the *bos1* gene. In most isolates, we found mutations, from nonsense mutations (G81STOP) – leading to the expected loss-of-function phenotype – to amino-acid replacements in the HAMP domains potentially involved in ST. In most cases, strains with identical mutations showed similar resistance profiles to the dicarboximide iprodione and to the phenylpyrroles suggesting that the mutations could be responsible for the phenotypes. Secondly, using site directed mutagenesis of the *bos1* gene, we were able to test eight of the twelve sequenced point mutations. By comparing these resistance profiles to those of the corresponding *in vivo* selected mutants, we confirmed the role of the following mutations: I365S leading to an ImiR1 phenotype and G278D, G415D, I365S T581P, A493T leading to loss-of-function phenotypes. The *in vitro* mutants with the genotypes I365S V239F, I365S M427T, I365S E529K displaying low to moderate resistance to iprodione, are not in agreement with the phenotypes observed for the *in vivo* mutants (BC1G20I3, H6G20I2/I3, BC26G20I1/I3), which are highly resistant to iprodione associated with moderate resistance to phenylpyrroles. We suspect that additional mutations responsible for higher resistance levels within these genetic backgrounds were selected on high iprodione concentrations.

We confirmed that the A493T mutation confers moderate levels of resistance to phenylpyrroles and iprodione, but as the strains

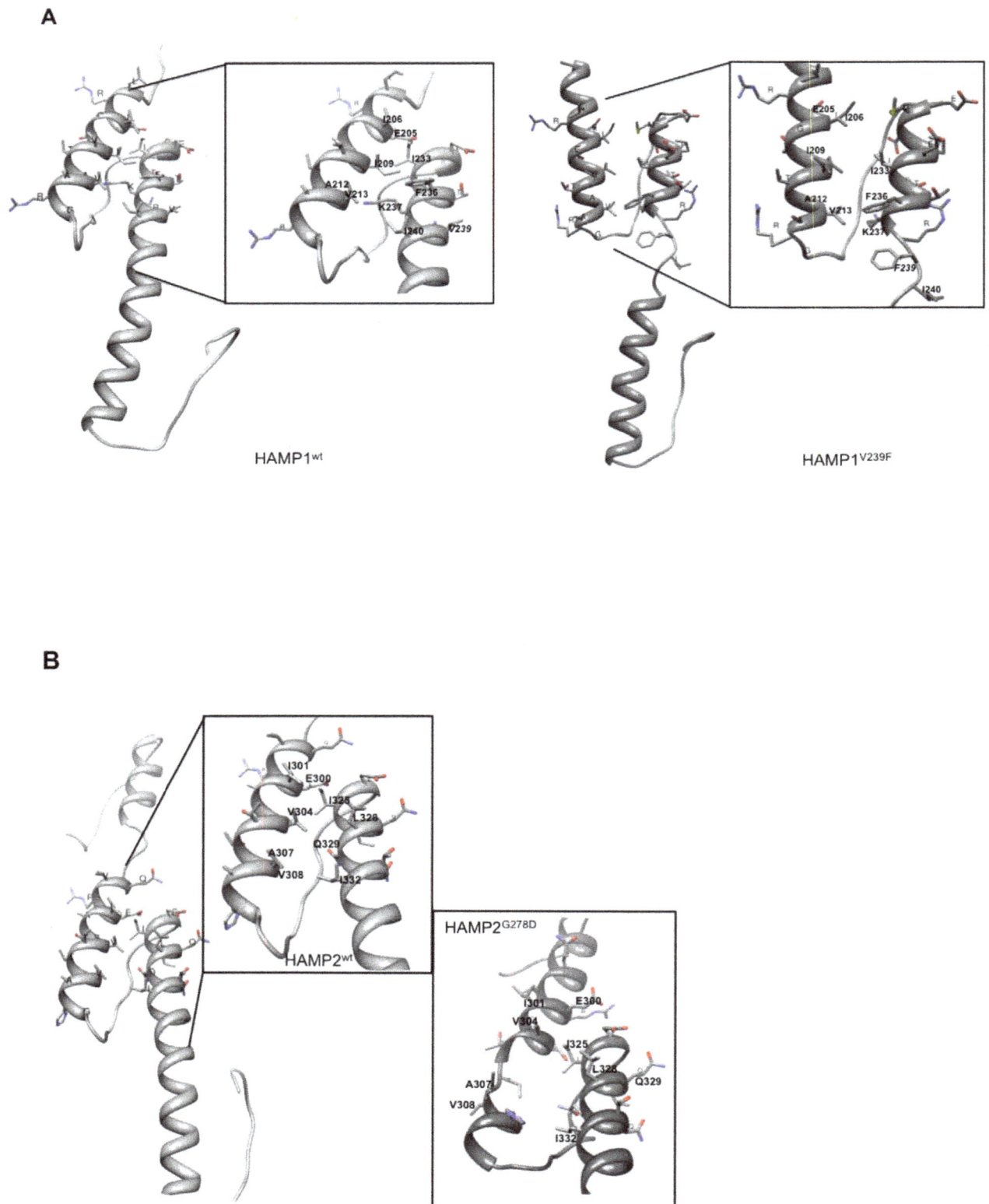

Figure 2. Homology based models of HAMP domains with focus on interactions between AS1 and AS2. A/ HAMP1 wild-type (wt) and mutant peptides. B/ HAMP2 wt and mutant peptides. The orientation of the peptides is from up (N-terminus) to down (C-terminus). Solely, the side chains from amino-acids of the connector proximal helices are displayed. Residues potentially involved in interactions are labelled; modified residues are written in italics.

A

B

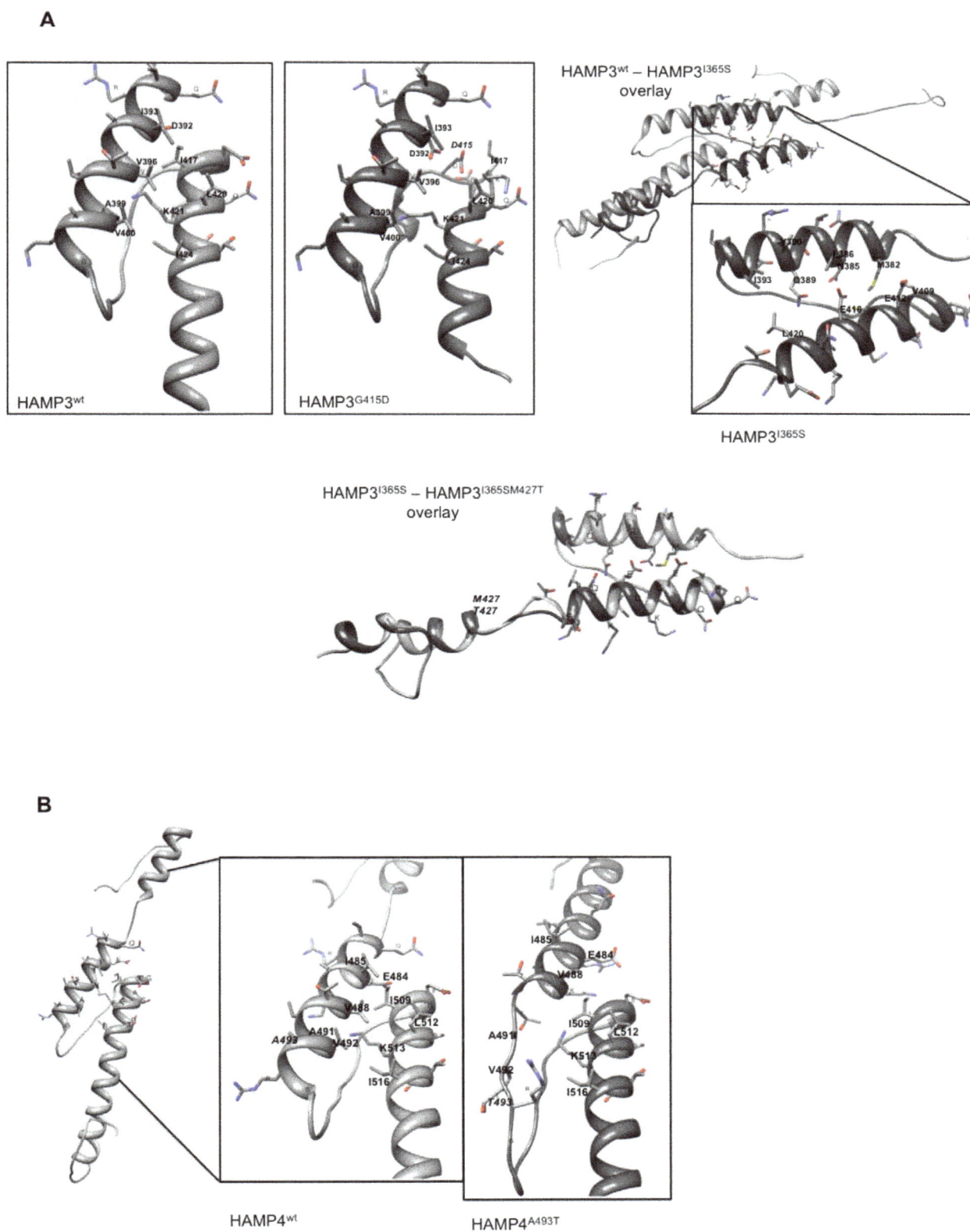

Figure 3. Homology based models of HAMP domains with focus on interactions between AS1 and AS2. A/ HAMP3 wt and mutant peptides. In the last panel I365S (light grey) and I365S M427T (dark grey) isoforms are shown in an overlay. B/ HAMP4 wt and mutant peptides. The orientation of the peptides is from up (N-terminus) to down (C-terminus) or from right to left (A). Solely, the side chains from amino-acids of the connector proximal helices are displayed. Residues potentially involved in interactions are labelled; modified residues are written in italics. In overlaid models, the wild-type peptide is shaded in light grey, the mutant peptide in dark grey.

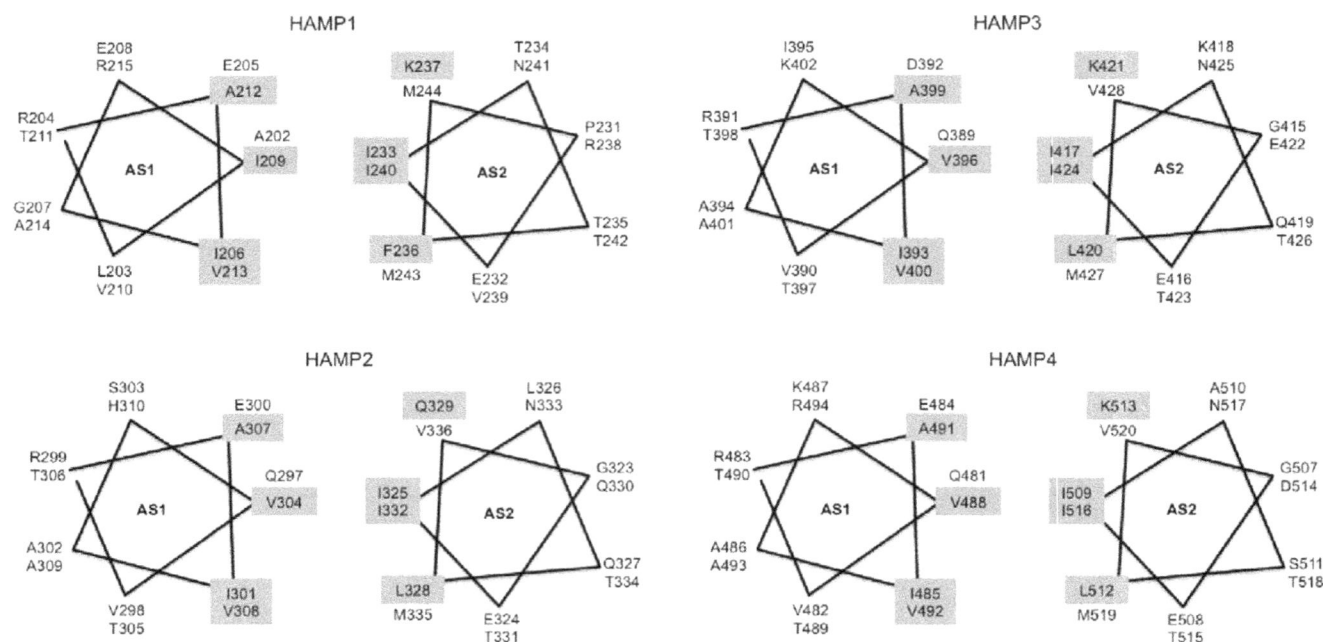

Figure 4. 2D-schematic representation of the helical arrangement of connector proximal helices from HAMP domains 1–4. Two helical turns of seven amino acids each are listed. Both helices are facing each other such as in the model of Figure 1. The grey shading highlights interacting residues. AS1 = amphipathic helix 1; AS2 = amphipathic helix 2.

BC25G20I2 and BC25G20I3 harboring this mutation show different resistance levels to phenylpyrroles. We can therefore hypothesize that at least strain BC25G20I3 carries an additional mutation leading to higher resistance levels to these compounds.

To sum up the molecular data of our mutational analysis: i/ all HAMP domains were affected by modifications inducing resistance. Only the mutation of V239F in HAMP1 did not modify the resistance profile in addition to the I365S mutation present in the initial strain; ii/ the replacement of conserved glycine residues (G278D, G415D) by charged amino acids led to ImiR4 (loss-of-function) phenotypes; iii/ mutations with a low impact on the resistance levels (highlighted in green or purple in Figure 1) localized outside AS1, whereas mutations leading to ImiR4 phenotypes affected either AS1 or AS2. Of particular interest are the mutations E529A and M427T (highlighted in purple in Figure 1), which led to osmosensitivity in the I365S background. They only weakly interfered with sensitivity to phenylpyrroles, but these replacements seem to abolish osmotic ST.

Concerning the structural changes predicted for the HAMP domains and the related phenotypes, our study provides first insights:

Generally, only the modifications of hydrophobic residues impact the helical structures of the HAMP domains, whereas the replacements of polar residues do not seem to interfere with them. We suspect that these perturbations of the helical structures abolish or modify the supposed interactions between AS1 and AS2. The strongest phenotypes observed (loss of function) correlate with the loss of two interacting residues in HAMP2 (mutant G278D), one in HAMP3 (mutant G415D) or two in HAMP4 (mutant A493T). The mutants displaying an ImiR1 phenotype (mutation I365S alone or associated with V239F) have an HAMP3 domain with modified interactions between AS1 and AS2 compared to the wt. The fact that the I365S mutant is not affected for phenylpyrroles and osmotic ST suggest that these newly created interactions in HAMP3 are sufficient to transduce

these signals. Concerning the modifications that did not alter the helical structure, but lead to a modification in signal transduction (T581P, M427T, E529A), we suspect that the affected residues could be important for ST or that the replacement residues might hinder ST – at least in combination with I365S. This holds true particularly for T581P leading to a loss-of-function phenotype in association with I365S, whereas E529A and M427T could be involved principally in osmotic signal transduction.

In summary, our results suggest that mutations resulting in Bos1-loss-of-function phenotypes (those highlighted in red in Figure 1) i.e. completely disrupting signal perception and/or transduction, either abolish important interactions between AS1 and AS2 of HAMPs 2–4 or affect potential key residues in Bos1 ST, such as T581.

In order to better understand differential ST through Bos1 we were particularly interested in mutations conferring resistance to only one chemical family, dicarboximides or phenylpyrroles, or modifying osmosensitivity, because they may have a partially functional Bos1 protein. Using site-directed mutagenesis, we obtained two categories of partially functional Bos1, one leading only to low resistance to dicarboximides (I365S, ImiR1) and the other affecting also osmosensitivity (I356S E529A, I365S M427T). These modifications do not seem to abolish interactions between AS1 and AS2, potentially essential to Bos1 ST.

Altogether our data reveal Bos1 modifications that lead to loss-of-signal-transduction, principally in HAMP domains 2–5. Only changes outside AS1 and the connector domains maintained partial Bos1 function. Some modifications interfere only with dicarboximide (I365S) or osmotic ST (E529A, M427T).

Our study gives a first glimpse on structure-function relationship for differential ST through an eukaryotic HAMP-containing histidine kinase. We analyzed the HAMP structures individually, although the structure of the model protein used for our analyses involves protein dimers [35], we cannot exclude that some mutations affect intra- or inter-molecule interactions different from those we analyzed. It would be interesting to resolve the crystal structure of eukaryotic HAMP containing proteins – especially of histidine kinases with successive HAMP domains – in order to better understand the signal transduction processes regulated by these proteins.

Supporting Information

Figure S1   Predicted models for the HAMP domains of the histidine-kinase Bos1. (A) front, (B) back. Model predictions were performed on the Swiss-model server [45] by alignment to the crystal structure 3lnrA of the aerotaxis receptor Aer2 of *Pseudomonas aeroginosa* [35]. The orientation of the peptides is from up (N-terminus) to down (C-terminus). The side chains of amino acids located in helical regions (A) and in the connector (B) facing the neighbouring helices are presented.

Table S1   *In vitro* response of pyrrolnitrin- and iprodione-induced mutants to pyrrolnitrin, iprodione, and phenylpyrroles.

Table S2   Primers used in this study for amplification and sequencing of the *bos1* gene.

Acknowledgments

The authors gratefully acknowledge Claire Troulet, Gisèle Riqueau, Johann Confais and Christiane Auclair for excellent technical assistance and Anne-Sophie Walker for fruitful discussions.

Author Contributions

Conceived and designed the experiments: SF SA PN MB. Performed the experiments: SF SA MB. Analyzed the data: SF SA PN PL MB. Contributed reagents/materials/analysis tools: SF SA PN PL MB. Wrote the paper: MB SA SF.

References

1. Jarvis RW (1977) *Botryotinia* and *Botrytis* species: taxonomy, physiology, and pathogenicity. A guide to the literature. Ottawa, Canada: Canada Department of Agriculture. 195 p.
2. Leroux P (2004) Chemical control of *Botrytis* and its resistance to chemical fungicides. In: Elad Y, Williamson B, Tudzynski P, Delen N, editors. *Botrytis*: Biology, Pathology and Control. Dordrecht, The Netherlands: Kluwer Academic Press. pp. 195–222.
3. Myresiotis CK, Karaoglanidis GS, Tzavella-Monari K (2007) Resistance of *Botrytis cinerea* isolates from vegetable crops to anilinopyrimidine, phenylpyrrole, hydroxyanilide, benzimidazole, and dicarboximide fungicides. Plant Disease 91: 407–413.
4. Duffy B, Schouten A, Raaijmakers JM (2003) Pathogen self-defense: mechanisms to counteract microbial antagonism. Annual Review of Phytopathology 41: 501–538.
5. Elad Y, Stewart A (2004) Microbial control of *Botrytis* spp. In: Elad Y, Williamson B, Tudzynski P, Delen N, editors. *Botrytis*: Biology, Pathology and Control. Dordrecht, The Netherlands: Kluwer Academic Press. pp. 223–241.
6. Elmer PAG, Reglinski T (2006) Biosuppression of *Botrytis cinerea* in grapes. Plant Pathology 55: 155–177.
7. Raaijmakers JM, Vlami M, de Souza JT (2002) Antibiotic production by bacterial biocontrol agents. Antonie Van Leeuwenhoek International (Journal of General and Molecular Microbiology) 81: 537–547.
8. Arima K, Imanaka H, Kousaka M, Fukuta A, Tamura G (1964) Pyrrolnitrin, a new antibiotic substance, produced by Pseudomonas. Agricultural and Biological Chemistry 28: 275–276.
9. Hammer PE, Burd W, Hill DS, Ligon JM, van Pee KH (1999) Conservation of the pyrrolnitrin biosynthetic gene cluster among six pyrrolnitrin-producing strains. FEMS Microbiology Letters 180: 39–44.
10. Chernin L, Brandis A, Ismailov Z, Chet I (1996) Pyrrolnitrin production by an *Enterobacter agglomerans* strain with a broad spectrum of antagonistic activity towards fungal and bacterial phytopathogens. Current Microbiology 32: 208–212.
11. Janisiewicz WJ, Roitman J (1988) Biological control of blue mold and grey mold on apple and pear with *Pseudomonas cepacia*. Phytopathology 78: 1697–1700.
12. Ajouz S, Walker AS, Fabre F, Leroux P, Nicot PC, et al. (2011) Variability of *Botrytis cinerea* sensitivity to pyrrolnitrin, an antibiotic produced by biological control agents. BioControl 56: 353–363.
13. Ajouz S, Nicot PC, Bardin M (2010) Adaptation to pyrrolnitrin in *Botrytis cinerea* and cost of resistance. Plant Pathology 59: 556–566.
14. Leroux P, Lanen C, Fritz R (1992) Similarities in the antifungal activities of fenpiclonil, ipridione and tolclofos-methyl against Botrytis cinerea and Fusarium nivale. Pesticide Science 36: 255–261.
15. Leroux P, Chapeland F, Desbrosses D, Gredt M (1999) Patterns of cross-resistance to fungicides in *Botryotinia fuckeliana* (*Botrytis cinerea*) isolates from French vineyards. Crop Protection 18: 687–697.
16. Faretra F, Pollastro S (1993) Isolation, characterization, and genetic analysis of laboratory mutants of *Botryotinia fuckeliana* resistant to the phenylpyrrole fungicide CGA 173506. Mycological Research 97: 620–624.
17. Ziogas BN, Markoglou AN, Spyropoulou V (2005) Effect of phenylpyrrole-resistance mutations on ecological fitness of *Botrytis cinerea* and their genetical basis in *Ustilago maydis*. European Journal of Plant Pathology 113: 83–100.
18. Cui W, Beever RE, Parkes SL, Templeton MD (2004) Evolution of an osmosensing histidine kinase in field strains of *Botryotinia fuckeliana* (*Botrytis cinerea*) in response to dicarboximide fungicide usage. Phytopathology 94: 1129–1135.

19. Cui W, Beever RE, Parkes SL, Weeds PL, Templeton MD (2002) An osmosensing histidine kinase mediates dicarboximide fungicide resistance in *Botryotinia fuckeliana* (*Botrytis cinerea*). Fungal Genetics and Biology 36: 187–198.
20. Liu W, Leroux P, Fillinger S (2008) The HOG1-like MAP kinase Sak1 of *Botrytis cinerea* is negatively regulated by the upstream histidine kinase Bos1 and is not involved in dicarboximide- and phenylpyrrole-resistance. Fungal Genetics and Biology 45: 1062–1074.
21. Oshima M, Fujimura M, Banno S, Hashimoto C, Motoyama T, et al. (2002) A point mutation in the two-component histidine kinase BcOS-1 confers dicarboximide resistance in field isolates of *Botrytis cinerea*. Phytopathology 92: 75–80.
22. Viaud M, Fillinger S, Liu W, Polepalli JS, Le Pecheur P, et al. (2006) A class III histidine kinase acts as a novel virulence factor in *Botrytis cinerea*. Molecular Plant-Microbe Interactions 19: 1042–1050.
23. Avenot H, Simoneau P, Iacomi-Vasilescu B, Bataille-Simoneau N (2005) Characterization of mutations in the two-component histidine kinase gene AbNIK1 from *Alternaria brassicicola* that confer high dicarboximide and phenylpyrrole resistance. Current Genetics 47: 234–243.
24. Dry IB, Yuan KH, Hutton DG (2004) Dicarboximide resistance in field isolates of *Alternaria alternata* is mediated by a mutation in a two-component histidine kinase gene. Fungal Genetics and Biology 41: 102–108.
25. Kanetis L, Forster H, Jones CA, Borkovich KA, Adaskaveg JE (2008) Characterization of genetic and biochemical mechanisms of fludioxonil and pyrimethanil resistance in field isolates of *Penicillium digitatum*. Phytopathology 98: 205–214.
26. Ma ZH, Luo Y, Michailides T (2006) Molecular characterization of the two-component histidine kinase gene from *Monilinia fructicola*. Pest Management Science 62: 991–998.
27. Nathues E, Jorgens C, Lorenz N, Tudzynski P (2007) The histidine kinase CpHK2 has impact on spore germination, oxidative stress and fungicide resistance, and virulence of the ergot fungus *Claviceps purpurea*. Molecular Plant Pathology 8: 653–665.
28. Okada A, Banno S, Ichiishi A, Kimura M, Yamaguchi I, et al. (2005) Pyrrolnitrin interferes with osmotic signal transduction in *Neurospora crassa*. Journal of Pesticide Science 30: 378–383.
29. Orth AB, Rzhetskaya M, Pell EJ, Tien M (1995) A serine (threonine) protein kinase confers fungicide resistance in the phytopathogenic fungus *Ustilago maydis*. Applied And Environmental Microbiology 61: 2341–2345.
30. Pillonel C, Meyer T (1997) Effect of phenylpyrroles on glycerol accumulation and protein kinase activity of *Neurospora crassa*. Pesticide Science 49: 229–236.
31. Catlett NL, Yoder OC, Turgeon BG (2003) Whole-genome analysis of two-component signal transduction genes in fungal pathogens. Eukaryotic Cell 2: 1151–1161.
32. Parkinson JS (2010) Signaling Mechanisms of HAMP Domains in Chemoreceptors and Sensor Kinases. Annual Review of Microbiology 64: 101–122.
33. Aravind L, Ponting CP (1999) The cytoplasmic helical linker domain of receptor histidine kinase and methyl-accepting proteins is common in many prokaryotic signalling proteins. FEMS Microbiol Lett 176: 111–116.
34. Hulko M, Berndt F, Gruber M, Linder JU, Truffault V, et al. (2006) The HAMP domain structure implies helix rotation in transmembrane signaling. Cell 126: 929–940.
35. Airola MV, Watts KJ, Bilwes AM, Crane BR (2010) Structure of Concatenated HAMP Domains Provides a Mechanism for Signal Transduction. Structure 18: 436–448.

36. Zhou Q, Ames P, Parkinson JS (2009) Mutational analyses of HAMP helices suggest a dynamic bundle model of input-output signalling in chemoreceptors. Molecular Microbiology 73: 801–814.

37. Ferris HU, Dunin-Horkawicz S, Mondejar LG, Hulko M, Hantke K, et al. (2011) The mechanisms of HAMP-mediated signaling in transmembrane receptors. Structure 19: 378–385.

38. Watts KJ, Johnson MS, Taylor BL (2008) Structure-function relationships in the HAMP and proximal signaling domains of the aerotaxis receptor Aer. Journal of Bacteriology 190: 2118–2127.

39. Meena N, Kaur H, Mondal AK (2010) Interactions among HAMP domain repeats act as an osmosensing molecular switch in group III hybrid histidine kinases from fungi. Journal of Biological Chemistry 285: 12121–12132.

40. Ajouz S, Decognet V, Nicot PC, Bardin M (2010) Microsatellite stability in the plant pathogen *Botrytis cinerea* after exposure to different selective pressures. Fungal Biology 114: 949–954.

41. Decognet V, Bardin M, Trottin-Caudal Y, Nicot PC (2009) Rapid change in the genetic diversity of *Botrytis cinerea* populations after the introduction of strains in a tomato glasshouse. Phytopathology 99: 185–193.

42. Dellaporta SL, Wood J, Hicks JB (1983) A plant DNA minipreparation: version 2. Plant Molecular Biology Reporter 1: 19–21.

43. Amselem J, Cuomo CA, van Kan JAL, Viaud M, Benito EP, et al. (2011) Genomic Analysis of the Necrotrophic Fungal Pathogens *Sclerotinia sclerotiorum* and *Botrytis cinerea*. PLoS Genet 7: e1002230.

44. Levis C, Fortini D, Brygoo Y (1997) Transformation of *Botrytis cinerea* with the nitrate reductase gene (niaD) shows a high frequency of homologous recombination. Current Genetics 32: 157–162.

45. Arnold K, Bordoli L, Kopp J, Schwede T (2006) The SWISS-MODEL workspace: a web-based environment for protein structure homology modelling. Bioinformatics 22: 195–201.

46. Zdobnov EM, Apweiler R (2001) InterProScan–an integration platform for the signature-recognition methods in InterPro. Bioinformatics 17: 847–848.

47. Thompson JD, Gibson TJ, Higgins DG (2002) Multiple sequence alignment using ClustalW and ClustalX. Curr Protoc Bioinformatics Chapter 2: Unit 2 3.

48. Kim KK, Yokota H, Kim S-H (1999) Four-helical-bundle structure of the cytoplasmic domain of a serine chemotaxis receptor. Nature 400: 787–792.

49. Edgar RC (2004) MUSCLE: multiple sequence alignment with high accuracy and high throughput. Nucleic Acids Research 32: 1792–1797.

50. Pettersen EF, Goddard TD, Huang CC, Couch GS, Greenblatt DM, et al. (2004) UCSF Chimera–a visualization system for exploratory research and analysis. Journal of Computational Chemistry 25: 1605–1612.

51. Ajouz S, Bardin M, Nicot PC, El Maataoui M (2011) Comparison of the development *in planta* of a pyrrolnitrin-resistant mutant of *Botrytis cinerea* and its sensitive wild-type parent isolate. European Journal of Plant Pathology 129: 31–42.

52. Alberoni G, Collina M, Lanen C, Leroux P, Brunelli A (2010) Field strains of *Stemphylium vesicarium* with a resistance to dicarboximide fungicides correlated with changes in a two-component histidine kinase. European Journal of Plant Pathology 128: 171–184.

53. Finn RD, Clements J, Eddy SR (2011) HMMER web server: interactive sequence similarity searching. Nucleic Acids Research 39: W29–W37.

54. Dunin-Horkawicz S, Lupas AN (2010) Comprehensive Analysis of HAMP Domains: Implications for Transmembrane Signal Transduction. Journal of Molecular Biology 397: 1156–1174.

55. Benkert P, Biasini M, Schwede T (2011) Toward the estimation of the absolute quality of individual protein structure models. Bioinformatics 27: 343–350.

56. Leroux P, Fritz R, Debieu D, Albertini C, Lanen C, et al. (2002) Mechanisms of resistance to fungicides in field strains of *Botrytis cinerea*. Pest Management Science 58: 876–888.

57. Vignutelli A, Hilber-Bodmer M, Hilber UW (2002) Genetic analysis of resistance to the phenylpyrrole fludioxonil and the dicarboximide vinclozolin in *Botryotinia fuckeliana* (*Botrytis cinerea*). Mycological Research 106: 329–335.

3

Widespread Occurrence of Chemical Residues in Beehive Matrices from Apiaries Located in Different Landscapes of Western France

Olivier Lambert[1,4,5]*, **Mélanie Piroux**[1,4,5], **Sophie Puyo**[1], **Chantal Thorin**[2], **Monique L'Hostis**[1], **Laure Wiest**[3], **Audrey Buleté**[3], **Frédéric Delbac**[4,5], **Hervé Pouliquen**[1]

[1] LUNAM Université, Oniris, Ecole Nationale Vétérinaire, Agroalimentaire et de l'Alimentation Nantes-Atlantique, Plateforme Environnementale Vétérinaire, Centre Vétérinaire de la Faune Sauvage et des Ecosystèmes des Pays de la Loire (CVFSE), Nantes, France, [2] LUNAM Université, Oniris, Ecole Nationale Vétérinaire, Agroalimentaire et de l'Alimentation Nantes-Atlantique, Unité de Physiopathologie Animale et Pharmacologie Fonctionnelle, Nantes, France, [3] Université de Lyon, Institut des Sciences Analytiques, Département Service Central d'Analyse, UMR 5280 CNRS, Université de Lyon1, ENS-Lyon, Villeurbanne, France, [4] Clermont Université, Université Blaise Pascal, Laboratoire Microorganismes: Génome et Environnement, BP 10448, Clermont-Ferrand, France, [5] CNRS, UMR 6023, LMGE, Aubière, France

Abstract

Background: The honey bee, *Apis mellifera*, is frequently used as a sentinel to monitor environmental pollution. In parallel, general weakening and unprecedented colony losses have been reported in Europe and the USA, and many factors are suspected to play a central role in these problems, including infection by pathogens, nutritional stress and pesticide poisoning. Honey bee, honey and pollen samples collected from eighteen apiaries of western France from four different landscape contexts during four different periods in 2008 and in 2009 were analyzed to evaluate the presence of pesticides and veterinary drug residues.

Methodology/Findings: A multi-residue analysis of 80 compounds was performed using a modified QuEChERS method, followed by GC-ToF and LC−MS/MS. The analysis revealed that 95.7%, 72.3% and 58.6% of the honey, honey bee and pollen samples, respectively, were contaminated by at least one compound. The frequency of detection was higher in the honey samples (n = 28) than in the pollen (n = 23) or honey bee (n = 20) samples, but the highest concentrations were found in pollen. Although most compounds were rarely found, some of the contaminants reached high concentrations that might lead to adverse effects on bee health. The three most frequent residues were the widely used fungicide carbendazim and two acaricides, amitraz and coumaphos, that are used by beekeepers to control *Varroa destructor*. Apiaries in rural-cultivated landscapes were more contaminated than those in other landscape contexts, but the differences were not significant. The contamination of the different matrices was shown to be higher in early spring than in all other periods.

Conclusions/Significance: Honey bees, honeys and pollens are appropriate sentinels for monitoring pesticide and veterinary drug environmental pollution. This study revealed the widespread occurrence of multiple residues in beehive matrices and suggests a potential issue with the effects of these residues alone or in combination on honey bee health.

Editor: Nicolas Desneux, French National Institute for Agricultural Research (INRA), France

Funding: This project was supported by the European Union through European Agricultural Guidance Guarantee fund projects, Nantes Métropole, le Conseil Général de Loire Atlantique, le Conseil Régional des Pays de la Loire and Pullman hotels by Bee My Friend. The funders had no role in study design, data collection and analysis, decision to publish, or preparation of the manuscript.

Competing Interests: The authors have declared that no competing interests exist.

* E-mail: olivier.lambert@oniris-nantes.fr

Introduction

Since the middle of the twentieth century, profound changes have occurred and damaged the ecological balance. Industrialization, growing urbanization, transportation and agricultural practices have led to overall ecosystem contamination and to major modifications in landscape structure and composition. These changes have had adverse effects on biodiversity, causing physiological and behavioral damage to living organisms and altering organism habitats and the quality and/or the quantity of food resources [1−5]. The consequences have been disastrous, particularly as supplementary stressors, such as infectious agents or invasive species, may be added [6−9].

Bees are at the center of this issue. First, although honey bee populations are globally increasing throughout the world, unprecedented colony losses have been reported in Europe and North America over the last decade, and the number of hives that must be replaced each year has drastically increased [10−11]. Multiple causes are suspected including (i) climate change; (ii) reduction of floral diversity and quality resources in relation to monocultural practices and fragmentation of natural habitats; (iii) infection by pathogens, including viruses, bacteria, fungi and parasites [12−15]; and (iv), poisoning by chemical compounds, including pesticides [16−17]. Even if each individual cause may have a real impact on honey bee health, no factor has emerged as the definitive and single stressor responsible for this decline. Many authors actually suggest that pesticides are not involved [12] because most field studies have demonstrated that pesticides have not been found at levels that would be harmful to bees. Thus, a combination of biological, chemical and physical stressors would be the most probable explanation for extensive colony losses. In particular, recent studies have reported that parasite-insecticide interactions can synergistically and negatively affect honey bee survival [18−22].

Second, honey bee and other beehive matrices are recognized as appropriate sentinels for monitoring anthropogenic contamination in the environment [23−24]. Honey bees are exposed to atmospheric pollutants during their foraging activities, their hairy bodies easily hold residues, and they may be exposed to contaminants *via* contaminated food resources such as nectar, pollen or water. Therefore, many studies have used honey bees, pollen or honey as relevant samples to assess the levels of heavy metals [25−26] and polycyclic aromatic hydrocarbons [27] in both wild and anthropogenic areas, but pesticides are also of concern in agricultural areas [28−31].

In this context, the Wildlife and Ecosystems Veterinary Center of Pays de la Loire (CVFSE/Oniris) conducted a program concerning the use of honey bees (*Apis mellifera* L.), honey and pollen for monitoring lead [32], polycyclic aromatic hydrocarbons [33] and pesticide environmental pollution in Pays de la Loire (western France). The aim of the present study was to investigate the contamination of 18 apiaries by pesticide residues through analyses of 3 different matrices over 2 years (2008 and 2009). The sampled apiaries were located in four different landscapes susceptible to various contamination levels due to different uses of pesticides and veterinary drugs (gardening, agricultural, herd breeding or apicultural practices). To our knowledge, this study is the first to compare the contamination of 3 matrices in 4 different landscape contexts. The temporal distribution of the pesticide and veterinary drug concentrations and the choice of the most relevant matrices for monitoring environmental pesticide and veterinary drug contamination are discussed.

Materials and Methods

Study sites

Apicultural matrices were collected from 18 apiaries located in four different landscapes from western France (Bretagne and Pays de la Loire) (**Fig. 1**). Two apiaries were located on small islands (Isle of Ouessant, I1, and Isle of Yeu, I2) that are free from high levels of anthropogenic activities. These islands were selected to represent landscapes with low levels of pesticides. Six apiaries (RG1 to RG6) were located in a rural-grassland landscape characterized by high length of hedges and numerous grassland plots. Five apiaries (RC1 to RC5) were located in a rural-cultivated landscape characterized by large plots of crops (permanent, oil seed, grain crops, and market gardening) and a low hedgerow network. The pesticide display in these 11 rural-sites is reflective of agricultural practices and veterinary treatments of farm animals. Finally, five apiaries (U1 to U5) were located in an urban landscape characterized by large urban areas and some rural areas. The observed pesticides in these apiaries are reflective of leisure gardening, and a small number of these pesticides emanate from agricultural treatments.

Sample collection

Three different biological matrices (foraging honey bees, trap pollen and honey) were collected from eight colonies randomly selected at each apiary. The samples were always collected from the same eight colonies during the survey. Otherwise, the number of hives sampled was kept similar by replacement of each dead colony. For the final assessment, this replacement was not subjected to a special statistical treatment. The apiaries were visited four times in both 2008 (periods AM8, JJ8, JA8 and SO8) and 2009 (periods AM9, JJ9, JA9 and SO9). In terms of seasons, the apiaries were visited in spring (late April-early May, periods AM8 and AM9), at the beginning of summer (late June-early July, periods JJ8 and JJ9), in summer (late July-early August, periods JA8 and JA9), and at the beginning of autumn (late September-early October, periods SO8 and SO9). During a single period, all samples were collected when possible within 10 days to minimize variations in climatic factors, flowering and pesticide treatments. The owners of the studied apiaries were present for each sampling, and their names are being kept confidential.

The honey bees were directly collected from the hive's flight board with a hand-held vacuum cleaner. The pollens were collected in pollen traps installed by beekeepers three days before the sampling. As foraging activities depend on meteorological conditions, some pollen samples were missing due to low temperatures or bad weather, especially at the beginning of autumn. Honey samples were collected from several honeycombs with a cutter or with a punch. Uncapped honeycombs were chosen (when possible in beehive rises) to collect fresh honey. In an apiary, each colony displays specific foraging activities that may not be representative of the whole apiary. However, for the purpose of this study, *i.e.*, the use of the apicultural matrices as sentinels for monitoring the contamination by pesticides and veterinary drugs around or in each apiary, samples of honey bees, honey and pollen collected in the hives of the same apiary and at the same period were pooled. These field-collected pools were immediately placed on ice after sampling and then stored at −20°C until analysis. In total, 141 honey bee samples, 141 honey samples and 128 pollen samples were collected.

Sample preparation, analysis and method performance

A multi-residue analysis was developed to identify and quantify 80 pesticides (gardening and agricultural) and veterinary drugs in the three beehive matrices. The 80 compounds covered 21 families of contaminants and corresponded to the majority of pesticides used for plant protection and some veterinary drugs used for treatments of farm animals or in apicultural practices to control the parasitic mite *Varroa destructor* (**Table 1**). The method consisted of a single extraction, based on a modified "QuEChERS method" ("Quick Easy Cheap Effective Rugged Safe method"), followed by gas chromatography coupled with time-of-flight mass spectrometry (GC-ToF) and liquid chromatography coupled with tandem mass spectrometry (LC−MS/MS) as previously described [34]. The combination of the "QuEChERS method" with the sensitive GC-ToF and LC−MS/MS analytical techniques enabled the detection of pesticide concentrations as low as 10 ng/g in honey bee, honey and pollen samples. The limit of detection (LOD) and limit of quantification (LOQ) of each chemical compound are given in **Table 1**.

Statistical analyses

The statistical parameters for the concentrations (mean, median and standard deviation) were calculated from all the analyzed samples of each matrix and not only from samples for which the residues were detected or quantified. When a compound was not detected (< LOD), the concentration used for statistical analysis was ½ LOD [35]. When a compound was not quantified (> LOD and < LOQ), the concentration used was ½ (LOD + LOQ).

For each matrix, statistical analyses were performed only for residues that were detected or quantified at least once. We

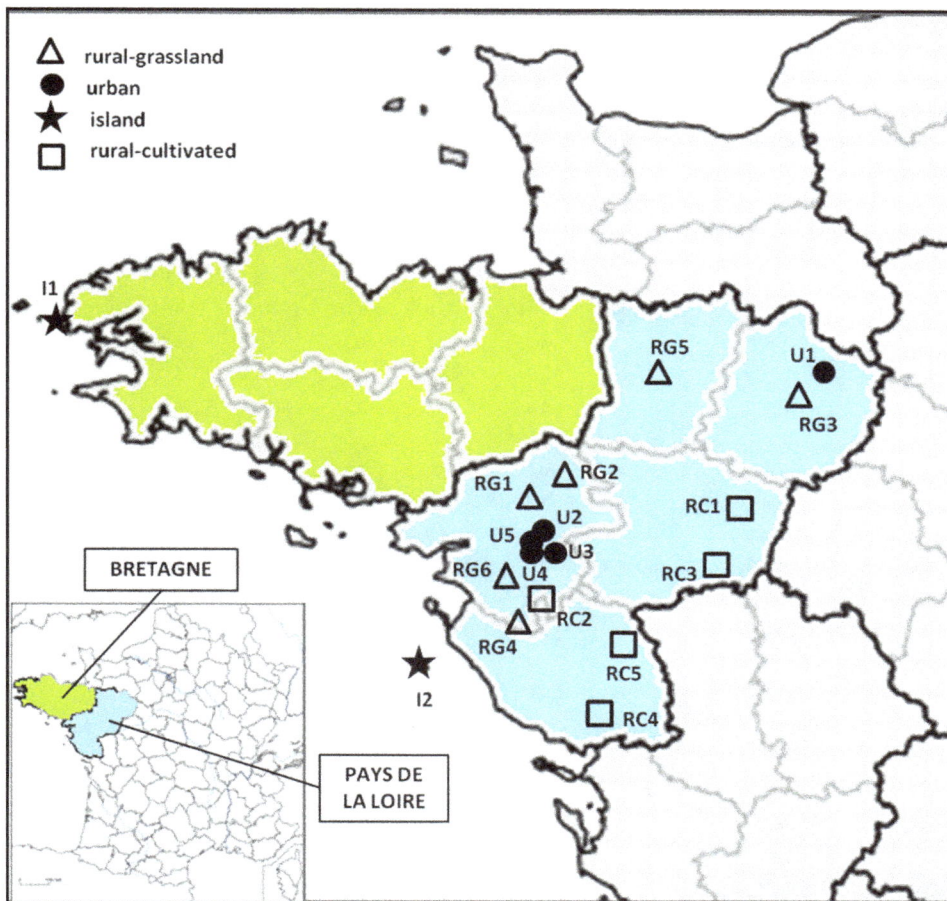

Figure 1. Location of the 18 surveyed apiaries. The apiaries are located in four different landscape contexts (rural-grassland landscapes: RG1 to RG6, rural-cultivated landscapes: RC1 to RC5, urban landscapes: U1 to U5, islands: I1 and I2) from two regions of western France (Bretagne in green and Pays de la Loire in blue).

transformed the data into a present/absent dataset and considered the number of compounds detected and/or quantified in each sample as an explicative variable in the following models.

Linear mixed effects models were used to perform a comparison between the number of residues detected or quantified (i) in honey bees (n = 141), honey (n = 141) and pollen (n = 128); (ii) in different landscape structures (rural-grassland, rural-cultivated, urban and island); and (iii) for different sampling periods (AM8, JJ8, JA8, SO8, AM9, JJ9, JA9 and SO9). These models were the best way to take into account the repeated measurements on each apiary. The linear mixed effect models are the theoretical presentation of ANOVA for repeated measurements.

For each of these three models, the assumption of independence and normality of the residues and random effects was checked through diagnostic graphs generated by the parametric estimation theory of mixed effects models (data not shown) [36]. Then, Tukey post-hoc tests (a specific version designed for mixed effects models) were used to implement multiple comparisons of the means in each model, (i) difference in matrix, (ii) difference in landscape and (iii) difference in sampling periods.

The statistical analyses were performed using R software with the "nlme package" for the mixed effects and the "multcomp package" for the post-hoc tests [37]. Significant differences were evaluated based on a 5% type one error ($\alpha = 5\%$).

Results

Multiple contaminant residues were detected in the honey bee, honey and pollen matrices

Among the 141 honey bee, 141 honey and 128 pollen samples collected from 18 apiaries during 2008 and 2009, 102 (72.3%), 135 (95.7%) and 75 (58.6%) of the samples, respectively, were contaminated by at least one contaminant.

Twenty compounds were detected in honey bees (**Table 2**), with up to 6 different residues in a single sample and a mean of 1.4 residues per analyzed sample. In this matrix, 36.2% of the samples contained at least 2 residues, and 18.4% contained at least 3 residues. Twenty-eight compounds were detected in honey (**Table 3**), with up to 8 different residues in a single sample and a mean of 2.9 residues per analyzed sample. In this matrix, 80.8% of the samples contained at least 2 residues, and more than 3 residues were detected in 57.4% of these samples. Twenty-three compounds were detected in pollen (**Table 4**), with up to 7 different residues in a single sample and a mean of 1.1 residues per analyzed sample. In this matrix, 30.5% of the samples contained at least 2 residues, and 11.7% of the samples contained at least 3 residues. The mixed effects models and the Tukey post-hoc tests indicated a significant difference between the number of residues in the honey and honey bee samples (Tukey test, z = 9.991,

Table 1. Method, limits of detection and limits of quantification for the 80 compounds analyzed for beehive matrices from western France honey bee colonies.

Compound	Method[1]	Class[2]	Effect[3]	Honey bees		Honey		Pollen	
				LOD[4]	LOQ[5]	LOD[4]	LOQ[5]	LOD[4]	LOQ[5]
4,4'-dichlorobenzophenone	GC	OC	A	3.6	9.0	3.6	17.9	3.1	11.2
Abamectin	LC	AVER	I	10.2	20.4	10.2	30.6	nd	nd
Aldrin	GC	OH	I	4.5	22.3	0.2	4.5	11.1	13.9
Amitraz I	LC	FORM	I	18.5	27.8	10.0	37.0	46.3	69.4
Amitraz II	LC	FORM	I	4.3	10.8	0.3	4.3	8.1	17.3
Benalaxyl	GC	PHENA	F	5.7	28.4	5.7	14.2	21.3	42.7
Bifenthrin	GC	PYRE	I	1.3	5.1	3.3	12.9	4.5	19.3
Bitertanol	GC	TRIA	F	1.1	4.4	11.0	16.5	3.9	16.5
Bromopropylate	GC	CARBI	A	0.2	3.9	0.3	3.9	1.0	14.5
Bupirimate	GC	PYRI	F	5.7	14.2	5.7	14.2	2.8	21.4
Buprofezine	GC	THIAD	I	23.9	71.8	23.9	35.9	29.9	59.9
Cadusaphos	GC	OP	N	1.0	8.9	3.6	8.9	8.9	22.3
Carbaryl	LC	CARB	I	0.4	3.8	0.1	3.8	0.7	1.2
Carbendazim	LC	CARB	F	0.6	4.0	0.5	4.0	0.1	1.0
Carbofuran	LC	CARB	I	0.1	3.8	0.03	3.8	0.4	1.0
Chlorothalonil	GC	ISOP	F	nd	nd	22.2	33.3	11.1	22.2
Chlorpyrifos	GC	OP	I	0.8	3.2	3.2	8.0	8.0	20.0
Chlorpyrifos-methyl	GC	OP	I	0.3	5.2	0.1	5.2	1.3	19.5
Clofentezine	LC	QUIN	A	1.0	3.9	1.0	3.9	9.7	48.6
Clothianidine	LC	NEO	I	0.9	10.6	0.3	4.3	1.4	17.0
Coumaphos	LC	OP	I	0.4	3.7	0.3	3.0	1.8	6.0
Cyfluthrin	GC	PYRE	I	12.3	61.5	12.3	30.8	76.9	230.7
Cyhalothrin-lambda	GC	PYRE	I	3.8	9.6	6.7	9.6	23.9	47.9
Cypermethrin	GC	PYRE	I	4.5	27.1	4.5	37.6	56.4	169.1
Cyproconazole	LC	TRIA	F	2.0	10.1	0.2	3.5	3.0	10.1
Deltamethrin	GC	PYRE	I	4.6	16.2	6.9	17.3	28.9	57.8
Diazinon	GC	OP	I	6.3	14.7	7.4	10.5	10.5	26.3
Dichloran	GC	OH	I	38.0	nd	19.0	57.0	47.5	nd
Dichlorvos	GC	OP	I	5.8	14.6	5.8	14.6	14.6	21.9
Dieldrin	GC	OH	I	3.9	9.8	3.9	29.5	9.8	24.6
Diethofencarbe	LC	CARB	F	0.2	3.8	0.04	3.8	0.6	1.9
Dimethoate	GC	OP	I	3.6	27.3	13.6	18.2	9.1	45.4
Endosulfan I	GC	OH	I	5.1	38.0	5.1	12.7	12.7	31.7
Endosulfan II	GC	OH	I	10.3	30.9	10.3	30.9	15.5	51.5
Endosulfan sulphate	GC	OH	I	5.1	8.4	1.2	3.4	8.4	21.1
Eprinomectin	LC	AVER	I	3.9	9.7	9.7	29.1	nd	nd
Esfenvalerate	GC	PYRE	I	10.1	30.2	10.1	30.2	25.1	150.9
Ethoprofos	GC	OH	I	0.6	3.6	1.3	6.4	3.2	13.7
Fenarimol	GC	CARBI	F	3.3	8.1	8.1	16.3	20.3	28.4
Fenitrothion	GC	OP	I	1.1	6.2	6.2	15.5	3.9	19.4
Fenoxycarbe	LC	CARB	I	0.6	4.1	0.1	4.1	1.0	3.3
Flusilazole	GC	TRIA	F	2.1	10.3	4.1	10.3	3.6	15.5
Hexachlorobenzene	GC	OH	F	0.8	3.9	0.2	3.9	9.7	24.3
Hexythiazox	LC	THIAZ	A	0.8	3.9	0.1	4.0	4.8	10.2
Imazalil	LC	IMI	F	1.4	10.2	0.7	4.1	6.9	25.5
Imidacloprid	LC	NEO	I	0.4	9.6	0.2	3.9	2.6	12.0
Iprodione	LC	DICA	F	9.7	19.5	9.7	19.5	15.6	48.7

Table 1. Cont.

Compound	Method[1]	Class[2]	Effect[3]	Honey bees		Honey		Pollen	
				LOD[4]	LOQ[5]	LOD[4]	LOQ[5]	LOD[4]	LOQ[5]
Ivermectin	LC	AVER	I	11.7	23.5	23.5	70.4	nd	nd
Lindane	GC	OH	I	1.0	5.2	1.2	3.4	8.6	17.2
Malathion	GC	OP	I	7.8	15.6	5.5	11.7	39.1	58.6
Metamidophos	LC	OP	I	0.8	10.0	10.0	40.1	2.2	25.1
Methiocarbe	LC	CARB	M	0.4	10.3	0.01	4.1	0.2	0.5
Methomyl	LC	CARB	I	0.3	10.5	0.1	10.5	0.8	3.2
Methoxychlor	GC	OH	I	1.2	3.9	3.9	9.8	2.0	9.8
Moxidectin	LC	AVER	I	3.7	9.4	18.7	nd	nd	nd
Myclobutanil	GC	TRIA	F	10.7	21.4	10.7	32.2	10.7	37.5
o,p DDD	GC	OH	I	3.7	9.2	0.3	3.7	4.6	13.9
p,p-DDT	GC	OH	I	1.3	4.4	21.9	65.8	11.0	27.4
Paclobutrazol	GC	TRIA	F	4.3	10.8	7.5	16.2	3.8	10.8
Parathion	GC	OP	I	1.6	8.0	4.6	11.4	11.4	17.1
Penconazole	GC	TRIA	F	1.9	13.5	5.4	13.5	6.7	16.9
Permethrin	GC	PYRE	I	4.3	10.7	4.3	10.7	5.3	32.1
Phenthoate	GC	OP	I	0.6	14.4	0.3	14.4	1.4	14.4
Phosalone	GC	OP	I	4.1	10.2	4.1	10.2	10.2	15.4
Phosmet	GC	OP	I	9.8	19.7	3.9	9.8	14.8	24.6
Phoxim	LC	OP	I	1.8	7.3	0.1	7.3	2.7	15.5
Piperonyl Butoxyde	LC	BENZ	I	0.1	3.6	0.2	9.0	6.8	22.6
Prochloraz	LC	IMI	F	0.7	4.6	0.2	11.4	4.9	14.8
Procymidone	GC	DICA	F	nd	nd	1.3	3.7	nd	nd
Propargite	GC	SULES	A	11.9	34.1	17.1	25.6	42.7	128.0
Propiconazole	GC	TRIA	F	2.6	17.0	11.1	42.5	4.3	85.1
Pyriproxyfen	LC	PHENP	I	2.1	4.3	1.5	4.3	2.1	8.6
Tau-fluvalinate	GC	PYRE	I	3.7	9.1	3.7	9.1	4.6	22.8
Tebuconazole	GC	TRIA	F	5.1	17.9	12.8	25.6	12.8	38.4
Tetradifon	GC	OH	I	3.3	8.2	3.3	5.7	8.2	20.4
Thiamethoxam	LC	NEO	I	0.6	4.0	0.3	4.0	2.0	8.5
Thiophanate-methyl	LC	CARB	F	4.1	10.3	0.3	10.3	16.5	51.5
Tolclofos-methyl	GC	OP	I	0.3	3.0	0.1	3.0	1.1	11.4
Triadimenol	LC	TRIA	F	9.6	16.0	1.0	6.4	5.6	19.2
Triphenylphosphate	GC	OP	I	0.4	9.3	0.7	9.3	0.5	9.3
Vinclozoline	GC	DICA	F	4.0	10.1	4.0	10.1	1.5	12.6

[1]Method: LC = liquid chromatography coupled with tandem mass spectrometry (LC−MS/MS); GC = gas chromatography coupled with Time of Flight mass spectrometry (GC-ToF).
[2]Class: AVER = avermectine; BENZ = benzodioxole; CARB = carbamate; CARBI = carbinole; DICA = dicarboximide; FORM = formamidin; IMI = imidazole; ISOP = isophtalonitrile; NEO = neonicotinoid; OC = organochloride; OH = organohalogenus; OP = organophosphorus; PHENA = phenylamide; PHENP = phenylpyrazole; PYRE = pyrethroid; PYRI = pyrimidin; QUIN = quinoxaline; SULES = sulfite ester; THIAD = thiadiazin; THIAZ = thiazolidinone; TRIA = triazole.
[3]Effect: A = acaricide; F = fungicide; I = insecticide; M = molluscicide; N = nematodicide.
[4]LOD = limit of detection in ng/g; nd = not determined.
[5]LOQ = limit of quantification in ng/g; nd = not determined.

p<0.0001) and between the number of residues in the honey and pollen samples (Tukey test, z = −11.578, p<0.0001).

A total of 37 different compounds were detected when considering all the matrices (**Tables 2, 3 and 4**), and only 12 compounds were detected in all three matrices: 3 acaricides (2 metabolites of amitraz, amitraz I and amitraz II, and coumaphos), 3 fungicides (carbendazim, flusilazole and thiophanate-methyl) and 6 insecticides (carbaryl, phosmet, piperonyl butoxide, pyriproxyfen, tau-fluvalinate and triphenylphosphate).

Most prevalent detected contaminant residues and their concentrations in the three beehive matrices

The most frequently detected residues (in more than 10% of the samples) were 1) carbendazim (41.1%), triphenylphosphate (24.8%), coumaphos (17.8%) and amitraz II (16.3%) in honey

Table 2. Summary of contaminant residues detections in honey bee samples from western France honey bee colonies.

Compound	Class[1]	Effect[2]	%[3]	Detections				
				Min[4]	Max[4]	Mean[5]	Median[5]	SD[5]
Amitraz I	FORM	A	5.0	> LOD and < LOQ	29.60	9.99	9.25	3.27
Amitraz II	FORM	A	16.3	> LOD and < LOQ	17.00	3.07	2.15	2.41
Benalaxyl	PHENA	F	1.4	> LOD	< LOQ	3.05	2.85	1.69
Carbaryl	CARB	I	2.1	> LOD	< LOQ	0.24	0.20	0.28
Carbendazim	CARB	F	41.1	> LOD and < LOQ	66.30	2.04	0.30	6.59
Chlorpyrifos	OP	I	3.5	> LOD and < LOQ	180.20	1.72	0.40	15.14
Coumaphos	OP	A	17.8	> LOD and < LOQ	47.30	1.04	0.20	4.33
Cypermethrin	PYRE	I	1.4	28.50	48.80	2.52	2.00	4.52
Diazinon	OP	I	0.7	> LOD	< LOQ	3.20	3.15	0.62
Fenoxycarb	CARB	I	0.7	20.10	20.10	0.44	0.30	1.67
Flusilazole	TRIA	F	1.4	> LOD	< LOQ	1.12	1.05	0.61
Hexythiazox	THIAZ	A	0.7	> LOD	< LOQ	0.41	0.40	0.16
Phosalone	OP	I	0.7	> LOD	< LOQ	2.09	2.05	0.43
Phosmet	OP	I	2.8	> LOD and < LOQ	62.20	5.52	4.90	5.01
Piperonyl Butoxide	BENZ	I	2.1	> LOD	< LOQ	0.09	0.05	0.26
Propiconazole	TRIA	F	1.4	> LOD and < LOQ	7.80	0.37	0.30	0.65
Pyriproxyfen	PHENP	I	1.4	> LOD	< LOQ	1.08	1.05	0.26
Tau-fluvalinate	PYRE	I	7.1	> LOD and < LOQ	52.90	3.41	1.85	7.20
Thiophanate-methyl	CARB	F	5.7	> LOD and < LOQ	2418.70	22.96	2.05	207.54
Triphenylphosphate	OP	I	24.8	> LOD and < LOQ	61.60	1.95	0.20	5.86

[1]Class: BENZ = benzodioxole; CARB = carbamate; FORM = formamidin; OP = organophosphorus; PHENA = phenylamide; PHENP = phenylpyrazole; PYRE = pyrethroid; THIAZ = thiazolidinone; TRIA = triazole.
[2]Effect: A = acaricide; F = fungicide; I = insecticide.
[3]$n = 141$ honey bee samples.
[4]Min = minimum in ng/g; LOD = limit of detection; LOQ = limit of quantification.
[5]Mean, Median and SD (standard deviation) were calculated taking into account all the analyzed samples.

bees (**Table 2**); 2) coumaphos (78.0%), amitraz II (68.8%), carbendazim (64.5%), phosmet (12.8%) and cyproconazole (11.3%) in honey (**Table 3**), and 3) carbendazim (34.4%) and amitraz II (14.8%) in pollen (**Table 4**).

The highest maximum concentrations were obtained in pollen for the fungicides thiophanate-methyl (max = 3674.00 ng/g) and carbendazim (max = 2595.00 ng/g). Thiophanate-methyl (max = 2418.70 ng/g) and the insecticide chlorpyrifos (max = 180.20 ng/g) were also quantified in high concentrations in honey bees, although they were rarely detected (5.7% and 3.5% for thiophanate-methyl and chlorpyrifos, respectively). In honey, the highest concentrations concerned the acaricide amitraz II (max = 116.10 ng/g) and the fungicide carbendazim (max = 87.90 ng/g). Although coumaphos was more frequently found in honey (78% of samples) compared with honey bees (17.8%) and pollen (3.9%), the maximal concentrations did not differ significantly between the three matrices (56.40, 47.30 and 40.40 ng/g).

The neonicotinoid imidacloprid was only detected in 3/141 honey samples (2.1%) and in 1/128 pollen samples (0.8%) and was not found among the 141 honey bee samples. Pyrethroids were also rarely detected, with tau-fluvalinate as the most frequent (7.1% in bees, 5.0% in honey and 3.1% in pollen) followed by cypermethrin (1.4% in bees and 0.7% in honey). In contrast, bifenthrin, cyfluthrin, deltamethrin, esfenvalerate, lambda cyhalothrin, permethrin and tefluthrin were never detected.

Honey was the matrix most contaminated by triazole fungicides, with 5 different compounds, including cyproconazole, which was

detected in 11.3% of the samples. The azole fungicide concentrations were the highest in pollen and were always higher than the LOQ. Three samples were contaminated by flusilazole (51.60, 35.88 and 19.9 ng/g), three samples by triadimenol (35.70, 35.4 and 34.3 ng/g), and one sample by cyproconazole (22.30 ng/g).

Coumaphos (17.8% in bees, 78.0% in honey and 3.9% in pollen), triphenylphosphate (24.8% in bees, 2.1% in honey and 9.4% in pollen) and phosmet (2.8% in bees, 12.8% in honey and 7.4% in pollen) were the three most prevalent cholinesterase inhibitor insecticides detected. The insecticides carbaryl and chlorpyrifos were rarely observed. Carbaryl was detected in 3/141, 9/141 and 10/128 of the honey bee, honey and pollen samples, respectively. Chlorpyrifos was detected in 5/141 honey bee samples and in 5/128 pollen samples, but the maximum concentrations were very high (180.20 ng/g and 139.50 ng/g in the honey bee and pollen samples, respectively). The apicultural matrices were less contaminated by other carbamate and organophosphorus compounds, such as carbofuran, diazinon, dimethoate, fenoxycarb and phoxim, with detection rates between 0% and 2.1% depending on the matrices and compounds.

Contamination according to the landscape context and the sampling period

Figure 2 shows the number of agricultural and veterinary residues (all matrices confounded) detected according to the landscape context. Even if the median for rural-cultivated

Table 3. Summary of contaminant residues detections in honey samples from western France honey bee colonies.

Compound	Class[1]	Effect[2]	%[3]	Detections Min[4]	Max[4]	Mean[5]	Median[5]	SD[5]
Amitraz I	FORM	A	4.2	> LOD and < LOQ	26.00	5.65	5.00	3.29
Amitraz II	FORM	A	68.8	> LOD and < LOQ	116.10	10.21	2.30	18.62
Bupirimate	PYRI	F	1.4	> LOD	< LOQ	2.95	2.85	0.84
Buprofezin	THIAD	I	1.4	> LOD and < LOQ	42.80	12.30	11.95	3.00
Carbaryl	CARB	I	6.4	> LOD and < LOQ	4.10	0.18	0.05	0.53
Carbendazim	CARB	F	64.5	> LOD and < LOQ	87.90	2.89	2.25	8.42
Carbofuran	CARB	I	2.1	> LOD	< LOQ	0.06	0.02	0.28
Chlorpyrifos-methyl	OP	I	1.4	> LOD	< LOQ	0.24	0.20	0.31
Coumaphos	OP	A	78.0	> LOD and < LOQ	56.40	2.48	1.65	5.69
Cypermethrin	PYRE	I	0.7	> LOD	< LOQ	2.38	2.25	1.58
Cyprocanozole	TRIA	F	11.3	> LOD and < LOQ	3.80	0.31	0.10	0.62
Diazinon	OP	I	2.1	> LOD and < LOQ	14.00	3.85	3.70	1.06
Diethofencarb	CARB	F	0.7	> LOD	< LOQ	0.03	0.02	0.16
Endosulfan-beta	OH	I	0.7	> LOD	< LOQ	5.26	5.15	1.30
Fenoxycarb	CARB	I	0.7	> LOD	< LOQ	0.06	0.05	0.17
Flusilazole	TRIA	F	2.1	> LOD	< LOQ	2.16	2.05	0.75
Hexythiazox	THIAZ	A	1.4	> LOD	< LOQ	0.08	0.05	0.24
Imazalil	IMI	F	4.2	> LOD	< LOQ	0.44	0.35	0.42
Imidacloprid	NEO	I	2.1	> LOD	< LOQ	0.14	0.10	0.28
Phosmet	OP	I	12.8	> LOD and < LOQ	42.10	3.57	1.95	5.07
Phoxim	OP	I	2.1	> LOD	< LOQ	0.13	0.05	0.53
Piperonyl Butoxide	BENZ	I	8.5	> LOD	< LOQ	0.48	0.10	1.26
Prochloraz	IMI	F	1.4	> LOD	< LOQ	0.18	0.10	0.68
Pyriproxyfen	PHENP	I	3.5	> LOD	< LOQ	0.83	0.75	0.40
Tau-fluvalinate	PYRE	I	5.0	> LOD and < LOQ	30.00	2.30	1.85	2.73
Tebuconazole	TRIA	F	0.7	> LOD	< LOQ	6.49	6.40	1.08
Thiophanate-methyl	CARB	F	1.4	4.00	5.30	0.21	0.15	0.54
Triphenylphosphate	OP	I	2.1	> LOD	< LOQ	0.45	0.35	0.67

[1]Class: BENZ = benzodioxole; CARB = carbamate; FORM = formamidin; IMI = imidazole; NEO = neonicotinoid; OH = organohalogenus; OP = organophosphorus; PHENP = phenylpyrazole; PYRE = pyrethroid; PYRI = pyrimidin; THIAD = thiadiazin; THIAZ = thiazolidinone; TRIA = triazole.
[2]Effect: A = acaricide; F = fungicide; I = insecticide.
[3]n = 141 honey samples.
[4]Min = minimum in ng/g; LOD = limit of detection; LOQ = limit of quantification.
[5]Mean, Median and SD (standard deviation) were calculated taking into account all the analyzed samples.

landscape appears to be higher than the others, the linear mixed effects models failed to reveal a significant difference. This was due to a high standard error of the different estimations (Tukey test, $-2.3 < z < 2.3$, $p > 0.05$).

Figure 3 shows the number of agricultural and veterinary residues (three matrices confounded) detected during 8 periods over 2008 and 2009. Contamination was higher during the period AM8 (late April-early May 2008) than during all other periods (Tukey test, $-z < -3.68$, $p < 0.01$).

Discussion

Most prevalent contaminants in beehive matrices

The fungicide carbendazim and the acaricides amitraz and coumaphos, which are commonly used in beehives to control the parasitic mite *Varroa destructor*, were the three most prevalent residues. Several other studies have previously demonstrated that

the chemicals used by beekeepers inside the hives are frequently found in the apicultural matrices [30,38−40]. Amitraz residues (amitraz I and amitraz II) were mainly detected in bees at the beginning of autumn (periods SO8 and SO9), which corresponds to periods of treatments against *Varroa destructor*. In the pollen matrix, this acaricide was detected during the periods AM8, SO8, AM9 and SO9, which correspond to the end of the beekeeping season with anti-*Varroa* treatments and the beginning of the next beekeeping season. The presence of amitraz in the pollen matrix might be the result of transfer from contaminated bees because its use as a plant-protective acaricide is no longer authorized in France. Surprisingly, amitraz residues were identified in honey samples for all the eight periods. In contrast, Maver and Poklukar [41] and Martel et al. [42] did not detect any amitraz residue in honey after treatment with this compound, which may be explained by their LOD and LOQ being at least ten-fold higher than ours. Coumaphos, another acaricide extensively used against

Table 4. Summary of contaminant residues detections in pollen samples from western France honey bee colonies.

Compound	Class[1]	Effect[2]	%[3]	Detections				
				Min[4]	Max[4]	Mean[5]	Median[5]	SD[5]
Amitraz I	FORM	A	1.6	> LOD and < LOQ	115.20	24.14	23.15	8.67
Amitraz II	FORM	A	14.8	> LOD and < LOQ	129.40	7.39	4.10	13.82
Bupirimate	PYRI	F	0.8	> LOD	< LOQ	1.48	1.40	0.95
Carbaryl	CARB	I	7.8	> LOD and < LOQ	14.67	0.70	0.35	1.68
Carbendazim	CARB	F	34.4	> LOD and < LOQ	2595.00	24.31	0.05	229.47
Carbofuran	CARB	I	1.6	> LOD and < LOQ	2.30	0.22	0.20	0.19
Chlorpyrifos	OP	I	3.9	> LOD and < LOQ	139.50	5.61	4.00	12.53
Coumaphos	OP	A	3.9	> LOD and < LOQ	40.40	1.95	0.90	5.10
Cyprocanozole	TRIA	F	0.8	22.30	22.30	1.66	1.50	1.84
Dieldrin	OH	I	0.8	> LOD	< LOQ	5.0	4.90	1.09
Diethofencarb	CARB	F	0.8	2.60	2.60	0.32	0.30	0.32
Dimethoate	OP	I	0.8	> LOD	< LOQ	4.73	4.55	2.01
Flusilazole	TRIA	F	2.3	19.90	51.60	2.60	1.80	5.53
Imidacloprid	NEO	I	0.8	> LOD	< LOQ	1.35	1.30	0.53
Iprodione	DICA	F	0.8	> LOD	< LOQ	7.99	7.80	2.15
Phosmet	OP	I	7.4	> LOD and < LOQ	78.10	9.38	7.40	8.80
Piperonyl Butoxide	BENZ	I	0.8	> LOD	< LOQ	3.49	3.40	1.00
Pyriproxyfen	PHENP	I	4.7	> LOD	< LOQ	5.85	5.35	2.28
Tau-fluvalinate	PYRE	I	3.1	> LOD and < LOQ	85.42	3.52	2.30	8.31
Thiophanate-methyl	CARB	F	1.6	1395.00	3674.00	47.72	8.25	345.52
Triadimenol	TRIA	F	2.3	34.30	35.70	8.64	8.00	4.12
Triphenylphosphate	OP	I	9.4	> LOD	< LOQ	0.69	0.25	1.36
Vinclozolin	DICA	F	0.8	70.31	70.31	1.29	0.75	6.15

[1]Class: BENZ = benzodioxole; CARB = carbamate; DICA = dicarboximide; FORM = formamidin; NEO = neonicotinoid; OH = organohalogenus; OP = organophosphorus; PHENP = phenylpyrazole; PYRE = pyrethroid; PYRI = pyrimidin; TRIA = triazole.
[2]Effect: A = acaricide; F = fungicide; I = insecticide.
[3]n = 128 honey bee samples.
[4]Min = minimum in ng/g; LOD = limit of detection; LOQ = limit of quantification.
[5]Mean, Median and SD (standard deviation) were calculated taking into account all the analyzed samples.

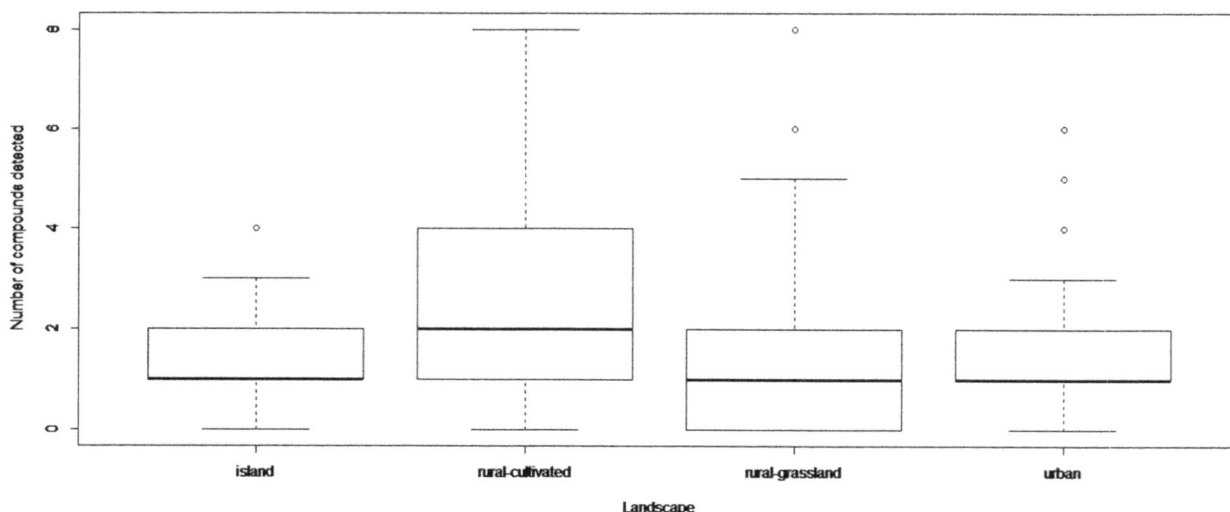

Figure 2. **Number of compounds detected according to the landscape context.** The number of compounds was calculated irrespective of the matrix (honey bees, honey and pollen) for each landscape context (rural-grassland, rural-cultivated, island and urban landscapes).

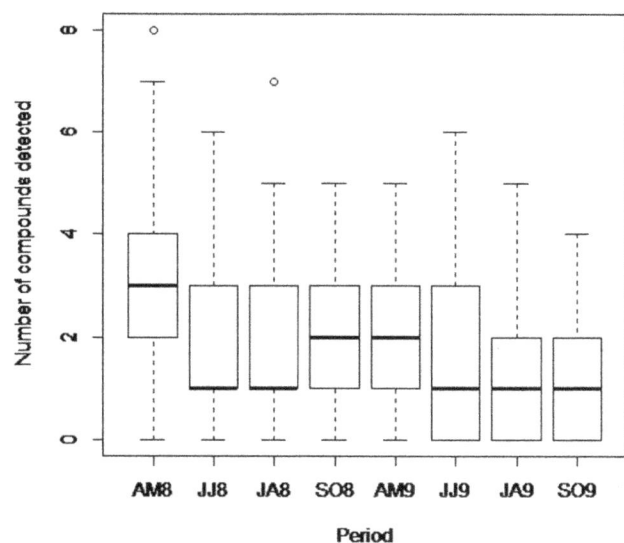

Figure 3. Number of compounds detected according to the period. The number of compounds was calculated irrespective of the matrix (honey bees, honey and pollen) for each period: late April-early May 2008 (period AM8) and 2009 (period AM9), late June-early July 2008 (period JJ8) and 2009 (period JJ9), late July-early August 2008 (period JA8) and 2009 (period JA9), and late September-early October 2008 (period SO8) and 2009 (period SO9).

Varroa in recent decades, was also frequently detected in apicultural matrices [28,30]. In addition, many studies indicate that coumaphos was persistent in wax and diffused from wax to honey in high proportions [43−44]. Although this acaricide is now banned in France, we detected this molecule in most honey samples (110/141 samples). This might be the result of past chronic use, accumulation and then transfer from contaminated wax, except for one apiary, where an illegal use of coumaphos was demonstrated after an investigation.

The high prevalence of the fungicide carbendazim in the three beehive matrices might result from its very wide use on orchards, vineyards, and grains, along with oleaginous and vegetable crops, both in agricultural and in gardening treatments. Otherwise, carbendazim is also a metabolite of thiophanate-methyl, and some detection of this substance might be linked to the field use of thiophanate-methyl. This is most likely the reason why all samples contaminated by thiophanate-methyl also contained carbendazim residues.

Among the five most frequently detected residues, Mullin et al. [30] identified two in-hive miticides, coumaphos and amitraz, and the fungicide chlorothalonil, widely applied for the control of fungal diseases in agricultural and gardening treatments [45]. Those results were consistent with those of the present study, even if the fungicide was not the same, which was most likely due to different agricultural treatment practices. The acaricide fluvalinate, which was the most frequent residue found by Mullin et al. [30], was detected in lower frequency in the present study and in a recent French study [31], reflecting different apicultural practices in Europe and North America. Systematic combinations of acaricides used to fight *Varroa* or fungicides and other pesticides have been reported in many studies [30,46], and the interaction between such compounds can induce lethal or adverse sublethal effects in honey bees [17,47−49].

Contamination according to the landscape context and to the sampling period

The spatial patterns of the contamination of beehive matrices by environmental pollutants such as polycyclic aromatic hydrocarbons or heavy metals have been previously studied [26,32−33,50−52]. Most of these studies compared the contamination in both urban and wild sites. To our knowledge, the pesticide spatial distribution in bees, honey or pollen was only studied in different areas within the same landscape context and, in particular, cultivated landscapes [28−29,31,53]. The present study is the first to monitor beehive matrix contamination by pesticides in 4 different landscape structures characterized by different pesticide and veterinary drug use patterns (private or professional uses, veterinary, apicultural or agricultural uses). Apiaries in rural-cultivated landscapes were more contaminated than apiaries in all other landscape types, but the differences were not significant.

Acaricide residues were observed in all the 18 apiaries, and the differences in spatial distribution were linked to the nature of the compounds used in agricultural treatments such as organophosphorus and carbamate insecticides and azole fungicides. In rural-cultivated landscapes (dominated by permanent, oleaginous, grain crops, market gardening), agricultural treatments are performed on large plots and often on melliferous plants (*Asteraceae, Brassicaceae, Fabaceae*). These treatments are more likely to contaminate honey bees than local treatments in an urban context, particularly if the urban district green services practice a zero pesticide policy. Despite this trend, the present results indicate that some apiaries located in urban landscapes, which are supposedly less exposed to pesticide pressure, were more contaminated than apiaries in rural landscapes.

Previous studies have demonstrated seasonal variation in beehive matrix contamination by polycyclic aromatic hydrocarbons [33] and heavy metals [26,32,51]. The contamination levels were linked to the meteorological conditions and were generally higher during the dry months. However, few studies have reported the evolution of pesticide contamination throughout the beekeeping season. In the present study, the most contaminated period corresponded to late April-early May in 2008 and was associated with agricultural uses and crop treatments [54] and with a high foraging activity. Ghini et al. [53] demonstrated that the maximum level and frequency of pesticides (organophosphate and carbamate residues) occurred in the late spring (May and June). Samplings of the present study were collected 4 times per year and reflected a single time point and not the contamination kinetics throughout the year. This sampling methodology might explain the differences with Ghini et al. [53], who performed samplings each month from April to October 2000. The higher contamination in spring 2008 was not observed in 2009, most likely due to differences in the meteorological conditions and treatment time.

Pesticide and veterinary drug contamination of beehive matrices

Honey was the most contaminated matrix in the present study when taking into account the number of residues observed by matrix and by sample. These results are not in agreement with the results of other studies in which pollen was observed to be the most contaminated matrix [30−31]. Indeed, Mullin et al. [30] demonstrated that among 140 honey bee and 350 pollen samples collected in the USA, in both 2007 and 2008, 91.4% and 99.1% of the samples were contaminated, respectively. However, the context of their study and the method used for pesticide detection and quantification were different: (i), samples were collected to investigate possible threats to colony health for the CCD (Colony Collapse Disorder) working group; (ii), more than twice as many compounds were searched, with an average of 171 pesticides and toxic metabolites studied per analysis; and (iii), the honey matrix was not analyzed for pesticide presence. Another recently published study on beehive matrices collected in France between 2002 and 2005 also demonstrated that pollen samples were the most contaminated (69.5%) compared with honey bee (44.3%) and honey (43.1%) samples [31]. Whereas LOD and LOQ were similar for the different matrices in these two previous studies [30–31], they were generally higher in pollen than in the honey bee or honey matrices in our survey. This could explain the different prevalence for residues among our matrices, and in particular, the lower frequency of detection in pollen samples. However, for most compounds, the maximum concentration and the mean concentration were higher in pollen than in honey bees and honey. If a lower LOD and LOQ could be achieved for pollen samples, it is likely that pollen would be the best matrix for assessing the presence of pesticides in foraging areas.

Other contaminants and implications for bee colony health

Systemic insecticides, including neonicotinoids, have been demonstrated to present a high acute and chronic toxicity in bees [16–17,55–59]. A survey in France from 2002 to 2005 indicated that imidacloprid was frequently detected in honey bee matrices: 11.2%, 40.5% and 21.8% of the honey bee, pollen and honey samples were found to be contaminated, respectively [31]. In contrast, this molecule was rarely detected in our study. A potential explanation is that our data were from a more recent field-study and might reflect changes in agricultural practices; for example, imidacloprid is being gradually replaced by new neonicotinoids such as thiamethoxam [60], even if this neonicotinoid was never detected in the present samples. The low presence of neonicotinoids in our survey was most likely linked to the multi-residue analysis, which is characterized by a high LOD. Indeed, these molecules are used at very low doses, and our detection method was not sensitive enough to detect them in most samples. However, no adverse effect has been established by previous studies at our concentrations and at other field-relevant doses for pollen and nectar.

Like neonicotinoids, pyrethroids can induce adverse acute sublethal effects in bees [16–17]. However, except for tau-fluvalinate and cypermethrin, other pyrethroids were never detected in our study. These contamination levels in the present study were consistent with those of Chauzat et al. [31] but were much lower than those of Mullin et al. [30], in relation to lower LODs than ours for these residues. In addition, it has been demonstrated that a synergistic action between pyrethroids and azole fungicides can occur [61] and may increase the risk to bees. Interestingly, one apiary located in a rural-cultivated landscape displayed pyrethroid (tau-fluvalinate) and azole fungicide (flusila-zole) contamination in the same period and in both pollen and bee matrices. Important levels of bee mortality were noticed in this apiary during and after the sampling, but no direct relationship with the co-contamination by tau-fluvalinate and flusilazole was demonstrated.

Carbaryl was frequently detected in pollen and honey samples. Ninety percent of carbaryl detections concerned the samples collected in 2008, which is consistent with the ban on the use of this compound in November 2008 in France, and confirmed that the beehive matrices are very sensitive in terms of reflecting agricultural practices and pesticide use. In addition, it was previously demonstrated that sublethal exposure to carbaryl has adverse effects on the longevity and foraging activities of honey bees [62]. Although the concentrations determined in the present study were lower than those in this semi-field study, chronic exposure to this insecticide might cause problems to bee colonies.

Except for coumaphos, phosmet and triphenylphosphate, few organophosphorus compounds were detected, which is contrast to the results of other studies. For example, Ghini et al. [53] found that twelve organophosphorus compounds were present in more than 10% of examined samples (58% for malathion and 53% for fenitrothion). The adverse effects of these compounds have been demonstrated, and their use is now controlled and limited. When considering the weight of one bee (0.1108 g, mean weight determined for the sampled honey bees in the present study), some maximum concentrations in the honey bees were very high and close to or above the LD_{50} (lethal dose 50) divided by the safety factor of 100 that is regularly used in environmental toxicology. Such was the case for the insecticides chlorpyrifos (LD_{50} = 122 ng/bee, maximum concentration determined = 19.97 ng/bee) and phosmet (LD_{50} = 803 ng/bee, maximum concentration determined = 6.89 ng/bee). Such concentration levels might have affected the bee health in the corresponding samples either by inducing direct mortality or via sublethal effects on bee physiology and behavior [16]. The maximum concentrations in pollen were similar to those measured in bees for chlorpyrifos and phosmet, and the sublethal effects of the residues in bees, especially in brood, were likely the result of delivery through consumption of contaminated pollen samples. The organophosphate triphenylphosphate is not allowed for agricultural use in France, but it is highly used in industrial settings as a flame retardant [63]. Thus, this ubiquitous environmental pollutant might contaminate apicultural matrices via atmospheric residues. Honey bees are more likely to be exposed to atmospheric pollutants during their foraging due to their flight, their hairy bodies and their ability to hold residues, which might explain the contamination levels found in the present study.

Because of the difference in residue LOD and LOQ for each matrix, no correlation was tested between the three beehive matrices collected from the same apiary and at the same period. Such differences might explain the fact that a residue found in one matrix was not systematically present in the two others. Other explanations might include the following features: 1) the sampling methodology at a single time point (honey was stored inside the hive for several days before the sampling, pollen was collected from a pollen trap set up 3 days before the sampling, and foraging honey bees were directly collected on the hive's flight board), 2) the specific biotransformations of residues in each matrix, and 3) the ability of each matrix to reflect contamination in time.

In conclusion, our field study revealed the widespread presence of multiple residues in honey, honey bees and pollen with different distribution patterns according to the landscape context and the sampling period. This contamination by multiple residues also raises the issue of the impact of the combinations of these pesticides and veterinary drugs and their potential synergistic effects on the health of bees and other pollinators [64,65].

Acknowledgments

We thank all the beekeepers who allowed samplings in their colonies and their apiaries. We also thank Allard J-M, Bastian S, and Pouleur B for their help in field sampling.

Author Contributions

Conceived and designed the experiments: OL MLH HP. Performed the experiments: OL MP MLH. Analyzed the data: OL MP SP CT HP FD MLH. Contributed reagents/materials/analysis tools: LW AB. Wrote the paper: OL HP FD.

References

1. McLaughlin A, Mineau P (1995) The impact of agricultural practices on biodiversity. Agric Ecosys Environ 55: 201−212.
2. Reidsma P, Tekelenburg T, van den Berg M, Alkemade R (2006) Impacts of land-use change on biodiversity: an assessment of agricultural biodiversity in the European Union. Agric Ecosys Environ 114: 86−102.
3. McKinney ML (2008) Effects of urbanization on species richness: a review of plants and animals. Urban Ecosys 11: 161−176.
4. Stoate C, Bàldi A, Beja P, Boatman ND, Herzon I, et al. (2009) Ecological impacts of early 21st century agricultural change in Europe – A review. J Environ Manag 91: 22−46.
5. Butler SJ, Boccaccio L, Gregory RD, Vorisek P, Norris K (2010) Quantifying the impact of land-use change to European farmland bird populations. Agric Ecosys Environ 137: 348−357.
6. Daszak P, Berger L, Cunningham AA, Hyatt AD, Green DE, et al. (1999) Emerging infectious diseases and amphibian population declines. Emerg Infect Dis 5: 735−748.
7. Crowl TA, Crist TO, Parmenter RR, Belovsky G, Lugo AE (2008) The spread of invasive species and infectious disease as drivers of ecosystem change. Front Ecol Environ 6: 238−246.
8. Hejda M, Pyšek P, Jarošik V (2009) Impact of invasive plants on the species richness, diversity and composition of invaded communities. J Ecol 97: 393−403.
9. Kenis M, Auger-Rozenberg M-AA, Roques A, Timms L, Péré C, et al. (2009) Ecological effects of invasive alien insects. Biol Invas 11: 21−45.
10. Aizen MA, Harder LD (2009) The global stock of domesticated honey bees is growing slower than agricultural demand for pollination. Curr Biol 19: 1−4.
11. vanEngelsdorp D, Meixner MD (2010) A historical review of managed honey bee populations in Europe and the United States and the factors that may affect them. J Invertebr Pathol 103: S80−S95.
12. Guzmán-Novoa E, Eccles L, Calvete Y, McGowan J, Kelly PG, et al. (2010) Varroa destructor is the main culprit for the death and reduced populations of overwintered honey bee (Apis mellifera) colonies in Ontario, Canada. Apidologie 41: 443−450.
13. Higes M, Martín-Hernández R, Meana A (2010) Nosema ceranae in Europe: an emergent type C nosemis. Apidologie 41: 375−392.
14. Rosenkranz P, Aumeier P, Ziegelmann B (2010) Biology and control of Varroa destructor. J Invertebr Pathol 103: S96−S119.
15. Core A, Runckel C, Ivers J, Quock C, Siapno T, et al. (2012) A new threat to honey bees, the parasitic phorid fly Apocephalus borealis. PLoS ONE 7: e29639.
16. Desneux N, Decourtye A, Delpuech JM (2007). The sublethal effects of pesticides on beneficial arthropods. Annu Rev Entomol 52: 81−106.
17. Belzunces LP, Tchamitchian S, Brunet J-L (2012) Neural effects of insecticides in the honey bee. Apidologie 43: 348−370.
18. Alaux C, Brunet J-L, Dussaubat C, Mondet F, Tchamitchian S, et al. (2010) Interactions between Nosema microspores and a neonicotinoid weaken honey bees (Apis mellifera). Environ Microbiol 12: 774−782.
19. Vidau C, Diogon M, Aufauvre J, Fontbonne R, Viguès B, et al. (2011) Exposure to sublethal doses of fipronil and thiacloprid highly increases mortality of honey bees previously infected by Nosema ceranae. PLoS ONE 6: e21550.
20. Aufauvre J, Biron DG, Vidau C, Fontbonne R, Roudel M, et al. (2012) Parasite-insecticide interactions: a case study of Nosema ceranae and fipronil synergy on honey bee. Scientific Reports 2 (326): 1−7.
21. Pettis JS, vanEngelsdorp D, Johnson J, Dively G (2012) Pesticide exposure in honey bees results in increased levels of the gut pathogen Nosema. Naturwissenschaften 99: 153−158.
22. Wu JY, Smart MD, Anelli CM, Sheppard WS (2012) Honey bees (Apis mellifera) reared in brood combs containing high levels of pesticide residues exhibit increased susceptibility to Nosema (Microsporidia) infection. J Invertebr Pathol 109: 326−329.
23. Celli G, Maccagnani B (2003) Honey bees as bioindicators of environmental pollution. Bull Insectol 56: 137−139.
24. Porrini C, Sabatini AG, Girotti S, Ghini S, Medrzycki P, et al. (2003) Honey bees and bee products as monitors of the environmental contamination. Apiacta 38: 63−70.
25. Kalbande DM, Dhadse SN, Chaudhari PR, Wate SR (2008) Biomonitoring of heavy metals by pollen in urban environment. Environ Monit Assess 138: 233−238.
26. Perugini M, Manera M, Grotta L, Abete MC, Tarasco R, et al. (2011) Heavy metals (Hg, Cr, Cd and Pb) contamination in urban areas and wildlife reserves: honey bees as bioindicators. Biol Trace Elem Res 140: 170−176.
27. Perugini M, Di Serafino G, Giacomelli A, Medrzyck P, Sabatini AG, et al. (2009) Monitoring of polycyclic aromatic hydrocarbons in bees (Apis mellifera) and honey in urban areas and wildlife reserves. J Agric Food and Chem 57: 7440−7444.
28. Balayiannis G, Balayiannis P (2008) Bee honey as an environmental bioindicator of pesticides' occurrence in six agricultural areas of Greece. Arch Environ Contam Toxicol 55: 462−470.
29. Barmaz S, Potts SG, Vighi M (2010) A novel method for assessing risks to pollinators from plant protection products using honey bees as a model species. Ecotoxicology 19: 1347−1359.
30. Mullin CA, Frazier M, Frazier JL, Ashcraft S, Simonds R, et al. (2010) High levels of miticides and agrochemicals in North American apiaries: implications for honey bee health. PLoS ONE 5: e9754.
31. Chauzat M-P, Martel A-C, Cougoule N, Porta P, Lachaize J, et al. (2011) An assessment of honey bee colony matrices, Apis mellifera (Hymenoptera: apidae) to monitor pesticide presence in continental France. Environ Toxicol Chem 30: 103−111.
32. Lambert O, Piroux M, Puyo S, Thorin C, Larhantec M, et al. (2012a) Bees, honey and pollen as sentinels for lead environmental contamination. Environ Pollut 170: 254−259.
33. Lambert O, Veyrand B, Durand S, Marchand P, Le Bizec B, et al. (2012b) Polycyclic aromatic hydrocarbons: bees, honey and pollen as sentinels for environmental chemical contaminants. Chemosphere 86: 98−104.
34. Wiest L, Buleté A, Giroud B, Fratta C, Amic S, et al. (2011) Multi-residue analysis of 80 environmental contaminants in honeys, honey and pollens by one extraction procedure followed by liquid and gas chromatography coupled with mass spectrometric detection. J Chromatogr A 1218: 5743−5756.
35. Office of Pesticide Programs, US Environmental Protection Agency (2000) Assigning values to non-detected/non-quantified pesticide residues in human health food exposure assesments, 33p.
36. Pinheiro JC, Bates DM (2000). Mixed-effects models in S and S-PLUS. Springer-Verlag, New York, 528p.
37. R Core Team (2012) R: a language and environment for statistical computing. R Foundation for Statistical Computing, Vienna, Austria. ISBN 3-900051-07-0.
38. Lodesani M, Costa C, Bigliardi M, Colombo R (2003) Acaricide residues in bee wax and organic beekeeping. Apiacta 38: 31−33.
39. Sabatini AG, Carpana E, Serra G, Colombo R (2003) Presence of acaricides and antibiotics in samples of Italian honey. Apiacta 38: 46−49.
40. Tsigouri A, Menkissoglu-Spiroudi U, Thrasyvoulou A, Diamantidis G (2003) Fluvalinate residues in Greek honey and beeswax. Apiacta 38: 50−53.
41. Maver L, Poklukar J (2003) Coumaphos and amitraz residues in Slovenian honey. Apiacta 38: 54−57.
42. Martel A-C, Zeggane S, Aurière C, Drajnudel P, Faucon J-P, et al. (2007) Acaricide residues in honey and wax after treatment of honey bee colonies with Apivar® and Asuntol®50. Apidologie 38: 534−544.
43. Van Buren NWM, Mariën J, Velthuis HHW, Oudejans RCHM (1992) Residues in beeswax and honey of perizin, an acaricide to combat the mite Varroa jacobsoni Oudemans (Acari: Mesostigmata). Environ Entomol 21: 860−865.
44. Wallner K (1999) Varroacides and their residues in bee products. Apidologie 30: 235−248.
45. Chaves A, Shea D, Danehower D (2008) Analysis of chlorothalonil and degradation products in soil and water by GC/MS and LC/MS. Chemosphere 71: 629−638.
46. Orantes-Bermejo FJ, Pajuelo AG, Megias MM, Fernández-Piñar CT (2010) Pesticide residues in beeswax and beebread samples collected from honey bee colonies (Apis mellifera L.) in Spain. Possible implications for bee losses. J Apic Res 48: 243−250.
47. Colin ME, Belzunces LP (1992). Evidence of synergy between prochloraz and deltamethrin in Apis mellifera L – A convenient biological approach. Pestic Sci 36: 115−119.
48. Johnson RM (2011) Managed pollinator coordinated agricultural project – miticide and fungicide interactions. Am Bee J 151: 975−978.
49. Johnson RM, Dahlgren L, Siegfried BD, Ellis MD (2013) Acaricide, fungicide and drug interactions in honey bees (Apis mellifera). PLoS ONE 8: e54092.
50. Tuzen M, Silici S, Mendil D, Soylak M (2007) Trace element levels in honeys from different regions of Turkey. Food Chem 103: 325−330.
51. Morgano MA, Teixeira Martins MC, Rabonato LC, Milani RF, Yotsuyanagi K, et al. (2010) Inorganic contaminants in bee pollen from Southeastern Brazil. J Agric Food and Chem 58: 6876−6883.
52. Bilandžić N, Dokić M, Sedak M, Kolanović BS, Varenina I, et al. (2011) Determination of trace elements in Croatian floral honey originating from different regions. Food Chem 128: 1160−1164.
53. Ghini S, Fernández M, Picó Y, Marín R, Fini F, et al. (2004) Occurrence and distribution of pesticides in the province of Bologna, Italy, using honey bees as bioindicators. Arch Environ Contam Toxicol 47: 479−488.
54. Krupke CH, Hunt GJ, Eitzer BD, Andino G, Given K (2012) Multiple routes of pesticide exposure for honey bees living near agricultural fields. PLoS ONE 7: e29268.

55. El Hassani AK, Dacher M, Gary V, Lambin M, Gauthier M, et al. (2008) Effects of sublethal doses of acetamiprid and thiamethoxam on the behavior of the honey bee (*Apis mellifera*). Arch Environ Contam Toxicol 54: 653−661.

56. Yang EC, Chuang YC, Chen YL, Chang LH (2008) Abnormal foraging behavior induced by sublethal dosage of imidacloprid in the honey bee (Hymenoptera: Apidae). J Econ Entomol 101: 1743–1748.

57. Han P, Niu C-Y, Lei C-L, Cui J-J, Desneux N (2010) Use of an innovative T-tube maze assay and the proboscis extension response assay to assess sublethal effects of GM products and pesticides on learning capacity of the honey bee *Apis mellifera* L. Ecotoxicology 19: 1612–1619.

58. Schneider CW, Tautz J, Grünewald B, Fuchs S (2012) RFID tracking of sublethal effects of two neonicotinoid insecticides on the foraging behavior of *Apis mellifera*. PLoSONE 7: e30023.

59. Henry M, Beguin M, Requier F, Rollin O, Odoux J-F, et al. (2012) A common pesticide decreases foraging success and survival in honey bees. Sciencexpress 10.1126, science.1215039, 4p.

60. Jeschke P, Nauen R, Schindler M, Elbert A (2011) Overview of the status and global strategy for neonicotinoids. J Agric Food Chem 59: 2897–2908.

61. Pilling ED, Bromley-Challenor KAC, Walker CH, Jepson PC (1995) Mechanism of synergism between the pyrethroid insecticide λ-cyhalothrin and the imidazole fungicide prochloraz, in the honey bee (*Apis mellifera* L.). Pestic Biochem Physiol 51: 1−11.

62. MacKenzie KE, Winston ML (1989) The effects of sublethal exposure to diazinon, carbaryl and resmethrin on longevity and foraging in *Apis mellifera* L. Apidology 20: 29−40.

63. Andresen JA, Grundmann A, Bester K (2004) Organophosphorus flame retardants and plasticisers in surface waters. Sci Total Environ 332: 155–166.

64. Brittain CA, Vighi M, Bommarco R, Setteled J, Potts SG (2010) Impacts of a pesticide on pollinator species richness at different spatial scales. Basic Appl Ecol 11: 106–115.

65. Whitehorn PR, O'Connor S, Wackers FL, Goulson D (2012) Neonicotinoid pesticide reduces bumble bee colony growth and queen production. Sciencexpress, 10.1126, science.1215025, 3p.

4

Assessing the Risk That *Phytophthora melonis* Can Develop a Point Mutation (V1109L) in CesA3 Conferring Resistance to Carboxylic Acid Amide Fungicides

Lei Chen[1], Shusheng Zhu[2], Xiaohong Lu[1], Zhili Pang[1], Meng Cai[1], Xili Liu[1]*

1 Department of Plant Pathology, College of Agriculture and Biotechnology, China Agricultural University, Beijing, China, 2 Key Laboratory of Agro-Biodiversity and Pest Management of Education Ministry of China, Yunnan Agricultural University, Kunming, Yunnan, China

Abstract

The risk that the plant pathogen *Phytophthora melonis* develops resistance to carboxylic acid amide (CAA) fungicides was determined by measuring baseline sensitivities of field isolates, generating resistant mutants, and measuring the fitness of the resistant mutants. The baseline sensitivities of 80 isolates to flumorph, dimethomorph and iprovalicarb were described by unimodal curves, with mean EC_{50} values of 0.986 ($\pm$0.245), 0.284 ($\pm$0.060) and 0.327 ($\pm$0.068) µg/ml, respectively. Seven isolates with different genetic background (as indicated by RAPD markers) were selected to generate CAA-resistance. Fifty-five resistant mutants were obtained from three out of seven isolates by spontaneous selection and UV-mutagenesis with frequencies of 1×10^{-7} and 1×10^{-6}, respectively. CAA-resistance was stable for all mutants. The resistance factors of these mutants ranged from 7 to 601. The compound fitness index (CFI = mycelial growth × zoospore production × pathogenicity) was often lower for the CAA-resistant isolates than for wild-type isolates, suggesting that the risk of *P. melonis* developing resistance to CAA fungicides is low to moderate. Among the CAA-resistant isolates, a negative correlation between EC_{50} values was found for iprovalicarb vs. flumorph and for iprovalicarb vs. dimethomorph. Comparison of the full-length cellulose synthase 3 (CesA3) between wild-type and CAA-resistant isolates revealed only one point mutation at codon position 1109: a valine residue (codon GTG in wild-type isolates) was converted to leucine (codon CTG in resistant mutants). This represents a novel point mutation with respect to mutations in CesA3 conferring resistance to CAA fungicides. Based on this mutation, an efficient allelic-specific PCR (AS-PCR) method was developed for rapid detection of CAA-resistance in *P. melonis* populations.

Editor: Joy Sturtevant, Louisiana State University, United States of America

Funding: This work was funded by the China National Science Foundation (NO. 30671390 and 30800731). The funders had no role in study design, data collection and analysis, decision to publish, or preparation of the manuscript.

Competing Interests: The authors have declared that no competing interests exist.

* E-mail: seedling@cau.edu.cn

Introduction

The oomycete *Phytophthora melonis* Katsura, which is conspecific with *P. sinesis*, causes a severe disease of cucumber (*Cucumis sativus*) which has been reported in China, Japan, Egypt, Turkey, Korea, India and Iran [1]. In addition to cucumber, *P. melonis* infects other cucurbits including zucchini (*Cucurbita pepo* L.), hami melon (*Cucumis melo* L.), wax gourd (*Benincasa hispida* (Thunb.) Cogn.) [2–5], and pointed gourd (*Trichosanthes dioica* Roxb.) [6]. It also infects pistachio (*Pistacia vera* L) [7], causing blight, dieback, root rot, foot rot and crown rot. The use of resistant cultivars and chemical fungicides are two efficient control methods [2,5,8]. Phenylamides (e.g. metalaxyl) have been widely used for *P. melonis* disease control. However, metalaxyl-resistance of *P. melonis* has been reported in China [9]. Since the early 1980s, the efficacy of phenylamides has declined due to the emergence of resistant populations of oomycete pathogens in fields [10,11].

The current study concerns resistance of *P. melonis* to the carboxylic acid amide (CAA) fungicides, which are divided into three different chemical groups based on differences in structure: the cinnamic acid amides (e.g., dimethomorph and flumorph), the valine amide carbamates (e.g., benthiavalicarb, benthiavalicarb-

isopropyl and iprovalicarb) and the mandelic acid amides (e.g., mandipropamid) (FRAC Code List, www.frac.info). These fungicides are used to control the pathogens in the families Peronosporaceae (e.g., *Plasmopara viticola* and *Bremia lactucea*) and Pythiaceae (e.g., *Phytophthora* spp., but not *Pythium* spp.) [12]. All CAA fungicides strongly inhibit all asexual stages of susceptible pathogens but do not inhibit zoospore release and mobility [13–16]. Inhibition by CAA fungicides results from the interruption of cellulose biosynthesis and the disruption of cell wall structure [17].

istance to phenylamide fungicides, resistance to CAA fungicides is an important problem. Since dimethomorph's introduction in the 1980s, CAA-resistant isolates of *P. viticola* have been detected in most areas of Europe (FRAC web). In China, flumorph-resistant isolates of *Pseudoperonospora cubensis* were obtained after successive applications of flumorph in a greenhouse [18]. *P. viticola* and *Ps. cubensis* are classified by FRAC as being at high risk to develop resistance to CAA fungicides, but *P. infestans* is considered to have a low risk of developing such resistance (FRAC pathogen risk list, www.frac.info). No resistant isolates of *P. infestans* have been detected in field since the introduction of CAA fungicides over 15 years ago (CAA Minutes

2010 final RG, www.frac.info). Using UV- and EMS- mutagenesis, researchers obtained stable CAA-resistant mutants of *P. infestans* only with difficulty [19–21] and an isolate of *P. capsici* failed to develop dimethomorph-resistance after repeated exposure to the fungicide [19]. However, *P. capsici* mutants with high CAA-resistance were obtained by mass selection from zoospores and oospores, and the resistance risk was considered low to moderate to CAAs in *P. capsici* [22,23].

Cross-resistance among CAA fungicides has been reported in *P. viticola* [24] and *P. capsici* [22]. The resistance mechanism of some pathogens to CAA fungicides has been elucidated in recent reports. Amino acid substitutions at codon position 1105 (G1105V or G1105A) in cellulose synthase (CesA3) were responsible for resistance to the CAA fungicide mandipropamid in EMS-generated mutants of *P. infestans* [17]. Changes of G1105V or G1105W led to CAA-resistance in *Ps. cubensis* [25]. In *P. viticola*, changes of G1105S in PvCesA3 conferred CAA-resistance [26]. Based on the point mutation, a PCR-RFLP method has been developed for detecting CAA-resistant isolates in *P. viticola* populations [27].

The objectives of the present study were to (i) determine the baseline sensitivity of *P. melonis* to the CAA fungicides flumorph, dimethomorph and iprovalicarb; (ii) assess the risk of resistance to the three CAA fungicides; (iii) investigate the CAA-resistance mechanism in *P. melonis*; and (iv) develop a rapid and reliable method for detection of CAA-resistant isolates in populations of *P. melonis*.

Results

Baseline Sensitivity of *P. melonis* to Flumorph, Dimethomorph and Iprovalicarb

For the 80 *P. melonis* isolates investigated, the frequency distribution of EC_{50} values for each of the three CAA fungicides were described by unimodal curves (Figure 1), indicating the absence of CAA-resistant subpopulations among these isolates. The mean and range of EC_{50} values were 0.986 ± 0.245 μg/ml and $0.410–1.577$ μg/ml for flumorph, 0.284 ± 0.060 μg/ml and $0.171–0.590$ μg/ml for dimethomorph, and 0.327 ± 0.068 μg/ml and $0.100–0.482$ μg/ml for iprovalicarb. The highest EC_{50} value was 3.45-, 3.85-, and 4.82-times greater than the smallest value for flumorph, dimethomorph and iprovalicarb, respectively.

Development of CAA-resistant Isolates of *P. melonis* in *vitro*

Random amplified polymorphic DNA (RAPD). RAPD analysis was used to identify isolates with different genetic backgrounds; these isolates would be used as the parents for the generation of CAA-resistant mutants. Sixteen primers (Table 1) that produced easily recognizable and consistent banding patterns were used for RAPD analysis of 16 isolates from different geographic origins (Table 2). RAPD analysis using primer combinations clearly separated these isolates. A dendrogram based on UPGMA analysis indicated that the 15 isolates of *P. melonis* formed three major groups (Figure 2). One isolate of *P. drechsleri* was separated from *P. melonis*. RAPD groups were not related to the geographic origins or sensitivity to CAA fungicides of these isolates. Seven isolates (TJ-58, TX-11, TJ-99, TJ-104, TJ-114, 63 and 70) from different RAPD groups were selected for generation of CAA-resistant mutants.

Generation of CAA-resistant isolates. When the seven isolates were exposed to the CAA fungicides, spontaneous mutation resulted in five isolates with flumorph-resistance, one with dimethomorph resistance and three with iprovalicarb resistance.

These nine isolates with spontaneous mutations were obtained from TJ-58, 63 and 70, and the survival frequency was approximately 1×10^{-7} for mutants exposed to the three fungicides (Table 3). No resistant isolates were derived from the other parent isolates. Following the UV treatment, however, the survival frequency was increased to approximately 1×10^{-6}, and 19 flumorph-resistant, 17 dimethomorph-resistant and 10 iprovalicarb-resistant mutants were obtained from the parent isolates TJ-58, 63 and 70 (Table 3).

Characteristics of CAA-resistant Mutants

Resistance stability and resistance factor. After 10 transfers on non-amended medium, the mutant isolates grew as well on fungicide-amended medium as on non-amended medium, indicating that the resistance to CAA fungicides was stable. The level of resistance, as indicated by the resistance factor (RF = EC_{50} of mutant at the 10th transfer/EC_{50} of its parent), ranged from 7 to 601 (Table 3).

Mycelial growth, sporulation and virulence *in vitro*. Compared to the mycelial growth of the corresponding parents (TJ-58, 63 and 70), mycelial growth was faster for some resistant isolates and slower for others. For example, the mycelial growth rate relative to the parent TJ-58 was significantly decreased for F58-1, F58-3, F58-4 and I58-2 ($p<0.05$), but significantly increased for D58-2 and I58-1 ($p<0.05$) on the non-amended medium (Table 4). Virulence also increased or decreased, depending on the resistant isolate. Lesions were significantly larger for the resistant isolates D58-3, D58-2, I58-1, I58-2 and F58-1 than for their parent TJ-58 ($p<0.05$) (Table 4). The resistant isolates F58-3, D58-3, D58-5 and I58-1, however, produced fewer zoospores than the wild-type isolate (Table 4). A compound fitness index (CFI) was calculated: CFI = *in vitro* mycelia growth × zoospore production × lesion area on cucumber leaves. CFI values of resistant isolates were significantly lower for two of nine mutants derived from parent isolate TJ-58, for five of nine mutants derived from parent isolate 63, and all eight isolates derived from parent isolate 70 (Table 4). CFI values were never significantly greater for the mutants than for the parents and were frequently lower but without statistical significance.

Cross-resistance. There was a high level of cross-resistance among all three CAA fungicides: the values of Spearman's rho (ρ) were all >0.8000 ($p<0.0001$) (Figure 3 A to C). Examination of the EC_{50} values (those for CAA-resistant isolates and clustered on the right side of Figure 3 A, B and C) once again indicated a positive correlation between resistance to dimethomorph and flumorph (Figure 3 D), but a negative correlation between resistance to iprovalicarb and flumorph (Figure 3 E) and between resistance to iprovalicarb and dimethomorph (Figure 3 F). No cross-resistance was detected between CAA fungicides and non-CAA fungicides such as metalaxyl, cymoxanil, azoxystrobin and cyazofamid ($p>0.05$) (data not shown).

Analysis of the *CesA3* Gene in *P. melonis*

The full-length of *PmCesA3* gene contained 3550 bp, with one intron of 130-bp located after nucleotide 143 (Figure 4 A). The *PmCesA3* gene coded for a polypeptide chain of 1139 amino-acids and had a predicted molecular weight of 126.5 kDa. The analysis of identities between the PmCesA3 amino acid sequence and those of the closest organisms found in the NCBI GenBank database revealed that homologies were higher with the CesA3 in oomycetes than with CesA3 in *Arabidopsis thaliana* (Table 5). Compared to the CesA3 in sensitive isolates, only one amino-acid substitutions was detected in the CesA3 in the CAA-resistant isolates: this substitution (a GTG to CTG mutation) occurred at

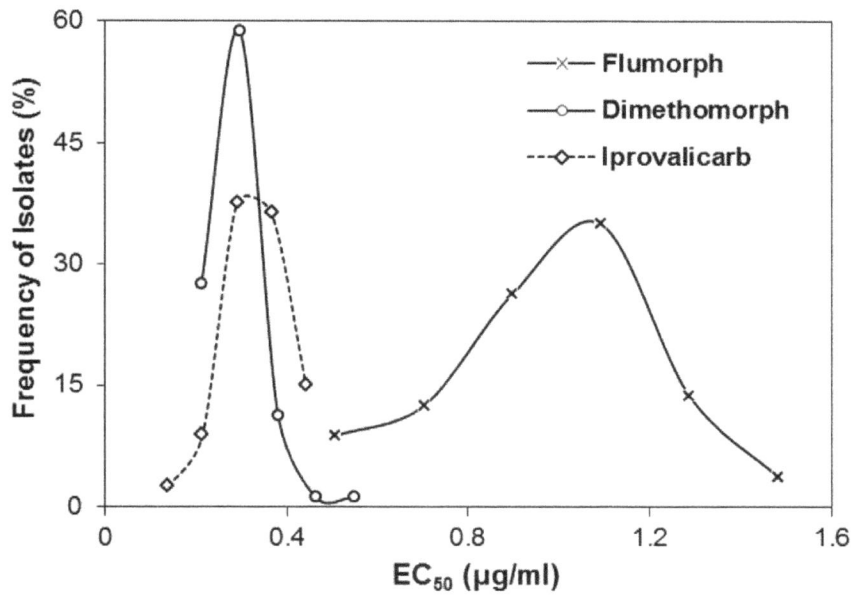

Figure 1. Frequency distributions of EC_{50} values (the effective concentration causing 50% inhibition of mycelial growth of *Phytophthora melonis*) **for flumorph, dimethomorph and iprovalicarb.** In total, 80 isolates of *P. melonis* were collected from areas never exposed to carboxylic acid amide fungicides.

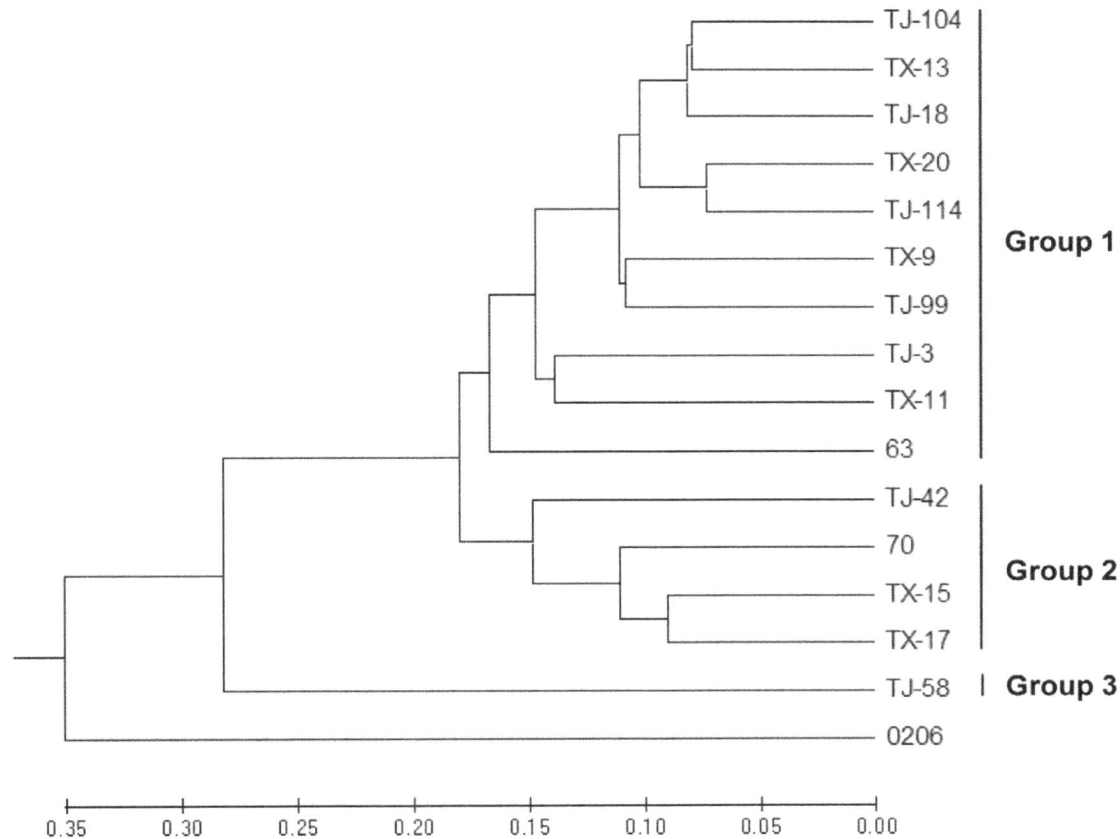

Figure 2. Genetic relationships among 15 isolates of *Phytophthora melonis.* The denrogram (UPGMA) shows the relationships among the isolates of *P. melonis* based on randomly amplified polymorphic DNA (RAPD) analysis with 16 decamer primers. Scale at the bottom depicts the genetic distance.

Table 1. Nucleotide sequences and characteristics of primers used in this study.

Primers	Sequence (5'–3')	Description	Source or reference
ABA3	AGTCAGCCAC	Primer for RAPD analysis	[42]
ABA7	GAAACGGGTG	Same as for ABA3	[42]
ABA9	GGGTAACGCC	Same as for ABA3	[42]
ABA10	GTGATCGCAG	Same as for ABA3	[42]
ABA13	CAGCACCCAC	Same as for ABA3	[42]
ABA17	GACCGCTTGT	Same as for ABA3	[42]
ABA18	AGGTGACCGT	Same as for ABA3	[42]
ABA20	GTTGCGATCC	Same as for ABA3	[42]
Y11	AGACGATGGG	Same as for ABA3	[43]
Y04	GGCTGCAATG	Same as for ABA3	[43]
OPG-16	AGCGTCCTCC	Same as for ABA3	[44]
OPS-14	AAAGGGGTCC	Same as for ABA3	[44]
OPX-12	TCGCCAGCCA	Same as for ABA3	[44]
OPG-11	TGCCCGTCGT	Same as for ABA3	[44]
OPG-14	GGATGAGACC	Same as for ABA3	[44]
OPG-15	ACTGGGACTC	Same as for ABA3	[44]
PmA3F1	TCTCGTGTCGGACGGACCAA	Primer for amplification and partial sequencing of *PmCesA3* gene	This study
PmA3S1	ATCATCGCGTGCTACCTGC	Sequencing primer for *PmCesA3* gene	This study
PmA3S2	CCGTCTTTGTTGTTGGCGGACTG	Same as for PmA3S1	This study
PmA3S3	TCGACGTACTCGATCGCCA	Same as for PmA3S1	This study
PmA3S4	TCACTACATGGAACCGGTGACG	Same as for PmA3S1	This study
PmA3S5	TGGACGGTGGAGGTCGTCAG	Same as for PmA3S1	This study
PmA3S6	AAGCCGTCGCTTGCGTTCC	Same as for PmA3S1	This study
PmA3R1	TCTGCATGTCCAGCCTTCC	Same as for PmA3F1	This study
PMF	ATCTACGCTCGCGGTACCAAG	Primer for rapid detection of resistance	This study
PMR1109A	CGAACACCACGATGTACACCAG	Same as for PMF	This study
PMR1109B	CGAACACCACGATGTACACCTG	Same as for PMF	This study
PMR1109C	CGAACACCACGATGTACACCCG	Same as for PMF	This study
PMR1109D	CGAACACCACGATGTACACCGG	Same as for PMF	This study

codon 1109 and resulted in the replacement of a valine residue with a leucine residue (Figure 4B).

AS-PCR for Rapid Detection of CAA-resistant Isolates of *P. melonis*

Four pairs of allele-specific primers, designed according to the single mutation in the *PmCesA3* gene, were used for PCR with DNA template from CAA-resistant and -sensitive isolates. Using the primer pair PMR1109A + PMF, a 500-bp fragment was amplified at different annealing temperatures whether the template DNA was from resistant or sensitive isolates (Figure 5A), indicating that primers designed by the traditional method could not discriminate between sensitive and resistant alleles. The introduction of an artificial mismatch base at the second nucleotide at the 3'-end of the primers improved specificity at various annealing temperatures (Figure 5 A). As the annealing temperature increased, the reverse primer with artificial mismatch 'T' at the second nucleotide showed more specificity than the primers with mismatch 'C' or 'G'. At the annealing temperature of 68.5°C, the primer PMR1109B was optimal for distinguishing the mutation at codon 1109. With the primer pairs PMF +

PMR1109B, the 500-bp fragment was amplified from CAA-resistant isolates F58-4, I63-2, D63-1, F63-11 and D70-3 but not from CAA-sensitive isolates TX21, TX33, TJ90 and TJ12 (Figure 5 B).

Discussion

The sensitivity of 80 *P. melonis* isolates (collected from 13 fields in China) to the CAA fungicides flumorph, dimethomorph and iprovalicarb was determined by measuring EC_{50} values. The frequency distributions of the EC_{50} values were described as unimodal curves with a narrow range for each fungicide, indicating the absence of CAA-resistant subpopulations among the 80 isolates. Therefore, these results can be used as baselines for tracking future sensitivity shifts of *P. melonis* populations to these three CAA fungicides. Mycelial growth was inhibited more strongly by dimethomorph than by iprovalicarb or flumorph. Similar results were reported for *P. capsici* [22,23,28], *Bremia lactucae* [29], *P. infestans* [21,30] and *Peronophythora litchi* [15], indicating that dimethomorph is generally more effective than iprovalicarb or flumorph for control of oomycete plant pathogens.

Table 2. Isolates of *Phytophthora melonis* used for RAPD analysis and their sensitivities to flumorph, dimethomorph and iprovalicarb.

Isolate	EC$_{50}$ ^{b} (µg/ml) for the three fungicides			Origin
	Flumorph	**Dimethomorph**	**Iprovalicarb**	
TX-9	1.240	0.280	0.230	Xiqing 1^c, Tianjin
TX-11	1.080	0.260	0.310	Xiqing 2, Tianjin
TX-13	1.193	0.280	0.330	Xiqing 3, Tianjin
TX-15	0.983	0.250	0.330	Xiqing 4, Tianjin
TX-17	1.161	0.319	0.340	Xiqing 5, Tianjin
TX-20	0.857	0.240	0.320	Xiqing 6, Tianjin
TJ-3	1.091	0.296	0.354	Hexi 1, Tianjin
TJ-18	1.182	0.281	0.311	Hexi 2, Tianjin
TJ-42	0.760	0.300	0.370	Hexi 3, Tianjin
TJ-58	1.010	0.360	0.240	Hexi 4, Tianjin
TJ-99	0.552	0.250	0.400	Nankai 1, Tianjin
TJ-104	0.775	0.270	0.360	Nankai 2, Tianjin
TJ-114	1.022	0.280	0.360	Nankai 3, Tianjin
63	0.440	0.360	0.290	UC Riverside, USA
70	1.410	0.590	0.370	UC Riverside, USA
0206^a	0.710	0.680	0.420	Nanjing

^{a}One isolate of *Phytophthora drechsleri* was used as an outgroup control.
^{b}EC$_{50}$ values, the effective concentration for causing 50% inhibition of mycelial growth inhibition of *P. melonis*.
^{c}Number represents a different field in the same district.

Although RAPD analysis revealed a high degree of genetic diversity in *P. melonis* collected from different geographical regions, the groups defined by RAPD markers did not share CAA-sensitivity. A likely reason for this lack of correlation is that RAPD markers could not reflected the defined loci responding to sensitivity to fungicides [31]. The RAPD results, however, made it possible to select isolates with different genetic backgrounds for resistance generation.

Isolates with resistance to CAA fungicides were generated from three of the seven isolates used, suggesting that the risk of *P. melonis* resistance to CAA fungicides may be associated with an isolate's genetic background. This would explain why dimethomorph-resistant mutants of *P. capsici* could not be obtained from only one isolate by taming [21], but why CAA-resistance could be obtained by mass selection from zoospores and sexual progeny [22,23].

The risk of fungicide resistance also depends on the pathogen species and its biological characteristics. Based on disease cycles, dispersal ability, frequency of sexual recombination and the competitive ability, *P. viticola* and *Ps. cubensis* have been considered high risk pathogens, while *P. infestans*, *P. capsici* and *P. melonis* have been considered low risk pathogens (FRAC, www.frac.info). Assessments of the risk of fungicide resistance are also based on field observations. Thus, CAA-resistant isolates that are stable and competitive have been detected among field populations of *P. viticola* [24] and *Ps. cubensis* [18] but not among field populations of *P. infestans* [20] (FRAC, www.frac.info), indicating a high risk of resistance to CAAs in *P. viticola* and *Ps. cubensis* but a low risk in *P. infestans*. Until now, no CAA-resistant isolates of *P. capsici* have been reported in the field, but *P. capsici* mutants with high CAA-resistance were obtained by mass selection from zoospores and oospores, and

the risk of resistance to CAAs was considered low to moderate in *P. capsici* [22,23]. For *P. melonis* in the current study, CAA-resistant mutants were generated *in vitro* with a frequency of 1×10^{-7} by spontaneous selection and 1×10^{-6} by UV-mutagenesis of zoospores. That the frequency was higher with UV-mutagenesis than with spontaneous selection suggests that UV radiation can increase the probability of CAA fungicide resistance in *P. melonis*. The CFIs (compound fitness indices) were often lower for the CAA-resistant isolates than the wild-type isolates, indicating that CAA-resistance in this study was generally associated with reduced fitness. This supports our inference that the risk of resistance to CAA fungicides in *P. melonis* is low to moderate.

Mutants resistant to one of the CAA fungicides in the current study were resistant to other CAA fungicides but not non-CAA fungicides, indicating that there was cross-resistance among flumorph, dimethomorph and iprovalicarb but not between the CAA and non-CAA fungicides. Similar results have been reported for *P. viticola* [24], *Ps. cubensis* [18] and *P. capsici* [22,23,32]. Although the cross-resistance suggests that the CAA-resistant isolates have a similar resistance mechanism, the negative correlation between higher EC$_{50}$ values for iprovalicarb and flumorph and between higher EC$_{50}$ values for iprovalicarb and dimethomorph but not between those for flumorph and dimethomorph suggests that the resistance mechanism may differ somewhat between the cinnamic acid amides (dimethomorph and flumorph) and the valine amide carbamates (iprovalicarb).

We amplified and sequenced the *CesA3* gene of *P. melonis*. Analysis of the CesA3 amino acid sequence revealed that the wild-type and CAA-resistant isolates of *P. melonis* differed only in the V1109L substitution (Figure 5B). Previous studies reported that resistance to CAA fungicides was conferred by G1105V or G1105A substitution in CesA3 of *P. infestans* [17], G1105S in CesA3 of *P. viticola* [26] and G1105V or G1105W in CesA3 of *Ps. cubensis* [25]. The substitution of V1109L in PmCesA3 would therefore represent a novel mutation causing resistance to CAA fungicides. The finding of only one mutation and the detailed cross-resistance results suggest that other genes might also be involved in CAA resistance. In addition, CAA resistance was considered to be controlled by a recessive gene in *P. infestans* [17] and *P. viticola* [26], but by two dominant genes in *P. capsici* [23]. In this study, we did not find any CAA-resistant isolates with a heterozygous mutation at codon position 1109 on PmCesA3, suggesting that CAA resistance in *P. melonis* may also be controlled by a recessive gene(s). Confirming this will require further genetic experiments, but genetic manipulation of *P. melonis* is difficult because it is homothallic.

Several methods such as AS-PCR and PCR-RFLP have been developed for detecting isolates with mutations associated with fungicides resistance [33]. A recent study described a PCR-RFLP method that rapidly detects CAA resistance in *P. viticola* populations [27]. In our study, AS-PCR primers were designed (based on the mutation of V1109L); these primers effectively identified CAA resistance in *P. melonis*. Compared with the traditional AS-PCR primers, the new reverse primer contained an additional mismatch at the second nucleotide of the 3'-end; the introduction of this mismatch was previously reported to increase specificity of the allele-specific primer [34–36]. In our trial, the mismatch nucleotide 'T' was more optimal than the mismatch nucleotides of 'C' and 'G'. However, different mismatches can increase or decrease the specificity of the primer, indicating that the most suitable mismatch must be tested in different cases [37]. The AS-PCR primers described

Table 3. Results of the experiments conducted to induce resistance against flumorph, dimethomorph, and iprovalicarb in *Phytophthora melonis*.

Parental isolates	Fungicides	Type of induction[a]	No. of mutants	Survival frequency[b] ($\times 10^{-6}$)	EC$_{50}$[c] (µg/ml)	RF[d]
TJ-58	Flumorph	SM	2	0.40	114~151	113~150
		UV	5	1.00	48~155	133~431
	Dimethomorph	SM	0	–	–	–
		UV	5	1.00	40~151	111~419
	Iprovalicarb	SM	1	0.25	101	421
		UV	2	0.40	46~193	192~804
63	Flumorph	SM	2	0.40	58~159	132~361
		UV	10	2.00	16~174	35~395
	Dimethomorph	SM	1	0.25	47	131
		UV	8	1.60	5~194	14~539
	Iprovalicarb	SM	2	0.40	74~104	206~289
		UV	3	0.60	43~174	119~483
70	Flumorph	SM	1	0.25	63	45
		UV	4	0.80	22~57	16~41
	Dimethomorph	SM	0	–	–	–
		UV	4	0.80	9~76	15~129
	Iprovalicarb	SM	0	–	–	–
		UV	5	1.00	8~42	22~114

[a]SM, spontaneous mutation. UV, UV-mutagenesis.
[b]Survival frequency, number of mutants/total number of zoospores used for mutant generation.
[c]EC$_{50}$, the effective concentration for causing 50% inhibition of mycelial growth inhibition of *P. melonis*.
[d]Resistance factor = EC$_{50}$ of resistant isolates at the 10th transfer/EC$_{50}$ of its parent.

here will be useful for detecting CAA-resistant isolates of *P. melonis* from field populations.

Materials and Methods

Isolates and Culture Conditions

Roots and stems of cucumber (*Cucumis sativus* Linn.) with typical signs and symptoms of infection by *P. melonis* were collected from 13 fields in Xiqing, Hexi and Nankai districts in Tianjin of China in 2005, where CAA fungicides had never been used. Tissue plugs were cut from the margin of lesions on stems and roots. The plugs were disinfested for 3 min in 0.5% (vol/vol) NaClO. After being rinsed three times with sterile water, these plugs were placed on white kidney bean agar (WKB) (60 g of white kidney bean, 7 g of agar and distilled water up to 1 liter) plates amended with 50 µg/ml of ampicillin (98% a.i., Tuoyingfang Biotech Co., Ltd., Beijing), 50 µg/ml of rifampicin (98% a.i., Tuoyingfang Biotech Co., Ltd., Beijing) and 50 µg/ml of pentachloronitrobenzene (PCNB) (40% a.i., Sanli Chemical Industry Co., Ltd., Shanxi, China). After the cultures had been incubated at 25°C in darkness for 5 d, mycelial plugs were cut from margin of the culture and transferred to a new WKB agar plate. In total, 80 isolates of *P. melonis* were obtained. The isolates were identified using specific primers and morphology as described previously [1,38].

For acquisition of single-zoospore isolates, a zoospore suspension was prepared by placing mycelial plugs into sterile soil extract (10 g of soil per liter of water). After incubation under light at 25°C for 72 h, sporangia formed. Following incubation at 4°C for 1 h and at 25°C for 40 min, zoospores were discharged from sporangia. A 0.2-ml volume of the zoospore suspension was placed

on water-agar plate at 25°C. After 12 h, single germinated zoospores and associated agar were transferred to fresh WKB agar plate. Two single-zoospore isolates of *P. melonis* (68 and 70) were kindly provided by Dr. Michael D. Coffey (University of California, Riverside), and one single-zoospore isolate of *P. drechsleri* (0206) was kindly provided by Dr. Zheng Xiaobo (Nanjing Agricultural University). These three isolates were also single-zoospore cultures. All isolates were maintained on WKB agar medium. For long-term storage, each culture was transferred to WKB agar slants, covered with sterile mineral oil, and stored at room temperature.

Fungicides

The following technical-grade fungicides were individually dissolved in dimethyl sulfoxide (DMSO) to prepare stock solutions (1×10^4 µg/ml) and stored at 4°C in the dark: flumorph (96% a.i., Research Institute of Chemical Industry, Shenyang, China), dimethomorph (95% a.i., Frey Agrochemicals Ltd), iprovalicarb (98% a.i.; Sigma-Aldrich Shanghai Trading Co. Ltd), metalaxyl (97% a.i., Agrolex P. Ltd., Beijing), azoxystrobin (96% a.i., Syngenta Biotechnology Co. Ltd., Shanghai, China), cyazofamid (96% a.i., Sigma-Aldrich Shanghai Trading Co. Ltd) and cymoxanil (98% a.i., Xinyi Agrochemicals Company, Jiangsu, China). The final concentration of DMSO in the WKB agar medium was adjusted to 0.1% (vol/vol) throughout this study. WKB agar plates amended with fungicides were prepared by adding the same volume of serially diluted solutions to the molten agar medium at ≈50°C. WKB agar medium without fungicide but with the same volume of DMSO was used as a control.

Table 4. Fitness of CAA-resistant and -sensitive isolates of *Phytophthora melonis in vitro*.

Isolates [a]	Mycelial growth (mm) [b]	Zoospore production ($\times 10^5/cm^2$)	Lesion area on cucumber leaves (mm^2)	CFI [c] ($\times 10^5$)
TJ-58	77 c	1.53 a	370 d	43566 ab
F58-1	73 e	1.30 abcd	398 bc	37728 bc
F58-3	72 e	1.22 bcd	380 cd	33684 c
F58-4	74 de	1.44 ab	383 cd	40980 abc
D58-2	81 a	1.41 abc	420 ab	47483 a
D58-3	76 cd	1.20 bcd	425 a	38712 abc
D58-5	77 c	1.16 cd	383 cd	34205 c
I58-1	80 ab	1.13 d	420 ab	37743 bc
I58-2	73 e	1.28 abcd	398 bc	36977 bc
I58-3	78 bc	1.35 abcd	368 d	38660 abc
63	78 c	0.76 a	555 c	32988 ab
F63-1	73 f	0.42 c	532 c	16296 e
F63-3	76 de	0.48 bc	473 d	17381 e
F63-5	75 e	0.75 a	476 d	27014 abc
D63-1	76 de	0.42 c	462 d	14893 e
D63-2	77 de	0.53 bc	457 d	18652 de
D63-8	81 b	0.51 bc	476 d	19648 cde
I63-2	83 a	0.39 c	583 b	18784 de
I63-5	77 cd	0.62 ab	542 c	25872 bcd
I63-9	72 f	0.76 a	621 a	34020 a
70	87 c	0.87 a	575 d	43719 a
F70-1	82 e	0.74 b	497 e	30125 cd
F70-5	91 a	0.58 de	619 bc	32809 bc
F70-11	83 e	0.70 bc	561 d	32795 bc
D70-1	87 cd	0.53 e	639 b	29280 cd
D70-5	79 f	0.68 bc	681 a	36690 b
I70-1	89 b	0.52 e	567 d	26310 d
I70-5	85 d	0.53 e	564 d	25505 d
I70-9	87 cd	0.63 cd	607 c	33155 bc

[a] Isolates in bold font are parents of the resistant isolates listed under them in regular font. Isolates starting with the letter F, D, and I, are flumorph-resistant mutants, dimethomorph-resistant mutants, and iprovalicarb-resistant mutants, respectively.
[b] For each parent and its resistant progeny, means followed by same letters are not significantly different according to Fisher's least significance difference ($\alpha = 0.05$).
[c] CFI (compound fitness index) = mycelial growth $\times$ zoospore production $\times$ lesion area on cucumber leaves.

Baseline Sensitivities of *P. melonis* to Flumorph, Dimethomorph and Iprovalicarb

The sensitivities of 80 *P. melonis* isolates to the CAA fungicides flumorph, dimethomorph and iprovalicarb were determined by measuring mycelium growth on fungicide-amended medium. Fresh mycelial plugs (5 mm in diameter) were cut from the edge of an actively growing colony and placed face up in the center of WKB agar medium plates, which were amended with flumorph (0, 0.50, 0.70, 0.90, 1.00, 1.25, 1.50 μg/ml), dimethomorph (0, 0.10, 0.15, 0.20, 0.25, 0.30, 0.35 μg/ml) or iprovalicarb (0, 0.20, 0.25, 0.30, 0.35, 0.40, 0.45 μg/ml). Each treatment was represented by four replicate plates. After incubation for 4 days at 25°C in darkness, colony diameter was measured at perpendicular angles, and the average of the two measurements (minus 5 mm for the mycelial plug) was used for data analysis. The percentage of inhibition was calculated for each concentration and the concentration of each fungicide causing 50% inhibition (EC_{50}) was estimated from the regression of the probit of the percentage of growth inhibition against the logarithmic value of fungicide concentration. For each of the three CAA fungicides, the frequency distribution of 80 EC_{50} values was plotted as a representation of baseline sensitivity.

Development of CAA-resistant Mutants of *P. melonis* in vitro

RAPD. To select *P. melonis* isolates with different genetic background for generation of CAA-resistant mutants, 15 isolates collected from different fields were randomly chosen for genetic relationship analysis by using RAPD, and one isolate of *P. drechsleri* was used as the outgroup control (Table 2). Mycelia were frozen in liquid nitrogen and ground into fine powder with mortar and pestle, which has been previously sterilized at 160°C for 2 h. Genomic DNA was extracted according to the modified Ristaino's CTAB protocol [39]. About 100 mg of mycelial powder was placed in a 1.5-ml centrifuge tube. A 150-μl volume of extraction buffer (0.35 M sorbitol, 0.1 M Tris, 0.005 M EDTA [pH 7.5], and 0.02 M sodium bisulfite) was added, and the tube was then mixed with a vortex mixer. A 150-μl volume of nuclear lysis buffer

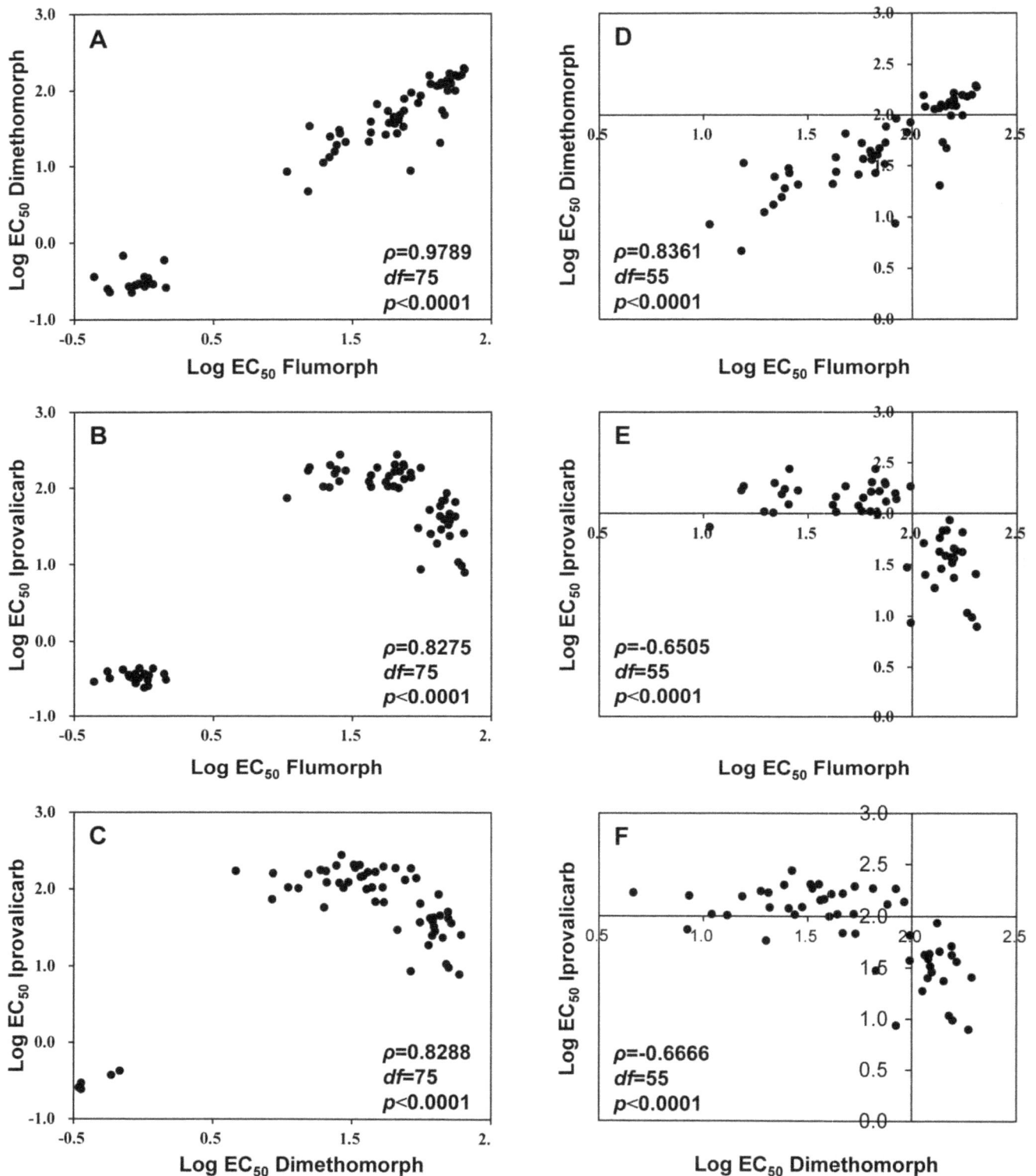

Figure 3. Cross-resistance among flumorph, dimethomorph and iprovalicarb. Log-transformed EC_{50} values (the effective concentration for causing 50% inhibition of mycelial growth inhibition of *Phytophthora melonis*) for isolates of *P. melonis* were compared among the three carboxylic acid amide fungicides using Spearman's rank correlation coefficients. (A), (B), and (C) indicate positive cross-resistance among flumorph, dimethomorph, and iprovalicarb; (D-F) include only the higher EC_{50} values from (A-C), i.e., EC_{50} values from CAA-resistant isolates. (D) reveals a positive correlation between the EC_{50} values for dimethomorph and flumorph among CAA-resistant isolates, while (E) and (F) reveals a negative correlation between iprovalicarb and flumorph and between iprovalicarb and dimethomorph among CAA-resistant isolates.

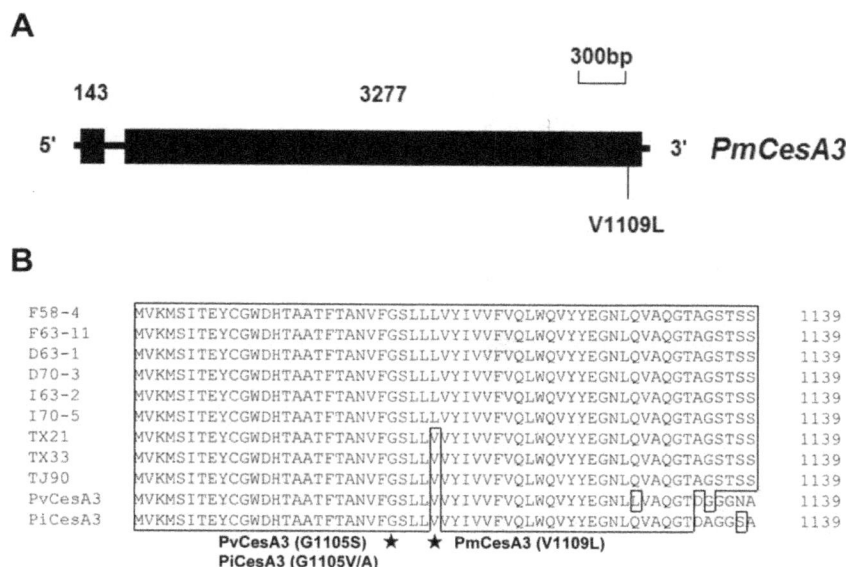

Figure 4. Structure and site of mutation in the *PmCesA3* gene associated with carboxylic acid amide (CAA) fungicide resistance. (A) Intron/exon structure of the *PmCesA3* gene. Numbers represent the size in base pairs. Point mutations in CAA-resistant mutants and the predicted amino acid substitution in the mutant gene products are indicated. (B) Alignment of partial amino acid sequences of CesA3 in *P. melonis* (PmCesA3), *P. infestans* (PiCesA3), and *P. viticola* (PvCesA3). TJ-90, TX-21, and TX-33 were wild-type isolates. D63-1 and D70-3 were dimethomorph-resistant mutants. F58-4 and F63-11 were flumorph-resistant mutants. I63-2 and I70-5 were iprovalicarb-resistant mutants. Mutations in CAA-resistant mutants of *P. infestans*, *P. viticola* and *P. melonis* are indicated by asterisks.

(0.2 M Tris, 0.05 M EDTA [pH 7.5], 2.0 M NaCl and 2% CTAB [pH 7.5]) and 60 μl of 20% SDS (20 g SDS per 100 ml water) was added, and the tube was mixed again. After incubation at 65°C for 30 min, an isopyknic mixture of chloroform-isoamyl alcohol (24:1, v/v) was added, and the tube was centrifuged for 15 min at 13,000 *g*. The aqueous phase was transferred to a new tube, and the chloroform extraction was repeated. After adding 0.1 volume of 3 M sodium acetate (pH 8.0) and 0.6 volume of cold isopropyl alcohol, DNA was precipitated at −20°C for 2 h. The tube was centrifuged at 13,000 *g* for 15 min, and the precipitate was washed with 75% ethanol and then dried at room temperature. DNA was resuspended using 50 μl of TE buffer (10 mM Tris-HCl, 0.1 mM EDTA [pH 8.0]) for PCR.

RAPD-PCR was performed with each of 16 decamer primers (Table 1). The primers were synthesized by Beijing Sunbiotech Co. Ltd. (Beijing, China). PCR was performed in a 25-μl volume containing 50 ng of template DNA, 1 μl of primer (10 μM), 2 μl of dNTP mixture (2.5 mM of each dNTP and 20 mM Mg^{2+}), 2.5 μl of Easy Taq DNA Polymerase Buffer (10×), and 2.5 U of EasyTaq

DNA Polymerse (TransGen Biotech, Beijing, China). Amplification was performed in a MyCycler™ Thermal Cycler (Bio-Rad) with the following parameters: 94°C for 6 min; followed by 40 cycles of denaturation at 94°C for 30 s, annealing at 36°C for 1 min, and extension at 72°C for 2 min and a final cycle of extension at 72°C for 10 min. Amplification products were separated on 1.5% agarose gels in Tris-acetate (TAE) buffer at 110 V for 2 h and were visualized under UV light after being stained with ethidium bromide. All PCRs were repeated at least twice.

Differences in fingerprinting patterns among isolates were assessed based on the clear and reproducible bands. Presumed homologous bands were scored zero (absent) or one (present) and then transformed into a binary matrix. Genetic distance coefficients were calculated for all pairwise comparisons by Nei's method [40]. The phylogenic tree was generated based on the genetic distance coefficients by using UPGMA (unweighted pair-group method arithmetic averages) and MEGA (molecular evolutionary genetics analysis) software (version 5).

Generation of CAA-resistant isolates. Based on the genetic analysis, seven isolates of *P. melonis* were selected for generation of CAA-resistant mutants. In the case of flumorph, zoospores suspensions were prepared as described above, and 100 μl of a zoospore suspension (approximately 1.0×10^6 zoospores/ml) was inoculated onto WKB plates amended with 10 μg/ml of flumorph (WKBF). After incubation at 25°C in darkness for 5 days, the emergent colonies were transferred to a fresh WKBF plate. Single-zoospore isolates were obtained. The same procedure was used for generation of resistant mutants to dimethomorph (10 μg/ml of medium) and iprovalicarb (5 μg/ml of medium). This selection procedure was performed twice.

Ultraviolet (UV)-mutagenesis of zoospores. Zoospore suspensions were continuously agitated while they were exposed to UV irradiation (TUV Philips, 15 W, 254 nm) for 1 min at a distance of 30 cm. The suspensions were then spread on WKB

Table 5. Predicted amino acid sequence identities (%) among known CesA3s from four *Phytophthora* species, *Plasmopara viticola*, and *Arabidopsis thaliana*.

	PmCesA3	PiCesA3	PrCesA3	PvCesA3	AtCesA3
PmCesA3	100	–	–	–	–
PiCesA3	82	100	–	–	–
PrCesA3	95	95	100	–	–
PvCesA3	81	95	94	100	–
AtCesA3	14	16	15	16	100

Values indicate identity expressed as a percentage.

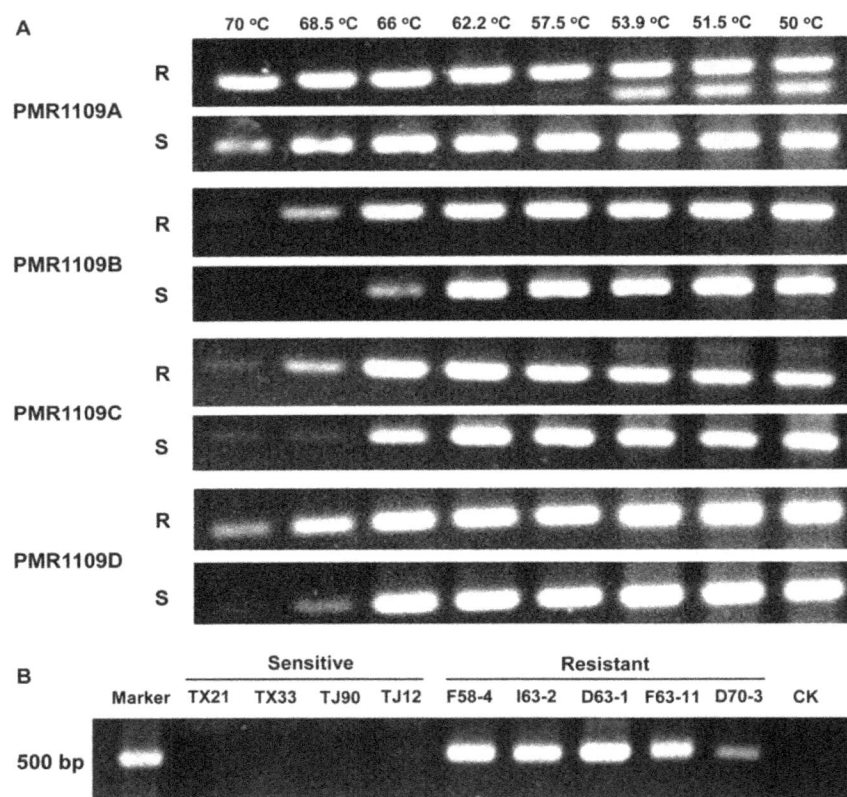

Figure 5. Specificity of four allele-specific PCR primer pairs for the detection of carboxylic acid amide (CAA)-resistant isolates of *Phytophthora melonis.* (A) Specificity of the four primer pairs for the CAA-sensitive isolate TJ-58 (S) and the CAA-resistant isolate F58-4 (R) at gradient annealing temperatures. (B) Specificity of primer pair (PMF + PMR1109B) for four CAA-sensitive and five CAA-resistant isolates at 68.5°C.

medium plates amended with the corresponding CAA fungicide as described in the previous section. These plates were incubated in the dark for 30 min to minimize light repair of DNA damage. This selection procedure was performed twice and included control plates that were not exposed to UV.

Biological Characteristics of CAA-resistant Mutants

Stability and level of resistance. For determination of the stability of CAA resistance of the mutants, the mutants were subjected to 10 successive transfers on fungicide-free medium before mycelium growth was measured on WKB agar medium amended with each corresponding fungicide at the concentrations described previously. The experiment was done twice. EC_{50} values of mutants were estimated by measuring mycelium growth on fungicide-amended medium at 0, 5, 10, 20, 40, 80 and 100 μg/ml of each CAA fungicide. The level of resistance was described by the resistance factor: $RF = EC_{50}$ of mutant at the 10th transfer/ EC_{50} of its parent.

In addition, one spontaneous and one UV-induced mutant resistant to each of fungicide was randomly selected for determination of resistance stability of zoospore progeny. At least 20 single-zoospore isolates randomly sampled from each mutant were grown on WKBF medium.If these single-zoospore isolates grew on WKBF medium, their resistance was considered to be stable; otherwise their resistance was unstable. The same procedure was followed with dimethomorph at 10 μg/ml and with iprovalicarb at 5 μg/ml. This experiment was performed twice.

Mycelial growth and zoospores production. For determination of mycelial growth, the 26 CAA-resistant isolates and their parents were transferred to WKB medium in plates as described in the section concerning baseline sensitivities except that the medium did not contain fungicide. Each isolate was represented by three replicated plates. After incubation in the darkness at 25°C for 5 days, the colony diameter was measured at perpendicular angles, and the average of the two measurements was used to compare the mycelial growth of each resistant isolate and its parent. For comparison of zoospore production, 10 plugs (5-mm in diameter) from the colony margin and 10 plugs from the area near the initial inoculum plug were harvested, and zoospore production was induced as described above and quantified with a hemacytometer. The number of zoospores per cm^2 of culture was calculated. These experiments were conducted at least twice.

Virulence. The second or third true leaf from a cucumber plant (cv. Changchunmici) at the fifth true leaf stage was used for virulence tests. The leaves were harvested and rinsed three times with sterile-distilled water. Zoospore suspensions were prepared as described earlier for each of the 26 CAA-resistant mutants and their parent isolates. Four 10-μl droplets of a zoospore suspension (1.0×10^5 zoospores/ml) were placed on the upper surface of leaves. One half of each leaf was inoculated with a resistant mutant and the other half was inoculated with the corresponding parent. Ten replicate leaves were used for each combination of mutant and parent. After 5 days in a moist chamber at 20°C with 12 h of light and 12 h of darkness, the lesion areas on each leaf were measured. This experiment was conducted at least twice for each combination of mutant and parent.

Cross-resistance. The 55 CAA-resistant mutants and 20 wild-type isolates were cultured on WKB agar medium amended with the non-CAA fungicides metalaxyl (0, 0.01, 0.02, 0.05, 0.10, and 0.20 µg/ml). azoxystrobin (0, 0.01, 0.05, 0.10, 0.50, and 1.00 µg/ml), cymoxanil (0, 10, 20, 40, 80, and 100 µg/ml), or cyazofamid (0.01, 0.02, 0.05, 1.00, and 2.00 µg/ml) or with the CAA fungicides at the concentrations described above. After incubation in darkness at 25°C for 4 days, the colony diameters were measured and the EC_{50} values were calculated as described above. Each treatment was represented by three replicate plates. The experiment was conducted at least twice for each isolate.

Amplification of the CesA3 gene of P. melonis

Based on the conserved sequence of the CesA3 genes in *P. infestans* (ABP96904), *P. ramorum* (ABP96912) and *P. sojae* (ABP96908) in the Genbank/EMBL data libraries, homologous primers were designed for amplification of the partial PmCesA3 gene fragment. The 5′and 3′end of the PmCesA3 gene were acquired using SiteFinding-PCR [41]. The full-length PmCesA3 gene was amplified and sequenced using primers listed in Table 1. All primers were synthesized by Beijing Sunbiotech Co. Ltd. (Beijing, China). Primers PmA3F1 and PmA3R1 were used to amplify the PmCesA3 gene. PCRs were performed in a 50-µl volume containing 50 ng of template DNA, 1 µl of each primer (10 µM), 4 µl of dNTP mixture (2.5 mM each dNTP), 1×Easy Taq DNA Polymerase Buffer, and 2.5 U of EasyTaq DNA Polymerase (TransGen Biotech, Beijing, China). The PCR was performed in a MyCycler™ Thermal Cycler (Bio-Rad) with the following parameters: an initial preheating for 5 min at 95°C; followed by 35 cycles of denaturation at 94°C for 30 s, annealing at 62°C for 30 s, and extension at 72°C for 4 min; and with a final extension at 72°C for 10 min. All PCR products were separated and purified by electrophoresis in a 1% agarose gel in Tris-acetate (TAE) buffer and were cloned into the pEASY-T3 Vector (TransGen Biotech, Beijing, China) and sequenced by Beijing Sunbiotech Co. Ltd. (Beijing, China). The programs in the DNAMAN software were used to predict the PmCesA3 amino acid sequences and to compare the amino acid sequences of the wild-type isolates with those of the CAA-resistant mutants.

Molecular Detection of Resistance Mutation in PmCesA3 by Allele-specific PCR

According to the single mutation in the PmCesA3 gene, allele-specific primers were designed with the match the nucleotide 'C' at the 3′-end of the reverse primers. The specificity of the primers was improved by introducing an artificial mismatch base at the second nucleotide at the 3′-end of the primers (Table 1). To test the specificity of the primers, all the primer pairs were used for gradient PCR using the DNA templates from wild-type isolate TJ-58 and CAA-resistant mutant F58-4. PCR amplification was performed in a MyCycler™ Thermal Cycler (Bio-Rad) with the following parameters: an initial preheating for 5 min at 95°C; followed by 30 cycles of denaturation at 95°C for 30 s, annealing at 50 to 70°C for 30 s, and extension at 72°C for 30 s; and terminated with a final extension at 72°C for 10 min. A 5-µl volume of PCR product from each sample was analyzed by electrophoresis using a 1.5% agarose gel in TAE buffer.

Statistical Analysis

Data were analyzed by using the general linear model (GLM) procedure with Statistical Analysis System software (version 9; SAS Inc., Cary, NC, USA). Means were separated using Fisher's protected least significant difference (LSD, $\alpha = 0.05$). Cross-resistance between fungicides was analyzed using Spearman's rank correlation coefficient for log-transformed EC_{50} values.

Acknowledgments

We thank M. D. Coffey and X. Zheng for kindly providing strains. We thank B. Jaffee and J. Hao for reviewing and providing professional opinions on this manuscript.

Author Contributions

Conceived and designed the experiments: LC SZ X. Liu. Performed the experiments: LC SZ X. Liu. ZP MC. Analyzed the data: LC SZ. Wrote the paper: LC SZ.

References

1. Ho HH, Gallegly ME, Hong CX (2007) Redescription of *Phytophthora melonis*. Mycotaxon 102: 339–345.
2. Mohaghegh P, Khoshgoftarmanesh AH, Shirvani M, Sharifnabi B, Nili N (2011) Effect of silicon nutrition on oxidative stress induced by *Phytophthora melonis* infection in cucumber. Plant Dis 95: 455–460.
3. Ho HH (1986) *Phytophthora melonis* and *P. sinensis* synonymous with *P. drechsleri*. Mycologia 78: 907–912.
4. Li G, Xue YL (1989) Research on induced immunization of hami melon against *Phytophthora melonis*. Chin Sci Bull 34: 253–256.
5. Wang R, Wang RP (2000) Induction of resistance to *Phytophthora melonis* in Hami melon (*Cucumis melo* L.). Zhi Wu Bao Hu 26: 9–11.
6. Guharoy S, Bhattacharyya S, Mukherjee SK, Mandal N, Khatua DC (2006) *Phytophthora melonis* associated with fruit and vine rot disease of pointed gourd in India as revealed by RFLP and sequencing of ITS region. J. Plant Pathol 154: 612–615.
7. Mirabolfathy M, Cooke DEL, Duncan JM, Williams NA, Ershad D, et al. (2001) *Phytophthora pistaciae* sp. nov. and *P. melonis*: The principal causes of pistachio gummosis in Iran. Mycol Res 105: 1166–1175.
8. Shaofeng L, Yongguan W, Siliang H, Lizhi L (2007) Reasearch Progress in Control of Wax Gourd *Phytophthora* Blight. Zhongguo Nong Xue Tong Bao 23: 301–304.
9. Wu Y, Lu S, Huang S, Fu G, Chen L, et al. (2011) Field resistance of *Phytophthora melonis* to metalaxyl in South China. Weishengwu Xuebao 51: 1078–1086.
10. Georgopoulos S, Grigoriu A (1981) Metalaxyl-Resistant Strains of *Pseudoperonospora cubensis* in Cucumber Greenhouses of Southern Greece. Plant Dis 65: 729–730.
11. Shattock RC (2002) *Phytophthora infestans*: populations, pathogenicity and phenylamides. Pest Manag Sci 58: 944–950.
12. Gisi U, Lamberth C, Mehl A, Seitz T (2007) Carboxylic acid amide (CAA) fungicides. Modern crop protection compounds: 651–674.
13. Dutzmann S (1999) Iprovalicarb (SZX 0722) - a novel fungicide with specific activity against oomycetes. Pflanzenschutz-Nachrichten Bayer 52: 15–32.
14. Reuveni M (2003) Activity of the new fungicide benthiavalicarb against *Plasmopara viticola* and its efficacy in controlling downy mildew in grapevines. Eur. J. Plant Pathol. 109: 243–251.
15. Wang HC, Sun HY, Stammler G, Ma JX, Zhou MG (2009) Baseline and differential sensitivity of *Peronophythora litchii* (lychee downy blight) to three carboxylic acid amide fungicides. Plant Pathol 58: 571–576.
16. Zhu SS, Chen L, Lu XH, Li JQ, Liu XL (2010) Effect of three carboxylic acid amide fungicides on different life stages of *Phytophthora melonis* and determination of the sensitivities. Nongyaoxue Xuebao 12: 168–172.
17. Blum M, Boehler M, Randall E, Young V, Csukai M, et al. (2010) Mandipropamid targets the cellulose synthase-like PiCesA3 to inhibit cell wall biosynthesis in the oomycete plant pathogen, *Phytophthora infestans*. Mol Plant Pathol 11: 227–243.
18. Zhu SS, Liu XL, Wang Y, Wu XH, Liu PF, et al. (2007) Resistance of *Pseudoperonospora cubensis* to flumorph on cucumber in plastic houses. Plant Pathol 56: 967–975.
19. Young DH, Spiewak SL, Slawecki RA (2001) Laboratory studies to assess the risk of development of resistance to zoxamide. Pest Manag Sci 57: 1081–1087.
20. Cohen Y, Rubin E, Hadad T, Gotlieb D, Sierotzki H, et al. (2007) Sensitivity of *Phytophthora infestans* to mandipropamid and the effect of enforced selection pressure in the field. Plant Pathol 56: 836–842.
21. Yuan SK, Liu XL, Si NG, Dong J, Gu BG, et al. (2006) Sensitivity of *Phytophthora infestans* to flumorph: *In vitro* determination of baseline sensitivity and the risk of resistance. Plant Pathol 55: 258–263.

22. Lu XH, Zhu SS, Bi Y, Liu XL, Hao JJ (2010) Baseline sensitivity and resistance-risk assessment of *Phytophthora capsici* to iprovalicarb. Phytopathology 100: 1162–1168.

23. Meng QX, Cui XL, Bi Y, Wang Q, Hao JJ, et al. (2011) Biological and genetic characterization of *Phytophthora capsici* mutants resistant to flumorph. Plant Pathol 60: 957–966.

24. Gisi U, Waldner M, Kraus N, Dubuis PH, Sierotzki H (2007) Inheritance of resistance to carboxylic acid amide (CAA) fungicides in *Plasmopara viticola*. Plant Pathol 56: 199–208.

25. Blum M, Waldner M, Olaya G, Cohen Y, Gisi U, et al. (2011) Resistance mechanism to carboxylic acid amide fungicides in the cucurbit downy mildew pathogen *Pseudoperonospora cubensis*. Pest Manag Sci 67: 1211–1214.

26. Blum M, Waldner M, Gisi U (2010) A single point mutation in the novel *PvCesA3* gene confers resistance to the carboxylic acid amide fungicide mandipropamid in *Plasmopara viticola*. Fungal Genet Biol 47: 499–510.

27. Aoki Y, Furuya S, Suzuki S (2011) Method for rapid detection of the *PvCesA3* gene allele conferring resistance to mandipropamid, a carboxylic acid amide fungicide, in *Plasmopara viticola* populations. Pest Manag Sci 67: 1557–1561.

28. Sun H, Wang H, Stammler G, Ma J, Zhou M (2010) Baseline Sensitivity of Populations of *Phytophthora capsici* from China to Three Carboxylic Acid Amide (CAA) Fungicides and Sequence Analysis of Cholinephosphotranferases from a CAA-sensitive Isolate and CAA-resistant Laboratory Mutants. J Plant Pathol 158: 244–252.

29. Cohen Y, Rubin A, Gotlieb D (2008) Activity of carboxylic acid amide (CAA) fungicides against *Bremia lactucae*. Eur J Plant Pathol 122: 169–183.

30. Cohen Y, Gisi U (2007) Differential activity of carboxylic acid amide fungicides against various developmental stages of *Phytophthora infestans*. Phytopathology 97: 1274–1283.

31. Mahuku G, Peters RD, Platt HW, Daayf F (2000) Random amplified polymorphic DNA (RAPD) analysis of *Phytophthora infestans* isolates collected in Canada during 1994 to 1996. Plant Pathol 49: 252–260.

32. Bi Y, Cui X, Lu X, Cai M, Liu X, et al. (2011) Baseline sensitivity of natural population and resistance of mutants in *Phytophthora capsici* to zoxamide. Phytopathology 101: 1104–1111.

33. Ma Z, Michailides TJ (2005) Advances in understanding molecular mechanisms of fungicide resistance and molecular detection of resistant genotypes in phytopathogenic fungi. Crop Prot 24: 853–863.

34. Drenkard E, Richter BG, Rozen S, Stutius LM, Angell NA, et al. (2000) A simple procedure for the analysis of single nucleotide polymorphism facilitates map-based cloning in Arabidopsis. Plant Physiol 124: 1483–1492.

35. Hayashi K, Hashimoto N, Daigen M, Ashikawa I (2004) Development of PCR-based SNP markers for rice blast resistance genes at the Piz locus. Theor Appl Genet 108: 1212–1220.

36. Zhu LX, Zhang ZW, Liang D, Jiang D, Wang C, et al. (2007) Multiplex asymmetric PCR-based oligonucleotide microarray for detection of drug resistance genes containing single mutations in Enterobacteriaceae. Antimicrob Agents Chemother 51: 3707–3713.

37. Yin Y, Liu X, Shi Z, Ma Z (2010) A multiplex allele-specific PCR method for the detection of carbendazim-resistant *Sclerotinia sclerotiorum*. Pestic Biochem Physiol 97: 36–42.

38. Wang Y, Ren Z, Zheng X (2007) Detection of Phytophthora melonis in samples of soil, water, and plant tissue with polymerase chain reaction. Can J Plant Pathol 29: 172–181.

39. Ristaino JB, Madritch M, Trout CL, Parra G (1998) PCR amplification of ribosomal DNA for species identification in the plant pathogen genus *Phytophthora*. Appl Environ Microbiol 64: 948–954.

40. Nei M (1972) Genetic distance between populations. Am Nat 106: 283–292.

41. Tan G, Gao Y, Shi M, Zhang X, He S, et al. (2005) SiteFinding-PCR: A simple and efficient PCR method for chromosome walking. Nucleic Acids Res 33: 1–7.

42. Silvar C, Merino F, Díaz J (2006) Diversity of *Phytophthora capsici* in Northwest Spain: Analysis of virulence, metalaxyl response, and molecular characterization. Plant Dis 90: 1135–1142.

43. Bagirova S, Li AZ, Dolgova A, Elansky S, Shaw D, et al. (2001) Mutants of *Phytophthora infestans* resistant to dimethomorph fungicide. J Russ Phytopathol Soc 2: 19–24.

44. Linde C, Soo SH, Drenth A (2001) Sexual recombination in *Phytophthora cinnamomi in vitro* and aggressiveness of single-oospore progeny to eucalyptus. Plant Pathol 50: 97–102.

5

Honeybee Colony Disorder in Crop Areas: The Role of Pesticides and Viruses

Noa Simon-Delso[1,2]*, Gilles San Martin[3], Etienne Bruneau[1], Laure-Anne Minsart[1], Coralie Mouret[1], Louis Hautier[3]

1 Beekeeping Research and Information Centre, Louvain la Neuve, Belgium, 2 Environmental Sciences, Copernicus Institute, Utrecht University, Utrecht, The Netherlands, 3 Plant Protection and Ecotoxicology Unit, Life Sciences Department, Walloon Agricultural Research Centre, Gembloux, Belgium

Abstract

As in many other locations in the world, honeybee colony losses and disorders have increased in Belgium. Some of the symptoms observed rest unspecific and their causes remain unknown. The present study aims to determine the role of both pesticide exposure and virus load on the appraisal of unexplained honeybee colony disorders in field conditions. From July 2011 to May 2012, 330 colonies were monitored. Honeybees, wax, beebread and honey samples were collected. Morbidity and mortality information provided by beekeepers, colony clinical visits and availability of analytical matrix were used to form 2 groups: healthy colonies and colonies with disorders (n = 29, n = 25, respectively). Disorders included: (1) dead colonies or colonies in which part of the colony appeared dead, or had disappeared; (2) weak colonies; (3) queen loss; (4) problems linked to brood and not related to any known disease. Five common viruses and 99 pesticides (41 fungicides, 39 insecticides and synergist, 14 herbicides, 5 acaricides and metabolites) were quantified in the samples. The main symptoms observed in the group with disorders are linked to brood and queens. The viruses most frequently found are Black Queen Cell Virus, Sac Brood Virus, Deformed Wing Virus. No significant difference in virus load was observed between the two groups. Three acaricides, 5 insecticides and 13 fungicides were detected in the analysed samples. A significant correlation was found between the presence of fungicide residues and honeybee colony disorders. A significant positive link could also be established between the observation of disorder and the abundance of crop surface around the beehive. According to our results, the role of fungicides as a potential stressor for honeybee colonies should be further studied, either by their direct and/or indirect impacts on bees and bee colonies.

Editor: Fabio S. Nascimento, Universidade de São Paulo, Faculdade de Filosofia Ciências e Letras de Ribeirão Preto, Brazil

Funding: The Walloon Region was the only funder of this study. The funders had no role in study design, data collection and analysis, or preparation of the manuscript.

Competing Interests: The authors have declared that no competing interests exist.

* Email: simon@cari.be

Introduction

The evolution of pollinator populations has been the subject of an increasing number of studies, most of them showing worrying negative trends [1–4]. Furthermore, beekeepers have long been notifying enhanced bee winter losses and disorders [5–8]. Bees contribute to ecosystem services and their decline thus threatens pollination of both wild and cultured plants, endangering biodiversity, food and fibre production [9,10]. This decline may also have an impact on the production of other goods with pharmacological uses (e.g. honey, propolis) [11,12], and scientific and technological inspiration (e.g. development of visual guided flight robotics) [13]. Bees are also part of our culture (e.g. culinary, hobby occupation, etc.), contributing to the dynamism of rural and urban areas [14] and providing a source of inspiration and well-being for many [15].

Belgium shares the trends observed worldwide both in terms of wild and reared pollinators [8,16], with enhanced winter mortality observed from 1999 [8,17,18]. However, apart from colony mortality, Walloon beekeepers have reported a number of imprecise symptoms: colony weakness, mainly in spring; fast renewal of young queens; rapid loss of individuals in the colony, mainly foragers, or slow loss of individuals in the colony. In many cases brood and food remains in the colony. Sometime a small cluster of bees with the queen survives [17]. Some beekeepers described unspecific brood abnormalities not characteristic to any known disease (Baudoin and Lequeux, pers. com.). Other studies describe similar mortality trends [8,19,20] as well as unspecific symptoms: increased colony mortality and/or weakening [21–23], queen failure [21,23–25] or honey yield reduction [21].

Multifarious factors have been proposed to provide a cause to such a phenomenon. Climate change is proposed as one of them, together with a decrease of genetic variability of the bee colonies, electromagnetic radiation, pathogens and parasites or the impact of intensive agricultural systems (nutritional lack, GM plants or pesticides) [26]. Previous studies developed in the region discard some of these factors as causes of bee losses, specifically *Nosema* spp and American Foulbrood [27]. However, pesticide residues and certain viruses were detected in bee colonies [28,27]. These elements were the most relevant to us.

Countless studies have shown lethal and sub-lethal effects of pesticides on bees [29–31]. Insecticides are often the most studied for obvious reasons. However other substances (fungicides, herbicides) deserve analysis for their specific toxicity, individually or in synergy with other substances [32,33] or pathogens [34–36], or their extensive exposure given their large scale and/or repeated use. Indeed, residue analyses show that these types of substance are found in the hives, even though the crops in which they were applied would suggest no bee exposure [37–40].

Beside pesticides, bee viruses are often mentioned as a putative cause of decline of the colonies, or at least to reside among the presumed multi-factorial causes [41]. In Europe, at least 12 viruses infecting bees have been compiled [21]. Many honeybee viruses commonly occur in seemingly healthy populations that continue to run well. Viruses may remain latent and confined within certain cells or tissues with no active replication and no disruption of cellular function. Likewise, they may be replicating at low level in permissive cells but in non-vital sites or in honeybee life stages that do not exhibit any symptoms or obvious pathology [21,42,43]. Nevertheless, two viruses, ABPV and DWV, are able of inducing serious disorders to honeybees and have been shown to cause -in association with *V. destructor* - winter losses in Germany [23].

In this paper, we first studied the relationship of in-hive viral prevalence as well as the presence of pesticide residues on honeybee colonies' health. As a next step, we focused on the relationship between the environment surrounding the monitored apiaries and the health condition of their colonies.

Materials and Methods

Field work – colony follow up

A group of voluntary beekeepers were requested to participate in the study. All of them shared the following criteria: (1) have a minimum of 5 production colonies per apiary in June 2011; (2) regularly follow up the health status and development of their colonies; and (3) monitor the varroa infestation level and carry out officially recommended varroa treatments (treatment in July-August with veterinary medicaments based on thymol and winter treatment with veterinary medicaments based on oxalic acid). A total of 330 colonies distributed among 66 apiaries (5 colonies/apiary) in Walloon and Brussels regions (Belgium) were followed. In those apiaries composed of 5 or more colonies, the 5 well-developed colonies at the beginning of the study were selected.

Colonies received three visits along the study, the first one from mid-July to mid-August 2011, the second one from mid-September to mid-October 2011 and the third one from March to April 2012. State official beekeeping technicians specialised in bee health were trained in the framework of the project in order to minimise as far as possible the variability of the results due to handling and observations. For each of the visits, beekeepers were requested to fulfil a form asking information about the health history of the apiary and of the followed-up colonies, as well as about their beekeeping practices. Honeybee colony disorder symptoms were reported in the questionnaire. These include the following symptoms for which no causal agent could obviously be identified:

(1) dead and disappeared colonies : (1a) death of part or the whole colony, where dead bees can be found close to the hive or inside it. Beekeepers also describe the (1b) disappearance of part or the whole colony, leaving behind food reserves and brood, a phenomenon similar to the one described in North-American apiaries (Colony Collapse Disorder (CCD) [24];

(2) weak colonies : weakening of the colony, showing in occasions a slow development in spring under optimal conditions (e.g. optimal weather, low varroa pressure, etc) with as consequence the loss of the spring production, but in occasions showing abnormally small colonies or colonies with low activity;

(3) queen loss : replacement of young queens sometimes leading to queen-less colonies or interruption of the egg-laying activity of the queen [17];

(4) problems linked to brood and not related to any known disease.

The form also included requests about other symptoms typically linked to known diseases (e.g. diarrhoea, mummified larvae, varroa presence, presence of deformed wing bees, etc). Each of the questions requested further characterisation of the symptoms (e.g. population size, population dynamics, bees behaviour, etc.).

In addition, a thorough clinical inspection was carried out for each colony in the mentioned period. Special attention was given to the strength of the colony in terms of bee numbers, brood surface and reserves content, the presence of the queen and the varroa infestation level. Finally, two different samples were collected before and after the winter: (1) in-hive bees (minimum of 100 bees); (2) a section of the frame containing beebread and honey (about 100 cm^2). Samples were collected in hermetic plastic bags, cooled after collection and stored at −20°C.

Case choice – hierarchical sample clustering

Information available from bee colonies comprised field observations, beekeeper answers to the questionnaire and results from the analyses of the samples of beekeeping matrices (honey, beebread, wax and bees). Colonies for which this information was available were considered for our analyses. Colonies with well identified problems (heavy varroa infestation, lack of food or drone-laying queens) were discarded. After that, two groups were made: a group with disorders and another with healthy colonies. In order to constitute these groups and to limit variations due to potential different bee management, a hierarchical classification was made. The criteria used were the amount of food stores before the winter, the year of colony creation, the subspecies and the age of the queen. This classification was made by using Ward aggregation method. This method allows to build group with the lower variation within the group and the higher variation between groups [44].

Virus analyses

The viruses under investigation were the Black Queen Cell Virus (BQCV), the Chronic Bee Paralysis Virus (CBPV), the Acute Bee Paralysis Virus (ABPV), the Deformed Wings Virus (DWV), and the Sac Brood Virus (SBV). Viral analyses were conducted on the honeybee workers collected before the winter, during the first and second periods (July–August and September–October). The samples were analysed with a quantitative RT-PCR by the National Bee Unit laboratory, Food and Environment Research Agency (Sand Hutton, York, United-Kingdom).

Total Nucleic acid (TNA) was extracted from 60 bees homogenised for 12 minutes with 20 ml GITC Lysis Buffer (5 M Guandine Thiocynate, 0.05 M Tris base, 0.02 M EDTA, pH 8.0) in a 30 ml bottle containing 3, 7/16″ ball bearings. GITC Lysis buffer also contained 17.3 mM SDS buffer (173 mM Sodium dodecyl sulphate (SDS) in 100 ml MGW). The SDS buffer is added prior to use in warmed GITC Lysis Buffer. The homogenate was then incubated at 65°C for 40 minutes and then spun at 6189 g for 5 minutes. Polypropylene 96 -deep well plates (DT850301 Elkay Laboratory Products Ltd) were prepared as follows (1 well/extract); plate A: 800 µl extract, and 100 µl MagneSil™ beads (MD1441, Promega); plate B: 1 ml of GITC wash buffer (5 M Guandine Thiocynate, 0.05 M Tris base); plates C, D, E: 1 ml 70%v/v ethanol (E/00665DF/17, Fisher Scientific); plate F: 1 ml 5 M Betaine (B2629, Sigma), plate G, 300 ul 1×TE

Buffer (stock 100X TE (EC-862 National Diagnostics). Plates were loaded onto the Kingfisher Flex and processed as follows: Plate A – Bind 5 mins (fast dual mix), Plate B – Wash 3 mins (fast dual mix), Plate C, D, E – Wash 2 mins (fast dual mix), collect beads at 1 min intervals, Plate F – Wash 20 secs (medium), without releasing beads, Plate G – Mix 1 min then incubate at 65°C for 5 mins with mixing. All steps of the process are looped through twice. TNA was collected from plate G of each reaction and stored at −80°C prior to use in real-time PCR.

Reactions were set up in 96 well reaction plates using Absolute Blue QPCR ROX mix (AB-4139, Thermo Scientific) following the protocols supplied. 0.1 mM of 0.1 M Dithiothreitol (165680250, Arcos Organics) and 0.33 unit of MMLV (EPO441: Fermentas) were added to each reaction. The primers (Table 1) were all used at 400 nM and probes at 200 nM final concentration. Total nucleic acid (5 µl) was added to each reaction, giving a final reaction volume of 20 µl. Plates were cycled for 48°C/30 min, 95°C/10 min and 40 cycles of 60°C/1 min, 95°C/15 sec within the 7900 HT Sequence Detection System (Applied Biosystems), using real time data collection. The results were recorded as the cycle threshold (Ct) or cycle number after which a significant accumulation of florescence over the baseline was observed; an average (of duplicate wells) Ct value below 40 was regarded as a positive result with a threshold ΔRn setting of 0.2. Given the absence of internal standard, we assumed that the extraction method had the same efficiency towards bee DNA/RNA and viral RNA.

Virus primers and probes used for all pathogens tested have been previously described in Chantawannakul et al., 2006 [45]. Additional APBV primers are described in Martin et al., 2012. [46].

Pesticide analyses

Samples collected before the winter were sent to Eurofins Chemiphar NV, Brugge, Belgium and analysed by SOFIA GmbH Chemisches Labor für Softwareentwicklung und Intelligente Analytik, Berlin, Germany. Two multi-residues methods (SF146 and SF150) were used searching for 99, 93 and 96 pesticides residues in wax (54 samples), beebread (108 samples) and honey (107 samples – one sample did not contain enough matrix) respectively (LOQ in Table 2).

For SF146 method, 10 ml of water was added to 10 g of samples. Methanol was added and the preparation was mixed. The mixture was filtered and centrifuged. Next, for analysis by gas chromatography coupled with mass spectrometry (GC-MS), the filtrate was mixed with sodium chloride and ethyl acetate solution 1:1 (v/v). All was dried with sodium sulfate and filtrated. This filtrate was then concentrated and analysed by GC-MS. For analysis with liquid chromatography coupled with tandem mass spectrometry (LC-MS/MS), 5 ml of sample were transferred in ChemElut column and eluted with dichloromethane. After concentration, 1 ml of water-methanol 1:1 (v/v) was added and this solution was analysed by LC-MS/MS. For the wax, some modifications were made. For GC-MS, sodium chloride was replaced by hexane to dissolve wax and the mixture was frozen until a precipitate appears. For LC-MS/MS, 0.2 g of sample were extracted with 30 ml of mixture of naphtha:acetonitrile 1:2 (v/v). 10 ml of acetonitrile phase was isolated and concentrated before the analysis by LC-MS/MS.

For SF150 method, 10 ml of water was added to 10 g of samples as well as methanol and hydrochloric acid. All was mixed, filtrated and centrifuged. A solution of sodium chloride (20%) was added and 5 ml of sample were transferred in ChemElut column and eluted with dichloromethane. After concentration, 1 ml of water-methanol 1:1 (v/v) was added and this solution was analysed by LC-MS/MS.

Descriptive analyses

A descriptive analysis of virus prevalence and pesticide residues was first carried out. The relationship between disorder and these stressing factors was then assessed. For each virus type, a linear model (two way ANOVA) was used, with the cycle threshold value as dependent variable and the visit (first or second), group (with disorder or healthy) and their interaction as explanatory variables. For the pesticides residues, we used a similar model with the total number of residues as dependent variable and type (fungicide or insecticide/acaricide), group (with disorder or healthy) and their interaction as explanatory variable. Based on these models, we constructed a contrast matrix to test explicit post-hoc hypotheses (e.g. compare virus load between groups within each visit...). As several dependent variables had a strongly asymmetric distribution (i.e. non-normal), we computed all p-values by permutation (n = 1000). We applied a Bonferoni correction on the post-hoc tests p-values to take into account the multiplicity of the tests.

Relationship between bee colony disorders and potential stressors or surrounding environment of the apiary

We used two separate generalized linear mixed models with a binomial distribution and logit link function. In both models, the "group" (with disorder or healthy) was used as dependent binary variable and the apiary was used as random effect (grouping

Table 1. List of primers used for virus analyses.

Target	Primer Name	Sequence (5′-3′)
BQCV	BQCV 9195F	GGT GCG GGA GAT GAT ATG GA
	BQCV 8265R	GCC GTC TGA GAT GCA TGA ATA C
	BQCV 8217T	TTT CCA TCT TTA TCG GTA CGC CGC C
SBV	SBV 311F	AAG TTG GAG GCG CGy AAT TG
	SBV 380R	CAA ATG TCT TCT TAC dAG AGG yAA GGA TTG
	SBV 331T (MGB)	CGG AGT GGA AAG AT
CPV	CPV 304F	TCT GGC TCT GTC TTC GCA AA
	CPV 371R	GAT ACC GTC GTC ACC CTC ATG
	CPV 325T	TGC CCA CCA ATA GTT GGC AGT CTG C

Table 2. Active ingredients or metabolite included in the multi-residue analyses by beekeeping matrix (A = Acaricide; F = Fungicide; H = Herbicide; I = Insecticide; S = Synergist).

Active ingredient or metabolite*	Class	LOQ (mg/kg)		
		Wax	Beebread	Honey
2,4-D	H	0.1	–	0.01
Abamectin	I	0.1	0.1	0.005
Acetamiprid	I	0.1	0.1	0.005
Aldicarb	I	0.1	0.1	0.005
Alpha-cypermethrin	I	0.1	0.1	0.005
Amitraz	A	0.1	0.1	0.01
Azoxystrobin	F	0.1	0.1	0.005
Bentazone	H	0.2	–	0.02
Benthiavalicarb	F	0.1	0.1	0.005
Beta-cyfluthrin	I	0.1	0.1	0.01
Bifenthrin	I	0.1	0.1	0.005
Boscalid	F	0.1	0.1	0.005
Captan	F	0.1	0.1	0.01
Carbaryl	I	0.1	0.1	0.005
Chlorpyriphos	I	0.1	0.1	0.005
Chlorpyriphos-methyl	I	0.1	0.1	0.005
Chlorothalonil	F	0.1	0.1	0.005
Clothianidin	I	0.1	0.1	0.01
Coumaphos	A	0.1	0.1	0.05
Coumaphos oxon *	A	0.1	0.1	–
Coumaphos phenolic*	A	0.1	0.1	–
Cyazofamid	F	0.2	0.2	0.02
Cyfluthrin	I	0.1	0.1	0.01
Cymoxanil	F	0.1	0.1	0.01
Cypermethrin	I	0.1	0.1	0.005
Cyproconazole	F	0.1	0.1	0.005
Cyprodinil	F	0.1	0.1	0.01
DDT	I	0.1	0.1	0.005
Deltamethrine	I	0.1	0.1	0.005
Dichlorprop-P	H	0.1	–	0.01
Difenoconazole	F	0.1	0.1	0.005
Diflubenzuron	I	0.2	0.2	0.02
Dimethenamid-P	H	0.1	0.1	0.005
Dimethoate	I	0.1	0.1	0.005
Dimethomorph	F	0.1	0.1	0.01
Dimoxystrobin	F	0.1	0.1	0.005
Epoxiconazole	F	0.1	0.1	0.005
Esfenvalerate	I	0.1	0.1	0.005
Ethofumesate	H	0.5	0.5	0.05
Famoxadone	F	0.1	0.1	0.01
Fenhexamid	F	0.1	0.1	0.005
Fenoxycarb	F	0.1	0.1	0.005
Fenpropidin	F	0.1	0.1	0.01
Fenpropimorph	F	0.1	0.1	0.005
Fipronil	I	0.1	0.1	0.01
Flonicamid	I	2	2	0.2
Fluazinam	F	0.1	0.1	0.01
Flufenacet	H	0.1	0.1	0.01

Table 2. Cont.

Active ingredient or metabolite*	Class	LOQ (mg/kg)		
		Wax	Beebread	Honey
Fluopicolide	F	0.1	0.1	0.01
Flusilazole	F	0.1	0.1	0.005
Heptenophos	I	0.1	0.1	0.01
Imidacloprid	I	0.1	0.1	0.005
Indoxacarb	I	0.1	0.1	0.005
Iprodione	F	0.1	0.1	0.005
Kresoxim-methyl	I	0.1	0.1	0.005
Lambda-cyhalothrin	I	0.1	0.1	0.005
Lindane	I	0.1	0.1	0.005
Linuron	H	0.1	0.1	0.005
MCPA	H	0.1	–	0.01
Mecoprop-P	H	0.1	–	0.01
Metalaxyl-M	F	0.1	0.1	0.005
Metamitron	H	0.1	0.1	0.005
Metconazole	F	0.2	0.2	0.02
Methiocarb	I	0.1	0.1	0.005
Methoxyfenozide	F	0.1	0.1	0.005
Metribuzin	H	0.1	0.1	0.005
Penconazole	F	0.1	0.1	0.005
Pendimethalin	H	0.1	0.1	0.005
Permethrin	I	0.2	0.2	0.02
Picoxystrobin	F	0.1	0.1	0.005
Piperonyl butoxide	S	0.1	0.1	0.01
Pirimicarb	I	0.1	0.1	0.005
Prochloraz	F	0.1	0.1	0.005
Propamocarb	F	0.1	0.1	0.005
Propiconazole	F	0.1	0.1	0.01
Propyzamide	F	0.1	0.1	0.01
Pymetrozine	I	0.1	0.1	0.005
Pyraclostrobin	F	0.1	0.1	0.005
Pyrethrin	I	0.2	0.2	0.02
Pyrimethanil	F	0.1	0.1	0.005
Rotenone	I	0.1	0.1	0.01
Spinosad	I	0.1	0.1	–
Spirodiclofen	I	0.1	0.1	0.01
Spirotetramat	F	0.1	0.1	0.01
Spiroxamine	F	0.1	0.1	0.005
Sulcotrione	H	0.1	–	0.01
Tau-fluvalinate	IA	0.1	0.1	0.01
Tebuconazole	F	0.1	0.1	0.005
Tebufenozide	I	0.1	0.1	0.005
Tefluthrine	I	0.1	0.1	0.005
Terbuthylazine	H	0.1	0.1	0.01
Tetraconazole	F	0.1	0.1	0.01
Thiabendazole	F	0.1	0.1	0.005
Thiacloprid	I	0.1	0.1	0.005
Thiamethoxam	I	0.1	0.1	0.005
Thiophanate-methyl	F	0.1	0.1	0.005
Trifloxystrobin	F	0.1	0.1	0.01

Table 2. Cont.

Active ingredient or metabolite*	Class	LOQ (mg/kg)		
		Wax	Beebread	Honey
Zeta-cypermethrin	I	0.1	0.1	0.005
Zoxamide	F	0.1	0.1	0.01

factor) to take into account the non-independence between colonies from the same apiary and allow the intercept to vary between apiaries [47].

With the first model, we explored the relationship between the probability of disorder in a colony and potential stress factors, i.e. pesticides or viruses total load into the colonies. We used as explanatory variables the total number of fungicides residues, the total number of insecticides or acaricides (pooled together) residues, the total number of viruses detected for both visits and all first level interactions between these three explanatory variables. The insecticides and acaricides were pooled together because the most frequently found active ingredient, namely tau-fluvalinate, is authorized in Belgium for both purposes.

With the second model, we investigated the relationship between the probability of disorder in a colony and the structure of the agricultural landscape around the beehive. The surfaces of all different kinds of agricultural soil occupancy in a circle with a radius of 1500 m around each apiary were calculated according to the information provided by the farmers to the Walloon administration (Land Parcel Identification System). The different soil occupancy categories were pooled into two categories defined according to the potential frequency of plant protection product use: (1) grasslands (low pesticides use), (2) crops *sensu lato* (potentially higher pesticides use), including major crops (mainly: cereals, potatoes, beet, oilseed rape, maize, flax, …) but also surfaces dedicated to fruit or vegetables production, fodder production (mainly legumes) and to horticulture. The surfaces of these two land use options, grasslands and crops (*sensu lato*), were used as explanatory variables in the model. The 1500 m radius was chosen accordingly to the mean pollen and nectar collecting distance for the honeybee [48,49]. As the agricultural soil occupancy data was available only for Wallonia, we eliminated from the analysis the apiaries whose buffer had less than 70% of their surface within the administrative boundaries of Wallonia (n = 2).

We used Likelihood Ratio (LR) Tests to evaluate the significance of the explanatory variables [47,50]. We respected the marginality rules, i.e. all main effects were tested after removing from the model the interactions in which they are involved [51].

All analyses were performed with R (R Core Team 2013) and the mixed models were fitted with the package lme4 [52].

Results

Description of honeybee colony disorders

We gathered all sources of information, i.e. questionnaires, analytical matrices and information from clinical visits, for 173 colonies. After the clustering, 54 colonies coming from 21 apiaries (Figure 1) were considered for the study: 25 presenting bee disorders and 29 not presenting them, defined as healthy group. Their data and samples were collected by 8 beekeeping technicians.

In the group presenting disorders, 6 colonies were dead or unviable -with only a handful of bees remaining with the queen- at the spring visit (Table 3). Pre-wintering weakness, winter worker losses, late-season re-queening were reported in these cases. One colony was dead due to queen failure without re-queening in spring. Furthermore, a number of colonies were weak in spring. They were characterised by a low number of individuals and a slow development. In total, nine colonies lost their queens during the study, five of them leading to queenless colonies. These were considered as "disorder colonies" in the framework of the study. Finally, brood abnormalities not linked to known disease were observed in 10 colonies, most of them both before and after the winter.

Symptoms of other diseases or bee parasite observation remained low within the selected colonies. Wax moths were observed in two cases, one in each group. *Varroa destructor* was present in all colonies of the study. All of them were treated according to official veterinary advice. Diarrhoea was observed in one of the colonies showing brood abnormalities. Anatomical modifications (i.e. deformed or undeveloped wings, small abdomens) were observed in 6 colonies, half of them being healthy and half of them showing disorders. No symptoms of any of the foulbroods were observed.

Virus content

One bees sample from the "disorder" group could not be analysed due to a lack of sufficient matrix. The three most abundantly found viruses were the Deformed Wings Virus (DWV), the Black Queen Cell Virus (BQCV) and the Sac Brood Virus (SBV) (Figure 2). The Acute Bee Paralysis Virus (ABPV) and the Chronic Bee Paralysis Virus (CBPV) were found only in a very

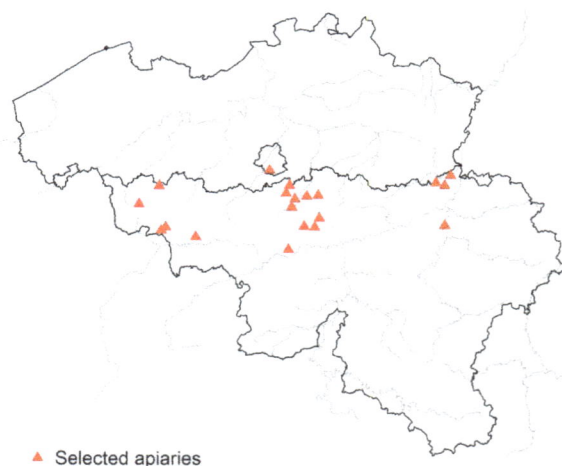

▲ Selected apiaries

Figure 1. Spatial distribution of selected apiaries.

Table 3. Symptoms observed in the group with disorders.

Symptoms of disorder	Frequency
Mortality	2*
Weakening	3
Queen failure	5
Brood problems	9
Mortality+Weakening	2
Mortality+Weakening+Queen failure	2
Mortality+Queen failure	1
Queen failure+Brood problems	1
Total	25

* One of these colonies not considered in the model due to a lack of viral results.

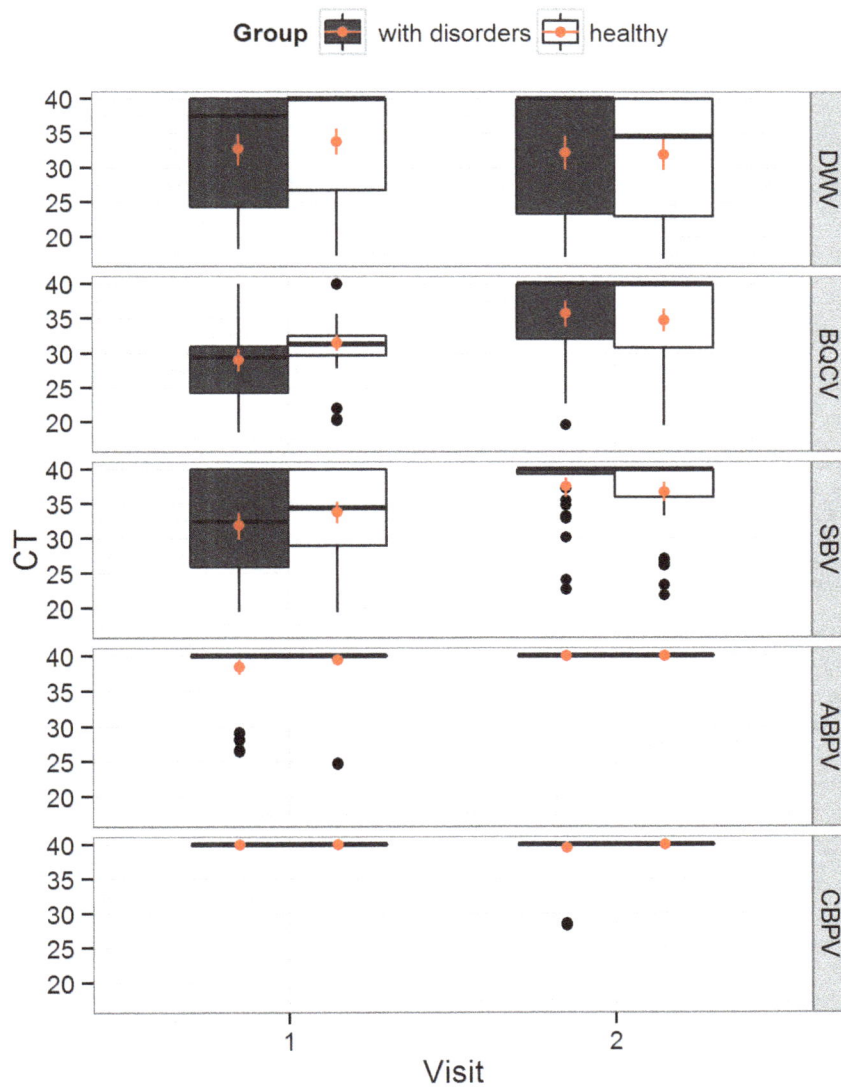

Figure 2. Virus content according to the Cycle Threshold (CT) for the groups "with disorder" and "healthy". Boxplot of Cycle Threshold for the first and second visits (visit 1 - mid-July to mid-august and visit 2 - mid-September to mid-October) and the group with disorders (grey, n = 24 colonies) and the healthy one (white, n = 29 colonies). Deformed Wings Virus (DWV), Black Queen Cell Virus (BQCV), SacBrood Virus (SBV), Acute Bee Paralysis Virus (ABPV), Chronic Bee Paralysis Virus (CBPV). In red, mean with confidence interval estimated by boostrap method. Note: CT values below 40 were regarded as positive results and the lowest CT values correspond to the higher virus contents.

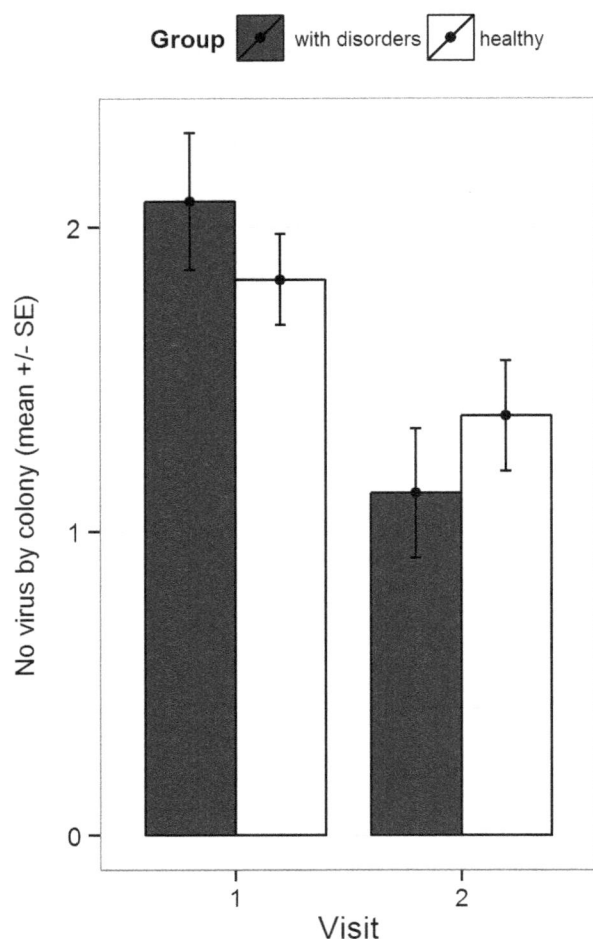

Figure 3. Average number of different viruses per colony. Data shown for the first and second visits (visit 1 - mid-July to mid-august and visit 2 - mid-September to mid-October) and the group with disorders (grey, n = 24 colonies) and the healthy one (white, n = 29 colonies). Whiskers show the standard error (SE).

limited number of samples and did not allowed particular statistical analyses on these two viruses.

There was no significant difference in viral content between the group with disorders or the healthy one, and independently of the visit for any of the three most abundant viruses (2-way ANOVA tested by permutations, DWV: p = 0.731, BQCV: p = 0.373, SBV: p = 0.54). We observed a decrease of virus content from summer to fall 2011, independently of the group, for BQCV (p<0.001) and SBV (p<0.001) but not for DWV (p = 0.271). For BQCV only, we observed a significant (p = 0.036) group×visit interaction indicating that the decrease of the virus abundance from visit 1 to visit 2 was higher in the group "with disorder" than in the group "healthy". Overall, the number of virus changed between visit 1 and visit 2 and a significant decrease was observed in the group with disorders (p<0.001) (Figure 3).

Pesticide analysis

172 agrochemical residues of 23 different active ingredients were detected in 94 out of 269 samples. Wax was the most contaminated matrix: 109 residues of 15 different active ingredients; while 39 and 24 residues of 10 and 8 substances were detected in beebread and honey, respectively (Figures 4 & 5).

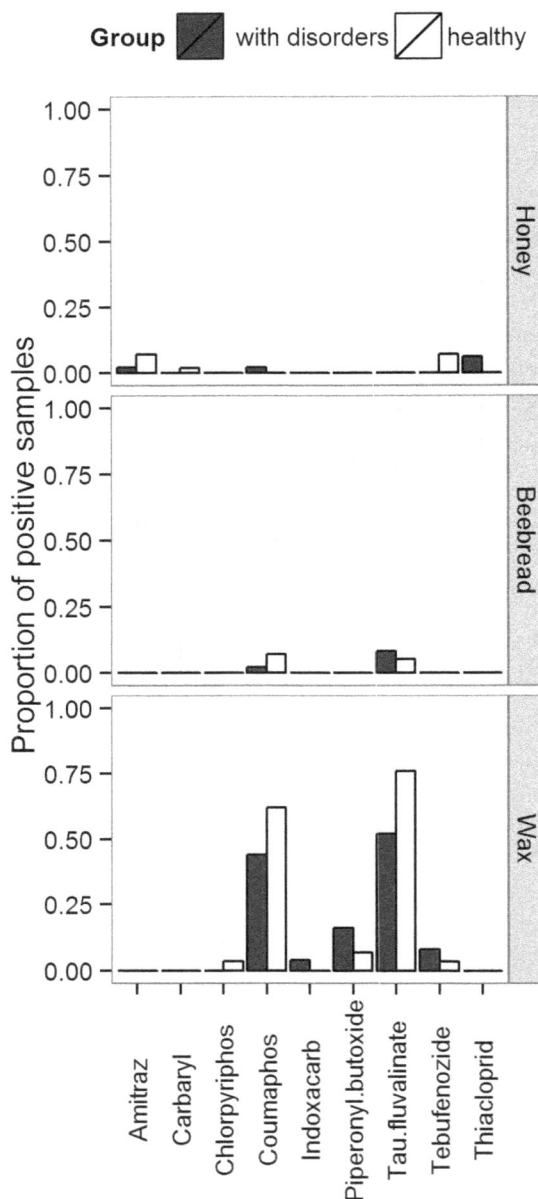

Figure 4. Proportion of samples containing residues of acaricides/insecticides in the different beekeeping matrices (honey, beebread and wax). Data shown for the group with disorders (grey, n = 25 colonies) and the healthy one (white, n = 29 colonies).

Residue levels contained in wax and beebread were higher (0.21–3.1 mg/kg) than those in honey (0.001–0.058 mg/kg).

For the subsequent statistical analysis, the results obtained for insecticide and acaricides residues were pooled together because the most frequently found active ingredient, tau-fluvalinate (n = 46), can be used in Belgium as an acaricide against varroa mite and as an insecticide to control *Meligethes aeneus* in rape. The second most frequently found residue was the coumaphos (n = 35), followed by two fungicides, boscalid (n = 33), iprodione (n = 13) (Table 4). Some residues of neonicotinoid insecticides, synergist and herbicide were also detected: thiacloprid (n = 3), piperonyl butoxide (n = 6), terbuthylazine (n = 1). The highest

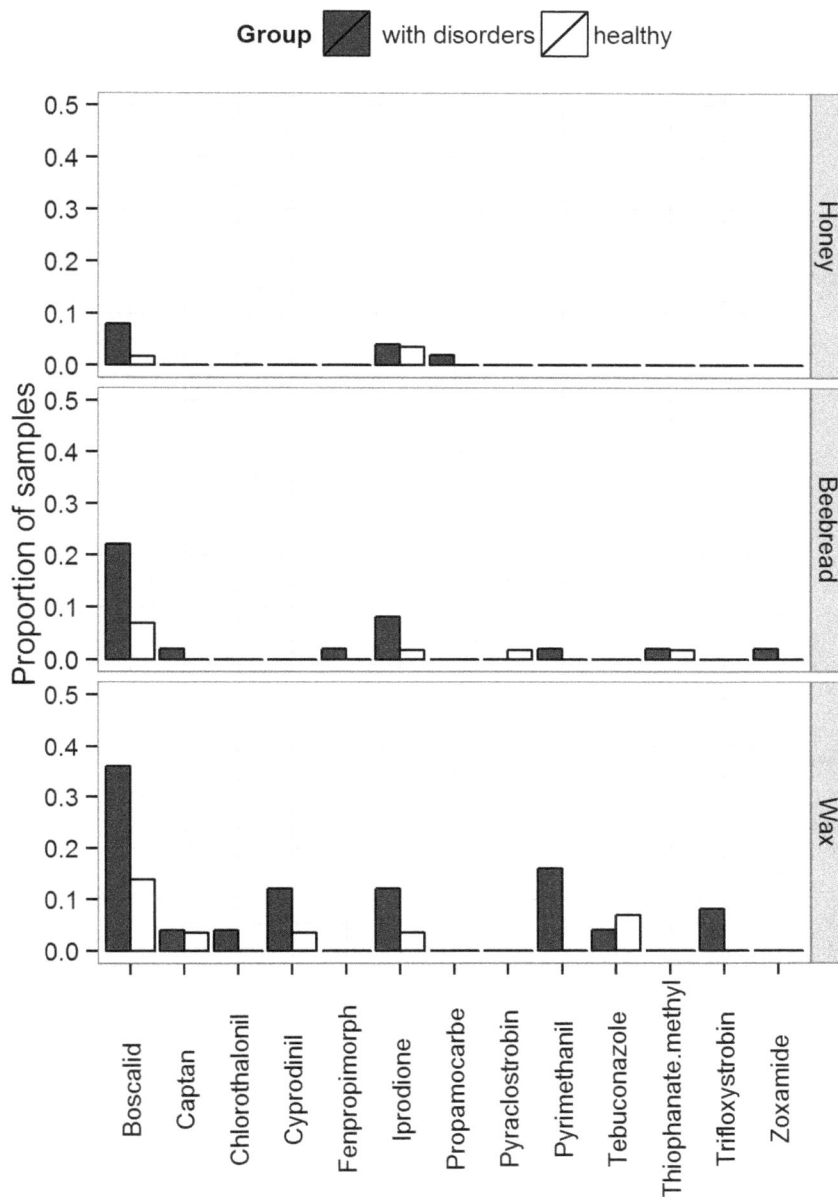

Figure 5. Proportion of samples containing residues of fungicides in the different beekeeping matrices (honey, beebread and wax). Data shown for the group with disorders (grey, n = 25 colonies) and the healthy one (white, n = 29 colonies).

residue level concerned captan with 3.1 and 1.9 mg/kg in the wax and bee bread of the same colony, respectively. Despite of being the most frequently found in matrices, tau-fluvalinate and coumaphos residues never exceeded 0.71 and 0.58 mg/kg, respectively. Boscalid, the most commonly found fungicide, ranged from 0.005 to 1.3 mg/kg.

There was a significant difference (p = 0.01) in terms of number of fungicide substances found between the "disorder" group (mean = 2.0) and the "healthy" one (mean = 0.7) as presented in Figure 6. The total number of insecticides/acaricides residues was slightly higher in the disorder group (mean = 2.1) than in the healthy group (mean = 1.64) but this difference is not statistically significant (p = 0.79). The most frequent fungicides in the group with disorders were boscalid and iprodione, detected in the three investigated matrices (Figure 5). However, for single active

substances, no significant difference of number of residues was observed between the groups "with disorder" and "healthy".

Probability of disorders in relation to potential stressors

There is clearly a significant positive relationship between the probability of a colony showing disorders and the total number of fungicides (Figure 7, LR = 7.128, df = 1, p = 0.008). The estimated probability for a colony to be in the "with disorder" group is 0.26 without fungicides residues, 0.60 with 2 fungicides residues and 0.88 with 4 fungicides residues when the insecticides-acaricides residues number and virus load are fixed to their observed mean value (Figure 7). None of the other variables (total number of viruses, total number of insecticides-acaricides) and none of the first level interactions seem to have any explanatory power on the probability of disorders in a colony (Table 5). No direct link could

Table 4. Residues of active ingredients found in wax, beebread and honey samples from colonies with disorders (Group D, n = 25 colonies) and healthy ones (Group H, n = 29 colonies).

Active ingredient	Group	Wax							Bee bread							Honey						
		LOQ (mg/kg)	<LOD (n=54)	Detected	Quantified	Range	Mean	s.d.	LOQ (mg/kg)	<LOD (n=108)	Detected	Quantified	Range	Mean	s.d.	LOQ (mg/kg)	<LOD (n=107)	Detected	Quantified	Range	Mean	s.d.
Amitraz	D	0.1	–	–	–	–	–	–	0.1	–	–	–	–	–	–	0.01	49	–	1	0.022	0.02	0.02
	H		–	–	–	–	–	–		–	–	–	–	–	–		53	–	4	0.01–0.028	0.02	0.01
Boscalid	D	0.1	16	8	1	0.29	0.29	–	0.1	39	7	4	0.4–1.3	0.68	0.42	0.005	46	–	4	0.005–0.026	0.01	0.01
	H		–	–	–	–	–	–		–	–	–	–	–	–		–	–	–	–	–	–
Captan	D	0.1	25	4	1	3.1	3.1	–	0.1	54	4	1	1.9	1.90	–	0.01	56	–	1	0.058	0.06	–
	H		–	–	–	–	–	–		–	–	–	–	–	–		–	–	–	–	–	–
Carbaryl	D	0.1	–	–	–	–	–	–	0.1	48	–	–	–	–	–	0.005	50	–	–	–	–	–
	H		–	–	–	–	–	–		–	–	–	–	–	–		–	–	–	–	–	–
Chlorpyriphos	D	0.1	25	–	–	–	–	–	0.1	58	–	–	–	–	–	0.005	56	–	1	0.02	0.02	–
	H		28	1	–	–	–	–		–	–	–	–	–	–		–	–	–	–	–	–
Chlorothalonil	D	0.1	24	1	–	–	–	–	0.1	–	–	–	–	–	–	0.005	–	–	–	–	–	–
	H		29	–	–	–	–	–		–	–	–	–	–	–		–	–	–	–	–	–
Coumaphos	D	0.1	14	9	2	0.32–0.34	0.33	0.01	0.1	48	1	1	–	–	–	0.05	49	–	1	0.012	0.01	–
	H		11	10	8	0.23–0.58	0.37	0.01		54	4	–	–	–	–		57	–	–	–	–	–
Cyprodinil	D	0.1	22	3	–	–	–	–	0.1	–	–	–	–	–	–	0.01	–	–	–	–	–	–
	H		28	1	–	–	–	–		57	1	–	–	–	–		–	–	–	–	–	–
Fenpropimorph	D	0.1	–	–	–	–	–	–	0.1	49	1	–	–	–	–	0.005	–	–	–	–	–	–
	H		–	–	–	–	–	–		58	–	–	–	–	–		–	–	–	–	–	–
Indoxacarb	D	0.1	24	1	–	–	–	–	0.1	–	–	–	–	–	–	0.005	–	–	–	–	–	–
	H		29	–	–	–	–	–		–	–	–	–	–	–		–	–	–	–	–	–
Iprodione	D	0.1	22	1	2	0.24–1.5	0.87	0.89	0.1	45	–	4	0.34–1.5	0.90	0.48	0.005	48	–	2	0.017–0.022	0.02	0.00
	H		28	1	–	–	–	–		57	1	–	–	–	–		55	–	2	0.022–0.04	0.03	0.01
Piperonyl butoxide	D	0.1	21	4	–	–	–	–	0.1	–	–	–	–	–	–	0.01	–	–	–	–	–	–
	H		27	2	–	–	–	–		–	–	–	–	–	–		–	–	–	–	–	–
Propamocarbe	D	0.1	–	–	–	–	–	–	0.1	–	–	–	–	–	–	0.005	49	–	1	0.008	0.01	0.00

Table 4. Cont.

Active ingredient	Group	Wax LOQ (mg/kg)	Wax <LOD (n=54)	Wax Detected	Wax Quantified	Wax Range (mg/kg)	Wax Mean	Wax s.d.	Bee bread LOQ (mg/kg)	BB <LOD (n=108)	BB Detected	BB Quantified	BB Range (mg/kg)	BB Mean	BB s.d.	Honey LOQ (mg/kg)	Honey <LOD (n=107)	Honey Detected	Honey Quantified	Honey Range (mg/kg)	Honey Mean	Honey s.d.
	H	–	–	–	–	–	–	–									57	–	–	–	–	–
Pyraclostrobin	D	0.1	–	–	–	–	–	–	0.1	50	–	–				0.005	–	–	–	–	–	–
	H		–	–	–	–	–	–		57	1	–					–	–	–	–	–	–
Pyrimethanil	D	0.1	21	4	–				0.1	49	1	–				0.005	–	–	–	–	–	–
	H		29	–	–					58	–	–					–	–	–	–	–	–
Tau-fluvalinate	D	0.1	12	9	4	0.29–0.46	0.4	0.08	0.1	45	4	–				0.01	50	–	–	–	–	–
	H		7	13	9	0.21–0.71	0.5	0.08		55	3	–					53	–	4	0.011–0.02	0.02	0.00
Tebuconazole	D	0.1	24	1	–				0.1	–	–	–				0.005	–	–	–	–	–	–
	H		27	2	–					–	–	–					–	–	–	–	–	–
Tebufenozide	D	0.1	23	2	–				0.1	–	–	–				0.005	–	–	–	–	–	–
	H		28	1	–					–	–	–					–	–	–	–	–	–
Terbuthylazine	D	0.1	24	1	–				0.1	–	–	–				0.01	–	–	–	–	–	–
	H		29	–	–					–	–	–					–	–	–	–	–	–
Thiacloprid	D	0.1	–	–	–				0.1	–	–	–				0.005	47	–	3	0.009–0.013	0.01	0.00
	H		–	–	–					–	–	–					–	–	–	–	–	–
Thiophanate-methyl	D	0.1	–	–	–				0.1	49	1	1	0.38	0.38	–	0.005	57	–	–	–	–	–
	H		–	–	–					57	–	–					–	–	–	–	–	–
Trifloxystrobin	D	0.1	23	2	–				0.1	–	–	–				0.01	–	–	–	–	–	–
	H		29	–	–					–	–	–					–	–	–	–	–	–
Zoxamide	D	0.1	–	–	–				0.1	49	1	–				0.01	–	–	–	–	–	–
	H		–	–	–					58	–	–					–	–	–	–	–	–

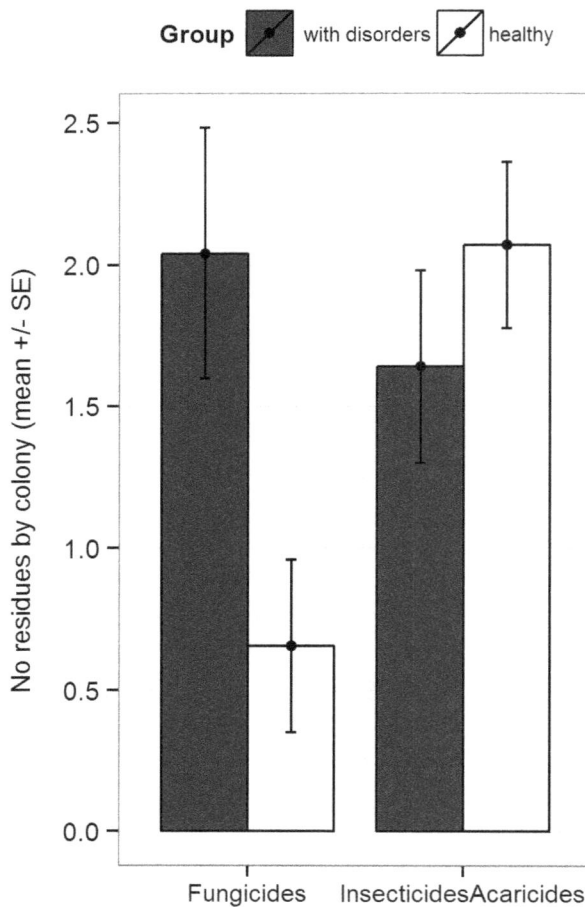

Figure 6. Average number of residues per colony. Data shown for the group with disorders (grey, n = 25 colonies) and the healthy ones (white, n = 29 colonies). Whiskers show the standard error (SE).

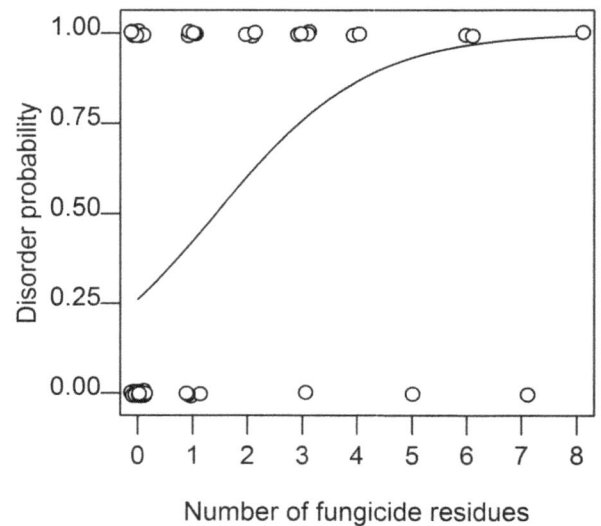

Figure 7. Probability of honeybee colony disorders depending on the number of fungicide residues detected. Model based on averaged coefficients and median value both for the number of insecticides-acaricides residues and total number of virus detected for both visits (n = 53 colonies).

Discussion

The bee disorders observed after the winter in the Walloon region happened despite of normal climatic conditions. The autumn of 2011 was dry (140.4 mm rainfall water from October to December, average being 219.9 mm), sunny and warm for Belgian conditions (12.4°C on average, which is normally 10.9°C for this period), followed by average winter and spring 2012 in terms of temperature, rainfall, with the exception of a cold wave lasting twelve days in February [53]. These conditions would not induce, *a priori*, a risk for honeybee colonies. Furthermore, based on studies carried out on the palynological diversity of the pollens collected by these colonies before the winter (Table S1), the hypothesis of nutritional lack is unlikely. All samples analysed contained at least 8 different botanic sources of pollen, including rich protein content as oilseed rape, ivy and *Phacelia*.

Viruses infections has often been mentioned as a source of stress for honeybee colonies [7]. However no significant difference in quantity or occurrence was observed between the healthy and the "with disorder" groups. DWV is one of the most commonly detected virus in *A. mellifera*. The prevalence of this virus is even more important at colonies infested by *Varroa destructor*, a well-established vector of this virus [7]. In accordance with the present study, further studies run on Belgian apiaries show DWV, BQCV and SBV as the most frequently found viruses [54]. Unlike DWV, we observed that the amount of SBV and BQCV has dropped significantly between the first and second visits for each group. In the case of BQCV, this is of no surprise as the cycle of incidence of this virus has been shown to increase in late winter, with a peak in May or June followed by a rapid decline [55]. The observed decrease of SBV has also been reported in other studies [42] in which the authors suggested that bees could develop a molecular defensive mechanism to reduce virus multiplication, or that the change in bee susceptibility to the virus could result from environmental conditions such as the quality of pollen.

We cannot prove any causal relationship between any of these viruses analysed and bee disorders, nor does the interaction

be established between bee colony disorders and the amount (in μg/kg) of residues found in the matrices.

Probability of disorders and crop/grassland surface

Our data clearly show a significant increase in disorder probability with the increase of crop surfaces (*sensu lato*, i.e. including fruit, vegetables, fodder production and horticulture) in the surrounding of the apiary (Figure 8, LR = 8.052, df = 1, p = 0.0045). The predicted probability of disorders is close to 0.1 for a crop surface of 0 ha in the radius of 1500 m and increases up to 0.8 for a surface of 500 ha of crops when fixing the grasslands surface to its observed mean.

On the contrary, the probability of disorders strongly decreases when the grassland surfaces increases (Figure 8, LR = 14.527, df = 1, p = 0.0001) after controlling for the crop surface. The predicted probability of disorders is close to 1 for a grassland surface of 0 ha in a radius of 1500 m and drops to ~0.1 for a surface of 150 ha of grasslands when fixing the crops surface to its observed mean.

Very similar results were found with a 3000 m buffer around the apiary (results not shown).

Table 5. Analysis of deviance table for generalized linear binomial models describing the relationships between the colony disorder probability and three variables: the total number of (1) fungicide residues (totfungicides), (2) insecticide-acaricide residues (totinsaca) and (3) virus detected for both visits (totvirus).

	LR	df	p(>Chisq)
totfungicides	7.13	1	0.008
totinsaca	0.005	1	0.943
totvirus	0.136	1	0.712
totfungicides:totvirus	2.222	1	0.136
totinsaca:totvirus	0.975	1	0.323
totfungicides:totinsaca	0.901	1	0.342

LR = likelihood ratio.

between these two factors. Cox-Foster et al., 2007 [41] found no clear correlation between a variety of pathogens, including *Nosema* spp., DWV, CBPV, ABPV, BQCV, *Mellisococcus pluton* and *Paenibacillus larvae* ssp and CCD. No specific spore counts were carried out in our study. However, no signs of nosemosis or foulbrood could be linked to colonies presenting bee disorders. It is noteworthy that a monitoring run at Belgian level found *Nosema* spp spores in one out of four colonies [56]. Nevertheless, Cox-Foster et al., 2007 [41], show a positive correlation between IAPV and CCD, which *a priori* would not be relevant in our conditions given the low prevalence of IAPV in this country [27,54]. A recent publication [36] shows a positive correlation between the presence of fungicides in pollen loads and the presence of spores of *Nosema ceranae*. We do not exclude the potential involvement of *Nosema* spp. in the case of bee disorders. However, in the framework of our study, *N. ceranae* seems either to play a role in the development of this weakening, while remaining asymptomatic, or not to play a decisive role in it.

When considering pesticide residues, no direct link could be established between bee colony disorders and the amount (in µg/ kg) of residues found in the matrices. Neither could we identify any specific molecule as cause of bee disorders. Nevertheless, the study of the residue load of pesticides, specifically fungicides, opens new avenues for a better understanding of honeybee colony disorders.

Even if insecticides/acaricides were the most numerous residues detected in hives mainly in wax, no significant difference in the number of accumulated residues was observed between colonies with disorders and the healthy ones. Indeed the two most abundant active ingredients, tau-fluvalinate and coumaphos, came most probably from varroa control measures even if tau-fluvalinate is used as an insecticide (Mavrik 2F) against *Meligethes aeneus* in oilseed rape. These active ingredients seem to be a frequent outcome of residue analyses studies [57,58,39,38,40,59]. Synergistic effects have been shown between acaricides and other molecules [33,60]. Nevertheless, our modelling did not suggest any synergistic effects in the appearance of bee disorders occurring between residues of fungicides and insecticides-acaricides. Residues of synergist, piperonyl butoxide, were also detected that indicates a prior exposure to synthetic or natural pyrethroids even though residues were not found in the analysed matrix, probably due to a fast degradation of these active ingredients [61]. A nonauthorised active ingredient was also detected: carbaryl, forbidden in Belgium since 2007, indicating an illegal agricultural or gardening use.

The only neonicotinoid found, thiacloprid, was detected in honey during the sampling period of July-August in two colonies. However, the limits of detection achieved in our study do not allow to determine the presence of neonicotinoids -with the exception of acetamiprid- or fipronil at levels in the range of the acute toxicity of these substances (30–40 µg/Kg). Residues of these substances could be present at lower levels and thus an exposure to these substances cannot be excluded.

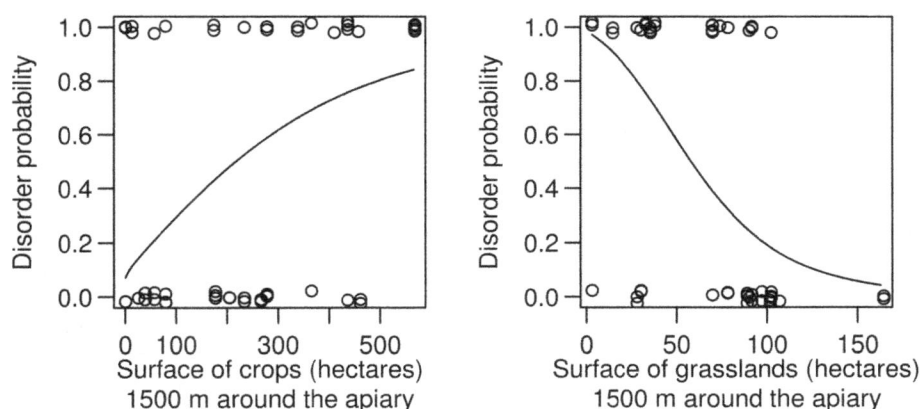

Figure 8. Probability of honeybee colony disorders depending on the apiary's environment. Consideration of crop surface *vs.* grassland surface in a radius of 1500 m around the apiary. Crops include fruit, vegetables, fodder production and horticulture. For each graph, the value of the variable not displayed is fixed to its observed mean (n = 53 colonies).

According to our results, the number of fungicide residues seems to plays a role in the appearance of honeybee colony disorder.

Significantly, more fungicides residues were detected in colonies with disorders than in the healthy ones. Mainly four active ingredients, frequently used in plant protection, were detected: boscalid, iprodione, pyrimethanil and cyprodinil. Fungicides are often considered safe for honeybees based on their acute toxicity. However some direct toxic effects on bees either by the mother compounds or their metabolites have been reported in the past [62]. For example, boscalid shows low acute toxicity to bees [63], despite the fact that beekeepers in the USA have already reported losses and adverse effects on bee brood development related to the foliar application of this systemic active ingredient [64]. Incidents were often linked to the use of a co-formulation boscalid – pyraclostrobin. However, both molecules appeared above detection levels in only two colonies of our study and could not explain the general trend. Interestingly, one metabolite of boscalid, the 2-chloronicotinic acid, is similar to the 6-chloronicotinic acid, a metabolite of imidacloprid. The latter has proved to be lethal to bees at low concentrations (0.1 µg/L) following chronic exposure [65]. Further research should be carried out in order to clarify the mechanism of bee toxicity of boscalid. Indeed, boscalid has proved toxic to other aquatic invertebrates, reducing daphnid fecundity and *Chironomid* emergence [64]. In addition, synergism with other active ingredients like insecticides are possible and increase the toxicity for honeybee [66,32,67]. Two other fungicides were also detected at very high levels in wax and beebread: iprodione and captan. The former is known for its synergistic effects in collaboration with insecticidal compounds [33]. The latter has been shown to induce effects on growth and development of larval honeybees [68,69]. Chronic and larval toxicity studies would be interesting at this stage in order to evaluate possible direct toxic effects on bee individuals and their behaviour. Indeed, a recent study showed increased larval mortality following chronic exposure to tau-fluvalinate, coumaphos, chlorpyriphos and chlorotalonyl or some of their combinations [70]. All these substances were found in our study. Increased mortality rates in the fall may compromise the size and age structure of the wintering cluster, which could lead to winter losses.

The indirect effects of fungicides on bees or on bee colonies are relatively little-known to date.

Fungicides may have an impact on the colony by modifying the existing microflora present in the food stores or in the bee intestinal tract [71]. Studies have already shown the possible modification of microbial composition both at beebread level [62,72] and at intestinal level [73]. This modification in the composition of microbiota may lead to dysbiosis [74]. The impact that such an unbalance in the bee gut microflora may have on bee health has already been considered. The link between the unspecific symptoms observed in our study and a possible microbial alteration could be subject of further research.

In parallel, the potential impact of microbial modification on digestibility and availability of nutrients should be a target for further research. Indeed, the content of essential amino acids might be altered when beebread is contaminated with fungicides (DeGrandi-Hoffman, 2013, pers com.). Given the importance of nutrition, especially pollen, in the good development of the colony [75] alterations in composition or lack of essential nutrients would put the homeostasis of the colony at stake. Some studies have already shown the impact of nutritional lack on bee development and health [76]. Provided that pollen is the unique source of amino acids for honeybees, royal jelly production could also be affected [77,78] with unexpected potential consequences for its main consumers, larvae and the queen. A poor nutrition of the queen, could have as a result an impact in its activity. Likewise, a poor nutrition of the larvae has been shown to impact their development [76]. As a result, the presence of fungicides on beebread and honey may have both a direct effect on their health, but also an indirect one on the colony development.

Fungicides are widely used in agriculture and are broadly present in bee matrices, sometimes at high concentrations at levels of mg/kg [79,80,38,59]. Boscalid, cyprodinil, iprodione are used to control a broad range of fungal pathogens including *Botrytis* spp., *Alternaria* spp. and *Sclerotinia* spp. on a wide range of crops including fruits, vegetables and ornamentals. Pyrimethanil is more specifically used to control grey mould on fruits, vegetables and ornamentals, and leaf scab on apple trees [81]. These active ingredients were already reported in bee matrices in Europe and in the USA [23,39,62]. Their frequent presence in bee matrices might be an indication of chemical plant protection intensity in all agricultural landscapes. Fungicides could also be markers of exposure to other active ingredients with higher toxicity to bees. Mixes of products like fungicides – insecticides are often applied to reduce the number of spray applications. As a result, other pesticides often used in combination with the fungicides found or applied in the same crops could have been at the origin of the effects observed in this study. However, the sensibility of the residue analyses used in our study might explain the lack of detection of such components. Further intensive monitoring and a thorough record of the agricultural practices on pesticide application in tank mixes would help clarifying this alternative explanation.

Factors different from fungicides are most likely involved in the development of bee disorders. According to our model, the disorder probability was not absent when fungicides residues were not detected in presence of insecticide residues and virus. Non identified pathogens, chemicals or factors of different nature could be operating as silent stressors. Ravoet and colleagues (2013) [82] reported the presence of new pathogens in Belgium that were not taken into consideration in our analyses (i.e. *Crithidia mellificae* or the Lake Sinai Virus (LSV)). Other stressing factors could be also linked to the intensive agriculture. In fact, we observed an increase of colony disorders in the area with high density of crops in comparison with areas with grassland. Some studies and beekeeper claims go in line with our outcome regarding the concentration of bee problems in areas with intensive agriculture [83–86]. Furthermore, the same negative trends on pollinators and biodiversity in areas with intensive agriculture have already been described as the result of as increased pesticide use, decreased landscape heterogeneity, loss and fragmentation of natural habitat [87,88].

In conclusion, the five virus studied (ABPV, CBPV, QBCV, SBV, DWV) do not seem determinant in the appearance of bee disorders in our study. These disorders seem clearly linked to the environment of the apiaries and were observed mainly in agricultural crop areas. We observed also that the number of fungicide residues appears as the main potential stress factor linked to bee disorder. However other stressing factors could be acting or interacting at the same time: insecticides exposure, a lack of amino acids and oligo-elements, etc. Our results open new avenues for future research in order to better understand the side effects of fungicides on the bee colony and questions the sustainability of intensive agriculture model and its impact on bees. Specific toxicological studies on both adult bees and larvae would be recommendable in order to better characterise the toxicity of fungicides.

Acknowledgments

We would like to salute and to thank the beekeepers and beekeeping technicians for their dedication and active contribution to our study. Likewise, we thank Gloria DiGrandi-Hoffman, Gerard Arnold, Szaniszlo Szöke and Martin Dermine for their critical and constructive review of the article. We would like also to thank the Direction Générale Opérationnelle Agriculture, Ressources Naturelles et Environnement (DGO3), Département des Aides (D4), Direction des Surfaces (D42), Service 42/3 - LPIS (Land Parcel Identification System) - Service Public de Wallonie for the spatial data. We would also like to thank the Molecular Technology Unit at The Food and Environment Research Agency, Sand Hutton, York, YO41 1LZ for Bee Pathogen Screening Services. Finally, we would like to thank the reviewers for their helpful comments.

Author Contributions

Conceived and designed the experiments: NS LH EB. Performed the experiments: NS LH CM. Analyzed the data: GSM LH. Wrote the paper: LH NS LAM GSM.

References

1. Biesmeijer JC, Roberts SPM, Reemer M, Ohlemüller R, Edwards M, et al. (2006) Parallel declines in pollinators and insect-pollinated plants in Britain and the Netherlands. Science 313: 351–354.

2. Potts SG, Biesmeijer JC, Kremen C, Neumann P, Schweiger O, et al. (2010) Global pollinator declines: trends, impacts and drivers. Trends Ecol Evol 25: 345–353.

3. Goulson D, Lye GC, Darvill B (2008) Decline and conservation of bumble bees. Annu Rev Entomol 53: 191–208.

4. Carvalheiro LG, Kunin WE, Keil P, Aguirre-Gutiérrez J, Ellis WN, et al. (2013) Species richness declines and biotic homogenisation have slowed down for NW-European pollinators and plants. Ecol Lett 16: 870–878. doi:10.1111/ele.12121.

5. Kluser S, Peduzzi P (2007) Global pollinator decline: a literature review. UNEP/GRID- Europe. © UNEP 2007. Switzerland.

6. Maxim L, van der Sluijs JP (2010) Expert explanations of honeybee losses in areas of extensive agriculture in France: Gaucho ® compared with other supposed causal factors. Environ Res Lett 5: 014006. doi:10.1088/1748-9326/5/1/014006.

7. vanEngelsdorp D, Meixner MD (2010) A historical review of managed honey bee populations in Europe and the United States and the factors that may affect them. J Invertebr Pathol 103: S80–S95.

8. Van der Zee R, Pisa L, Andonov S, Brodschneider R, Charriere J-D, et al. (2012) Managed honey bee colony losses in Canada, China, Europe, Israel and Turkey, for the winters of 2008–9 and 1009–10. J Apic Res Bee World 51: 100–114.

9. Klein A-M, Müller C, Hoehn P, Kremen C (2009) Understanding the role of species richness for crop pollination services. Biodiversity, ecosystem function and human wellbeing. New York. 195–208.

10. Lautenbach S, Seppelt R, Liebscher J, Dormann CF (2012) Spatial and temporal trends of global pollination benefit. PloS One 7: e35954.

11. Banskota AH, Tezuka Y, Kadota S (2001) Recent progress in pharmacological research of propolis. Phytother Res 15: 561–571. doi:10.1002/ptr.1029.

12. Jull AB, Rodgers A, Walker N (2008) Honey as a topical treatment for wounds. Cochrane Database of Systematic Reviews. Chichester, UK: John Wiley & Sons, Ltd. 1–43.

13. Srinivasan MV (2011) Honeybees as a Model for the Study of Visually Guided Flight, Navigation, and Biologically Inspired Robotics. Physiol Rev 91: 413–460. doi:10.1152/physrev.00005.2010.

14. Bradber N (1990) Beekeeping in rural development. International Bee Research Association. 15 pp.

15. UNEP (2005) Ecosystems and Human Well-Being: Our Human Planet: Summary for Decision Makers. Island Press. Washington: Island Press.

16. Rasmont P, Pauly A, Terzo M, Patiny S, Michez D, et al. (2005) The survey of wild bees (Hymenoptera, Apoidea) in Belgium and France. Food Agric Organ Rome: 18.

17. Lefevbre M, Bruneau E (2004) État des lieux du phénomène de dépérissement des ruchers en Région wallonne. Louvain la Neuve.

18. Nguyen BK, Mignon J, Laget D, de Graaf D, Jacobs F, et al. (2010) Honey bee colony losses in Belgium during the 2008–9 winter. J Apic Res 49: 337–339. doi:10.3896/IBRA.1.49.4.07.

19. Neumann P, Carreck NL (2010) Honey bee colony losses. J Apic Res 49: 1–6.

20. Vanengelsdorp D, Hayes J, Underwood RM, Caron D, Pettis J (2011) A survey of managed honey bee colony losses in the USA, fall 2009 to winter 2010. J Apic Res 50: 1–10.

21. Aubert M, Faucon J-P, Chauzat M-P (2008) Enquête prospective multi-factorielle: influence des agents microbiens et parasitaires, et des résidus de pesticides sur le devenir de colonies d'abeilles domestiques en conditions naturelles. AFSSA.

22. Higes M, Martín-Hernández R, Garrido-Bailón E, González-Porto AV, García-Palencia P, et al. (2009) Honeybee colony collapse due to *Nosema ceranae* in professional apiaries. Environ Microbiol Rep 1: 110–113. doi:10.1111/j.1758-2229.2009.00014.x.

23. Genersch E, von der Ohe W, Kaatz H, Schroeder A, Otten C, et al. (2010) The German bee monitoring project: a long term study to understand periodically high winter losses of honey bee colonies. Apidologie 41: 332–352.

24. vanEngelsdorp D, Speybroeck N, Evans JD, Nguyen BK, Mullin C, et al. (2010) Weighing Risk Factors Associated With Bee Colony Collapse Disorder by Classification and Regression Tree Analysis. J Econ Entomol 103: 1517–1523. doi:10.1603/EC09429.

25. Pettis JS (2013) The Role Of Pesticides In Queen Health & Sperm Viability. Apimondia 2014, Kiev.

26. Farooqui T (2013) A potential link among biogenic amines-based pesticides, learning and memory, and colony collapse disorder: A unique hypothesis. Neurochem Int 62: 122–136. doi:10.1016/j.neuint.2012.09.020.

27. Nguyen BK, Ribiere M, Vanengelsdorp D, Snoeck C, Saegerman C, et al. (2011) Effects of honey bee virus prevalence, Varroa destructor load and queen condition on honey bee colony survival over the winter in Belgium. J Apic Res 50: 195–202.

28. Nguyen BK, Saegerman C, Pirard C, Mignon J, Widart J, et al. (2009) Does imidacloprid seed-treated maize have an impact on honey bee mortality? J Econ Entomol 102: 616–623.

29. Desneux N, Decourtye A, Delpuech J-M (2007) The Sublethal Effects of Pesticides on Beneficial Arthropods. Annu Rev Entomol 52: 81–106. doi:10.1146/annurev.ento.52.110405.091440.

30. Belzunces LP, Tchamitchian S, Brunet J-L (2012) Neural effects of insecticides in the honey bee. Apidologie 43: 348–370. doi:10.1007/s13592-012-0134-0.

31. Van der Sluijs JP, Simon-Delso N, Goulson D, Maxim L, Bonmatin J-M, et al. (2013) Neonicotinoids, bee disorders and the sustainability of pollinator services. Curr Opin Environ Sustain 5: 293–305. doi:10.1016/j.cosust.2013.05.007.

32. Pilling ED, Jepson PC (1993) Synergism between EBI fungicides and a pyrethroid insecticide in the honeybee (Apis mellifera). Pestic Sci 39: 293–297.

33. Fischer R, Wachendorff-Neumann U (2003) Active ingredient combinations with insecticidal, fungicidal and acaricidal properties.

34. Alaux C, Brunet J-L, Dussaubat C, Mondet F, Tchamitchan S, et al. (2010) Interactions between *Nosema* microspores and a neonicotinoid weaken honeybees (*Apis mellifera*). Environ Microbiol 12: 774–782. doi:10.1111/j.1462-2920.2009.02123.x.

35. Vidau C, González-Polo RA, Niso-Santano M, Gómez-Sánchez R, Bravo-San Pedro JM, et al. (2011) Fipronil is a powerful uncoupler of oxidative phosphorylation that triggers apoptosis in human neuronal cell line SHSY5Y. NeuroToxicology 32: 935–943. doi:10.1016/j.neuro.2011.04.006.

36. Pettis JS, Lichtenberg EM, Andree M, Stitzinger J, Rose R (2013) Crop pollination exposes honey bees to pesticides which alters their susceptibility to the gut pathogen Nosema ceranae. PloS One 8: e70182.

37. Bogdanov S (2005) Contaminants of bee products. Apidologie 37: 1–18. doi:10.1051/apido:2005043.

38. Johnson RM, Ellis MD, Mullin CA, Frazier M (2010) Pesticides and honey bee toxicity – USA. Apidologie 41: 312–331. doi:10.1051/apido/2010018.

39. Mullin CA, Frazier M, Frazier JL, Ashcraft S, Simonds R, et al. (2010) High Levels of Miticides and Agrochemicals in North American Apiaries: Implications for Honey Bee Health. PLoS ONE 5: e9754. doi:10.1371/journal.pone.0009754.

40. Chauzat M-P, Martel A-C, Cougoule N, Porta P, Lachaize J, et al. (2011) An assessment of honeybee colony matrices, Apis mellifera (Hymenoptera: Apidae) to monitor pesticide presence in continental France. Environ Toxicol Chem 30: 103–111. doi:10.1002/etc.361.

41. Cox-Foster DL, Conlan S, Holmes EC, Palacios G, Evans JD, et al. (2007) A metagenomic survey of microbes in honey bee colony collapse disorder. Science 318: 283–287.

42. Tentcheva D, Gauthier L, Zappulla N, Dainat B, Cousserans F, et al. (2004) Prevalence and seasonal variations of six bee viruses in Apis mellifera L. and Varroa destructor mite populations in France. Appl Environ Microbiol 70: 7185–7191.

43. Gauthier L, Tentcheva D, Tournaire M, Dainat B, Cousserans F, et al. (2007) Viral load estimation in asymptomatic honey bee colonies using the quantitative RT-PCR technique. Apidologie 38: 426–435.

44. Zuur AF, Ieno EN, Smith GM (2007) The Analysis of Ecological Data. Springer-Verlag.

45. Chantawannakul P, Ward L, Boonham N, Brown M (2006) A Scientific Note on the Detection of Honeybee Viruses Using real-time PCR (TaqMan®) in Varroa

Mites Collected from a Thai Honeybee (Apis mellifera) Apiary. Journal of Invertebrate Pathology: 91: 69–73.

46. Martin SJ, Highfield AC, Brettell L, Villalobos EM, Budge GE, et al. (2012) Global honey bee viral landscape altered by a parasitic mite. Science 336: 1304–1306.

47. Zuur AF, Ieno EN, Elphick CS (2010) A protocol for data exploration to avoid common statistical problems. Methods Ecol Evol 1: 3–14.

48. Beekman M, Ratnieks FLW (2000) Long-range foraging by the honey-bee, Apis mellifera L. Funct Ecol 14: 490–496.

49. Steffan-dewenter I, Kuhn A (2003) Honeybee foraging in differentially structured landscapes. Proc R Soc Lond B Biol Sci 270: 569–575.

50. Pinheiro JC, Bates DM (2000) Mixed-Effects Models in S and S-PLUS. Springer. 560 p.

51. Fox J (2002) An R and S-Plus companion to applied regression. Sage Publications, Inc. Available: http://books.google.be/books?hl = en&lr = &id = xWS8kgRjGcAC &oi = fnd&pg = PR9&dq = Fox+2002+An+R+and+S+Plus+&ots = o5qCjIMtOb &sig = lvi2jQFv4iDIBrv0QtaWiRIIRBM. Accessed 2012 Sep 18.

52. Bates D (2013) Linear mixed model implementation in lme4. Ms Univ Wis. Available: http://debianmirror.wwi.dk/cran/web/packages/lme4/vignettes/ Implementation.pdf. Accessed 2014 Feb 3.

53. IRM IR de M (2013) Bilan climatologique annuel. Available: http://www. meteo.be/meteo/view/fr/1317239-Bilan+climatologique+annuel.html. Accessed 2013 Dec 15.

54. De Smet L, Ravoet J, de Miranda JR, Wenseleers T, Mueller MY, et al. (2012) BeeDoctor, a versatile MLPA-based diagnostic tool for screening bee viruses. PloS One 7: e47953.

55. Bailey L, Ball BV, Perry JN (1981) The prevalence of viruses of honey bees in Britain. Ann Appl Biol 97: 109–118.

56. Verhoeven B (2014) Projet pilote de surveillance de la mortalité des abeilles 2012–2013 Premiers résultats. AFSCA. Journé apicole 26 janvier 2014, Namur.

57. Chauzat M-P, Faucon J-P, Martel A-C, Lachaize J, Cougoule N, et al. (2006) A survey of pesticide residues in pollen loads collected by honey bees in France. J Econ Entomol 99: 253–262.

58. Chauzat M-P, Faucon J-P (2007) Pesticide residues in beeswax samples collected from honey bee colonies (Apis mellifera L.) in France. Pest Manag Sci 63: 1100–1106. doi:10.1002/ps.1451.

59. Wiest L, Buleté A, Giroud B, Fratta C, Amic S, et al. (2011) Multi-residue analysis of 80 environmental contaminants in honeys, honeybees and pollens by one extraction procedure followed by liquid and gas chromatography coupled with mass spectrometric detection. J Chromatogr A 1218: 5743–5756. doi:10.1016/j.chroma.2011.06.079.

60. Johnson RM, Pollock HS, Berenbaum MR (2009) Synergistic interactions between in-hive miticides in Apis mellifera. J Econ Entomol 102: 474–479.

61. Yu SJ (2008) The toxicology and biochemistry of insecticides. CRC N Y NY USA. Available: http://www.lavoisier.fr/livre/notice.asp?ouvrage = 1795057. Accessed 2014 Jan 31.

62. DeGrandi-Hoffman G, Chen Y, Simonds R (2013) The Effects of Pesticides on Queen Rearing and Virus Titers in Honey Bees (Apis mellifera L.). Insects 4: 71–89. doi:10.3390/insects4010071.

63. Pesticide Properties Database (PPDB) (2014) Boscalid (Ref: BAS 510F). Available: http://sitem.herts.ac.uk/aeru/ppdb/en/86.htm. Accessed 2014 Jan 3.

64. Aubee C, Lieu D (2010) Environmental Fate and Ecological Risk Assessment for Boscalid New Use on Rapeseed, Including Canola (Seed Treatment). U.S. EPA - Washington.

65. Suchail S, Guez D, Belzunces LP (2001) Discrepancy between acute and chronic toxicity induced by imidacloprid and its metabolites in Apis mellifera. Environ Toxicol Chem SETAC 20: 2482–2486.

66. Belzunces LP, Colin ME (1992) Les phénomènes d'affaiblissement de ruchers.

67. Meled M, Thrasyvoulou A, Belzunces LP (1998) Seasonal variations in susceptibility of Apis mellifera to the synergistic action of prochloraz and deltamethrin. Environ Toxicol Chem 17: 2517–2520.

68. Thompson HM (2003) Behavioural effects of pesticides in bees–their potential for use in risk assessment. Ecotoxicology 12: 317–330.

69. Mussen EC, Lopez JE, Peng CY (2004) Effects of selected fungicides on growth and development of larval honey bees, Apis mellifera L.(Hymenoptera: Apidae). Environ Entomol 33: 1151–1154.

70. Zhu W, Schmehl DR, Mullin CA, Frazier JL (2014) Four Common Pesticides, Their Mixtures and a Formulation Solvent in the Hive Environment Have High Oral Toxicity to Honey Bee Larvae. PloS One 9: e77547.

71. Batra LR, Batra SWT, Bohart GE (1973) The mycoflora of domesticated and wild bees (Apoidea). Mycopathol Mycol Appl 49: 13–44.

72. Yoder JA, Jajack AJ, Rosselot AE, Smith TJ, Yerke MC, et al. (2013) Fungicide Contamination Reduces Beneficial Fungi in Bee Bread Based on an Area-Wide Field Study in Honey Bee, *Apis mellifera*, Colonies. J Toxicol Environ Health A 76: 587–600. doi:10.1080/15287394.2013.798846.

73. Anderson KE, Sheehan TH, Eckholm BJ, Mott BM, DeGrandi-Hoffman G (2011) An emerging paradigm of colony health: microbial balance of the honey bee and hive (Apis mellifera). Insectes Sociaux 58: 431–444. doi:10.1007/ s00040-011-0194-6.

74. Sartor RB (2008) Therapeutic correction of bacterial dysbiosis discovered by molecular techniques. Proc Natl Acad Sci 105: 16413–16414.

75. Di Pasquale G, Salignon M, Le Conte Y, Belzunces LP, Decourtye A, et al. (2013) Influence of Pollen Nutrition on Honey Bee Health: Do Pollen Quality and Diversity Matter? PLoS ONE 8: e72016. doi:10.1371/journal.pone. 0072016.

76. Brodschneider R, Crailsheim K (2010) Nutrition and health in honey bees. Apidologie 41: 278–294.

77. Standifer LN (1967) A comparison of the protein quality of pollens for growth-stimulation of the hypopharyngeal glands and longevity of honey bees, Apis mellifera L. (Hymenoptera: Apidae). Insectes Sociaux 14: 415–425. doi:10.1007/BF02223687.

78. Liming W, Jinhui Z, Xiaofeng X, Yi L, Jing Z (2009) Fast determination of 26 amino acids and their content changes in royal jelly during storage using ultra-performance liquid chromatography. J Food Compos Anal 22: 242–249.

79. Kubik M, Nowacki J, Pidek A, Warakomska Z, Michalczuk L, et al. (1999) Pesticide residues in bee products collected from cherry trees protected during blooming period with contact and systemic fungicides. Apidologie 30: 12. doi:10.1051/apido:19990607.

80. Kubik M, Nowacki J, Pidek A, Warakomska Z, Michalczuk L, et al. (2000) Residues of captan (contact) and difenoconazole (systemic) fungicides in bee products from an apple orchard. Apidologie 31: 531–542.

81. Pesticide Properties Database (PPDB) (2014) Pyrimethanil (Ref: SN 100309). Available: http://sitem.herts.ac.uk/aeru/ppdb/en/573.htm. Accessed 2014 Jan 3.

82. Ravoet J, Maharramov J, Meeus I, De Smet L, Wenseleers T, et al. (2013) Comprehensive bee pathogen screening in Belgium reveals Crithidia mellificae as a new contributory factor to winter mortality. PloS One 8: e72443.

83. Oomen PA, Belzunces L, Pélissier C, Lewis G (2001) Honey bee poisoning incidents over the last ten years, as reported by bee keepers in The Netherlands. Les Colloques 98: 129–136.

84. Brittain CA, Vighi M, Bommarco R, Settele J, Potts SG (2010) Impacts of a pesticide on pollinator species richness at different spatial scales. Basic Appl Ecol 11: 106–115.

85. CRA-API (2011) Effects of coated maize seed on honey bees. Report based on results obtained from the first year of activity of the APENET project. Italy.

86. Maxim L, van der Sluijs J (2013) Seed-dressing systemic insecticides and honeybees. Late Lessons from Early Warnings: Science, Precaution, Innovation. Copenhagen. 401–438. Available: http://www.beekeeping.com/articles/us/ late_lessons_from_early_warnings_2.pdf. Accessed 2014 Feb 10.

87. Holzschuh A, Steffan-Dewenter I, Tscharntke T (2008) Agricultural landscapes with organic crops support higher pollinator diversity. Oikos 117: 354–361.

88. Kleijn D, Kohler F, Báldi A, Batáry P, Concepcion ED, et al. (2009) On the relationship between farmland biodiversity and land-use intensity in Europe. Proc R Soc B Biol Sci 276: 903–909.

6

Crop Pollination Exposes Honey Bees to Pesticides Which Alters Their Susceptibility to the Gut Pathogen *Nosema ceranae*

Jeffery S. Pettis[1], Elinor M. Lichtenberg[2], Michael Andree[3], Jennie Stitzinger[2], Robyn Rose[4], Dennis vanEngelsdorp[2]*

[1] Bee Research Laboratory, USDA-ARS, Beltsville, Maryland, United States of America, [2] Department of Entomology, University of Maryland, College Park, College Park, Maryland, United States of America, [3] Cooperative Extension Butte County, University of California, Oroville, California, United States of America, [4] USDA-APHIS, Riverdale, Maryland, United States of America

Abstract

Recent declines in honey bee populations and increasing demand for insect-pollinated crops raise concerns about pollinator shortages. Pesticide exposure and pathogens may interact to have strong negative effects on managed honey bee colonies. Such findings are of great concern given the large numbers and high levels of pesticides found in honey bee colonies. Thus it is crucial to determine how field-relevant combinations and loads of pesticides affect bee health. We collected pollen from bee hives in seven major crops to determine 1) what types of pesticides bees are exposed to when rented for pollination of various crops and 2) how field-relevant pesticide blends affect bees' susceptibility to the gut parasite *Nosema ceranae*. Our samples represent pollen collected by foragers for use by the colony, and do not necessarily indicate foragers' roles as pollinators. In blueberry, cranberry, cucumber, pumpkin and watermelon bees collected pollen almost exclusively from weeds and wildflowers during our sampling. Thus more attention must be paid to how honey bees are exposed to pesticides outside of the field in which they are placed. We detected 35 different pesticides in the sampled pollen, and found high fungicide loads. The insecticides esfenvalerate and phosmet were at a concentration higher than their median lethal dose in at least one pollen sample. While fungicides are typically seen as fairly safe for honey bees, we found an increased probability of *Nosema* infection in bees that consumed pollen with a higher fungicide load. Our results highlight a need for research on sub-lethal effects of fungicides and other chemicals that bees placed in an agricultural setting are exposed to.

Editor: Fabio S. Nascimento, Universidade de São Paulo, Faculdade de Filosofia Ciências e Letras de Ribeirão Preto, Brazil

Funding: Funding for this study was provided by the National Honey Board (http://www.honey.com/) and the USDA-ARS Areawide Project on Bee Health (http://www.ars.usda.gov/research/projects/projects.htm?accn_no=412796). Neither the Honey Board nor USDA-ARS Program Staff had a role in study design, data collection and analysis, decision to publish, or preparation of the manuscript.

* E-mail: dennis.vanengelsdorp@gmail.com

Introduction

Honey bees, *Apis mellifera*, are one of the most important pollinators of agricultural crops [1]. Recent declines in honey bee populations in many North American and European countries [2–4] and increasing cultivation of crops that require insects for pollination [5] raise concerns about pollinator shortages [5,6]. Habitat destruction, pesticide use, pathogens and climate change are thought to have contributed to these losses [2,7,8]. Recent research suggests that honey bee diets, parasites, diseases and pesticides interact to have stronger negative effects on managed honey bee colonies [9,10]. Nutritional limitation [11,12] and exposure to sub-lethal doses of pesticides [13–16], in particular, may alter susceptibility to or severity of diverse bee parasites and pathogens.

Recent research is uncovering diverse sub-lethal effects of pesticides on bees. Insecticides and fungicides can alter insect and spider enzyme activity, development, oviposition behavior, offspring sex ratios, mobility, navigation and orientation, feeding behavior, learning and immune function [9,13,14,16–22]. Reduced immune functioning is of particular interest because of recent disease-related declines of bees including honey bees [3,23]. Pesticide and toxin exposure increases susceptibility to and mortality from diseases including the gut parasite *Nosema* spp. [14,15]. These increases may be linked to insecticide-induced alterations to immune system pathways, which have been found for several insects, including honey bees [22,24–26].

Surveys of colony food reserves and building materials (i.e. wax) have found high levels and diversity of chemicals in managed colonies [18,27,28]. These mixtures have strong potential to affect individual and colony immune functioning. However, almost all research to-date on pesticides' effects on pathogen susceptibility fed a single chemical to test bees [16]. Because pesticides may have interactive effects on non-target organisms (e.g. [29]), it is crucial to determine how real world combinations and loads of pesticides affect bee health.

One pathogen of major concern to beekeepers is *Nosema* spp. The endoparasitic fungal infections of *N. apis* and *N. ceranae* adversely affect honey bee colony health, and can result in complete colony collapse [30]. Infection with *Nosema* in the autumn leads to poor overwintering and performance the following spring [31], and queens can be superseded soon after becoming infected with *Nosema* [32]. We chose *Nosema* as a model pathogen because earlier work [13,14] had demonstrated an interaction with pesticide exposure.

This study addresses two important questions. 1) What types of pesticides might bees be exposed to in major crops? While multiple studies have characterized the pesticide profile of various materials inside a honey bee nest [27,28], few have looked at the pollen being brought back to the nest. 2) How do field-relevant pesticides blends affect bees' susceptibility to infection by the *Nosema* parasite?

Methods

Ethics Statement

Pollen was collected from honey bees with permission of the beekeepers and the land owners.

Hive Selection and Pollen Collection

We collected pollen carried by foraging honey bees returning to the hive for nine hives in seven crops: almond, apple, blueberry, cranberry, cucumber, pumpkin, and watermelon (Table 1). For each crop, we selected three fields that were separated by at least 3.2 km. Hives were deployed in these fields for pollination services based on growers' needs. Within each selected field, we chose the three honey bee hives with the strongest foraging forces by observing flight in the bee yard for 5–10 min, and attached plastic pollen traps (Brushy Mountain Bee Farm, Moravian Falls, NC) to these hives. Pollen traps collect the pollen pellets bees carry on their hind tibiae in flattened regions called corbiculae. Bees use this pollen to make food for larvae inside the nest. We checked traps after three days, and removed them if they contained at least 5 g of pollen. Traps with less than 5 g remained on hives until they contained 5 g of pollen or for 10 days. We placed pollen removed from traps in 50 mL centrifuge tubes and stored the samples on ice until they could be transferred to a $-29°C$ freezer in the lab.

Because our first round of pollen trapping in cranberry fields yielded little pollen, we collected pollen from each hive in cranberry fields twice: early in the flowering season and late in the season. We separate these samples in data analyses, referring to them as "Cranberry early" and "Cranberry late."

We measured the wet weight of each pollen sample, and compared the quantity of pollen collected by hives in different crops via a Kruskal-Wallis test followed by a post-hoc non-parametric Tukey-type test (using the R package nparcomp [33]). We then divided each sample into three portions. A 5 g subsample was sorted by color and then each group of similarly colored pollen pellets were identified (see below); a 3 g subsample was sent to the USDA's Agricultural Marketing Service Laboratory in Gastonia, NC for pesticide analysis; and a 10 g subsample was sent to the USDA-ARS Bee Research Laboratory (Beltsville, MD) for the *Nosema* infection study. Because almond pollen was collected after all other pollens, we were unable to include it in the pesticide analysis and *Nosema* infection study. In cases where the total amount of pollen collected from a single colony was less than 6 g all the pollen was used for pesticide analysis.

Pollen Identification

Each 5 g pollen subsample was dehydrated in a drying oven at 40°C. We considered a sample to be dry when its weight did not change between two consecutive time points (measured every 4–6 h). Typically pollen dried in 12–18 h. To identify pollen types collected by the bees, we sorted the pollen in each subsample by color, quantified each color by comparing to Sherwin-Williams® color palettes, re-weighed after color separation and fixed each color from each subsample on a separate slide. We prepared each slide by grinding 2 pollen pellets in 2 mL water and letting them dissolve to form a slurry. We placed a small amount of slurry on a slide with a drop of silicon oil, and covered slides and sealed with clear nail polish after letting air bubbles escape for 48 h. We visually identified each pollen type under 400x magnification by comparing with published reference collections [34–36]. Visual identification of pollen grains through comparison with voucher or reference specimens is standard in pollination ecology [37,38]. Similarities between closely related pollens, however, sometimes prevent identification to genus or species with this method [39]. Because of this limitation, we assumed that all pollen collected in apple (*Malus domestica*) orchards that was identified as *Malus* sp. was from apple trees, and that all pollen in the Cucurbitaceae family collected in cucumber (Cucurbitaceae, *Cucumis sativus*) fields was from cucumber flowers.

For each subsample, we estimated pollen diversity as the number of different pollen colors collected from that bee hive. We also calculated the proportion, by weight, of the pollen that was identified as belonging to the target crop's genus. Many samples could only be identified to genus, so assessing target genus rather than target crop permitted a more inclusive analysis. We used

Table 1. Quantity and diversity of pollen collected in pollen traps on individual honey bee hives.

Crop	Location	Mean grams of pollen collected (se)	Mean number of pollen types (se)
Almond	Rosedale, CA; Kern County	42.0 (9.1)[a,b]	1.7 (0.2)[a,b]
Apple	York Springs, PA; Adams County	26.7 (2.6)[a]	4.9 (0.5)[c]
Blueberry	Deblois, ME; Washington County	4.1 (1.5)[b]	6.0 (1.0)[c]
Cranberry (early season)	Hammonton, NJ; Atlantic County	13.0 (2.5)[a,b]	4.0 (1.0)[b,c]
Cranberry (late season)	Hammonton, NJ; Atlantic County	13.9 (3.8)[a,b]	4.1 (0.6)[b,c]
Cucumber	Cedarville, NJ; Cumberland County	8.1 (2.7)[b]	5.5 (1.3)[b,c]
Watermelon	Seaford, DE; Sussex County	27.1 (11.2)[a,b]	7.1 (1.2)[c]
Pumpkin	Kutztown, PA; Berks County	98.6 (29.0)[a,b]	3.7 (0.6)[b,c]

Letters indicate statistically different groups.

Kruskal-Wallis tests to determine whether either of these measures differed with the crop in which sampled bee hives were placed.

Pesticide Analysis

We determined the identity and load of pesticide residues present in pollen samples collected from all crops (except almond). For each field sampled ($n = 19$), we pooled pollen from the three hives for analysis. One early-season cranberry field and one cucumber field did not yield sufficient pollen in traps for pesticide analysis. Methods follow the LC/MS-MS and GC/MS methods for pollen analysis described in Mullin et al. [27]. We used these data to determine the total number of pesticides detected in each sample, each sample's total pesticide load, and the diversity and load of pesticides in each of 10 categories: insecticides, fungicides, herbicides, and several insecticide types (carbamates, cyclodienes, formamidines, neonicotinoids, organophosphates, oxadiazines and pyrethroids). To permit comparison between categories with different numbers of elements, we calculated diversity as the proportion of pesticides from a category found in a given sample, and load as the total load divided by the number of chemicals in that category. We only calculated diversity for categories with at least three chemicals.

The total number of pesticides present and total load did not meet parametric assumptions. We thus analyzed how these variables differ between crops using non-parametric Kruskal-Wallis tests. When separated by category and log-transformed, pesticide loads did meet parametric assumptions. We thus determined whether load varied by pesticide category using a general linear mixed model with sample as a random effect, to control for the fact that our regression included one data point per category from each sample. Insufficient degrees of freedom prevented us from expanding this model to include crop. We thus asked whether the pesticide load and diversity varied with crop for each category using one Kruskal-Wallis test per category and applying a sequential Bonferroni correction [40] across pesticide categories to control for multiple comparisons.

Nosema Infection

The *Nosema* infection experiment is similar to published methods [26]. We obtained 210 disease-free honey bees from each of three healthy colonies at the Bee Research Laboratory. Each bee was placed into one of 21 groups upon emergence, with the ten bees in the same group and from the same colony housed together in a wooden hoarding cage ($12 \times 12 \times 12$ cm). Each group of bees was fed 1 g of pollen mixed with 0.5 mL of syrup (1:1 sucrose to water by weight), which they fully consumed in 2–4 days. These pollen cakes were placed in small petri dishes with the laboratory cages. Pollen from either one of the crop fields or one of two control diets were used. The pollen control group ("BRL") was fed a mixed pollen diet prepared by the USDA-ARS Bee Research Laboratory. This pollen was collected in the desert Southwest (Arizona Bee Products, Tucson, AZ) and tested as pesticide-free by the USDA Agricultural Marketing Service prior to use. A protein control group was fed an artificial honey bee pollen substitute, MegaBee®. The *Nosema* inoculum was freshly prepared by mixing *Nosema* spores isolated from an infected colony (details provided in [26]) with 50% sucrose solution to obtain a concentration of ca. 2 million spores per 5 mL. We fed 5 mL of the *Nosema* inoculum to each cage during the first two days of adult life, then provided bees with *ad libitum* access to clean 50% (w/v) sucrose solution. We collected bees 12 days after infection and examined them for the presence or absence of *N. ceranae* spores by homogenizing individual abdomens in 1 mL distilled water. Here we focus only on infection prevalence, the number of individuals with *Nosema* spores.

To look for potential effects of individual pesticides on susceptibility to *Nosema* infection, we calculated the relative risk and its 95% confidence interval for bees becoming infected after consuming pollen with a specific pesticide. Relative risk measures the chance of developing a disease after a particular exposure [41], here each pesticide. A relative risk value of one indicates that the probability of infection is equal between exposed and non-exposed groups.

We further tested effects of pesticides in pollen on measured *Nosema* prevalence using a generalized linear mixed model with a bee's *Nosema* status as the response variable, the source hive and pesticide variables as fixed effects, and the pollen sample fed to the bee as a random effect. Collinearity prevented developing a full model to investigate in detail how pesticides and pollen source affect bees' susceptibility to *Nosema* infection. We thus selected for analysis two measures that vary with crop and are not nested: total pesticide diversity and fungal load. To graph logistic regression results in a meaningful manner, we followed recent recommendations [42,43] and a modification of the logi.hist.plot function in the R popbio package [44] that shows our mixed model output.

Results

Pollen Collection

Bee colonies collected different amounts of pollen in the different crops (Table 1; Kruskal-Wallis test: $H_7 = 29.6$, $p = 0.0001$). Pollen diversity, estimated by quantifying the number of differently colored pollen pellets collected in pollen traps, varied by crop (Table 1; Kruskal-Wallis test: $H_7 = 23.5$, $p = 0.0014$). The proportion of pollen that bees collected from the target crop, except for almond and apple, was low (mean$\pm$se $= 0.33\pm0.05$; Table S1). Like pollen weights, this proportion dramatically differed between crops (Fig. 1; $H_7 = 44.86$, $p<0.0001$). Notably, none of the pollen trapped from hives in blueberry, cranberry (early and late), pumpkin or watermelon fields was from the target crop.

Pesticide Analysis

All pollen collected in this study contained pesticides (Table 2; mean $\pm$ se $= 9.1\pm1.2$ different chemicals, range 3–21). Pesticide loads ranged from 23.6 to 51,310.0 ppb ($11,760.0\pm3,734.2$ ppb). The maximum pesticide concentration in any single pollen sample exceeded the median lethal dose (LD_{50}, the dose required to kill half a population within 24 or 48 h) for esfenvalerate and phosmet (Table 2). The number of pesticides detected in trapped pollen varied by the crop in which the bee hives were located (Kruskal-Wallis test: $H_6 = 12.96$, $p = 0.04$), but the total pesticide load did not ($H_6 = 11.21$, $p = 0.08$)(Fig. 2).

We found insecticides and fungicides in all 19, and herbicides in 23.6% of, pollen samples. Insecticides present in pollen collected by the bees came from seven categories. We found oxadiazines in 10.5%, neonicotinoids in 15.8%, carbamates in 31.6%, cyclodienes in 52.6%, formamidines in 52.6%, organophosphates in 63.2%, and pyrethroids in 100% of pollen samples. Both neonicotinoids and oxadiazines were present only in pollen collected by bees in apple orchards (Figs. 3, S1). Within a sample, pollen fungicide loads were significantly higher than loads of herbicides or any of the insecticide categories (Fig. 4; GLMM, likelihood ratio test: $\chi^2 = 121.9$, df $= 8$, $p<0.0001$).

After adjusting for multiple comparisons, pesticide loads did not vary by crop for any pesticide category (Fig. S1). We calculated

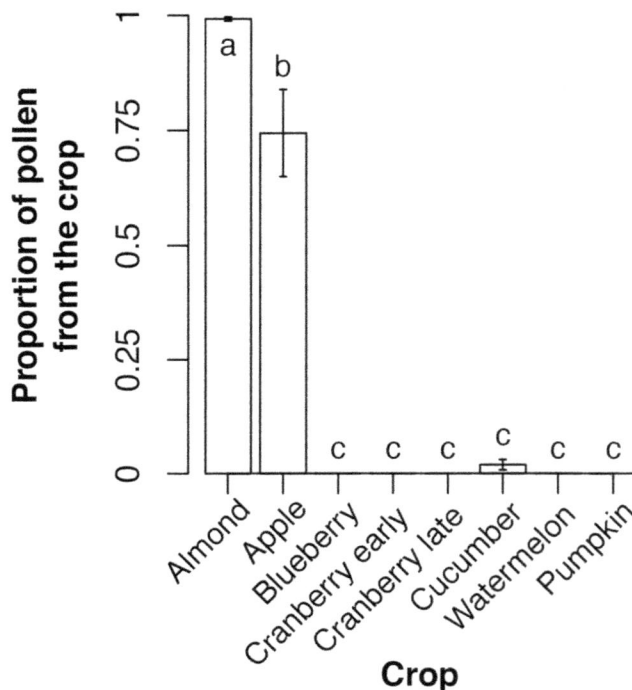

Figure 1. Pollen collection from the crop where a hive was located was low for most crops. Bars show mean ± se. Letters indicate statistically significant differences ($p<0.05$).

pesticide diversity within only those categories containing three or more chemicals. Fungicide and neonicotinoid diversities varied by crop, but diversities of other pesticide categories did not (Fig. 3).

Nosema Infection

147 of the 630 bees (23.3%) fed *Nosema* spores became infected. 22 of the 35 pesticides (62.9%) found in our pollen samples had relative risk values significantly different from 1 (Table 2). 8 pesticides (22.9%) were associated with increased *Nosema* prevalence, while the remaining 14 were associated with decreased *Nosema* prevalence. Two of the three detected pesticides applied by beekeepers to control hive mites (marked with a * in Table 2) had a relative risk larger than two, indicating *Nosema* prevalence in bees fed pollen containing those chemicals (DMPF and fluvalinate) was more than double the *Nosema* prevalence in bees that did not consume these chemicals. Of the seven pesticides found in pollen from over half, or at least four, of the crops, the majority were associated with higher *Nosema* prevalence in bees that consumed them. Both control diets had relative risk values not significantly different from one.

A pollen sample's fungicide load significantly affected *Nosema* prevalence among bees fed that pollen (Fig. 5; GLMM, likelihood ratio test: $\chi^2 = 5.8$, df $= 1$, $p = 0.02$), but pesticide diversity did not ($\chi^2 = 1.7$, df $= 1$, $p = 0.19$). A bee's source colony, included as a blocking variable, also did not affect *Nosema* prevalence ($\chi^2 = 2.0$, df $= 2$, $p = 0.36$). Replacing fungicide load with chlorothalonil load obtained the same result (chlorothalonil load: $\chi^2 = 5.3$, df $= 1$, $p = 0.02$; pesticide diversity: $\chi^2 = 1.5$, df $= 1$, $p = 0.23$; source colony: $\chi^2 = 2.0$, df $= 2$, $p = 0.36$; fungicide load model AIC $= 612.71$, chlorothalonil load model AIC $= 613.15$). Chlorothalonil was also the most abundant fungicide in our samples, and comprised $50.0 \pm 10.2\%$ (mean ± se) of the per sample total fungicide load.

Discussion

The results from this study highlight several patterns that merit further attention. First, despite being rented to pollinate specific crops, honey bees did not always return to the nest with corbicular pollen from those crops. These findings support other research with honey bees and native bees indicating that in some crops native bees may be more efficient pollinators [45]. Second, fungicides were present at high levels in both crop and non-crop pollen collected by bees. Third, two fungicides (chlorothalonil and pyraclostrobin), and two miticides used by beekeepers to control varroa infestation (amitraz and fluvalinate) had a pronounced effect on bees' ability to withstand parasite infection. Research on pesticides' effects on bee health has focused almost exclusively on insecticides (e.g. fipronil [15] and the neonicotinoids imidacloprid [13,14] and thiacloprid [15]). Finally, several individual pollen samples contained loads higher than the median lethal dose for a specific pesticide. While multiple studies have shown negative effects of specific pesticides on honey bee individual and colony health [14,15,22,26] and high pesticide exposure [27,28], ours is the first to demonstrate how real world pollen-pesticide blends affect honey bee health.

Our results show that beekeepers need to consider not only pesticide regimens of the fields in which they are placing their bees, but also spray programs near those fields that may contribute to pesticide drift onto weeds. The bees in our study collected pollen from diverse sources, often failing to collect any pollen from the target crop (Fig. 1). All of the non-target pollen that we were able to identify to genus or species was from wildflowers (Table S1), suggesting the honey bees were collecting significant amounts of pollen from weeds surrounding our focal fields. The two exceptions to this were hives placed in almond and apple orchards. Almond flowers early in the year, and almond orchards are large, thus providing honey bees with little access to other flowers. Honey bees rarely collect pollen from blueberry or cranberry flowers, which only release large quantities of pollen after being vibrated by visiting bees (buzz pollination) [46,47]. Honey bees are not capable of buzz pollination and thus are unlikely to collect large amounts of pollen from these plants to bring back to the colony. Bumble bees, which can buzz pollinate, collect mainly blueberry pollen when placed in blueberry fields [48]. Interestingly, the two crops that saw high levels of pollen collection by honey bees are Old World crops that evolved with honey bees as natural pollinators. Crops native to the New World, where honey bees have been introduced, yielded little or no pollen in our samples.

It is possible that bees were exposed to pesticides while collecting nectar from our focal crops, even when we detected no pollen from those crops. Because pollen traps collect only corbicular pollen intended for consumption by the colony, our data indicate only flowers from which bees are actively collecting pollen and not all flowers they visited. Several studies have detected pesticides in floral nectar and pollen [49,50], sometimes in concentrations with sublethal effects on honey and bumble bees [51,52]. Honey bees may collect nectar from blueberry and cranberry flowers via legitimate visits or "robbing" through slits cut at the base of flower corollas [53]. However, exposure to pesticides via nectar may be unlikely in cucumber, pumpkin and watermelon. Beekeepers often report poor honey production when their hives are placed in these crops (pers. obs.).

The combination of high pesticide loads and increased *Nosema* infection rates in bees that consumed greater quantities of the fungicides chlorothalonil and pyraclostrobin suggest that some fungicides have stronger impacts on bee health than previously

Table 2. Pesticides found in pollen trapped off honey bees returning to the nest.

Pesticide	Insecticide family	LD$_{50}$ (ppm)[a]	Crops in which detected[c]	Detections	Quantity detected, mean±se (max) (ppb)	Relative risk (95% CI)
Fungicides						
Azoxystrobin		>1,562.5 [64]	Cr, Cu, Wa	10	60.3±25.6 (332)	0.75 (0.56, 1.02)
Captan		>78.13 [65]	Ap, Cr, Cu, Wa	9	976.9±734.4 (13,800)	0.59 (0.42, 0.81)†
Chlorothalonil		>1,414.06 [66]	Ap, Bl, Cr, Cu, Pu, Wa	17	4,491.2±2,130.7 (29,000)	2.31 (1.35, 3.94)†
Cyprodinil		>6,125 [67]	Ap	3	996.9±707.5 (12,700)	0.31 (0.15, 0.65)†
Difenoconazole		>781.25 [68]	Ap	3	171.4±119.4 (2,110)	0.31 (0.15, 0.65)†
Fenbuconazole		>2,282.65 [69]	Ap, Cr, Cu	10	227.3±89.2 (1,420)	0.33 (0.23, 0.48)†
Pyraclostrobin		573.44 [70]	Cr, Pu	4	2,787.1±1,890.1 (27,000)	2.85 (2.16, 3.75)†
Quintozene (PCNB)		>0.78 [71]	Cr	2	0.3±0.3 (4.7)	0.97 (0.59, 1.61)
THPI	Captan metabolite		Cr, Cu	3	832.1±531.8 (9,470)	0.42 (0.21, 0.82)†
Herbicides						
Carfentrazone ethyl		>217.97 [72]	Cr	1	0.1±0.08 (1.6)	1.05 (0.54, 2.05)
Pendimethalin		>388.28 [73]	Ap, Cr, Pu	5	5.1±3.7 (69.5)	1.47 (1.08, 1.99)†
Insecticides						
2,4 Dimethylphenyl formamide (DMPF)*	Amitraz (formamidine) metabolite		Bl, Cu, Pu, Wa	10	171.5±117.0 (2,060)	2.13 (1.56, 2.92)†
Acetamiprid	Neonicotinoid	55.47 [60]	Ap	3	59.1±32.2 (401)	0.31 (0.15, 0.65)†
Bifenthrin	Pyrethroid	0.11 [74]	Pu, Wa	3	6.6±3.8 (53.1)	2.08 (1.53, 2.83)†
Carbaryl	Carbamate	8.59 [75]	Ap, Cu, Wa	6	57.8±30.0 (403)	0.42 (0.27, 0.66)†
Chlorpyrifos	Organophosphate	0.86 [16]	Ap, Cr, Cu, Pu	7	3.1±1.1 (15.5)	0.89 (0.64, 1.23)
Coumaphos*	Organophosphate	35.94 [16]	Bl, Cr, Cu	6	2.2±1.0 (17.5)	0.62 (0.43, 0.91)†
Cyfluthrin	Pyrethroid	<0.31 [76]	Cr, Wa	2	0.6±0.4 (5.4)	1.31 (0.85, 2.02)
Cyhalothrin	Pyrethroid	0.30 [77]	Ap, Pu, Wa	7	14.6±7.9 (131)	0.94 (0.69, 1.29)
Cypermethrin	Pyrethroid	0.18–4.38 [78]	Cr	1	0.4±0.4 (6.9)	1.05 (0.54, 2.05)
Deltamethrin	Pyrethroid	0.39 [79]	Cr	1	4.5±4.5 (85.3)	1.05 (0.54, 2.04)
Diazinon	Organophosphate	1.72 [80]	Ap, Cr	3	1.4±1.0 (19.8)	0.56 (0.32, 0.97)†
Endosulfan I	Cyclodiene	54.69 [16]	Ap, Cr, Cu, Pu, Wa	8	1.5±0.7 (12.9)	1.60 (1.20, 2.14)†
Endosulfan II	Cyclodiene	54.69 [16]	Ap, Cr, Cu, Pu	6	0.8±0.3 (5.3)	1.41 (1.04, 1.91)†
Endosulfan sulfate	Endosulfan metabolite		Cr, Cu	4	0.3±0.2 (2.1)	0.79 (0.52, 1.19)
Esfenvalerate	Pyrethroid	0.13 [81]	Ap, Cr, Cu	7	16.9±12.0 (216)	0.51 (0.35, 0.75)†
Fluvalinate*	Pyrethroid	1.56 [82]	Bl, Cr, Cu, Pu, Wa	16	42.4±29.7 (570)	2.43 (1.49, 3.96)†
Heptachlor epoxide	Heptachlor[b] (cyclodiene) metabolite		Cr	1	0.6±0.6 (12)	1.05 (0.54, 2.04)
Imidacloprid	Neonicotinoid	0.23 [83]	Ap	3	2.8±2.0 (36.5)	0.31 (0.15, 0.65)†
Indoxacarb	Oxadiazine	1.41 [84]	Ap	2	0.5±0.5 (9)	0.28 (0.11, 0.73)†
Methidathion	Organophosphate	1.85 [85]	Cr	1	1.6±1.6 (31)	1.05 (0.54, 2.04)
Methomyl	Carbamate	<3.91 [86]	Wa	1	13.6±13.6 (259)	1.54 (0.91, 2.61)
Phosmet	Organophosphate	8.83 [85]	Ap, Cr, Cu	5	798.7±772.4 (14,700)	0.36 (0.21, 0.61)†
Pyrethrins	Pyrethroid	0.16 [16]	Cr	1	5.1±5.1 (97.4)	1.05 (0.54, 2.05)
Thiacloprid	Neonicotinoid	114.06 [60]	Ap	2	1.1±0.8 (12.4)	0.35 (0.15, 0.82)†
Control diets						
BRL	NA	NA	NA	NA	NA	0.58 (0.23, 1.48)
MegaBee	NA	NA	NA	NA	NA	0.74 (0.33, 1.67)

[a]We divided LD$_{50}$ values given as μg/bee (g) by 0.128 (equivalent to multiplying by 7.8) to obtain ppm when necessary [85]. If multiple values have been published, we include only the smallest.
[b]Heptachlor has been banned for use on cranberries since 1978 [87], but can persist in the soil for extended periods of time.
[c]Ap = apple, Bl = blueberry, Cr = cranberry, Cu = cucumber, Pu = pumpkin, Wa = watermelon.
*Used by beekeepers within the hive for parasitic mite control.
†Relative risk different from 1 at the 95% confidence level.
NA indicates information that is not relevant to control diets.

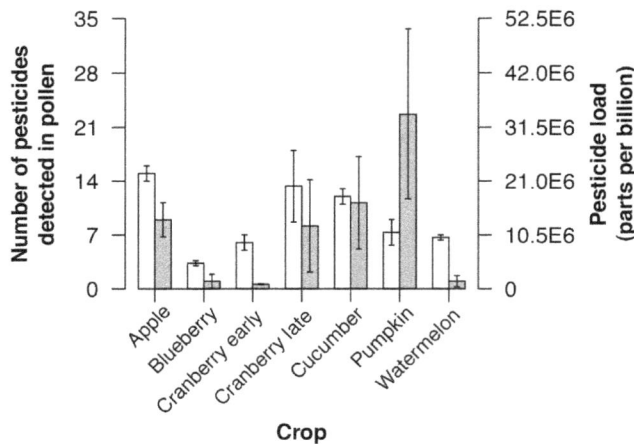

Figure 2. Pesticide diversity found in pollen samples, but not pesticide load, varied by crop. White bars show pesticide diversity, gray bars show pesticide load (mean ± se). Post-hoc testing found the following groups, where letters indicate statistically significant differences: apple a, b; blueberry c; cranberry_early d; cranberry_late b, d, e, f; cucumber e; pumpkin c, d, f; and watermelon d.

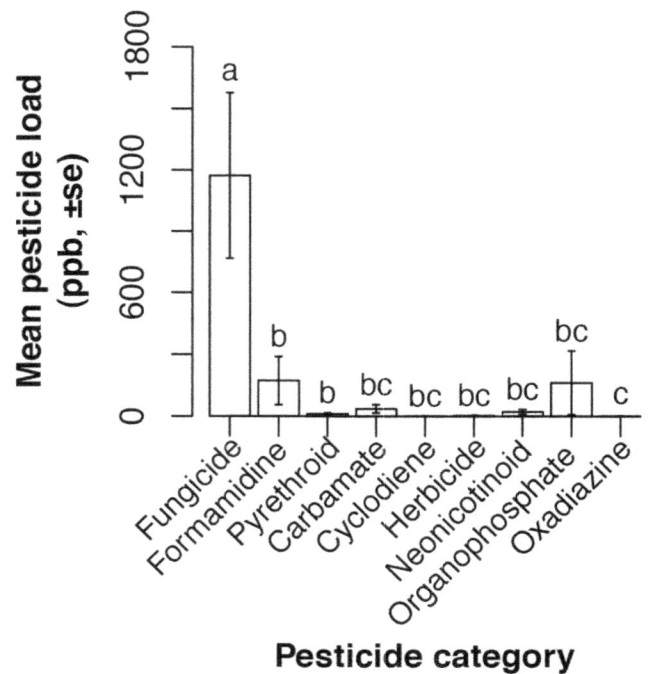

Figure 4. Load varied by pesticide category. Letters indicate statistically significant differences. The total load for each category is weighted by the number of chemicals in that category, to facilitate comparison across categories.

thought. *Nosema* infection was more than twice as likely (relative risk >2) in bees that consumed these fungicides than in bees that did not. Research on the sub-lethal effects of pesticides on honey bees has focused almost entirely on insecticides, especially neonicotinoids [54]. In our study, neonicotinoids entered the nest only via apple pollen. However, we found fungicides at high loads in our sampled crops. While fungicides are typically less lethal to bees than insecticides (see LD_{50} values in Table 2), these chemicals still have potential for lethal [55] and sub-lethal effects. Indeed, the fungicides chlorothalonil (found at high concentrations in our pollen samples) and myclobutanil increases gut cell mortality to the same degree as imidacloprid [56], an insecticide with numerous sub-lethal effects (e.g. [21,57]). Exposure to fungicides can also

make bees more sensitive to acaricides, reducing medial lethal doses [58]. In our study, consuming pollen with higher fungicide loads increased bees' susceptibility to *Nosema* infection. This result is likely driven by chlorothalonil loads. The pesticide with the highest relative risk was the fungicide pyraclostrobin. Bees that consumed pollen containing pyraclostrobin were almost three times as likely (relative risk = 2.85, 95% CI 2.16–3.75; Table 2)

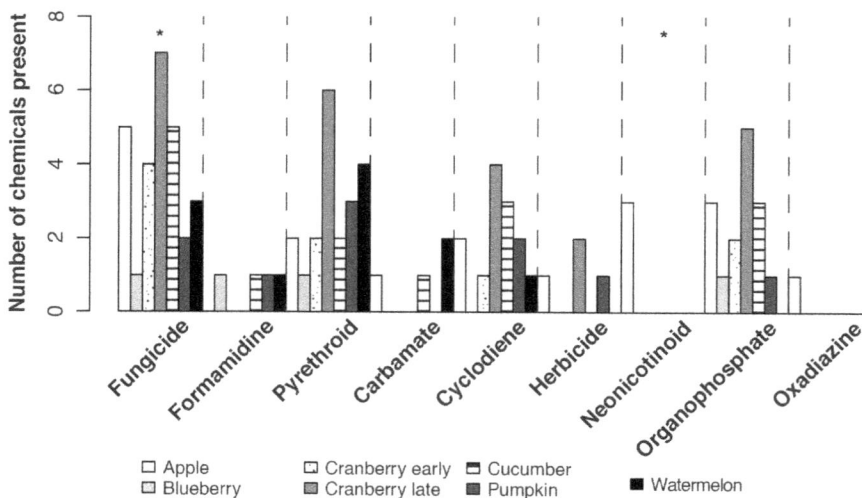

Figure 3. Fungicide and neonicotinoid diversities varied by crop. Bars show the total number of pesticides in each category found in each crop. Kruskal-Wallis test statistics comparing pesticide diversity between crops are: fungicides, $H_6 = 16.1$, $p = 0.01$; cyclodienes, $H_6 = 6.9$, $p = 0.33$; neonicotinoids, $H_6 = 17.9$, $p = 0.007$; organophosphates, $H_6 = 14.3$, $p = 0.03$; pyrethroids, $H_6 = 7.8$, $p = 0.26$. We only compared pesticide diversities for categories containing at least three chemicals. Sequential Bonferroni adjusted critical values are: 0.01, 0.0125, 0.0167, 0.025, 0.05. A * indicates that the total number of pesticides varied between crops within that pesticide category.

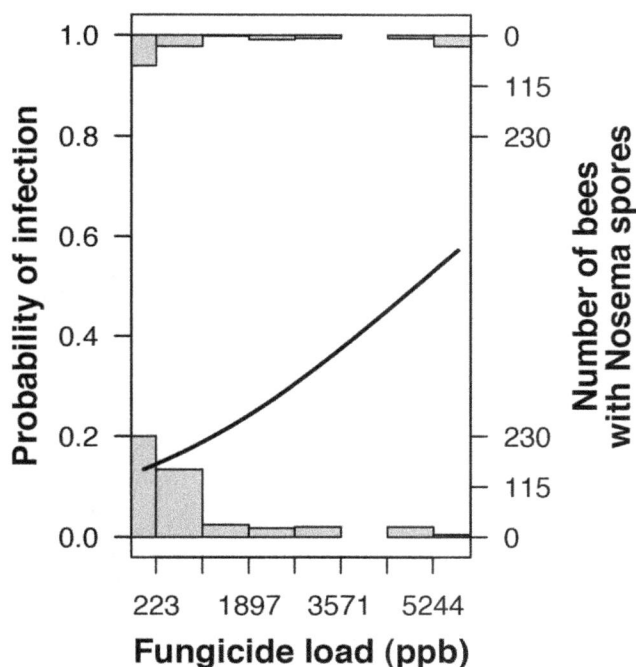

Figure 5. Probability of *Nosema* infection increased with fungicide load in consumed pollen. Histograms show the number of bees with (top) and without (bottom) *Nosema* spores as a function of the fungicide load in the pollen they were fed. The curve shows the predicted probability of *Nosema* infection.

than bees consuming pollen without this chemical to become infected after *Nosema* exposure. Our results show the necessity of testing for sub-lethal effects of pesticides on bees, and advocate for testing more broadly than the insecticides that are the targets of most current research.

A similarly large increased risk of *Nosema* infection was associated with consumption of DMPF and fluvalinate, miticides applied by beekeepers to help control the highly-destructive *Varroa* mite [3]. The path from in-hive application of these miticides to pollen on foragers returning to the hive is unclear. An increasingly popular practice, rotating combs out of hives to remove accumulated pesticides, is expected to reduce miticide levels in hives, and will hopefully decrease spread of these chemicals to the environment. Potential extra-nest sources, however, would slow efforts to reduce miticide accumulation and slow the development of resistance to these chemicals.

Insecticide relative risk values showed an interesting pattern: directional separation by insecticide family. Within a family, relative risk values significantly different than one were almost all in the same direction. The formamidine (DMPF) and two of the three the pyrethroids (bifenthrin and fluvalinate, but not esfenvalerate) were associated with an increased risk of *Nosema* infection. The carbamate (carbaryl), all neonicotinoids (acetamiprid, imidacloprid and thiacloprid), organophosphates (coumaphos, diazinon and phosmet) and the oxadiazine (indoxacarb) were associated with reduced risk of *Nosema* infection. Esfenvalerate and coumaphos have previously been found to be associated with apiaries without Colony Collapse Disorder [59]. These patterns suggest that insecticides' modes of action have differential effects on honey bee immune functioning. Because of the relatively small number of pesticides we found in each insecticide family,

however, additional sampling is necessary to determine how robust this pattern is.

The large numbers of pesticides found per sample and the high concentrations of some pesticides are concerning. First, two pollen samples contained one pesticide each at a concentration higher than the median lethal dose. Esfenvalerate ($LD_{50} = 0.13$ ppm) was measured at 0.216 ppm in pollen collected by bees in a cucumber field, and phosmet ($LD_{50} = 8.83$ ppm) at 14.7 ppm in one apple orchard. While the mean loads for these pesticides are well below their respective median lethal doses (0.0169 ppm for esfenvalerate, 0.7987 ppm for phosmet), our data indicate some bee colonies are being exposed to incredibly high levels of these chemicals. Second, research suggests that simultaneous exposure to multiple pesticides decreases lethal doses [58,60] or increases supersedure (queen replacement) rate [61]. Our pollen samples contained an average of nine different pesticides, ranging as high as 21 pesticides in one cranberry field. Thus published LD_{50} values may not accurately indicate pesticide toxicity inside a hive containing large numbers of pesticides. Research looking at additive and synergistic effects between multiple pesticides is clearly needed. Third, pesticides can have sub-lethal effects on development, reproduction, learning and memory, and foraging behavior. The mean and maximum imidacloprid loads in our samples (0.0028 and 0.0365 ppm, respectively) are higher than some published imidacloprid concentrations with sub-lethal effects on honey and bumble bees (0.001–0.0098 ppm [21,54,62]).

It is not surprising that total pollen collection varied by crop. Bee foraging activity levels vary with weather [63], thus outcomes of short-term measurements may be sensitive to temperature, cloud cover or humidity during data collection. Because we collected pollen samples from different parts of the country and on different days, weather conditions undoubtedly differed between crops. Crop flowering timing and landscape-level floral availability can also affect bee activity levels. We focused our analyses on variables less affected by these factors, such as the diversity of pollen types found in samples and the proportion of a sample that was from the target crop.

Our results are consistent with previously published pesticide analyses of pollen collected by honey bees or honey bee nest material [16,18,27]. The more intensive and geographically more diverse sampling of Mullin et al. [27] resulted in almost triple the number of pesticides we found, but the average number of pesticides per sample (7.1) is slightly lower than our 9.1. In our study and those listed above, pesticides applied by beekeepers to control hive pests were present in a large proportion of the samples, often in quantities higher than most of the pesticides that are applied to crops.

Our results combined with several recent studies of specific pesticides' effects on *Nosema* infection dynamics [13–15] indicate that a detrimental interaction occurs when honey bees are exposed to both pesticides and *Nosema*. Specific results vary, and may depend on the pesticide or dose used. For example, bees exposed to imidacloprid and *Nosema* can have lower spore counts than bees only infected with the pathogen but also exhibit hindered immune functioning [13]. Our study improves on previous methodologies by feeding pollen with real-world pesticide blends and levels that truly represents the types of exposure expected with pollination of agricultural crops. The significant increase in *Nosema* infection following exposure to the fungicides in pollen we found therefore indicates a pressing need for further research on lethal and sub-lethal effects of fungicides on bees. Given the diverse routes of exposure to pesticides we show, and increasing evidence that pesticide blends harm bees [16,18,58], there is a pressing need for further research on the mechanisms underlying pesticide-pesticide and pesticide-disease synergistic effects on honey bee health.

Supporting Information

Figure S1   Pesticide loads did not differ by crop for any pesticide category. Kruskal-Wallis test statistics comparing pesticide loads between crops are: fungicides, $H_6 = 10.6$, $p = 0.10$; herbicides, $H_6 = 8.3$, $p = 0.22$; carbamates, $H_6 = 13.4$, $p = 0.04$; cyclodienes, $H_6 = 6.7$, $p = 0.35$; formamidines, $H_6 = 13.6$, $p = 0.03$; neonicotinoids, $H_6 = 17.8$, $p = 0.007$; organophosphates, $H_6 = 14.5$, $p = 0.02$; oxadiazines, $H_6 = 11.3$, $p = 0.08$; pyrethroids, $H_6 = 9.6$, $p = 0.14$. Sequential Bonferroni adjusted critical values are: 0.0055, 0.0063, 0.0071, 0.0083, 0.01, 0.0125, 0.0167, 0.025, 0.06.

Table S1   Plant sources of pollens collected by bees placed in seven crops.

Acknowledgments

We thank David Hackenberg and David Mendes for letting us work with their bee hives, John Baker and Rob Snyder for field assistance, Roger Simonds for pesticide identification, and Vic Levi and Nathan Rice for assistance with *Nosema* assays.

The views expressed in this article are those of the authors and do not necessarily represent the policies or positions of the US Department of Agriculture (USDA).

Author Contributions

Conceived and designed the experiments: JSP RR DV. Performed the experiments: JSP MA JS DV. Analyzed the data: EML DV. Wrote the paper: JP EML DV.

References

1. Klein A-M, Vaissiere BE, Cane JH, Steffan-Dewenter I, Cunningham SA, et al. (2007) Importance of pollinators in changing landscapes for world crops. Proceedings of the Royal Society B-Biological Sciences 274: 303–313.
2. vanEngelsdorp D, Meixner MD (2010) A historical review of managed honey bee populations in Europe and the United States and the factors that may affect them. Journal of Invertebrate Pathology 103: S80–S95.
3. vanEngelsdorp D, Caron D, Hayes J, Underwood R, Henson M, et al. (2012) A national survey of managed honey bee 2010–11 winter colony losses in the USA: results from the Bee Informed Partnership. Journal of Apicultural Research 51: 115–124.
4. van der Zee R, Pisa L, Andonov S, Brodschneider R, Charriere JD, et al. (2012) Managed honey bee colony losses in Canada, China, Europe, Israel and Turkey, for the winters of 2008–9 and 2009–10. Journal of Apicultural Research 51: 91–114.
5. Aizen MA, Garibaldi LA, Cunningham SA, Klein AM (2009) How much does agriculture depend on pollinators? Lessons from long-term trends in crop production. Annals of Botany 103: 1579–1588.
6. Gallai N, Salles J-M, Settele J, Vaissiere BE (2009) Economic valuation of the vulnerability of world agriculture confronted with pollinator decline. Ecological Economics 68: 810–821.
7. NRC (2007) Status of pollinators in North America. Washington, DC: The National Academies Press. 312 p.
8. Biesmeijer JC, Roberts SPM, Reemer M, Ohlemuller R, Edwards M, et al. (2006) Parallel declines in pollinators and insect-pollinated plants in Britain and the Netherlands. Science 313: 351–354.
9. vanEngelsdorp D, Speybroeck N, Evans JD, Nguyen BK, Mullin C, et al. (2010) Weighing risk factors associated with bee colony collapse disorder by classification and regression tree analysis. Journal of Economic Entomology 103: 1517–1523.
10. Cornman RS, Tarpy DR, Chen Y, Jeffreys L, Lopez D, et al. (2012) Pathogen webs in collapsing honey bee colonies. PloS ONE 7: e43562.
11. Foley K, Fazio G, Jensen AB, Hughes WOH (2012) Nutritional limitation and resistance to opportunistic *Aspergillus* parasites in honey bee larvae. Journal of Invertebrate Pathology 111: 68–73.
12. Alaux C, Ducloz F, Crauser D, Le Conte Y (2010) Diet effects on honeybee immunocompetence. Biology Letters 6: 562–565.
13. Alaux C, Brunet J-L, Dussaubat C, Mondet F, Tchamitchan S, et al. (2010) Interactions between *Nosema* microspores and a neonicotinoid weaken honeybees (*Apis mellifera*). Environmental Microbiology 12: 774–782.
14. Pettis JS, vanEngelsdorp D, Johnson J, Dively G (2012) Pesticide exposure in honey bees results in increased levels of the gut pathogen *Nosema*. Naturwissenschaften 99: 153–158.
15. Vidau C, Diogon M, Aufauvre J, Fontbonne R, Vigues B, et al. (2011) Exposure to sublethal doses of fipronil and thiacloprid highly increases mortality of honeybees previously infected by *Nosema ceranae*. PLoS ONE 6: e21550.
16. Wu JY, Smart MD, Anelli CM, Sheppard WS (2012) Honey bees (*Apis mellifera*) reared in brood combs containing high levels of pesticide residues exhibit increased susceptibility to *Nosema* (Microsporidia) infection. Journal of Invertebrate Pathology 109: 326–329.
17. Desneux N, Decourtye A, Delpuech J-M (2007) The sublethal effects of pesticides on beneficial arthropods. Annual Review of Entomology 52: 81–106.
18. Wu JY, Anelli CM, Sheppard WS (2011) Sub-lethal effects of pesticide residues in brood comb on worker honey bee (*Apis mellifera*) development and longevity. PLoS ONE 6: e14720.
19. Eiri D, Nieh JC (2012) A nicotinic acetylcholine receptor agonist affects honey bee sucrose responsiveness and decreases waggle dancing. Journal of Experimental Biology 215: 2022–2029.
20. Gregorc A, Evans JD, Scharf M, Ellis JD (2012) Gene expression in honey bee (*Apis mellifera*) larvae exposed to pesticides and Varroa mites (*Varroa destructor*). Journal of Insect Physiology 58: 1042–1049.
21. Whitehorn PR, O'Connor S, Wackers FL, Goulson D (2012) Neonicotinoid pesticide reduces bumble bee colony growth and queen production. Science 336: 351–352.
22. Boncristiani H, Underwood R, Schwarz R, Evans JD, Pettis J, et al. (2012) Direct effect of acaricides on pathogen loads and gene expression levels in honey bees *Apis mellifera*. Journal of Insect Physiology 58: 613–620.
23. Cameron SA, Lozier JD, Strange JP, Koch JB, Cordes N, et al. (2011) Patterns of widespread decline in North American bumble bees. Proceedings of the National Academy of Sciences 108: 662–667.
24. George PJE, Ambrose DP (2004) Impact of insecticides on the haemogram of *Rhynocoris kumarii* Ambrose and Livingstone (Hem., Reduviidae). Journal of Applied Entomology 128: 600–604.
25. Delpuech JM, Frey F, Carton Y (1996) Action of insecticides on the cellular immune reaction of *Drosophila melanogaster* against the parasitoid *Leptopilina boulardi*. Environmental Toxicology and Chemistry 15: 2267–2271.
26. Chaimanee V, Chantawannakul P, Chen Y, Evans JD, Pettis JS (2012) Differential expression of immune genes of adult honey bee (*Apis mellifera*) after inoculated by *Nosema ceranae*. Journal of Insect Physiology 58: 1090–1095.
27. Mullin CA, Frazier M, Frazier JL, Ashcraft S, Simonds R, et al. (2010) High levels of miticides and agrochemicals in North American apiaries: implications for honey bee health. PLoS ONE 5: e9754.
28. Chauzat MP, Faucon JP, Martel AC, Lachaize J, Cougoule N, et al. (2006) A survey of pesticide residues in pollen loads collected by honey bees in France. Journal of Economic Entomology 99: 253–262.
29. Johnson RM, Wen Z, Schuler MA, Berenbaum MR (2006) Mediation of pyrethroid insecticide toxicity to honey bees (Hymenoptera : Apidae) by cytochrome P450 monooxygenases. Journal of Economic Entomology 99: 1046–1050.
30. Higes M, Martin-Hernandez R, Botias C, Garrido Bailon E, Gonzalez-Porto AV, et al. (2008) How natural infection by *Nosema ceranae* causes honeybee colony collapse. Environmental Microbiology 10: 2659–2669.
31. Higes M, Martin-Hernandez R, Meana A (2010) *Nosema ceranae* in Europe: an emergent type C nosemosis. Apidologie 41: 375–392.
32. Farrar CL (1947) Nosema losses in package bees as related to queen supersedure and honey yields. Journal of Economic Entomology 40: 333–338.
33. Konietschke F (2011) nparcomp v. 1.0–1. R package.
34. Crompton CW, Wojtas WA (1993) Pollen grains of Canadian honey plants. Ottawa: Agriculture Canada. 228 p.
35. Hodges D (1964) Pollen grain drawings from the pollen loads of the honeybee. London: Bee Research Association. 51 p.
36. Kirk WDJ (2006) A colour guide to pollen loads of the honey bee. Cardiff, UK: International Bee Research Association. 54 p.
37. Kearns CA, Inouye DW (1993) Techniques for Pollination Biologists. Niwot, CO: University Press of Colorado. 583 p.
38. Wilson EE, Sidhu CS, LeVan KE, Holway DA (2010) Pollen foraging behaviour of solitary Hawaiian bees revealed through molecular pollen analysis. Molecular Ecology 19: 4823–4829.
39. Cane JH, Snipes S (2006) Characterizing floral specialization by bees: analytical methods and a revised lexicon for oligolecty. In: Waser NM, Ollerton J, editors. Plant-Pollinator Interactions: From Specialization to Generalization. Chicago: University of Chicago Press. 99–122.
40. Holm S (1979) A simple sequentially rejective multiple test procedure. Scandinavian Journal of Statistics 6: 65–70.
41. vanEngelsdorp D, Lengerich E, Spleen A, Dainat B, Cresswell J, et al. (2013) Standard epidemiological methods to understand and improve *Apis mellifera*

health. The COLOSS BEEBOOK: Volume II: Standard methods for *Apis mellifera* pest and pathogen research. Journal of Apicultural Research 52: 1–16.

42. de la Cruz Rot M (2005) Improving the presentation of results of logistic regression with R. Bulletin of the Ecological Society of America 86: 41–48.

43. Smart J, Sutherland WJ, Watkinson AR, Gill JA (2004) A new means of presenting the results of logistic regression. Bulletin of the Ecological Society of America 85: 100–102.

44. Stubben C, Milligan B, Nantel P (2012) popbio: Construction and analysis of matrix population models v. 2.4. R package.

45. Garibaldi LA, Steffan-Dewenter I, Winfree R, Aizen MA, Bommarco R, et al. (2013) Wild pollinators enhance fruit set of crops regardless of honey bee abundance. Science 339: 1608–1611.

46. Delaplane KS, Mayer DF (2000) Crop Pollination by Bees. New York: CABI Publishing. 344 p.

47. Kremen C (2008) Crop pollination services from wild bees. In: James RR, Pitts-Singer TL, editors. Bee Pollination in Agricultural Ecosystems. New York: Oxford University Press. 10–26.

48. Whidden TL (1996) The fidelity of commercially reared colonies of *Bombus impatiens* Cresson (Hymenoptera: Apidae) to lowbush blueberry in southern New Brunswick. The Canadian Entomologist 128: 957–958.

49. Choudhary A, Sharma DC (2008) Dynamics of pesticide residues in nectar and pollen of mustard (*Brassica juncea* (L.) Czern.) grown in Himachal Pradesh (India). Environmental Monitoring and Assessment 144: 143–150.

50. Krischik VA, Landmark AL, Heimpel GE (2007) Soil-applied imidacloprid is translocated to nectar and kills nectar-feeding *Anagyrus pseudococci* (Girault) (Hymenoptera : Encyrtidae). Environmental Entomology 36: 1238–1245.

51. Stoner KA, Eitzer BD (2012) Movement of soil-applied imidacloprid and thiamethoxam into nectar and pollen of squash (*Cucurbita pepo*). PLoS ONE 7: e39114.

52. Dively GP, Kamel A (2012) Insecticide residues in pollen and nectar of a cucurbit crop and their potential exposure to pollinators. Journal of Agricultural and Food Chemistry 60: 4449–4456.

53. Sampson BJ, Danka RG, Stringer SJ (2004) Nectar robbery by bees *Xylocopa virginica* and *Apis mellifera* contributes to the pollination of rabbiteye blueberry. Journal of Economic Entomology 97: 735–740.

54. Blacquiere T, Smagghe G, van Gestel CAM, Mommaerts V (2012) Neonicotinoids in bees: a review on concentrations, side-effects and risk assessment. Ecotoxicology 21: 973–992.

55. Hooven LA (2013) Effect of fungicides on development and behavior of honey bees. American Bee Research Conference. Hershey, PA.

56. Gregorc A, Ellis JD (2011) Cell death localization in situ in laboratory reared honey bee (*Apis mellifera* L.) larvae treated with pesticides. Pesticide Biochemistry and Physiology 99: 200–207.

57. Gill RJ, Ramos-Rodriguez O, Raine NE (2012) Combined pesticide exposure severely affects individual- and colony-level traits in bees. Nature 491: 105–108.

58. Johnson RM, Dahlgren L, Siegfried BD, Ellis MD (2013) Acaricide, fungicide and drug interactions in honey bees (*Apis mellifera*). PLoS ONE 8: e54092.

59. vanEngelsdorp D, Evans JD, Saegerman C, Mullin C, Haubruge E, et al. (2009) Colony Collapse Disorder: a descriptive study. PLoS ONE 4: e6481.

60. Iwasa T, Motoyama N, Ambrose JT, Roe RM (2004) Mechanism for the differential toxicity of neonicotinoid insecticides in the honey bee, *Apis mellifera*. Crop Protection 23: 371–378.

61. Drummond FA, Aronstein K, Chen YP, Eitzer B, Ellis J, et al. (2013) Colony losses in stationary apiaries related to pesticide contamination of pollen. American Bee Research Conference. Hershey, PA.

62. Laycock I, Lenthall KM, Barratt AT, Cresswell JE (2012) Effects of imidacloprid, a neonicotinoid pesticide, on reproduction in worker bumble bees (*Bombus terrestris*). Ecotoxicology 21: 1937–1945.

63. Danka RG, Sylvester HA, Boykin D (2006) Environmental influences on flight activity of USDA-ARS Russian and Italian stocks of honey bees (Hymenoptera : Apidae) during almond pollination. Journal of Economic Entomology 99: 1565–1570.

64. US EPA (1997) Fact Sheet for Azoxystrobin. Washington, DC: US EPA. 23 p.

65. US EPA (1999) Reregistration Eligibility Decision (RED) for Captan. Washington, DC: US EPA. 250 p.

66. US EPA (1999) Reregistration Eligibility Decision (RED) for Chlorothalonil. Washington, DC: US EPA. 337 p.

67. US EPA (1998) Pesticide Fact Sheet for Cyprodinil. Washington, DC: US EPA. 13 p.

68. OPP Pesticide Toxicity Database: Ecotox Results for Difenoconazole website. Available: http://www.ipmcenters.org/Ecotox/Details.cfm?RecordID = 10104. Accessed 2013 Jan 15.

69. Atkins EL (1988) RH-7592 technical: bee adult toxicity dusting test. Spring House, PA: Rohm and Haas Company. 6 p.

70. ECCO-Team (2012) Full Report on Pyraclostrobin. York, UK: Pesticides Safety Directorate. 208 p.

71. US EPA (2006) Reregistration Eligibility Decision for Pentachloronitrobenzene. Washington, DC: US EPA. 127 p.

72. Carleton J (1999) Carfentrazone-ethyl Herbicide Environmental Fate and Ecological Effects Assessment and Risk Characterization for a Section 3 for Use on Sweet Corn, Sorghum, and Rice. Washington, DC: US EPA. 25 p.

73. US EPA (1997) Reregistration Eligibility Decision (RED) for Pendimethalin. Washington, DC: US EPA. 239 p.

74. Atkins EL, Kellum D (1981) Effect of pesticides on agriculture: maximizing the effectiveness of honey bees as pollinators. 1981 Report of research to California Alfalfa Seed Production Research Board. Philadelphia, PA: FMC Corporation. 2 p.

75. US EPA (2004) Interim Reregistration Eligibility Decision (IRED) for Carbaryl. Washington, DC: US EPA. 305 p.

76. Cox C (1994) Insecticide fact sheet: cyfluthrin. Journal of Pesticide Reform 14: 28–34.

77. Gough HJ, Collins IG, Everett CJ, Wilkinson W (1984) PP321: acute contact and oral toxicity to honey bees (*Apis mellifera*). Wilmington, DE: ICI Americas Inc. 7 p.

78. US EPA (2008) Reregistration Eligibility Decision (RED) for Cypermethrin. Washington, DC: US EPA. 113 p.

79. Pesticide Management Education Program: Pesticide Information Profile: Deltamethrin website. Available: http://pmep.cce.cornell.edu/profiles/extoxnet/carbaryl-dicrotophos/deltamethrin-ext.html. Accessed 2013 Jan 15.

80. US EPA (2004) Interim Reregistration Eligibility Decision (IRED) for Diazinon. Washington, DC: US EPA. 134 p.

81. Hoxter KA, Smith GJ (1990) H-#18, 151 (Technical Asana): an acute contact toxicity study with the honey bee. Wildlife International Ltd. 4 p.

82. US EPA (2005) Reregistration Eligibility Decision (RED) for Tau-fluvalinate. Washington, DC: US EPA. 85 p.

83. Decourtye A, Lacassie E, Pham-Delegue MH (2003) Learning performances of honeybees (*Apis mellifera* L) are differentially affected by imidacloprid according to the season. Pest Management Science 59: 269–278.

84. Abel S (2005) Environmental fate & effects division risk assessment for proposed new users of indoxacarb on grapes, fire ants, mole crickets, alfalfa, peanuts, soybeans. Washington, DC: US EPA. 263 p.

85. Johansen CA, Mayer DF (1990) Pollinator Protection: A Bee & Pesticide Handbook. Cheshire, CN: Wicwas Press. 212 p.

86. US EPA (1998) Reregistration Eligibility Decision (RED) for Methomyl. Washington, DC: US EPA. 266 p.

87. US EPA (1992) R.E.D. Facts for Heptachlor. Washington, DC: US EPA. 4 p.

7

The Synthetic Amphipathic Peptidomimetic LTX109 Is a Potent Fungicide That Disturbs Plasma Membrane Integrity in a Sphingolipid Dependent Manner

Rasmus Bojsen[1][9], **Rasmus Torbensen**[2][9], **Camilla Eggert Larsen**[2], **Anders Folkesson**[1,3], **Birgitte Regenberg**[2]*

1 Department of Systems Biology, Technical University of Denmark, Kgs. Lyngby, Denmark, 2 Department of Biology, University of Copenhagen, Copenhagen, Denmark, 3 Section for Bacteriology, Pathology and Parasitology, National Veterinary Institute, Frederiksberg C, Denmark

Abstract

The peptidomimetic LTX109 (arginine-tertbutyl tryptophan-arginine-phenylethan) was previously shown to have antibacterial properties. Here, we investigated the activity of this novel antimicrobial peptidomimetic on the yeast *Saccharomyces cerevisiae*. We found that LTX109 was an efficient fungicide that killed all viable cells in an exponentially growing population as well as a large proportion of cells in biofilm formed on an abiotic surface. LTX109 had similar killing kinetics to the membrane-permeabilizing fungicide amphotericin B, which led us to investigate the ability of LTX109 to disrupt plasma membrane integrity. *S. cerevisiae* cells exposed to a high concentration of LTX109 showed rapid release of potassium and amino acids, suggesting that LTX109 acted by destabilizing the plasma membrane. This was supported by the finding that cells were permeable to the fluorescent nucleic acid stain SYTOX Green after a few minutes of LTX109 treatment. We screened a haploid *S. cerevisiae* gene deletion library for mutants resistant to LTX109 to uncover potential molecular targets. Eight genes conferred LTX109 resistance when deleted and six were involved in the sphingolipid biosynthetic pathway (*SUR1*, *SUR2*, *SKN1*, *IPT1*, *FEN1* and *ORM2*). The involvement of all of these genes in the biosynthetic pathway for the fungal-specific lipids mannosylinositol phosphorylceramide (MIPC) and mannosyl di-(inositol phosphoryl) ceramide (M(IP)$_2$C) suggested that these lipids were essential for LTX109 sensitivity. Our observations are consistent with a model in which LTX109 kills *S. cerevisiae* by nonspecific destabilization of the plasma membrane through direct or indirect interaction with the sphingolipids.

Editor: Christopher Beh, Simon Fraser University, Canada

Funding: Funding provided by the Danish Agency for Science Technology and Innovation (FTP 10-084027). The funders had no role in study design, data collection and analysis, decision to publish, or preparation of the manuscript.

Competing Interests: The authors have declared that no competing interests exist.

* E-mail: bregenberg@bio.ku.dk

[9] These authors contributed equally to this work.

Introduction

Infections caused by pathogenic yeast such as *Candida spp.* affect a large number of immunosuppressed patients and are an increasing medical problem [1,2]. Fungal infections are currently treated with one of four major classes of antifungals. Azoles target ergosterol synthesis [3], polyenes bind to ergosterol in the cell membrane and form pores [4,5], echinocandins inhibit cell wall synthesis [6], and 5-fluorocytosine interferes with protein and DNA synthesis [7].

Decreased susceptibility to the most frequently used antifungal, fluconazole, has recently been reported, and the number of nonsusceptible *C. glabrata* isolates from humans is increasing [8,9]. Resistance towards 5-fluorocystosine is also rapidly developing [10]. Polyenes can be toxic [11] and echinocandins have a narrow spectrum of activity [12]. An additional complication in the treatment of nosocomial fungal infections is the frequent formation by fungi of sessile communities called biofilms in association with medical implants [13]. Limited nutrient access leads to slow-growing, antibiotic tolerant cells in biofilms that can serve as a reservoir for infection [14,15]. Most systemic antifungals are fungistatic against yeasts, so they are primarily effective against actively growing cells and have poor activity against cells in biofilms.

The limited number of antifungal classes and drugs with fungicidal properties raises the need for novel drugs with activity against slow-growing and biofilm-forming pathogenic fungi [16,17]. Antimicrobial peptides (AMPs) and modified forms of AMPs offer an attractive alternative to conventional antifungal drugs. AMPs are cationic and amphipathic peptides of 12–50 amino acids that are produced by species in almost every kingdom and phylum of life [18]. The amphipathic structure of AMPs suggests that they might have targets that are different from conventional antifungals [19,20]. The high degradation rate of many natural AMPs can be circumvented by backbone and side chain alterations that create structural analogs that mimics natural peptides [21]. A number of synthesized peptidomimetics have *in vitro* antifungal activity, making these compounds attractive candidates for novel antifungal drugs [22–24].

We tested the antifungal activity of the short, antibacterial peptidomimetic LTX109 (arginine-tertbutyl tryptophan-arginine-phenylethan). LTX109 is based on an Arg-Trp-Arg sequence found in the AMP bovine lactoferrin and was originally developed as an antibacterial [25–27].

We used killing kinetics to describe the antimicrobial effect of LTX109 and investigated its mode of action by measuring transport of H^+, K^+, amino acids and a fluorescent dye across the cell membrane. To uncover potential molecular targets that would explain the fungicidal activity of LTX109, we screened a *Saccharomyces cerevisiae* gene deletion library for mutants resistant to LTX109. Most mutations that led to LTX109 resistance were in genes involved in the synthesis of the sphingolipids mannosyl-inositol phosphorylceramide, MIPC, and mannosyl di-(inositol phosphoryl) ceramide, $M(IP)_2C$. These results indicate that $M(IP)_2C$ and/or MIPC in the plasma membrane are essential for the action of LTX109.

Materials and Methods

Strains, growth media and antifungal drugs

The S288c *S. cerevisiae* strain M3750 (*MATa ura3-52*) [28] was used as the reference strain in all experiments unless otherwise indicated, while the barcode-tagged deletion-mutant library was from Johnston and coworkers [29]. Σ1278b (10560-2B; *MATa ura3-52 leu2:hisG his3:hisG*) was used for biofilm susceptibility experiments [30]. Complex YPD medium [31] was used in all experiments except for amino acid release and biofilm where cells were grown in synthetic complete medium [31]. LTX109 (Lytixar; LytixBiopharmaAS, Oslo, Norway) and amphotericin B (Sigma) were dissolved in water and stock solutions were kept at −20°C.

Broth microdilution minimal inhibitory concentrations

Minimal inhibitory concentration (MIC) values were measured under static conditions in polystyrene microtiter plates. Two-fold dilution series of antifungal drug were prepared in fresh YPD medium and distributed to microtiter-plate wells. Overnight cultures of wild type (WT) S288c were diluted and added to antifungal-containing wells to a final concentration of 2×10^5 cells/ml. Growth inhibition was recorded with absorbance at 600 nm after 24 hours at 30°C. The lowest drug concentration resulting in 90% growth inhibition was the MIC. MIC values of LTX109 were determined three times with triplicate measurements, while MIC values of amphotericin B was determined once with triplicate measurements.

Killing kinetics

Overnight cultures of *S. cerevisiae* were diluted in fresh, preheated YPD to 4×10^5 cells/ml and incubated at 30°C with aeration. Exponential growth phase cells were challenged with LTX109 or amphotericin B at concentrations that were five times the MIC. Control samples were treated with water to ensure that cells applied in the time-kill experiment were in exponential growth phase. Samples were taken at the indicated time points, diluted 10-fold, and plated on YPD agar to determine colony forming units (CFUs). The time-kill experiment was conducted in triplicates.

Acidification assay

Glucose-induced acidification was measured as previously described [32] with modifications. Exponentially growing *S. cerevisiae* cells were washed and resuspended in sterile water to a final concentration of 10^8 cells/ml. Cells were subsequently incubated with LTX109 (100 µg/ml) or water (control) for 10 minutes before the assay was initiated by addition of glucose to a final concentration of 2% (w/v). The assay was conducted in triplicate at room temperature with continuous magnetic stirring. The assay was stopped by sampling at indicated time points, followed by immediate centrifugation ($2000 \times g$ for two minutes). pH of the resulting supernatants was measured and changes in extracellular H^+ concentration were calculated by applying the obtained values to the equation $pH = -\log [H^+]$.

Potassium release

Exponentially growing *S. cerevisiae* were harvested and resuspended in sterile water as described above. The potassium release assay was initiated by addition of LTX109 to a final concentration of 10 times the MIC. Samples treated with water instead of LTX109 served as control. The assay was stopped by centrifugation of samples ($13,000 \times g$ for 1 min) at indicated time points. Supernatants were transferred to sterile microtubes for spectrometric analysis. Potassium concentrations were measured with a FLM3 flame photometer (Radiometer). A standard concentration curve was generated from diluted S3336 urine flame standard (Radiometer). For spectrometric analysis, 5 µl of sample was added to 1000 µl of S3336 lithium solution (Radiometer). Experiments were carried out in triplicates at room temperature.

SYTOX Green uptake

SYTOX Green uptake was measured as previously described [33] with modifications. Exponentially growing *S. cerevisiae* cells were centrifuged, washed and suspended in 5 µM SYTOX Green (Life Technologies) to a final concentration of 10^8 cells/ml. LTX109 or water (control) was added to cell suspensions and SYTOX Green uptake was recorded microscopically after 4, 8, 16, 32, 64 and 128 minutes. Fluorescence was recorded with a Nikon Eclipse (Tokyo, Japan) fluorescence microscope equipped with a F36–525 EGFP HC-filter set (AHF Analysentechnik). Experiments were carried out at room temperature. SYTOX green uptake upon LTX109 treatment was observed in three independent experiments.

Amino acid release

Exponentially growing *S. cerevisiae* were harvested and suspended in sterile water or water with 10 times the MIC of LTX109 to a final concentration of 2×10^6 cells/ml. Loss of free amino acids from cells was recorded at room temperature after 16 minutes LTX109 exposure by instant centrifugation and subsequent HPLC of the cell free supernatant. Amino acids were detected and quantified by reverse-phase HPLC using an LKB-Alpha Plus amino-acid analyzer and a mixture of L-α-amino acids, 1 nmol each, as standards. The experiment was repeated three times.

Identification of LTX109-resistant mutants

Haploid knockout mutants of approximately 4000 nonessential genes in the S288c deletion mutant library [29] were pooled. About 10^6 cells from the mutant pool were transferred to YPD agar containing 10 times the MIC of LTX109. After 72 hours at 30°C, 17 colonies were picked from the LTX109 plates. LTX109-resistant clones were identified by PCR amplification and Sanger sequencing of the unique barcode tag of each mutant. PCR templates were DNA from clonal isolates of LTX109-resistant mutants. Primers were 5′-GATGTCCACGAGGTCTCT and 5′-CTGCAGCGAGGAGCCGTAAT. Gene deletions were identified using barcode sequences in the Stanford SGD deletion database.

Spot test

Exponentially growing *S. cerevisiae* cells were diluted to 10^7 cells/ml and 6 µl of a 10-fold dilution series was spotted on YPD agar and YPD agar containing 10 times the LTX109 MIC. Plates were incubated for 24 hours at 30°C and growth results were recorded.

Biofilm susceptibility

S. cerevisiae (Σ1278*b*) cells were grown in Lab-Tek™ Chamber Slide™ System; Permanox® (NUNC, Denmark) [34,35] in 1 ml synthetic complete medium. Cells were initially allowed to form biofilm for 12 hours before LTX109 was added for 5 hour in a concentration of 0 or 70 µg/ml. The biofilm was subsequently stained 15 minutes with Syto 9 (Invitrogen, Irvine, CA) for life cell staining and propidium iodine for dead cell staining before confocal laser scanning microscopy (CLSM). Imaging was carried out using a 63x/0.95NA water immersion lens. CLSM was performed with a Zeiss LSM510 microscope. Staining of biofilm treated with LTX109 was repeated in four independent experiments.

Results

Fungicidal properties of LTX109

We tested the antifungal properties of the peptidomimetic LTX109 on the yeast *S. cerevisiae* using microdilution. LTX109 had antifungal activity against *S. cerevisiae* at 2×10^5 cells/ml with a MIC value of 8 µg/ml compared to 2 µg/ml amphotericin B. Assay to determine the killing kinetics of LTX109 against *S. cerevisiae* revealed rapid and efficient fungicidal properties resulting in a 3-log reduction in viable cells within one hour, while amphotericin B required 90 minutes to achieve a similar fungicidal effect when using drug concentrations in multiples of MIC (Fig. 1). Additionally, LTX109 reduced the yeast population to the detection limit within only 2 hours, an effect that was not achieved by amphotericin B in the first 3 hours of exposure.

Exposure to LTX109 disrupts plasma membrane integrity

The speed with which LTX109 killed *S. cerevisiae* suggested that the compound was acting directly on the plasma membrane. To investigate the effect of LTX109 on plasma membrane integrity,

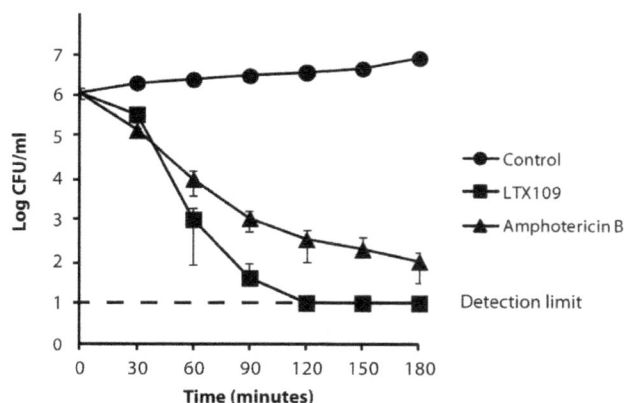

Figure 1. Fungicidal properties of LTX109 and amphotericin B. Time-kill kinetics of exponentially growing yeast cells exposed to water (circles) or five times the MIC of LTX109 (40 µg/ml) (squares) or amphotericin B (10 µg/ml) (triangles). Viability was examined every half hour as CFUs. Each data point is the average of three individual measurements ± standard deviation.

we measured H⁺ efflux, ion loss, loss of amino acids and uptake of the fluorophore SYTOX Green across the plasma membrane.

Yeast cells treated with glucose acidify their surroundings primarily by active transport of H⁺ by the plasma membrane H⁺-ATPase [36]. We found that glucose-induced acidification was completely absent when cells were treated with LTX109 for 10 minutes before glucose addition (Fig. 2A). These results suggested that LTX109 decoupled the plasma membrane potential directly or indirectly by inhibition of e.g. ATP synthesis.

We next tested the loss of potassium from cells treated with LTX109. Potassium release occurred immediately and increased during the first 16 minutes of exposure to LTX109, reaching a steady state that was more than four times higher than the maximum of the untreated control (Fig. 2B). Much of the K⁺ that was lost was detected within the first two minutes of challenge with a high LTX109 concentration. These results suggested that LTX109 acted by direct interaction with and disturbance of the plasma membrane rather than through indirect inhibition of metabolism or another intracellular pathway.

To investigate if LTX109 treatment also led to loss of other small molecules, cells were treated with LTX109 for 16 minutes and free amino acids measured in the extract. Yeast cells treated with LTX109 lost substantial amounts of at least 14 different amino acids whereas cells treated with water only leaked aspartate (Fig. 3). The loss corresponds well to the pool of intracellular amino acids found in other experiments [37], suggesting that most if not all free amino acids are depleted from cells treated with LTX109.

We finally investigated if the membrane potential decoupling was a consequence of plasma membrane destabilization by monitoring the uptake of the 600-Dalton nucleic acid stain SYTOX Green. SYTOX is an inorganic compound that fluoresces upon DNA binding and SYTOX Green can only enter a cell and fluoresce if the plasma membrane is compromised [38]. We found that untreated cells were impermeable to SYTOX Green, while cells treated with LTX109 became permeable. The dye was visible in the nucleus of LTX109-treated cells after only eight minutes (Fig. 2C) and fluorescence increased with LTX109 exposure time.

Defects in sphingolipid synthesis lead to LTX109 resistance

To gain further insight into the LTX109 mode of action, we screened a collection of haploid *S. cerevisiae* knockout mutants for LTX109 resistance. We isolated 17 mutants that were resistant to LTX109 at 10 times the MIC on YPD agar. Eight genes conferred LTX109 resistance when deleted (Table 1; Fig. 4A). Six of the identified genes (*SUR1*, *FEN1*, *SUR2*, *IPT1*, *SKN1*, *ORM2*) were involved in the biosynthesis of sphingolipids, which are a major plasma membrane component. Fen1p and Sur2p are involved in synthesis of ceramides, which are precursors for inositol phosphoceramide (IPC), the first complex sphingolipid in the synthesis pathway [39] (Fig 4B). Fen1p elongates long-chain fatty acids that are linked to a sphingoid base to form ceramides [40] and Sur2p hydroxylates dihydrosphingosine (DHS) to form phytosphingosine (PHS) [41], which is the most abundant sphingoid base in yeast ceramides [39]. Sur1p mannosylates IPC to form the intermediate sphingolipid mannose inositol phosphoceramide (MIPC) [42] and Skn1p and Ipt1p have similar functions in the biosynthesis of the terminal sphingolipid mannosyl di-inositol phosphorylceramide (M(IP)₂C) [43]. Orm2p is a regulator of the sphingolipid biosynthesis that links the biosynthesis to the regulatory Target Of Rapamycin pathway [44]. Mutants that fail to activate Orm2p have reduced levels of sphingolipids as do *fen1*, *sur1*, *ipt1* and *skn1*

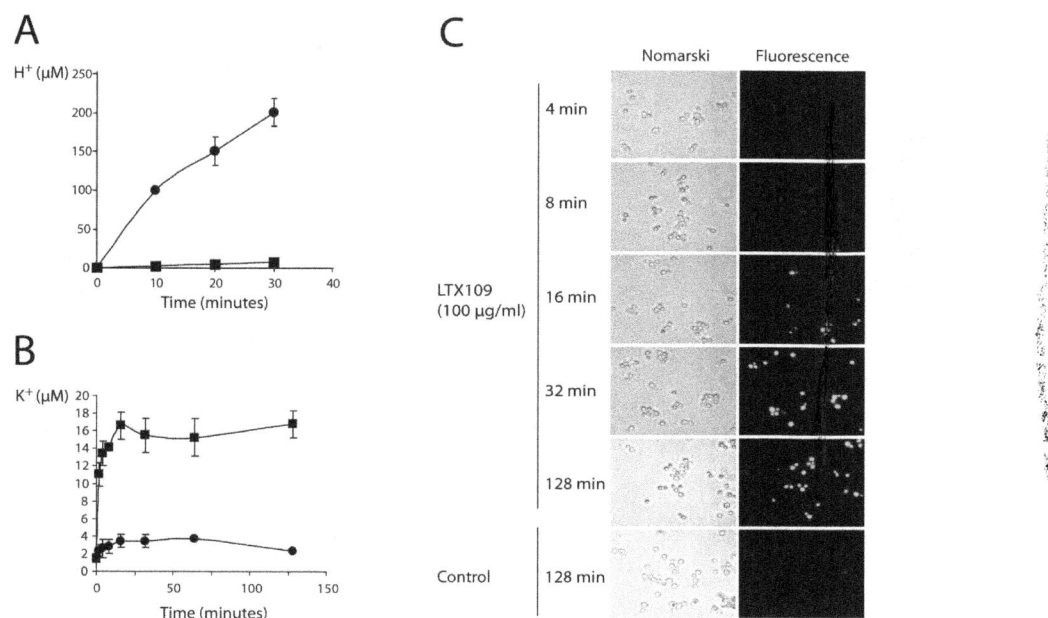

Figure 2. Transport of H⁺, K⁺ and a fluorescent dye by cells treated with LTX109. (A) Glucose-induced acidification of medium by yeast cells. Exponentially growing *S. cerevisiae* was washed and suspended in sterile water and exposed to 100 μg/ml LTX109 (squares) or water (circles) before glucose addition at time zero. Medium pH was measured and H⁺ concentration calculated from pH = −log [H⁺]. Each data point is the average of three individual measurements with standard deviations as error bars. (B) Potassium release from yeast cells. Exponentially growing yeast cells were washed, resuspended in water, and challenged with 100 μg/ml LTX109 (squares) or water (circles) at time zero. Potassium release was measured using flame atomic absorption spectrometry in binary increasing intervals. Each data point is the average of three individual measurements ± standard deviation. (C) Nomarski (left) and fluorescent (right) microscopy of SYTOX Green-stained yeast cells. Exponential growing cells were exposed to 100 μg/ml LTX109 and SYTOX Green uptake was monitored. Cells treated with SYTOX Green and 0 μg/ml LTX109 served as control. SYTOX green uptake upon LTX109 treatment was observed in three independent experiments.

mutants [40,43,45,46], suggesting a role of sphinolipids in sensitivity to LTX109.

One LTX109-resistant mutant was affected in the *YSP2* gene, which is involved in apoptosis, and another was affected in *OPI9*.

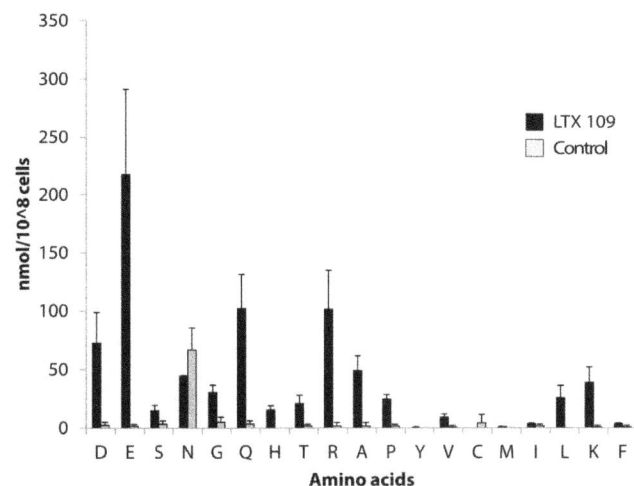

Figure 3. Efflux of amino acids from cells treated with LTX109. Exponentially growing yeast cells were washed, resuspended in water, and challenged with 70 μg/ml LTX109 (black bars) or water (grey bars) for 16 minutes. Amino acids (one letter code) in the extracellular medium were subsequently measured by HPLC. Each data point is the average of three individual measurements ± standard deviation.

OPI9 has an unknown function but partly overlaps with *VRP1*, which encodes an actin-associated protein with a role in actin filament organization. The *opi9* mutant therefore also has a partial deletion of *VRP1*, so the LTX109-resistance phenotype could be caused by loss of Vrp1p activity. Resistance of each mutant was confirmed by spot-testing diluted yeast suspensions on YPD agar containing LTX109 (Fig. 4A). Five of the mutants affected in sphingolipid biosynthesis showed similar, high resistance towards LTX109 (*sur1*, *fen1*, *sur2*, *ipt1* and *skn1*).

LTX109 efficiently kill S. cerevisiae growing as biofilm

Because S288c is incompetent of biofilm growth [47,48] we used the Σ1278*b* strain background to test the antifungal activity of LTX109 against *S. cerevisiae* biofilm. To visualize the antifungal properties of LTX109, we used CLSM in combination with Syto 9 DNA viability stain and propidium iodide that only penetrates damaged cell membranes. Intermediate (12 h) *S. cerevisiae* biofilm grown in batch culture slides were treated with 10 times MIC LTX109 for 5 hours before LIVE/DEAD staining and CLSM (Fig. 5). The LTX109 treatment killed the majority of the biofilm population as indicated by uptake and staining of dead cells with propidium iodide (Fig. 5), suggesting that LTX109 is also an efficient anti-biofilm agent in addition to its fungicidal activity against planktonic cells in exponential growth phase.

Discussion

This study demonstrated the fungicidal activity of the peptidomimetic LTX109. Antimicrobial peptidomimetics are peptide-like compounds, of which most are bactericides [49–53].

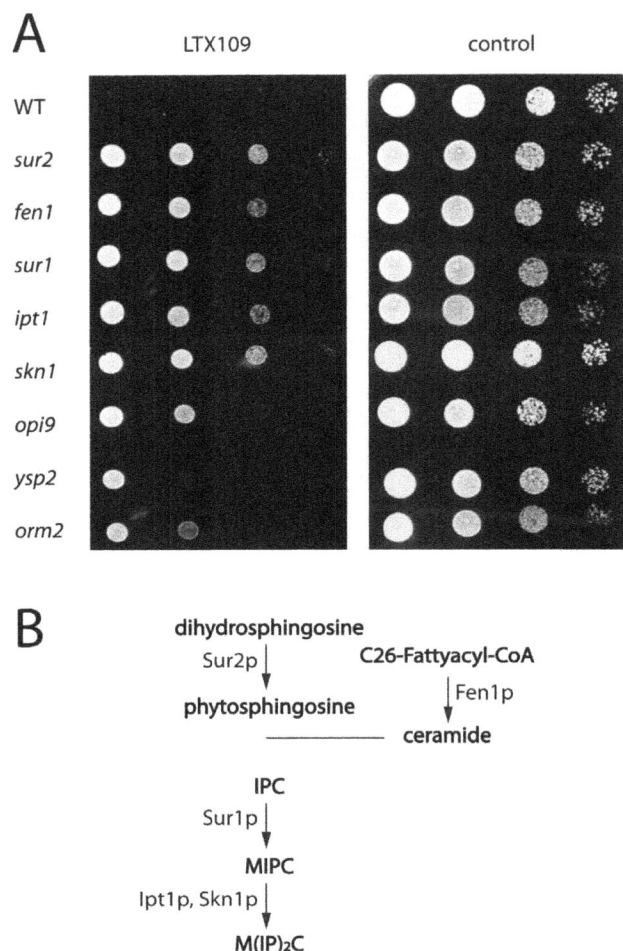

Figure 4. Mutants in sphingolipid biosynthesis are resistant to LTX109. (A) Spot test of wild type (WT) *S. cerevisiae* and eight deletion mutants identified by screening a deletion collection for LTX109 resistance. Exponentially growing yeast was resuspended in YPD to 10^7 cells/ml and serially diluted 1:10. Aliquots (6 μl) were spotted on solid YPD plates containing 70 μg/ml LTX109 (left panel), and without LTX109 (right panel). (B) *S. cerevisiae* sphingolipid biosynthetic pathway.

LTX109 was previously shown to have bactericidal activity [27]. The arginine–tertbutyl tryptophan–arginine sequence of LTX109 makes it amphipathic, with two bulky side groups and two cationic side groups that are proposed to interact with negatively charged bacterial cell membranes [26].

We found similar killing kinetic for LTX109 and the membrane permeabilizing drug amphotericin B, suggesting that the two compounds could have a related mode of action. The rapid uptake of the fluorescent dye SYTOX Green, potassium and amino acid efflux from cells exposed to a high concentration of LTX109 suggest that this drug disturbs the plasma membrane by direct interaction with one or several components in the plasma membrane. Furthermore, inability of LTX109 treated cells to acidify their surrounding media support an effect on the cell membrane.

These results are similar to results with amphotericin B, which also causes yeast cells to inhibit glucose-induced acidification [32] and to release potassium as a consequence of general membrane disorganization [54,55].

The high concentration of drug could have obscured other toxic effects of LTX109 on *S. cerevisiae*, so we cannot exclude that LTX109 has other effects in addition to membrane disruption as previously discussed for peptide drugs [56].

To gain further insight into the mode of function of LTX109, we screened for resistant mutants. Six of eight resistance mutants were affected in sphingolipid biosynthesis, and five of these showed similar, high resistance towards LTX109 (Fig. 4). *fen1*, *sur1*, *ipt1* and *skn1* mutants all have reduced amount of sphingolipids [40,43,45,46] as do mutants that reduce Orm2p activity [44], suggesting an essential role of complex sphingolipids in sensitivity to LTX109. Lack of Sur2p lead to decreased sphinganine hydroxylation, but does not prevent formation of MIPC [57]. Furthermore, the *fen1* mutant produce reduced amount of sphingolipids containing the C26 acyl group [40]. The resistance phenotype of the *sur2* and *fen1* mutants therefore suggested that it is not only the quantity, but also the structural modifications that occur during sphingolipid synthesis that is required for optimal LTX109 activity. The terminal steps of sphingolipid biosynthesis in yeasts are MIPC and M(IP)$_2$C. The fact that these lipids are reduced in the resistant mutants suggests that MIPC and M(IP)$_2$C are essential for the fungicidal activity of LTX109, either by direct interaction with LTX109 or by interaction with another mem-

Table 1. *S. cerevisiae* genes that confer LTX109 resistance upon deletion.

Deleted gene and function	ORF	Gene product	n
Sphingolipid biosynthesis			
SUR1	YPL057C	Mannosylinositol phosphorylceramide (MIPC) synthase catalytic subunit	2
SUR2	YDR297W	Sphinganine C4-hydroxylase	8
ORM2	YLR350W	Sphingolipid homeostasis. Interacts with serine palmitoyl transferase (SPT)	1
IPT1	YDR072C	Inositolphosphotransferase, involved in synthesis of mannose-(inositol-P)2-ceramide (M(IP)2C)	1
FEN1	YCR034W	Involved in membrane-bound fatty acid elongation up to 24 C (ceramide precursor)	1
SKN1	YGR143W	Involved in the terminal M(IP)C →M(IP)2C process	2
Apoptosis			
YSP2	YDR326C	Mitochondrial protein in programmed cell death.	1
Unknown function			
OPI9	YLR338W	Dubious ORF unlikely to encode a protein. Partly overlaps VRP1	1

n, number of mutants identified.

Control
(0 µg/ml 6h)

LTX109
(70 µg/ml 6 hours)

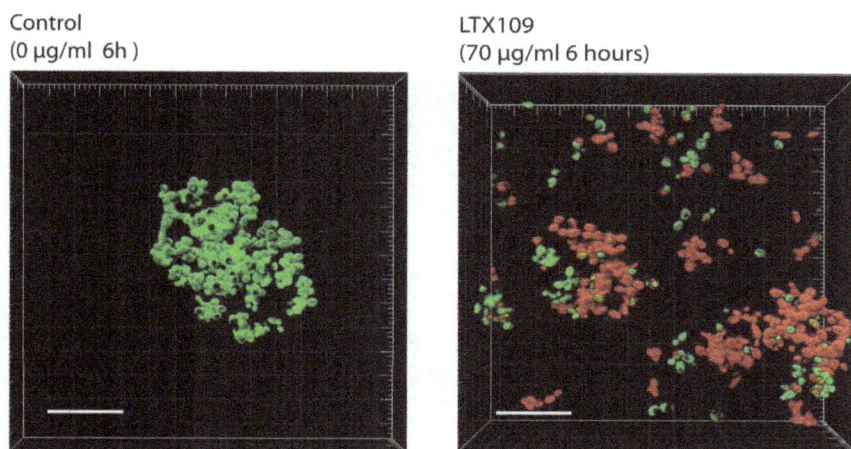

Figure 5. Activity of LTX109 against yeast biofilm. Confocal Laser Scanning Microscopy of *S. cerevisiae* (Σ1278b) biofilm. Cells were grown in Lab-Tek^TM Chamber Slide^TM System; Permanox - (NUNC, Denmark) in 1 ml synthetic complete medium After 12 hours, the cells were exposed to 0 µg/ml LTX109 (control) or 70 µg/ml LTX109 for another 5 hours. The biofilm cells were then stained with Syto9 (green) and propidium iodide (red) LIVE/DEAD stain before confocal laser scanning microscopy. Images are 3D reconstructions of biofilm made from 2 µm thick images in stacks of 20 individual images. CLSM was perform with a Zeiss LSM510 microscope using a 63x/0.95NA a water immersion lens. Life dead staining of biofilm treated with LTX109 was repeated in four independent experiments. White bar is 30 µm.

brane components that is the target for LTX109. It does however seem less likely that a component other than sphingolipids is the target for LTX109 for two reasons, (i) mutants depleted of the target would be expected to appear in the screen for mutants resistant to LTX109. (ii) Alternatively, the target could depend on sphingolipids for optimal activity, be essential for growth and thus not appear in the screen, but then *fen1, sur1, ipt1, skn1* and *orm2* mutants would be expected to have reduced growth rates which they do not (Fig. 4).

Sphingolipids are located primarily in the plasma membrane [58] and are often clustered together with ergosterol in lipid rafts [59]. Sphingolipids are not only a structural component of the cell membrane, but serve vital functions in the heat-shock response, cell cycle arrest, signaling pathways, endocytosis and protein trafficking [60,61]. Fungal sphingolipids are highly similar to each other [62,63], and the biosynthesis of complex fungal sphingolipids is distinctly different from mammals [64]. This makes the fungal sphingolipids attractive antifungal drug targets and several natural compounds with anti-IPC synthase activity have been identified [65–67].

The terminal M(IP)$_2$C is the major sphingolipid in the fungal plasma membrane [46] and has previously been suggested as a target for the plant defensin *Dahlia merckii* antimicrobial peptide 1 (DmAMP1) [43,68,69]. DmAMP1 is a 50 amino acid amtimicrobial peptide that leads to nonselective passage of potassium, calcium [70] and SYTOX Green [33]. Hence, DmAMP1 and LTX109 could have similar modes of action, although DmAMP1 does not contain the Arg-Trp-Arg sequence that serves as basis for LTX109.

Amphotericin B is currently the last in line treatment option for severe fungal infections [71]. Alternative drug candidates might therefore be developed for treatment in cases where use of amphotericin B becomes limited due to resistance. Biofilm

formation on medical devices is a major nosocomial problem and causes multidrug resistance [13]. Only a few of the current systemic antifungals have activity against yeast biofilms [72,73], but often it requires removal of the implant for effective treatment [74]. Peptide antibiotics including LTX109 analogues have been shown to be efficient drugs to kill bacterial biofilm cells [25,50]. This study shows for the first time a peptidomimetic with activity against yeast biofilm. This observation suggests antifungal peptidomimetics with rapid killing kinetics and membrane permeabilizing activities are attractive drugs for yeast biofilm treatment.

In conclusion, we have shown the efficient fungicidal properties of a synthetic peptidomimetic, LTX109, that killed the yeast *S. cerevisiae* with fast killing kinetics and complete eradication of viable cells in exponential growth phase. We found that yeast cells treated with a high concentration of LTX109 became permeable to free amino acids, potassium and SYTOX Green and prevented proton extrusion in response to a pulse of glucose. Fungal susceptibility to LTX109 depended on biosynthesis of sphingolipids. The sphingolipids M(IP)$_2$C and its precursor MIPC are found in fungal, but not human membranes, making LTX109 and derivatives attractive drug candidates for fungal infection treatment as alternatives to amphotericin B.

Acknowledgments

LTX109 was kindly provided by LytixBiopharma AS, Tromsø, Norway.

Author Contributions

Conceived and designed the experiments: RKB RT BR. Performed the experiments: RKB RT CEL BR. Analyzed the data: RKB RT AF BR. Wrote the paper: RKB RT BR.

References

1. Tortorano AM, Peman J, Bernhardt H, Klingspor L, Kibbler CC, et al. (2004) Epidemiology of candidaemia in Europe: results of 28-month European Confederation of Medical Mycology (ECMM) hospital-based surveillance study. Eur J Clin Microbiol Infect Dis 23: 317–322.

2. Pfaller MA, Diekema DJ (2007) Epidemiology of invasive candidiasis: a persistent public health problem. Clin Microbiol Rev 20: 133–163.

3. Vanden Bossche H, Koymans L, Moereels H (1995) P450 inhibitors of use in medical treatment: focus on mechanisms of action. Pharmacol Ther 67: 79–100.

4. Teerlink T, de Kruijff B, Demel RA (1980) The action of pimaricin, etruscomycin and amphotericin B on liposomes with varying sterol content. Biochim Biophys Acta 599: 484–492.

5. Ermishkin LN, Kasumov KM, Potzeluyev VM (1976) Single ionic channels induced in lipid bilayers by polyene antibiotics amphotericin B and nystatine. Nature 262: 698–699.

6. Deresinski SC, Stevens DA (2003) Caspofungin. Clin Infect Dis 36: 1445–1457.

7. Waldorf AR, Polak A (1983) Mechanisms of action of 5-fluorocytosine. Antimicrob Agents Chemother 23: 79–85.

8. Arendrup MC, Bruun B, Christensen JJ, Fuursted K, Johansen HK, et al. (2011) National surveillance of fungemia in Denmark (2004 to 2009). J Clin Microbiol 49: 325–334.

9. Oxman DA, Chow JK, Frendl G, Hadley S, Hershkovitz S, et al. (2010) Candidaemia associated with decreased in vitro fluconazole susceptibility: is Candida speciation predictive of the susceptibility pattern? J Antimicrob Chemother 65: 1460–1465.

10. Vermes A, Guchelaar HJ, Dankert J (2000) Flucytosine: a review of its pharmacology, clinical indications, pharmacokinetics, toxicity and drug interactions. J Antimicrob Chemother 46: 171–179.

11. Bates DW, Su L, Yu DT, Chertow GM, Seger DL, et al. (2001) Mortality and costs of acute renal failure associated with amphotericin B therapy. Clin Infect Dis 32: 686–693.

12. Kauffman CA (2006) Fungal infections. Proc Am Thorac Soc 3: 35–40.

13. Ramage G, Martinez JP, Lopez-Ribot JL (2006) Candida biofilms on implanted biomaterials: a clinically significant problem. FEMS Yeast Res 6: 979–986.

14. Brown MR, Allison DG, Gilbert P (1988) Resistance of bacterial biofilms to antibiotics: a growth-rate related effect? J Antimicrob Chemother 22: 777–780.

15. Baillie GS, Douglas LJ (1998) Effect of growth rate on resistance of Candida albicans biofilms to antifungal agents. Antimicrob Agents Chemother 42: 1900–1905.

16. Butts A, Krysan DJ (2012) Antifungal drug discovery: something old and something new. PLoS Pathog 8: e1002870.

17. van der Weerden NL, Bleackley MR, Anderson MA (2013) Properties and mechanisms of action of naturally occurring antifungal peptides. Cell Mol Life Sci.

18. Hancock RE, Scott MG (2000) The role of antimicrobial peptides in animal defenses. Proc Natl Acad Sci U S A 97: 8856–8861.

19. Yeaman MR, Yount NY (2003) Mechanisms of antimicrobial peptide action and resistance. Pharmacol Rev 55: 27–55.

20. Li Y, Xiang Q, Zhang Q, Huang Y, Su Z (2012) Overview on the recent study of antimicrobial peptides: Origins, functions, relative mechanisms and application. Peptides 37: 207–215.

21. Marr AK, Gooderham WJ, Hancock RE (2006) Antibacterial peptides for therapeutic use: obstacles and realistic outlook. Curr Opin Pharmacol 6: 468–472.

22. Maurya IK, Pathak S, Sharma M, Sanwal H, Chaudhary P, et al. (2011) Antifungal activity of novel synthetic peptides by accumulation of reactive oxygen species (ROS) and disruption of cell wall against Candida albicans. Peptides 32: 1732–1740.

23. Trabocchi A, Mannino C, Machetti F, De Bernardis F, Arancia S, et al. (2010) Identification of inhibitors of drug-resistant Candida albicans strains from a library of bicyclic peptidomimetic compounds. J Med Chem 53: 2502–2509.

24. Murillo LA, Lan CY, Agabian NM, Larios S, Lomonte B (2007) Fungicidal activity of a phospholipase-A2-derived synthetic peptide variant against Candida albicans. Rev Esp Quimioter 20: 330–333.

25. Flemming K, Klingenberg C, Cavanagh JP, Sletteng M, Stensen W, et al. (2009) High in vitro antimicrobial activity of synthetic antimicrobial peptidomimetics against staphylococcal biofilms. J Antimicrob Chemother 63: 136–145.

26. Isaksson J, Brandsdal BO, Engqvist M, Flaten GE, Svendsen JS, et al. (2011) A synthetic antimicrobial peptidomimetic (LTX 109): stereochemical impact on membrane disruption. J Med Chem 54: 5786–5795.

27. Saravolatz LD, Pawlak J, Johnson L, Bonilla H, Saravolatz LD 2nd, et al.(2012) In vitro activities of LTX-109, a synthetic antimicrobial peptide, against methicillin-resistant, vancomycin-intermediate, vancomycin-resistant, daptomy-cin-nonsusceptible, and linezolid-nonsusceptible Staphylococcus aureus. Antimicrob Agents Chemother 56: 4478–4482.

28. Jorgensen MU, Gjermansen C, Andersen HA, Kielland-Brandt MC (1997) STP1, a gene involved in pre-tRNA processing in yeast, is important for amino-acid uptake and transcription of the permease gene BAP2. Curr Genet 31: 241–247.

29. Giaever G, Chu AM, Ni L, Connelly C, Riles L, et al. (2002) Functional profiling of the Saccharomyces cerevisiae genome. Nature 418: 387–391.

30. Rupp S, Summers E, Lo HJ, Madhani H, Fink G (1999) MAP kinase and cAMP filamentation signaling pathways converge on the unusually large promoter of the yeast FLO11 gene. EMBO J 18: 1257–1269.

31. Sherman F (1991) Getting started with yeast. Methods Enzymol 194: 3–21.

32. Tanaka T, Nakayama K, Machida K, Taniguchi M (2000) Long-chain alkyl ester of AMP acts as an antagonist of glucose-induced signal transduction that mediates activation of plasma membrane proton pump in Saccharomyces cerevisiae. Microbiology 146 (Pt 2): 377–384.

33. Thevissen K, Terras FR, Broekaert WF (1999) Permeabilization of fungal membranes by plant defensins inhibits fungal growth. Appl Environ Microbiol 65: 5451–5458.

34. Bojsen RK, Andersen KS, Regenberg B (2012) Saccharomyces cerevisiae--a model to uncover molecular mechanisms for yeast biofilm biology. FEMS Immunol Med Microbiol 65: 169–182.

35. Haagensen JA, Regenberg B, Sternberg C (2011) Advanced microscopy of microbial cells. Adv Biochem Eng Biotechnol 124: 21–54.

36. Serrano R (1980) Effect of ATPase inhibitors on the proton pump of respiratory-deficient yeast. Eur J Biochem 105: 419–424.

37. Torbensen R, Moller HD, Gresham D, Alizadeh S, Ochmann D, et al. (2012) Amino acid transporter genes are essential for FLO11-dependent and FLO11-independent biofilm formation and invasive growth in Saccharomyces cerevisiae. PLoS One 7: e41272.

38. Roth BL, Poot M, Yue ST, Millard PJ (1997) Bacterial viability and antibiotic susceptibility testing with SYTOX green nucleic acid stain. Appl Environ Microbiol 63: 2421–2431.

39. Funato K, Vallee B, Riezman H (2002) Biosynthesis and trafficking of sphingolipids in the yeast Saccharomyces cerevisiae. Biochemistry 41: 15105–15114.

40. Oh CS, Toke DA, Mandala S, Martin CE (1997) ELO2 and ELO3, homologues of the Saccharomyces cerevisiae ELO1 gene, function in fatty acid elongation and are required for sphingolipid formation. J Biol Chem 272: 17376–17384.

41. Grilley MM, Stock SD, Dickson RC, Lester RL, Takemoto JY (1998) Syringomycin action gene SYR2 is essential for sphingolipid 4-hydroxylation in Saccharomyces cerevisiae. J Biol Chem 273: 11062–11068.

42. Beeler TJ, Fu D, Rivera J, Monaghan E, Gable K, et al. (1997) SUR1 (CSG1/BCL21), a gene necessary for growth of Saccharomyces cerevisiae in the presence of high Ca2+ concentrations at 37 degrees C, is required for mannosylation of inositolphosphorylceramide. Mol Gen Genet 255: 570–579.

43. Thevissen K, Idkowiak-Baldys J, Im YJ, Takemoto J, Francois IE, et al. (2005) SKN1, a novel plant defensin-sensitivity gene in Saccharomyces cerevisiae, is implicated in sphingolipid biosynthesis. FEBS Lett 579: 1973–1977.

44. Shimobayashi M, Oppliger W, Moes S, Jeno P, Hall MN (2013) TORC1-regulated protein kinase Npr1 phosphorylates Orm to stimulate complex sphingolipid synthesis. Mol Biol Cell 24: 870–881.

45. Stock SD, Hama H, Radding JA, Young DA, Takemoto JY (2000) Syringomycin E inhibition of Saccharomyces cerevisiae: requirement for biosynthesis of sphingolipids with very-long-chain fatty acids and mannose- and phosphoinositol-containing head groups. Antimicrob Agents Chemother 44: 1174–1180.

46. Dickson RC, Nagiec EE, Wells GB, Nagiec MM, Lester RL (1997) Synthesis of mannose-(inositol-P)2-ceramide, the major sphingolipid in Saccharomyces cerevisiae, requires the IPT1 (YDR072c) gene. J Biol Chem 272: 29620–29625.

47. Reynolds TB, Fink GR (2001) Bakers' yeast, a model for fungal biofilm formation. Science 291: 878–881.

48. Gimeno CJ, Ljungdahl PO, Styles CA, Fink GR (1992) Unipolar cell divisions in the yeast S. cerevisiae lead to filamentous growth: regulation by starvation and RAS. Cell 68: 1077–1090.

49. Violette A, Fournel S, Lamour K, Chaloin O, Frisch B, et al. (2006) Mimicking helical antibacterial peptides with nonpeptidic folding oligomers. Chem Biol 13: 531–538.

50. Liu Y, Knapp KM, Yang L, Molin S, Franzyk H, et al. (2013) High in vitro antimicrobial activity of beta-peptoid-peptide hybrid oligomers against plank-tonic and biofilm cultures of Staphylococcus epidermidis. Int J Antimicrob Agents 41: 20–27.

51. Niu Y, Padhee S, Wu H, Bai G, Qiao Q, et al. (2012) Lipo-gamma-AApeptides as a new class of potent and broad-spectrum antimicrobial agents. J Med Chem 55: 4003–4009.

52. Makobongo MO, Kovachi T, Gancz H, Mor A, Merrell DS (2009) In vitro antibacterial activity of acyl-lysyl oligomers against Helicobacter pylori. Antimicrob Agents Chemother 53: 4231–4239.

53. Niu Y, Wang RE, Wu H, Cai J (2012) Recent development of small antimicrobial peptidomimetics. Future Med Chem 4: 1853–1862.

54. Beggs WH (1994) Physicochemical cell damage in relation to lethal amphotericin B action. Antimicrob Agents Chemother 38: 363–364.

55. Chen WC, Chou DL, Feingold DS (1978) Dissociation between ion permeability and the lethal action of polyene antibiotics on Candida albicans. Antimicrob Agents Chemother 13: 914–917.

56. Theis T, Stahl U (2004) Antifungal proteins: targets, mechanisms and prospective applications. Cell Mol Life Sci 61: 437–455.

57. Haak D, Gable K, Beeler T, Dunn T (1997) Hydroxylation of Saccharomyces cerevisiae ceramides requires Sur2p and Scs7p. J Biol Chem 272: 29704–29710.

58. Patton JL, Lester RL (1991) The phosphoinositol sphingolipids of Saccharomyces cerevisiae are highly localized in the plasma membrane. J Bacteriol 173: 3101–3108.

59. Bagnat M, Keranen S, Shevchenko A, Simons K (2000) Lipid rafts function in biosynthetic delivery of proteins to the cell surface in yeast. Proc Natl Acad Sci U S A 97: 3254–3259.

60. Cowart LA, Obeid LM (2007) Yeast sphingolipids: recent developments in understanding biosynthesis, regulation, and function. Biochim Biophys Acta 1771: 421–431.

61. Dickson RC, Lester RL (2002) Sphingolipid functions in Saccharomyces cerevisiae. Biochim Biophys Acta 1583: 13–25.

62. Vincent VL, Klig LS (1995) Unusual effect of myo-inositol on phospholipid biosynthesis in Cryptococcus neoformans. Microbiology 141 (Pt 8): 1829–1837.

63. Wells GB, Dickson RC, Lester RL (1996) Isolation and composition of inositolphosphorylceramide-type sphingolipids of hyphal forms of Candida albicans. J Bacteriol 178: 6223–6226.

64. Dickson RC (1998) Sphingolipid functions in Saccharomyces cerevisiae: comparison to mammals. Annu Rev Biochem 67: 27–48.

65. Nagiec MM, Nagiec EE, Baltisberger JA, Wells GB, Lester RL, et al. (1997) Sphingolipid synthesis as a target for antifungal drugs. Complementation of the inositol phosphorylceramide synthase defect in a mutant strain of Saccharomyces cerevisiae by the AUR1 gene. J Biol Chem 272: 9809–9817.

66. Mandala SM, Thornton RA, Milligan J, Rosenbach M, Garcia-Calvo M, et al. (1998) Rustmicin, a potent antifungal agent, inhibits sphingolipid synthesis at inositol phosphoceramide synthase. J Biol Chem 273: 14942–14949.

67. Mandala SM, Thornton RA, Rosenbach M, Milligan J, Garcia-Calvo M, et al. (1997) Khafrefungin, a novel inhibitor of sphingolipid synthesis. J Biol Chem 272: 32709–32714.

68. Thevissen K, Cammue BP, Lemaire K, Winderickx J, Dickson RC, et al. (2000) A gene encoding a sphingolipid biosynthesis enzyme determines the sensitivity of Saccharomyces cerevisiae to an antifungal plant defensin from dahlia (Dahlia merckii). Proc Natl Acad Sci U S A 97: 9531–9536.

69. Aerts AM, Francois IE, Bammens L, Cammue BP, Smets B, et al. (2006) Level of M(IP)2C sphingolipid affects plant defensin sensitivity, oxidative stress resistance and chronological life-span in yeast. FEBS Lett 580: 1903–1907.

70. Thevissen K, Ghazi A, De Samblanx GW, Brownlee C, Osborn RW, et al. (1996) Fungal membrane responses induced by plant defensins and thionins. J Biol Chem 271: 15018–15025.

71. Chandrasekar P (2011) Management of invasive fungal infections: a role for polyenes. J Antimicrob Chemother 66: 457–465.

72. Bachmann SP, VandeWalle K, Ramage G, Patterson TF, Wickes BL, et al. (2002) In vitro activity of caspofungin against Candida albicans biofilms. Antimicrob Agents Chemother 46: 3591–3596.

73. Kuhn DM, George T, Chandra J, Mukherjee PK, Ghannoum MA (2002) Antifungal susceptibility of Candida biofilms: unique efficacy of amphotericin B lipid formulations and echinocandins. Antimicrob Agents Chemother 46: 1773–1780.

74. Lynch AS, Robertson GT (2008) Bacterial and fungal biofilm infections. Annu Rev Med 59: 415–428.

8

Repeated Evolution of Fungal Cultivar Specificity in Independently Evolved Ant-Plant-Fungus Symbioses

Rumsaïs Blatrix[1]*, Sarah Debaud[1], Alex Salas-Lopez[1], Céline Born[1], Laure Benoit[1], Doyle B. McKey[1,2], Christiane Attéké[3], Champlain Djiéto-Lordon[4]

1 Centre d'Ecologie Fonctionnelle et Evolutive (CEFE), CNRS/CIRAD-Bios/Université Montpellier 2, Montpellier, France, 2 Institut Universitaire de France, Montpellier, France, 3 Département de Biologie, Université des Sciences et Techniques de Masuku (USTM), Franceville, Gabon, 4 Laboratory of Zoology, University of Yaoundé I, Yaoundé, Cameroun

Abstract

Some tropical plant species possess hollow structures (domatia) occupied by ants that protect the plant and in some cases also provide it with nutrients. Most plant-ants tend patches of chaetothyrialean fungi within domatia. In a few systems it has been shown that the ants manure the fungal patches and use them as a food source, indicating agricultural practices. However, the identity of these fungi has been investigated only in a few samples. To examine the specificity and constancy of ant-plant-fungus interactions we characterised the content of fungal patches in an extensive sampling of three ant-plant symbioses (*Petalomyrmex phylax/Leonardoxa africana* subsp. *africana*, *Aphomomyrmex afer/Leonardoxa africana* subsp. *letouzeyi* and *Tetraponera aethiops/Barteria fistulosa*) by sequencing the Internal Transcribed Spacers of ribosomal DNA. For each system the content of fungal patches was constant over individuals and populations. Each symbiosis was associated with a specific, dominant, primary fungal taxon, and to a lesser extent, with one or two specific secondary taxa, all of the order Chaetothyriales. A single fungal patch sometimes contained both a primary and a secondary taxon. In one system, two founding queens were found with the primary fungal taxon only, one that was shown in a previous study to be consumed preferentially. Because the different ant-plant symbioses studied have evolved independently, the high specificity and constancy we observed in the composition of the fungal patches have evolved repeatedly. Specificity and constancy also characterize other cases of agriculture by insects.

Editor: Nicole M. Gerardo, Emory University, United States of America

Funding: This project was supported by the network "Bibliothèque du Vivant" (http://bdv.ups-tlse.fr/), funded by the CNRS, the Muséum National d'Histoire Naturelle, the INRA and the CEA (Centre National de Séquençage). It was also funded by grants to R. Blatrix and D. McKey from three programmes of the French Agence Nationale de la Recherche (www.agence-nationale-recherche.fr): "Young scientists" (research agreement no. ANR-06-JCJC-0127), "Biodiversity" (IFORA project) and "Sixth extinction" (C3A project). The funders had no role in study design, data collection and analysis, decision to publish, or preparation of the manuscript.

Competing Interests: The authors have declared that no competing interests exist.

* E-mail: rumsais.blatrix@cefe.cnrs.fr

Introduction

Ant-plants, or myrmecophytes, are plants that provide symbiotic ants with nesting cavities (specialized hollow structures, called domatia). Ant-plant symbioses involve about 100 plant and 40 ant genera in the tropics and have evolved many times independently [1]. Domatia originate from diverse modified plant structures: twigs, petioles, leaf laminae, stipules, rhizomes or tubers. The symbiotic ants usually obtain a large part of their food from plant products, either directly (extrafloral nectar and food bodies) or indirectly (honeydew produced by hemipterans reared in domatia) [2]. Most associated ant species protect the plant against herbivores, pathogens and competing vegetation [3,4,5]. They also often provide their host plant with nutrients [6]. In most cases, each individual plant is occupied by a single colony. In some species, a single colony can occupy several adjacent plants of the same species.

It has become evident that ant-plant symbioses should be considered not as bipartite interactions but as symbiotic communities involving, in many cases, plants, ants, hemipterans, fungi, bacteria and possibly nematodes [7,8,9,10]. This conceptual shift applies to all mutualistic interactions and proves useful for a better understanding of the functioning and evolution of ecosystems [11,12]. Microorganisms such as fungi have long been noticed within domatia [13,14,15], but their identities and roles are just beginning to be understood [10,16]. They have been detected in most ant-plant symbioses investigated and form a whole set of new species of the order Chaetothyriales (Ascomycota) [10]. They form dense and well delimited mats of hyphae covering a small area on the inner wall of the domatium. Fungal patches occur in limited number, but are present in each domatium of a single plant. The true symbiotic nature of the ant-plant-fungus association was first demonstrated in the African symbiosis between the ant *Petalomyrmex phylax* and the plant *Leonardoxa africana* subsp. *africana* (Fabaceae, subfamily Caesalpinioideae) [17]. Nutrient flux from ants to fungal patches was also demonstrated in this system [18], suggesting a manuring process. Although the role of these fungi remains largely unexplored, ants have been shown to use them as a food source in three ant-plant symbioses, *Pseudomyrmex penetrator/Tachigali* sp. (Fabaceae, subfamily Caesalpinioideae), *Petalomyrmex phylax/Leonardoxa africana* subsp. *africana* and *Tetraponera aethiops/Barteria fistulosa* (Passifloraceae) [16]. Considered together, along with more anecdotal observations, these studies strongly suggest that plant-ants farm these fungi for food. As ant-plant-fungus symbioses have evolved many times independently, they could represent multiple cases of parallel evolution of agriculture.

Fungiculture has been thoroughly investigated in three widely separated insect lineages [19]: fungus-growing ants (tribe Attini), fungus-growing termites (subfamily Macrotermitinae) and ambrosia beetles (Scolytidae, subfamily Scolytinae, including the Platypodinae). In contrast, very few data exist on other potential cases of agriculture conducted by animals. These cases involve damselfish and *Polysiphonia* algae [20], a marine snail and ascomycete fungi [21], gall midges (cecidomyiid flies) and dothideomycete fungi [22], and plant-ants and chaetothyrialean fungi [16,23,24]. Investigation of a greater range of agricultural systems is needed to obtain a more comprehensive understanding of the global pattern of the evolution of agriculture by animals, and to compare the features of these diverse and parallel coevolved systems.

In most ant-plant symbioses, the pattern of specificity between ants and plants is well known. However, the extent of specificity of their domatia-inhabiting fungal symbionts has never been assessed. We focussed on three ant-plant symbioses for which evidence strongly suggests that they are new cases of fungiculture by ants [7,16,17,18]: *Petalomyrmex phylax*/ *Leonardoxa africana* subsp. *africana*, *Aphomomyrmex afer*/ *Leonardoxa africana* subsp. *letouzeyi* and *Tetraponera aethiops*/ *Barteria fistulosa*. We aimed at characterising (i) the fungal community within domatia over a large number of samples in order to test for ant-plant-fungus specificity and (ii) the geographic pattern of variation in the occurrence of the specific fungal taxa in order to assess the degree of interdependency among the associated species. Sexual structures of domatia fungi have never been observed in fungal patches tended by ants and identification of species from hyphae is not possible. We thus used a stepwise DNA barcode approach, using polymerase chain reaction (PCR) with universal and then specific primers, to characterise the identity and distribution of the fungal partners associated with each of these ant-plant symbioses.

Methods

The symbiosis between the ant *Petalomyrmex phylax* and the plant *Leonardoxa africana* subsp. *africana* is obligatory, highly specific and endemic to the coastal rain forest of southern Cameroon [25]. A total of 98 fungal patches were sampled from 80 individual trees distributed along an 85-km transect of coastal forest, covering almost half the distribution area of this symbiosis. For 11 trees we sampled several domatia (two to five).

The ant-plant *Leonardoxa africana* subsp. *letouzeyi* can be occupied by non-specific ants at the sapling stage, but when trees are mature, the obligatory plant-ant *Aphomomyrmex afer* is by far the most common inhabitant [25,26]. This symbiosis is restricted to the lowland rain forests near the Bight of Biafra, across the boundary between Cameroon and Nigeria [25]. A single fungal patch was sampled from each of 17 individual trees occupied by *A. afer*, in a single site, around Iriba Inene camp in Korup National Park, Cameroon.

The ant-plant *Barteria fistulosa*, whose lateral branches are hollow throughout their length, and its ant symbiont *Tetraponera aethiops* are widely distributed over the whole Lower Guinea - Congo basin forest block [27]. They are considered highly dependent on each other because the ant has never been found nesting outside a *Barteria*, and unoccupied plants do not grow well [14]. However, *B. fistulosa* can also be found in association with the ant *Tetraponera latifrons*, and both *Tetraponera* spp. can also colonise the related and morphologically similar plant *Barteria dewevrei* [27,28]. A total of 440 fungal patches were collected in Cameroon and Gabon from 411 individual trees of *B. fistulosa* occupied by *T. aethiops*. Samples were collected over an area of nearly 100 000 km². For 13 trees we sampled several domatia (two to five).

The following authorities provided research permits and permitted sample collection: Ministry of Scientific Research and Innovation of the Republic of Cameroon, the conservator of Korup National Park (Cameroon), Université des Sciences et Techniques de Masuku (Gabon), Ministère de l'Education Nationale, de l'Enseignement Supérieur et Technique et de la Formation Professionnelle de la Recherche Scientifique chargé de la Culture, de la Jeunesse et des Sports (Gabon).

Fungal samples were either dried under silica gel or stored in extraction buffer immediately upon collection in the field. DNA was extracted using either the modified CTAB method described in [10] or the REDExtract-N-Amp Plant PCR Kit (Sigma–Aldrich, St. Louis, USA). Fungus identities were assessed by sequencing approximately 600 bp of the Internal Transcribed Spacer (ITS) of the nuclear ribosome, which comprises ITS1, 5.8S and ITS2, and is considered to be the best universal DNA barcode marker for fungi [29].

The first step for each biological system studied was to sequence ITS using fungal universal primers ITS1f [30] and ITS4 [31] for all fungal patches from the systems *Petalomyrmex*/*Leonardoxa* (98 samples) and *Aphomomyrmex*/*Leonardoxa* (17 samples), and for 78 fungal patches (out of 440) from the system *Tetraponera*/*Barteria*. This step allowed identifying the fungal taxa associated with each symbiosis. For the *Petalomyrmex*/*Leonardoxa* and *Aphomomyrmex*/*Leonardoxa* systems we performed molecular cloning respectively on nine and one first-step PCR products for which the sequence could not be read. This first step indicated that two specific fungal taxa occurred in each of the two systems *Petalomyrmex*/*Leonardoxa* and *Tetraponera*/*Barteria* and one specific taxon in the system *Aphomomyrmex*/*Leonardoxa*. However, this method did not allow determining whether two fungal taxa could co-occur in a single patch (except for the few samples on which we performed molecular cloning). We thus applied a second step, which involved only the two systems *Petalomyrmex*/*Leonardoxa* and *Tetraponera*/*Barteria*. This step consisted in testing for the presence of each specific fungal taxon in each fungal patch. For this, we designed primers specific to each of the Chaetothyriales taxa detected in the first step (primer sequences are given in Table 1).

To control whether the specific primers amplified the taxa they were respectively designed for, we sequenced all PCR products for the *Petalomyrmex*/*Leonardoxa* system. For the *Tetraponera*/*Barteria* system, 40 PCR products obtained with the two specific ITS primer pairs yielded sequences of the targeted species, confirming the high specificity of the primers in this system. As a consequence, 362 samples out of 440 were simply screened for the presence of each specific fungal taxon through success or failure of amplification with the specific primers (but no sequencing of PCR product).

Amplifications were performed in a 25 μl solution containing 1 × PCR mix (multiplex kit, Qiagen, Venlo, Netherlands), 0.5 μM of each primer and 1 μl of DNA template. They took place in a thermal cycler programmed for an initial denaturation step of 15 min at 95°C, followed by 35 cycles of 60 s at 94°C, 60 s at 53°C and 60 s at 72°C, and a final elongation step of 20 min at 60°C. Molecular cloning of PCR product was performed using the kit pGEM-T easy vector (Promega, Madison, USA) and following the manufacturer's instructions.

ITS sequences were first searched for relatives using the Basic Local Alignment Search Tool in GenBank (http://blast.ncbi.nlm.

Table 1. Sequences of primers developed in this study to amplify specifically the ITS region of fungal Molecular Operational Taxonomic Units (MOTU) detected in two focal ant-plant symbioses.

MOTU targeted	Name of primer	Sequence 5' - 3'	associated ant-plant symbiosis
La1	its1La1	GAGTGAGGGTCTCTGTGCCC	Petalomyrmex/Leonardoxa
	its4La1	TACAACTCGGACCCCAAGGGGC	
La2	its1La2	GTTAGGGTTCCTCTCACGGG	Petalomyrmex/Leonardoxa
	its4La2	AAATTACAACTCGGGCCGTG	
Y1	its1Y1	GGCTGCCGGGGGGTTCTATT	Tetraponera/Barteria
	its4Y1	GTCAACCTTAGATAAAACTA	
Y9	ITS1f is used as forward primer		Tetraponera/Barteria
	its4Y9	TCAACCTTTAGATATAAGA	

nih.gov/Blast.cgi). This allowed detecting which sequences belonged to species of the order Chaetothyriales. Sequences of Chaetothyriales were then classified into haplotypes. One sequence for each haplotype was deposited in GenBank. All haplotypes from the three systems were aligned with Muscle [32] and a maximum likelihood tree was constructed using PhyML [33] in order to guide our choice for delimitation of Molecular Operational Taxonomic Units (MOTUs, sensus [34]). In addition, we used the conservative cut-off value of 95% ITS sequence similarity for delimitation of MOTUs. Although higher cut-off values have been proposed in previous studies considering ITS as fungal barcodes [29,35], we prefer to use a conservative value because intra-specific sequence variation can vary across taxonomic groups and Chaetothyriales fungi are poorly known in this respect. Thus, the splitting into taxonomic units that we propose in this paper is likely to remain valid in the future.

Results

Out of a total of 311 ITS sequences, 305 were sequences of Chaetothyriales for the systems *Petalomyrmex/Leonardoxa* (208 sequences), *Aphomomyrmex/Leonardoxa* (10 sequences) and *Tetraponera/Barteria* (87 sequences), and only six sequences were of a different order. According to Blast results, these last six sequences most likely belonged to *Candida* (Ascomycota, Saccharomycetales), *Cryptococcus* (Basidiomycota, Tremellales), *Neurospora* (Ascomycota, Sordariales), *Fusarium* (Ascomycota, Hypocreales) and a Capnodiales (Ascomycota). From the Chaetothyriales sequences, we detected a total of 42 haplotypes (GenBank accession numbers KC951221 to KC951262) that were grouped into eight likely MOTUs (Fig. 1). For 7% of the sequences we could not determine the haplotype because of low sequence quality at determinant positions. We found four, two and two Chaetothyriales MOTUs in the *Petalomyrmex/Leonardoxa*, *Aphomomyrmex/Leonardoxa* and *Tetraponera/Barteria* systems respectively. Each symbiosis had its own set of MOTUs. Within each MOTU, haplotypes had more than 98.6% similarity in ITS sequence. The two most closely related MOTUs, La2 and La3, had ITS sequences similar at 95%. For the two systems with multiple sampling sites (*Petalomyrmex/Leonardoxa* and *Tetraponera/Barteria*), distribution of MOTUs did not seem to show spatial structure (Fig. 2 and Fig. 3).

For the *Petalomyrmex/Leonardoxa* system, samples for which a readable sequence was obtained using universal primers yielded La1 and La2 in 96% and 4% of the cases respectively (Table 2). We designed specific ITS primers for La1 and La2 (Table 1). In this system, all PCR products obtained with specific primers were sequenced. Sequences obtained with primers specific to La1 always yielded La1, whereas sequences obtained with primers specific to La2 yielded either La2 or La3. Molecular cloning of PCR products allowed detection of up to five haplotypes of a single MOTU in a single fungal patch. The number of haplotypes detected only when PCR products were cloned was 14 (out of 18), two (out of four) and zero (out of three) for La1, La2 and La3 respectively, showing that diversity within MOTUs is underestimated without molecular cloning. However, cloning revealed only one additional MOTU (La8). When we combine the results from all methods (PCR with universal or specific primers and molecular cloning of PCR products obtained with universal primers) La1, La2, La3 and La8 were detected in 97%, 47%, 32% and 2% of the samples respectively (Table 2). In 76% of the samples we detected both La1 and either La2 or La3. We cannot rule out the possibility that La2 and La3 co-occur in the same samples because we did not test diagnostic primer pairs. In 21% of the samples we detected La1 only. In 3% of the samples we detected either La2 or La3 only. In half of the plant individuals for which we sampled several fungal patches (one per domatium) we found exactly the same MOTUs in all patches from the same individual. In the other half we found patches with La1 only and patches with La1 and either La2 or La3 in the same individual.

In the *Aphomomyrmex/Leonardoxa* system, for 41% of the samples we did not obtain a readable sequence with universal primers. The other samples yielded Ll1 in eight out of 10 (80%) of the cases (Table 2). In the other two cases, the sequences revealed fungi that did not belong to the Chaetothyriales and that were likely contaminants or non-symbiotic competitors (*Candida*, *Fusarium*). For the one sample on which molecular cloning was performed, we detected Ll1, Kh1 and a Capnodiales. Kh1 is similar to the Chaetothyriales strain KhNk4-2a that has previously been isolated from the symbiosis between the African plant *Keetia hispida* (Rubiaceae) and ants of the genus *Crematogaster* [10], which can be found in the forest where we sampled the *Aphomomyrmex/Leonardoxa* system.

For the *Tetraponera/Barteria* system, samples for which a readable sequence was obtained using universal primers yielded Y1 and Y9 in 69% and 27% of the cases respectively (Table 2). In the other cases (two out of 78), the sequences revealed fungi that did not belong to the Chaetothyriales and that were likely contaminants or non-symbiotic competitors (*Neurospora*, *Cryptoccocus*). Specific primers were designed for both Y1 and Y9 (Table 1). When we combine the results from all methods (PCR with universal or specific primers), Y1 and Y9 were detected in 84% and 21% of the samples respectively (Table 2). In 17% of the samples we detected

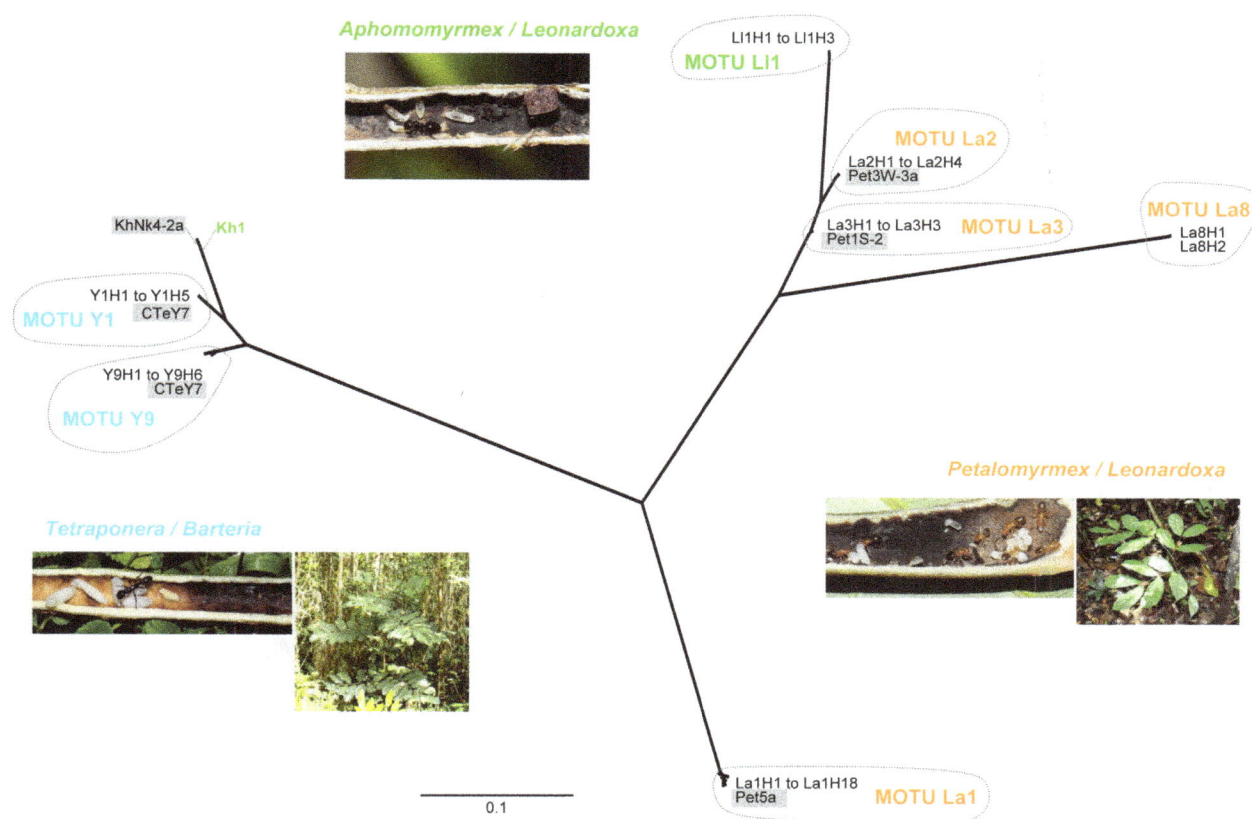

Figure 1. Maximum likelihood tree of ITS fungal haplotypes from three ant-plant-fungus symbioses. A total of 42 haplotypes (based on sequences of 647 aligned nucleotides) of Chaetothyriales were detected in fungal patches of the following ant-plant symbioses: *Petalomyrmex phylax/Leonardoxa africana* subsp. *africana* (MOTUs in orange), *Aphomomyrmex afer/Leonardoxa* subsp. *letouzeyi* (MOTU in green) and *Tetraponera aethiops/Barteria fistulosa* (MOTUs in blue). Note that haplotype Kh1 is labelled in green because it was detected in the system *Aphomomyrmex/ Leonardoxa*, although it is phylogenetically most closely related to MOTU Y1. The position of each MOTU on the tree is indicated by the intersection of the branches and the dotted lines. Branch tip labels highlighted in grey correspond to fungal strains obtained in previous studies following a culturing approach. For each symbiosis, the image on the left displays a domatium cut longitudinally to expose ants and a fungal patch (dark area on the inner surface). MOTU: Molecular Operational Taxonomic Unit.

both Y1 and Y9. In 67% of the samples we detected Y1 only. In 4% of the samples we detected Y9 only. In the other samples (12%) neither Y1 nor Y9 were detected. This high proportion of amplification failure is likely due to low quality of DNA extraction. Moreover, we did not repeat unsuccessful PCR for the 362 samples (out of 440) that were screened with specific primers and for which PCR product was not sequenced. In several plant individuals for which we sampled several fungal patches we found differences in MOTU composition among patches of a single individual. A young *B. fistulosa* individual (BF365) contained four founding queens, each in a separate domatium. A fungal patch was associated with each of these queens. For two patches amplifications failed. For the two others amplification was successful with Y1-specific primers and failed with Y9-specific primers.

Details on each individual sample of fungal patch, including collection information and detected MOTUs, are available in Table S1.

Discussion

The DNA barcode approach that we used on direct extracts of DNA fungal patches from three ant-plant symbioses detected mostly taxa belonging to the Chaetothyriales. In fact, fewer than 2% of the sequences were from other orders. From microscopic observation and culturing of fungal patches in previous studies [10,17] we know that many different fungi are present as spores or fragments of hyphae but do not grow in the natural conditions of domatia occupied by mutualist ants. The few sequences of non-Chaetothyriales taxa most likely represent such fungi that may have reached the domatia opportunistically. Previous studies showed that many ant-plant symbioses are associated with Chaetothyriales [7,10,36] and the present results confirm for three symbioses that this type of association is consistent over a large sampling.

Although intra-specific ITS sequence variability varies across taxa, it averages 2.5% in fungi, and more specifically, less than 2% in Ascomycota [29,35]. Applying mean cut-off values for species delimitation in poorly known groups, such as Chaetothyriales, is likely to bring erroneous conclusions. However, classification of the sequences from this study (42 haplotypes of Chaetothyriales) into MOTUs was rather straightforward through visual inspection of the Maximum Likelihood phylogenetic tree (Fig. 1). Moreover, sequence variability was less than 2% within and more than 5% between defined MOTUs. Although delimited MOTUs are likely to correspond to species, we are reluctant to use this term before more molecular data are available.

In our study models successful direct amplification of fungal patches with universal primers yielded one main MOTU in each

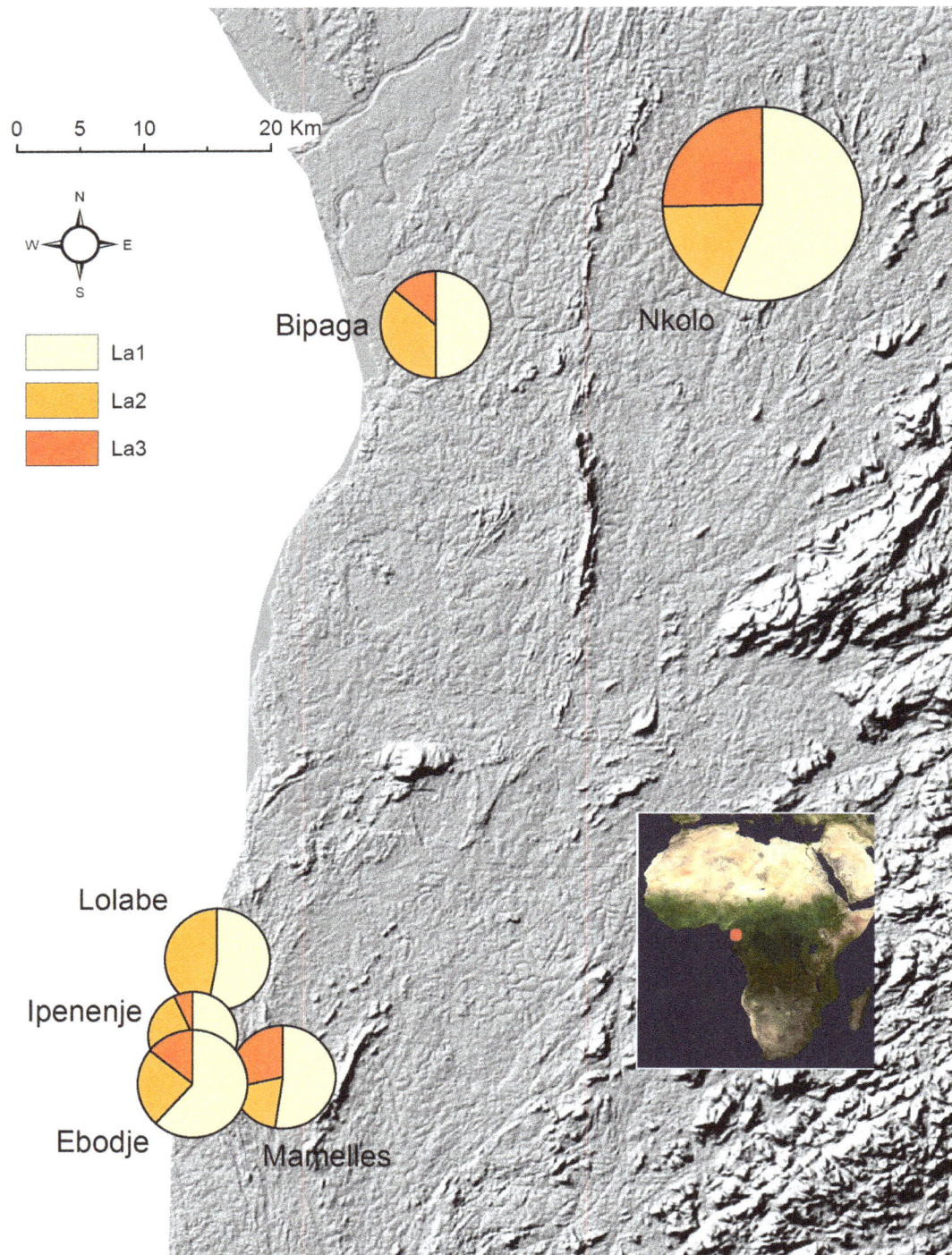

Figure 2. Spatial distribution of Chaetothyriales MOTUs of the *Petalomyrmex phylax/Leonardoxa africana* **subsp.** *africana* **system.** Sectors represent the proportion of each Molecular Operational Taxonomic Unit in each sampling site. MOTUs were detected using universal and/or specific ITS primers. Size of pie charts is proportional to sample size (i.e., the number of fungal samples for which at least one Chaetothyriales MOTU was detected).

system investigated. Amplification with primers specific for this primary MOTU revealed that it was also present in most samples in which it was not detected with universal primers. In the *Petalomyrmex/Leonardoxa* system two other secondary MOTUs were commonly detected using specific primers but only very rarely when using universal primers. This suggests that the primary MOTU is quantitatively the most abundant in fungal patches but that another MOTU occurs alongside. Moreover, the secondary MOTUs were very rarely detected alone, without the primary one. We did not note any particularity that was shared by the

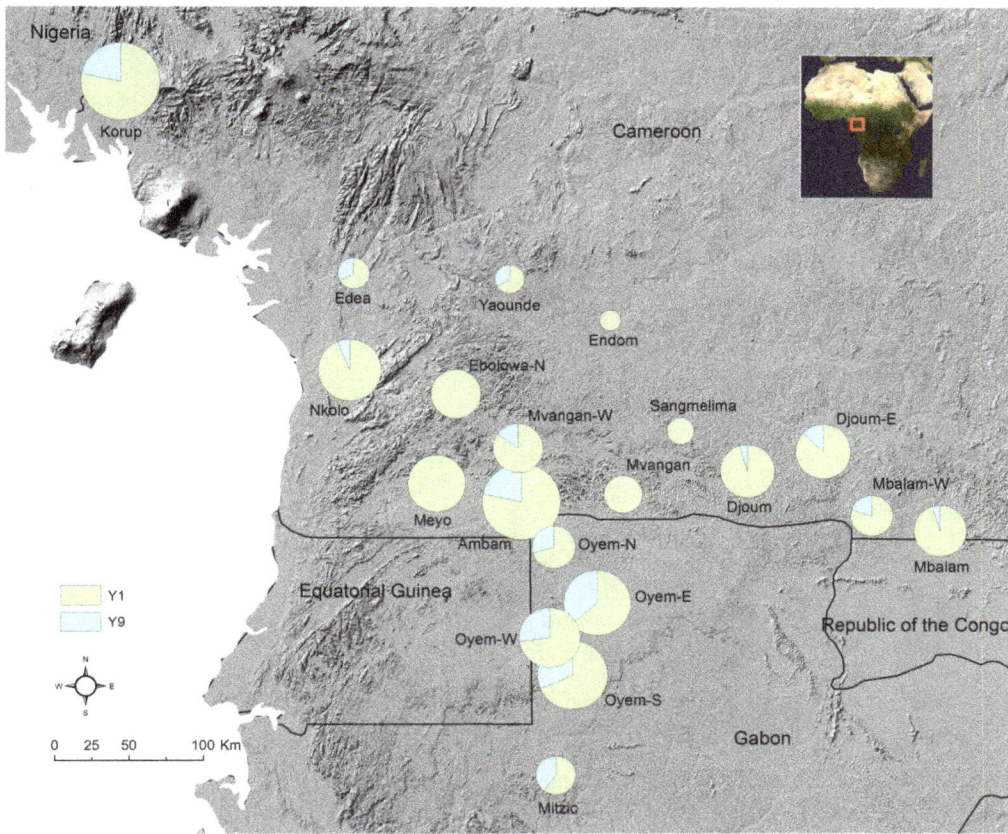

Figure 3. Spatial distribution of Chaetothyriales MOTUs of the *Tetraponera aethiops/Barteria fistulosa* **system.** Sectors represent the proportion of each Molecular Operational Taxonomic Unit in each sampling site. MOTUs were detected using universal and/or specific ITS primers. When specific ITS primers were used, PCR products were not always sequenced. Size of pie charts is proportional to sample size (i.e., the number of fungal samples for which at least one Chaetothyriales MOTU was detected).

samples in which only the secondary MOTUs were detected. A likely explanation for these cases is amplification failure of the primary MOTU due to poor quality of DNA extracts for these samples. In the *Tetraponera/Barteria* system, we detected only one secondary MOTU. In this system, both primary and secondary MOTUs were detected using universal primers and the proportion of samples that had only one of them (detected with specific primers) was higher than in the *Petalomyrmex/Leonardoxa* system.

The nature of the relationship between ants and the secondary MOTUs might differ between the two systems, as the pattern of occurrence appears different. The fungi associated with the three ant-plant symbioses are different between the symbioses, even when they occur in sympatry. For instance, in one sampling site (Nkolo, Cameroon) we collected specimens of the two symbioses, *Petalomyrmex/Leonardoxa* and *Tetraponera/Barteria*, that were only a few tens of meters apart, and still they did not share fungal

Table 2. Number of fungal samples in which the different MOTUs were detected using sequencing of the ITS region (ITS1, 5.8S, ITS2) of ribosomal DNA.

	Universal primers[a]					All methods[b]					
Petalomyrmex/Leonardoxa	La1	La2	NS[c]		Total	La1	La2	La3	La8	others[d]	Total
	80	3	15		98	95	46	31	2	1	98
Aphomomyrex/Leonardoxa	LI1	others[d]	NS[c]		Total	LI1	Kh1	others[d]	NS[c]		Total
	8	2	7		17	9	1	3	6		17
Tetraponera/Barteria	Y1	Y9	others[d]	NS[c]	Total	Y1	Y9	others[d]	NS[c]		Total
	31	12	2	33	78	369	91	2	53		440

[a]PCR was performed directly on the fungal patch using fungal universal primers ITS1f and ITS4, and thus only one species per sample can be detected.
[b]species were detected using either universal primers, molecular cloning of PCR product or species-specific primers, so that several species per sample can be detected.
[c]either no amplification, or the sequence was not readable.
[d]sequences that do not belong to Chaetothyriales (likely contaminants or non-symbiotic competitors).

MOTUs. Clearly, MOTUs are consistently the same between individuals within each study model. For two of these symbioses the sampling covered a substantial part of the distribution, and showed no geographic variation in the identity of fungal symbionts. Altogether, this information shows that Chaetothyriales symbionts associated with our focal ant-plant symbioses are specific to the symbiosis. However, it seems possible that some Chaetothyriales move more freely among systems. For instance, in the *Aphomomyrmex/Leonardoxa* system, in addition to the primary MOTU, we obtained one sequence (through molecular cloning of PCR product) that was very similar to the sequence of a strain isolated and cultured previously from the symbiosis between the small tree *Keetia hispida* (Rubiaceae) and an ant of the genus *Crematogaster* [10], found in the same forest. This may be explained either by dispersal of the fungus or by contamination between samples during lab processing. The extent of sharing of fungal taxa and the presence of these taxa in the environment still remain to be investigated.

The *Petalomyrmex/Leonardoxa* and *Tetraponera/Barteria* systems involve ants and plants that belong to different subfamilies and families respectively. These symbioses are phylogenetically independent. The similarity in their global pattern of specificity with the fungal symbionts thus reflects a repeated pattern in evolution. These two ant-plant symbioses are highly specialised. Further work should describe the pattern of specificity of fungal symbionts in less specialised systems, to test whether cultivar specificity is correlated with ant-plant specialisation. Patterns of specificity are known to vary in other cases of agriculture by insects. In Attine ants, for instance, species in the genera *Acromyrmex* and *Atta* (higher Attines, or leaf-cutting ants) share a unique species of fungal symbiont that they grow in pure culture [37,38]. In contrast, in lower Attines a single species can use various fungal symbionts, because each species exchanges cultivars horizontally with neighbouring colonies of different ant species [39,40]. In fungus-growing termites, although most species seem to be associated with a single fungal strain, some can associate with different species of the symbiotic *Termitomyces* fungi [41].

Our study revealed that the two different fungal symbionts that are associated with a single ant-plant symbiosis frequently co-occur in each ant colony. In contrast, in higher Attine ants and Macrotermitinae, each colony seems to rely on the monoculture of a single fungal species [39,40,42,43], even in ant species that can use various fungi. Plant-ants may thus have a mode of agriculture more similar to that of ambrosia beetles, whose fungal gardens are composed of several species of fungi and bacteria [44,45]. Interestingly, these gardens contain a primary, dominant, fungal strain, along with secondary strains [46], a pattern very similar to that we describe in ant-plant-fungi symbioses. The nature of the interaction between ambrosia beetles and their secondary symbionts is not always understood and secondary symbionts, along with bacteria, may play roles in the agricultural process. Even in higher Attine ants, in which the agricultural process was first considered to involve a limited number of coevolving symbionts [47], a whole community of recruited symbionts are now suspected to play roles [48,49,50].

Molecular cloning allowed the detection of several haplotypes of the same MOTU in a single fungal patch. These haplotypes could correspond to different ITS copies from a single individual, but rRNA gene clusters and their spacers are usually homogenised by the process of gene conversion. An alternative explanation is the occurrence of several individuals of the same MOTU in a single fungal patch. In addition, sequencing several fungal patches from a single ant colony (corresponding to one plant individual) revealed variation in their composition in some cases. For instance, one patch could contain the primary MOTU and another one the secondary MOTU. This shows that the ants do not grow a single cultivar that is propagated clonally, but instead combine different individuals and strains. A strict clonal propagation of cultivars is very unlikely to occur in insect agriculture. It has long been thought that this was the mode of propagation of the symbiont in higher Attine ants because each founding queen starts its new fungal garden from hyphae taken from its mother colony before the nuptial flight, and each colony grows a single strain [51,52]. However, it is now clear that recombination and horizontal transfers occur regularly [39,40,53], with monoculture being maintained owing to strain incompatibility [37,43].

As soon as they produce domatia, saplings of *Barteria fistulosa* are colonised by several founding queens of *Tetraponera*, each of which barricades itself in a single separate domatium by using debris to plug its entrance hole (claustral foundation). When one founding colony has reached a critical size, the workers begin to patrol outside of the domatium and kill all the other founding colonies present on the tree [54]. In the course of this study we collected from a single sapling four founding queens with brood that were still locked in their respective domatia. Each of the four domatia contained a fungal patch, and amplification was successful for two of them. Both yielded Y1 but not Y9. As we never found fungal patches in unoccupied domatia, this suggests that the fungal cultivar is brought by the founding queen either from her mother colony or passively from the environment. In the last case, founding queens would probably also introduce non-symbiotic fungi into domatia and the specificity of the association would likely be achieved through growth on an ant-specific medium that selectively favours particular Chaetothyriales fungi. Although the number of samples we were able to obtain from foundations was very low, the occurrence of the sole strain Y1 suggests that primary and secondary fungal symbionts may have different propagation dynamics. Interestingly, a previous experiment showed that *T. aethiops* ants feed preferentially on Y1, the primary symbiont, rather than on Y9 [16]. Whether Y9 represents a non-preferred symbiont, or a parasite of the system that queens avoid when founding a new colony, deserves further investigation. Ambrosia beetles of the tribe Xyleborini also treat primary and secondary fungal symbionts differently. In most cases, pseudo-vertical transmission by the beetles concerns only the primary symbiont, which is also the one that provides the highest nutritional benefits [19,46,55]. For a better understanding of agricultural practices in ant-plant-fungus interactions, further work should link the way ants manage primary and secondary fungal symbionts with the nature of their relationships.

Patterns and processes in agriculture by insects have been thoroughly investigated only in a very limited number of groups: Attine ants, Macrotermitinae and ambrosia beetles. Moreover, each of the first two groups arose from a single evolutionary event, followed by radiation. We thus need to study a much broader range of evolutionarily independent cases of agriculture to understand which mechanisms led repeatedly to successful exploitation of crops. In this context, ant-plant-fungus symbioses are promising models because they are diverse and have evolved many times independently. The very widespread occurrence of chaetothyrialean fungi-ant-plant symbioses suggests they may have a common evolutionary antecedent, such as looser associations of these fungi with non-symbiotic ants. The results presented in this study reveal consistency in patterns of species association. Further comparative analysis of agricultural processes in these symbioses will broaden our understanding of the evolution of agricultural practices by insects.

Supporting Information

Table S1 Detailed information on each individual sample of fungal patch used in this study: code used in the laboratory where genetic analyses were performed (CEFE), species of associated plant and ant, country where the sample was collected, name of the closest village or town, date of collection, geographical coordinates (WGS84, decimal degrees), name of collector, identity of the Molecular Operational Taxonomic Unit and the corresponding haplotype detected using universal or specific primers or molecular cloning.

References

1. Davidson DW, McKey D (1993) The evolutionary ecology of symbiotic ant-plant relationships. J Hymenopt Res 2: 13–83.
2. McKey D, Gaume L, Brouat C, Di Giusto B, Pascal L, et al. (2005) The trophic structure of tropical ant-plant-herbivore interactions: community consequences and coevolutionary dynamics. In: Burslem D, Pinard M, Hartley S, editors. Biotic interactions in the tropics: their role in the maintenance of species diversity. Cambridge: Cambridge University Press. pp. 386–413.
3. Frederickson ME, Greene MJ, Gordon DM (2005) "Devil's gardens" bedevilled by ants. Nature 437: 495–496.
4. Letourneau DK (1998) Ants, stem-borers, and fungal pathogens: experimental tests of a fitness advantage in Piper ant-plants. Ecology 79: 593–603.
5. Rosumek FB, Silveira FAO, Neves FD, Barbosa NPD, Diniz L, et al. (2009) Ants on plants: a meta-analysis of the role of ants as plant biotic defenses. Oecologia 160: 537–549.
6. Rico-Gray V, Oliveira PS (2007) The ecology and evolution of ant-plant interactions. Chicago: The University of Chicago Press. 331 p.
7. Blatrix R, Bouamer S, Morand S, Selosse MA (2009) Ant-plant mutualisms should be viewed as symbiotic communities. Plant Signal Behav 4: 554–556.
8. Eilmus S, Heil M (2009) Bacterial associates of arboreal ants and their putative functions in an obligate ant-plant mutualism. Appl Environ Microbiol 75: 4324–4332.
9. Gullan PJ (1997) Relationships with ants. In: Ben-Dov Y, Hodgson CJ, editors. Soft scale insects: their biology, natural enemies and control. Amsterdam: Elsevier Science. pp. 351–373.
10. Voglmayr H, Mayer V, Maschwitz U, Moog J, Djiéto-Lordon C, et al. (2011) The diversity of ant-associated black yeasts: Insights into a newly discovered world of symbiotic interactions. Fungal Biol 115: 1077–1091.
11. Rezende EL, Lavabre JE, Guimaraes PR, Jordano P, Bascompte J (2007) Non-random coextinctions in phylogenetically structured mutualistic networks. Nature 448: 925–926.
12. Vazquez DP, Bluthgen N, Cagnolo L, Chacoff NP (2009) Uniting pattern and process in plant-animal mutualistic networks: a review. Ann Bot 103: 1445–1457.
13. Bailey IW (1920) Some relations between ants and fungi. Ecology 1: 174–189.
14. Janzen DH (1972) Protection of Barteria (Passifloraceae) by Pachysima ants (Pseudomyrmecinae) in a Nigerian rain-forest. Ecology 53: 885–892.
15. Miehe H (1911) Untersuchungen über die javanische Myrmecodia. Abhandl Math Phys Kl K Sächs Gesell Wiss 32: 312–361.
16. Blatrix R, Djiéto-Lordon C, Mondolot L, La Fisca P, Voglmayr H, et al. (2012) Plant-ants use symbiotic fungi as a food source: new insight into the nutritional ecology of ant-plant interactions. Proc R Soc B 279: 3940–3947.
17. Defossez E, Selosse MA, Dubois MP, Mondolot L, Faccio A, et al. (2009) Ant-plants and fungi: a new threeway symbiosis. New Phytol 182: 942–949.
18. Defossez E, Djiéto-Lordon C, McKey D, Selosse MA, Blatrix R (2011) Plant-ants feed their host plant, but above all a fungal symbiont to recycle nitrogen. Proc R Soc B 278: 1419–1426.
19. Mueller UG, Gerardo NM, Aanen DK, Six DL, Schultz TR (2005) The evolution of agriculture in insects. Annu Rev Ecol Evol Syst 36: 563–595.
20. Hata H, Kato M (2006) A novel obligate cultivation mutualism between damselfish and Polysiphonia algae. Biol Lett 2: 593–596.
21. Silliman BR, Newell SY (2003) Fungal farming in a snail. Proc Natl Acad Sci U S A 100: 15643–15648.
22. Heath JJ, Stireman JO III (2010) Dissecting the association between a gall midge, Asteromyia carbonifera, and its symbiotic fungus, Botryosphaeria dothidea. Entomol Exp Appl 137: 36–49.
23. Lauth J, Ruiz-Gonzalez MX, Orivel J (2011) New findings in insect fungiculture. Have ants developed non-food, agricultural products? Commun Integr Biol 4: 728–730.
24. Mayer VE, Voglmayr H (2009) Mycelial carton galleries of Azteca brevis (Formicidae) as a multi-species network. Proc R Soc B 276: 3265–3273.
25. McKey D (2000) Leonardoxa africana (Leguminosae: Caesalpinioideae): a complex of mostly allopatric subspecies. Adansonia 22: 71–109.
26. Gaume L, McKey D, Terrin S (1998) Ant-plant-homopteran mutualism: how the third partner affects the interaction between a plant-specialist ant and its myrmecophyte host. Proc R Soc Lond B 265: 569–575.
27. Breteler FJ (1999) Barteria Hook. f. (Passifloraceae) revised. Adansonia 21: 306–318.
28. Peccoud J, Piatscheck F, Yockteng R, Garcia M, Sauve M, et al. (2013) Multilocus phylogenies of the genus Barteria (Passifloraceae) portray complex patterns in the evolution of myrmecophytism. Mol Phylogenet Evol 66: 824–832.
29. Schoch CL, Seifert KA, Huhndorf S, Robert V, Spouge JL, et al. (2012) Nuclear ribosomal internal transcribed spacer (ITS) region as a universal DNA barcode marker for Fungi. Proc Natl Acad Sci U S A 109: 6241–6246.
30. Gardes M, Bruns TD (1993) ITS primers with enhanced specificity for basidiomycetes - Application to the identification of mycorrhizae and rusts. Mol Ecol 2: 113–118.
31. White TJ, Bruns TD, Lee S, Taylor JW (1990) Amplification and direct sequencing of fungal ribosomal RNA genes for phylogenetics. In: Innis MA, Gelfand DH, Sninsky JJ, White TJ, editors. PCR protocols: a guide to methods and applications. San Diego: Academic Press. pp. 315–322.
32. Edgar RC (2004) MUSCLE: multiple sequence alignment with high accuracy and high throughput. Nucleic Acids Res 32: 1792–1797.
33. Guindon S, Dufayard JF, Lefort V, Anisimova M, Hordijk W, et al. (2010) New algorithms and methods to estimate maximum-likelihood phylogenies: assessing the performance of PhyML 3.0. Syst Biol 59: 307–321.
34. Blaxter M, Mann J, Chapman T, Thomas F, Whitton C, et al. (2005) Defining operational taxonomic units using DNA barcode data. Philosophical Transactions of the Royal Society B-Biological Sciences 360: 1935–1943.
35. Nilsson RH, Kristiansson E, Ryberg M, Hallenberg N, Larsson KH (2008) Intraspecific ITS variability in the kingdom Fungi as expressed in the international sequence databases and its implications for molecular species identification. Evol Bioinform 4: 193–201.
36. Ruiz-González MX, Male PJG, Leroy C, Dejean A, Gryta H, et al. (2011) Specific, non-nutritional association between an ascomycete fungus and Allomerus plant-ants. Biol Lett 7: 475–479.
37. Mueller UG, Scott JJ, Ishak HD, Cooper M, Rodrigues A (2010) Monoculture of leafcutter ant gardens. Plos One 5.
38. Silva-Pinhati ACO, Bacci M, Hinkle G, Sogin ML, Pagnocca FC, et al. (2004) Low variation in ribosomal DNA and internal transcribed spacers of the symbiotic fungi of leaf-cutting ants (Attini: Formicidae). Braz J Med Biol Res 37: 1463–1472.
39. Green AM, Mueller UG, Adams RMM (2002) Extensive exchange of fungal cultivars between sympatric species of fungus-growing ants. Mol Ecol 11: 191–195.
40. Mueller UG, Rehner SA, Schultz TR (1998) The evolution of agriculture in ants. Science 281: 2034–2038.
41. Aanen DK, Eggleton P, Rouland-Lefevre C, Guldberg-Froslev T, Rosendahl S, et al. (2002) The evolution of fungus-growing termites and their mutualistic fungal symbionts. Proc Natl Acad Sci U S A 99: 14887–14892.
42. Aanen DK, Licht HHD, Debets AJM, Kerstes NAG, Hoekstra RF, et al. (2009) High symbiont relatedness stabilizes mutualistic cooperation in fungus-growing termites. Science 326: 1103–1106.
43. Poulsen M, Boomsma JJ (2005) Mutualistic fungi control crop diversity in fungus-growing ants. Science 307: 741–744.
44. Batra LR (1966) Ambrosia fungi: extent of specificity to ambrosia beetles. Science 153: 193–195.
45. Haanstad JO, Norris DM (1985) Microbial symbiotes of the ambrosia beetle Xyloterinus politus. Microb Ecol 11: 267–276.
46. Batra LR (1985) Ambrosia beetles and their associated fungi: research trends and techniques. Proc Indian Acad Sci (Plant Sci) 94: 137–148.
47. Caldera EJ, Poulsen M, Suen G, Currie CR (2009) Insect symbioses: a case study of past, present, and future fungus-growing ant research. Environ Entomol 38: 78–92.
48. Barke J, Seipke RF, Gruschow S, Heavens D, Drou N, et al. (2010) A mixed community of actinomycetes produce multiple antibiotics for the fungus farming ant Acromyrmex octospinosus. BMC Biol 8: 109.
49. Mueller UG (2012) Symbiont recruitment versus ant-symbiont co-evolution in the attine ant-microbe symbiosis. Curr Opin Microbiol 15: 269–277.

Acknowledgments

We thank Marie-Pierre Dubois of the "Service des Marqueurs Génétiques en Ecologie (CEFE)" for her help with lab work. We thank Big John and the traditional chief for their hospitality at the village of Nkolo, Cameroon. The manuscript was improved by the comments of two reviewers.

Author Contributions

Conceived and designed the experiments: RB CB DM CA CDL. Performed the experiments: RB SD ASL CB LB CDL. Analyzed the data: RB SD ASL CB LB. Wrote the paper: RB CB DM.

50. Sen R, Ishak HD, Estrada D, Dowd SE, Hong EK, et al. (2009) Generalized antifungal activity and 454-screening of *Pseudonocardia* and *Amycolatopsis* bacteria in nests of fungus-growing ants. Proc Natl Acad Sci U S A 106: 17805–17810.
51. Chapela IH, Rehner SA, Schultz TR, Mueller UG (1994) Evolutionary history of the symbiosis between fungus-growing ants and their fungi. Science 266: 1691–1694.
52. Mueller UG (2002) Ant versus fungus versus mutualism: ant-cultivar conflict and the deconstruction of the attine ant-fungus symbiosis. Am Nat 160, supplement: S67–S98.
53. Mikheyev AS, Mueller UG, Abbot P (2006) Cryptic sex and many-to-one colevolution in the fungus-growing ant symbiosis. Proc Natl Acad Sci U S A 103: 10702–10706.
54. Yumoto T, Maruhashi T (1999) Pruning behavior and intercolony competition of *Tetraponera* (*Pachysima*) *aethiops* (Pseudomyrmecinae, Hymenoptera) in *Barteria fistulosa* in a tropical forest, Democratic Republic of Congo. Ecol Res 14: 393–404.
55. Gebhardt H, Bergerow D, Oberwinkler F (2004) Identification of the ambrosia fungus of *Xyleborus monographus* and *X. dryographus* (Curculionidae, Scolytinae). Mycol Prog 3: 95–101.

9

Development and Validation of a Weather-Based Model for Predicting Infection of Loquat Fruit by *Fusicladium eriobotryae*

Elisa González-Domínguez[1]*****, **Josep Armengol**[1], **Vittorio Rossi**[2]

1 Instituto Agroforestal Mediterráneo, Universidad Politécnica de Valencia, Valencia, Spain, **2** Istituto di Entomologia e Patologia vegetale, Università Cattolica del Sacro Cuore, Piacenza, Italy

Abstract

A mechanistic, dynamic model was developed to predict infection of loquat fruit by conidia of *Fusicladium eriobotryae*, the causal agent of loquat scab. The model simulates scab infection periods and their severity through the sub-processes of spore dispersal, infection, and latency (i.e., the state variables); change from one state to the following one depends on environmental conditions and on processes described by mathematical equations. Equations were developed using published data on *F. eriobotryae* mycelium growth, conidial germination, infection, and conidial dispersion pattern. The model was then validated by comparing model output with three independent data sets. The model accurately predicts the occurrence and severity of infection periods as well as the progress of loquat scab incidence on fruit (with concordance correlation coefficients >0.95). Model output agreed with expert assessment of the disease severity in seven loquat-growing seasons. Use of the model for scheduling fungicide applications in loquat orchards may help optimise scab management and reduce fungicide applications.

Editor: Erjun Ling, Institute of Plant Physiology and Ecology, China

Funding: This work was funded by Cooperativa Agrícola de Callosa d'En Sarrià (Alicante, Spain). Three months' stay of E. González-Domínguez at the Università Cattolica del Sacro Cuore (Piacenza, Italy) was supported by the Programa de Apoyo a la Investigación y Desarrollo (PAID-00-12) de la Universidad Politécnica de Valencia. The funders had no role in study design, data collection and analysis, decision to publish, or preparation of the manuscript.

Competing Interests: The authors have declared that no competing interests exist.

* Email: elgondo2@gmail.com

Introduction

Scab, caused by the plant-pathogenic fungus *Fusicladium eriobotryae* (Cavara) Sacc., is the main disease affecting loquat in Spain and in the whole Mediterranean basin [1,2]. The fungus affects young twigs, leaves and fruits, causing circular olive-colored spots that, on fruits, reduce their commercial value [1]. *Fusicladium* spp. are the anamorphic stages of the ascomycete genus *Venturia* but the sexual stage of *F. eriobotryae* has never been found in nature [2].

Although loquat scab is a well-known problem in the areas where loquat trees are cultivated, the biology of *F. eriobotryae* and the epidemiology of the disease have been seldom studied [1,3–8]. These studies have depicted *F. eriobotryae* as a highly rain-dependent pathogen that requires mild temperatures and long wet periods to infect loquat trees.

Environmental requirements for infection and the dispersion patterns have been studied in detail for other *Venturia* spp., such as *Venturia inaequalis* [9–16], *V. nashicola* [17–19], *V. pyrina* [20–24], *F. carpophilum* [25,26], *F. effusum* [27–29], and *F. oleagineum* [30–35]. These studies have been used to elaborate epidemiological models for some of these pathogens including *V. pyrina* [36], *V. nashicola*[37], *V. inaequalis* [38,39], *F. oleagineum* [40], and *F. effusum* [41]. For *V. inaequalis*, the use of epidemiological models to schedule fungicide applications has reduced the number of treatments [42–45]. To date, no epidemiological model has been developed for *F. eriobotryae*.

Disease modelling is an important step towards the implementation of sustainable agriculture [46,47]. Since the 1990 s, modern crop production has focused on the implementation of less intensive systems with reduced inputs of fertilizers and pesticides, and reduced use of natural resources [46]. Sustainable agriculture has its roots in Integrated Pest Management (IPM) [48]. IPM concepts originated as a reaction to the disruption of agro-ecosystems caused by massive applications of broad-spectrum pesticides in the middle of the last century [46] and also because of concern about the effects of excessive pesticide use on human health [49].

In Europe, the implementation of IPM has been legislatively mandated in recent years because of Directive 2009/128/CE regarding sustainable use of pesticides. Among other actions, the Directive encourages EU Member States to promote low pesticide-input pest control and the implementation of tools for pest monitoring and decision making, as well as advisory services (Art. 14 of the Directive). De facto, the "sustainable use" directive has made IPM mandatory in European agriculture as of 2014. As a consequence, there is an increased interest in the development and use of plant disease models to improve the timing of pesticide applications and to thus limit unnecessary treatments [46,50,51].

Our aims in this paper were (i) to develop a mechanistic, dynamic model to predict infection of loquat fruit by the scab fungus *F. eriobotryae*, and (ii) to evaluate the model against three independent data sets. The model was elaborated based on the principles of "systems analysis" [52,53] and by using recent data on the biology and epidemiology of *F. eriobotryae* obtained under environmentally controlled and field conditions [1,6,7].

Model Development

Based on the available information [1,3–8], the life cycle of *F. eriobotryae* under the Mediterranean climate is described in Figure 1. The fungus oversummers in lesions on branches and leaves and on mummified fruits that remain in the tree after harvest; during summer, high temperatures and low humidity may prevent sporulation on these lesions. Under favorable conditions in the fall, the conidia produced by the oversummering lesions serve as the primary inoculum and infect young leaves or loquat fruits. Conidia are dispersed by splashing rain to nearby fruits and leaves; with suitable temperature and wetness, conidia germinate and penetrate the tissue, probably directly through the cuticle or through stomata. Once infection has occurred and if the temperature is favorable, the fungus grows under the cuticle; conidiophores then erupt through the cuticle and produce new conidia. These conidia cause secondary infections during the entire fruiting season as long as rains disperse them and as long as temperature and wetness duration permit conidial germination, infection, and lesion growth.

Model description

The relational diagram of the model for loquat fruit infection by *F. eriobotryae* is shown in Figure 2, and the acronyms are explained in Table 1. The time step of the model is 1 hour.

The model starts at fruit set and ends at harvest because fruits are assumed to be always susceptible to infection. The model considers the lesions from the previous season on branches, old leaves, and mummified fruits as the sources of primary inoculum. Because the abundance of these lesions in an orchard may vary depending on several conditions–on, for instance, the level of disease or the fungicide treatments in the previous season–and because it is difficult to quantify these lesions, the model assumes that oversummered forms are present in the orchard and that they hold conidia at fruit set and onwards.

The model considers that any measurable rain (i.e., R≥0.2 mm in 1 hour) causes dispersal and deposition of conidia on loquat fruit [7] and triggers an infection process that potentially ends with the appearance of scab symptoms. Each site on the fruit that is occupied by a conidium or conidia is considered a potential infection site and is referred to as a lesion unit (LU). During the infection process, infection on any LU can fail because conidia may fail to germinate or may germinate but then die because of unfavorable conditions. Therefore, the proportion of LUs that become scabbed at the end of the infection process may be less than that occupied by splashing conidia at the beginning of the process.

The model predicts the progress of infection on single LUs, which are the surface unit of the fruit which can become occupied by a scab lesion. This approach is related to the concept of "carrying capacity". In ecology, the carrying capacity is interpreted broadly as the maximum population size that any area of land or water can sustain [54,55]. In plant pathology, the host's carrying capacity for disease is the maximum possible number of lesions that a plant (or an organ) can hold [56]. The carrying capacity is a common concept in plant disease modeling [57–60]. In the model described here, a LU is initially healthy (LUH) but then becomes occupied by: ungerminated conidia

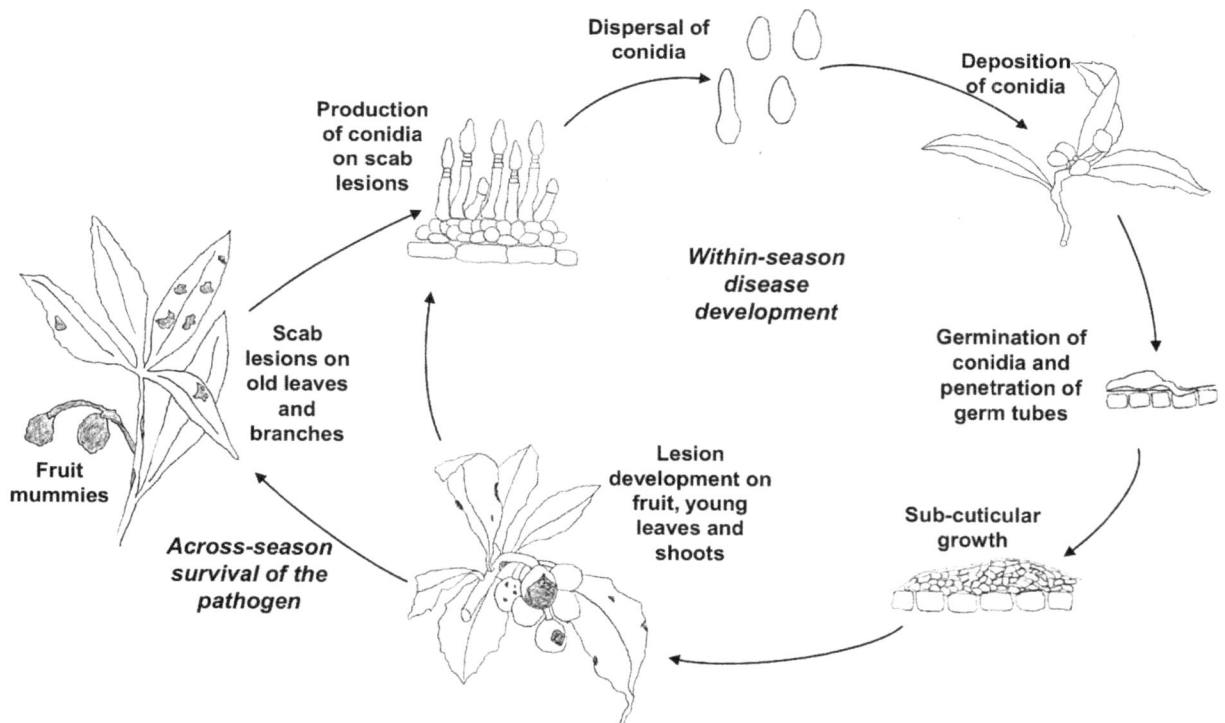

Figure 1. Disease cycle of loquat scab caused by *Fusicladium eriobotryae.*

Table 1. List of variables used in the model.

Acronym	Description	Unit
T	Air temperature	°C
RH	Relative humidity	%
R	Rainfall	mm
VPD	Vapour pressure deficit	hPa
WD	Wetness duration	hours
Teq	Temperature equivalent	°C
LUH	Unit of loquat fruit surface without conidia of *F. eriobotryae*	Number (0–1)
LUUC	Unit of loquat fruit surface with ungerminated conidia of *F. eriobotryae*	Number (0–1)
LUGC	Unit of loquat fruit surface with germinated conidia of *F. eriobotryae*	Number (0–1)
LULI	Unit of loquat fruit surface with latent infection by *F. eriobotryae*	Number (0–1)
LUVI	Unit of loquat fruit surface with visible scab lesions	Number (0–1)
GER	Cumulated conidial germination	Number (0–1)
INF	Cumulated infection	Number (0–1)
SUR	Cumulated conidial survival	Number (0–1)
GER'	Germination rate (first derivative of GERM)	Number (0–1)
INF'	Infection rate (first derivative of INF)	Number (0–1)
SUR'	Survival rate (first derivative of SUR)	Number (0–1)
C	Correction factor	Number (0–1)
DD	Degree days	Number

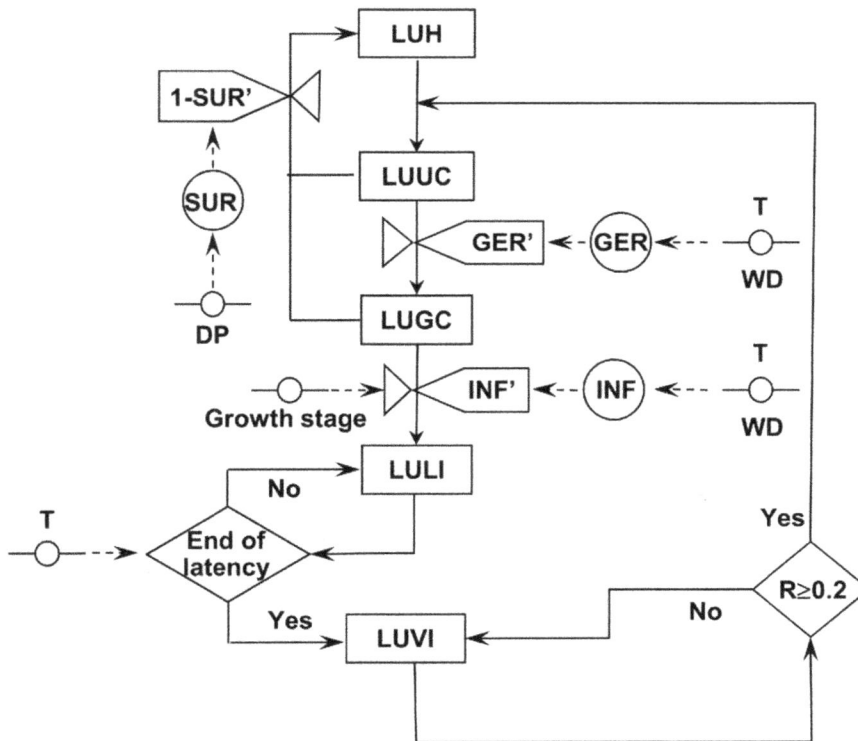

Figure 2. Relational diagram showing how the model simulates infection by *Fusicladium eriobotryae*. Legend: boxes are state variables; line arrows show fluxes and direction of changes from a state variable to the next one; valves define rates regulating these fluxes; diamonds show switches (i.e., conditions that open or close a flux); circles crossed by a line show parameters and external variables; dotted arrows show fluxes and direction of information from external variables or parameters to rates or intermediate variables; circles are intermediate variables. See Table 1 for acronym explanation.

(LUUC) at the time of conidial dispersal; germinated conidia (LUGC) at the time of conidial germination; latent infection after penetration (i.e., hyphae are invading the fruit cuticle; LULI); and visible and sporulating scab lesions at the end of latency (LUVI). Both LUUC and LUGC can fail to progress if ungerminated or germinated conidia die; these LUs then return to being LUHs because they can start a new infection process whenever new conidia are splashed on them.

At any dispersal event on hour h, the model considers that $LUUC_h = 1$. The rate at which $LUUC_h$ advances to $LUGC_h$ depends on a germination rate (GER'), and the rate at which $LUGC_h$ advances to $LULI_h$ depends on an infection rate (INF') (Figure 2). Both GER' and INF' are influenced by temperature (T in °C) and wetness duration (WD, in hours) (i.e., free water on the surface of the loquat fruit) caused by either rain or dew. Fruit surfaces are assumed to be wet on any hour when $R_h > 0$ mm, or $RH_h > 89\%$, or $VPD_h < 1$, where VPD is the vapour pressure deficit (in hPa) calculated using T_h and RH_h, following Buck [61]. The rate at which $LUUC_h$ and $LUGC_h$ returns to LUH_h depends on a survival rate (SUR'), which depends in turn on the length of the dry period (DP), i.e., the number of hours with no wetness on the fruit surface (Figure 2).

GER', INF', and SUR' are calculated at hourly intervals by using the first derivative of the equations described in González-Domínguez et al. [6] in the form:

$$
\begin{aligned}
GER' = \\
116.249 \times Teq^{4.347} \times \left(1 - Teq^{2.882}\right) \times \\
3.215 \times e^{(-0.376 \times WD)} \times e^{\left[-8.551 \times e^{(-0.376 \times WD)}\right]}
\end{aligned} \quad (1)
$$

$$
\begin{aligned}
INF' = \\
4.961 \times Teq^{1.700} \times (1 - Teq)^{0.771} \times \\
0.409 \times e^{-0.087 \times WD} \times e^{\left[4.704 \times e^{(-0.087 \times WD)}\right]}
\end{aligned} \quad (2)
$$

$$
SUR' = \frac{0.165}{DP} \quad (3)
$$

where: Teq is the temperature equivalent in the form $Teq = (T - T_{min})/(T_{max} - T_{min})$ where: T is the temperature regime, $T_{min} = 0°C$ and $T_{max} = 35°C$ in equation (1), and $T_{min} = 0°C$ and $T_{max} = 25°C$ in equation (2); WD = number of consecutive hours with wetness; DP = number of consecutive hours with no wetness. When $DP = 0$, $SUR' = 1$

At any time of the infection progress (i):

$$
LUGC_h = \sum_{i=1}^{i=t} GER'_i \times \left(1 - \sum_{i=1}^{i=t} SUR'_i\right) \times C_i
$$

$$
LUUC_h = \sum_{i=1}^{i=t} (1 - LUGC_h)_i \times \left(1 - \sum_{i=1}^{i=t} SUR'_i\right) \times C_i
$$

$$
LULI_h = \sum_{i=1}^{i=t} (INF'_i)
$$

$$
LUGC_h + LUUC_h + LULI_h + LUH_h = 1
$$

where C is a correction factor $(C = 1 - LULI_h)$.

Any infection period triggered by a conidial dispersal event ends when no viable conidia are present on any LUs, exactly when $LUUC \leq 0.01$. An example of model output for a single infection period is shown in Figure 3.

The model considers that any further rain event causes further dispersal and deposition of conidia if >5 hours have passed after the previous dispersal event. This is the time required by a lesion to produce new conidia.

Model output

The model output consists of: (i) the available inoculum on fruits (i.e., the frequency of LUs with ungerminated conidia on each day) as a measure of the potential for infection to occur; (ii) the dynamics of *LULI* for each infection process; and (iii) the seasonal dynamics of the accumulated values of *LULI* (*ΣLULI*) as an estimate of the disease in the orchard.

Examples of model output for the 2011 and 2012 loquat growing seasons are shown in Figures 4 and 5, respectively. The output is based on the weather data registered by a weather station of the Regional Agrometerological Service (http://riegos.ivia.es/) located in Callosa d'En Sarrià, Alicante Province, southeastern Spain.

Model Validation

Three data sets were used to validate the model: (i) incidence of affected fruits in a loquat orchard during growing seasons 2011 and 2012; (ii) disease occurrence on loquat fruits in single-exposure experiments in 2013; and (iii) expert assessment of the disease severity in seven loquat growing seasons.

To operate the model, hourly values of air temperature (T, °C), relative humidity (RH, %), and total rainfall (R, mm) were registered by the weather station of Callosa d'En Sarrià, which is ≤3.5 km from the orchards considered for validation.

Predicted vs. observed disease incidence in orchards

In data set (i), observations were carried out in a loquat orchard in Callosa d'En Sarrià, Alicante Province, southeastern Spain. Details on these data have been previously published [7]. Briefly, fruits from four shoots of each of 46 loquat trees were assessed weekly, and disease incidence was expressed as the percentage of fruits with scab symptoms. The disease incidence was lower in 2011 than 2012, with 27.3% and 97.6% of fruits affected by loquat scab at harvest, respectively [7]. This difference in disease incidence may be related to the fact that the orchard was treated with fungicides for scab control in 2010 but not in 2011 or 2012. Given that the inoculum sources for fruit infection in 2011 was very low because of effective disease control in 2010, a correction factor for LUUC was applied for the infection processes initiated in January 2011, i.e., LUUC = 0.1 instead of = 1 in January 2011.

Model validation was performed by comparing ΣLULI with observed data of disease incidence. Because there is a time lag (i.e., a latency period) between the predicted disease (as ΣLULI) and the disease incidence estimated in the orchard (DI), DI was shifted back by one latency period for comparison between predicted and observed disease. Sanchez-Torres et al. [1] observed a latency period of 21 days at a constant temperature of 20°C, which is a degree-day accumulation (DD base 0°C) of 420. Therefore, DI was shifted back by either 21 days or 420 DD. To calculate the

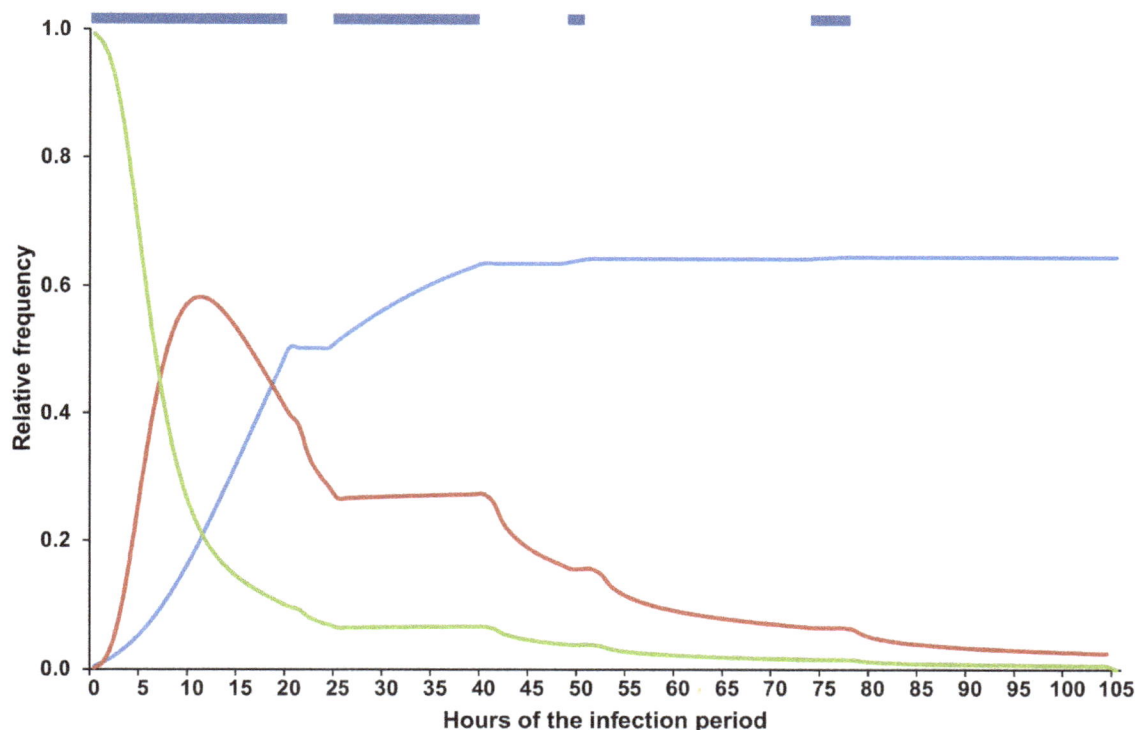

Figure 3. Dynamics of lesion units (LUs) during an infection period of *Fusicladium eriobotryae.* The graph shows the relative frequency of LUs occupied by ungerminated conidia (LUUC, in green), germinated conidia (LUGC, in red), and latent infections (LULI in blue). Blue bars at the top indicate hours with free water on the fruit surface. An infection period starts when a rain event splashes conidia on LUs and ends when no viable conidia are present on any LUs, i.e., when LUUC≤0.01.

DD, the average temperature of each day was considered with base temperature of 0°C.

Predicted vs. observed disease incidence in single-exposure experiments

In data set (ii), data were collected in an abandoned loquat orchard in Callosa d'En Sarrià from 4 February to 15 April 2013. On 25 January, 200 random shoots bearing fruits were covered with water-resistant paper bags (one shoot per bag) to prevent deposition of rain-splashed conidia. On 4 February, 10 random bags were opened to receive splashed inoculum; after seven additional days, the bags were closed again. Ten other randomly selected bags were opened on 11 February and closed again 7 days later. This operation was repeated until nine groups of shoots had been sequentially exposed to rain. At the end of the experiment (15 April 2013), disease incidence (percentage of fruits affected by loquat scab) and severity were assessed in each group of shoots. Disease severity refers to the percentage of fruit area covered by scab lesions and was measured as described by González-Domínguez et al. [8].

Model validation was performed by comparing the model output in the week when a group of shoots was exposed to splashing rain with final disease severity in that group.

Expert assessment

For data set (iii), Esteve Soler (technical advisor of the 'Cooperativa Agricola de Callosa d'En Sarrià') was asked to provide a subjective estimate of the severity (low, medium, or high) of loquat scab in the area for eight growing seasons (from 2005/2006 to 2012/2013). Mr. Soler's estimates were based on his

extensive experience in managing loquat orchards, on his scouting activities in the orchards of the cooperative, and on the number of fungicide treatments that were required to control the disease in the area.

For each season, the model was operated from 1 November to 31 March, and the numbers of disease outbreaks predicted by the model were counted. A disease outbreak was defined as $\Sigma LULI > 0.1$ in 1 day, when no outbreaks were predicted in the previous 5 days. Average and standard error of the number of predicted outbreaks were calculated for each category (low, medium, or high) of scab severity derived from the expert assessment.

Data analysis

Linear regression was used to compare the predicted and observed data of data sets (i) and (ii). To make data homogeneous, $\Sigma LULI$ values at the time of each disease assessment in the orchards were rescaled to the $\Sigma LULI$ at the end of the season; disease incidence was also rescaled to the final disease incidence. A t-test was used to test the null hypotheses that "a" (intercept of regression line) was equal to 0 and that "b" (slope of regression line) was equal to 1 [62]. The distribution of residuals of predicted versus observed values was examined to evaluate the goodness-of-fit. The concordance correlation coefficient (CCC) was calculated as a measure of model accuracy [63]; CCC is the product of two terms: the Pearson product-moment correlation coefficient between observed and predicted values and the coefficient Cb (bias estimation factor), which is an indication of the difference between the best fitting line and the perfect agreement line (CCC = 1 indicates perfect agreement). The following indexes of goodness-of-fit were also calculated [64]: NS model-efficacy

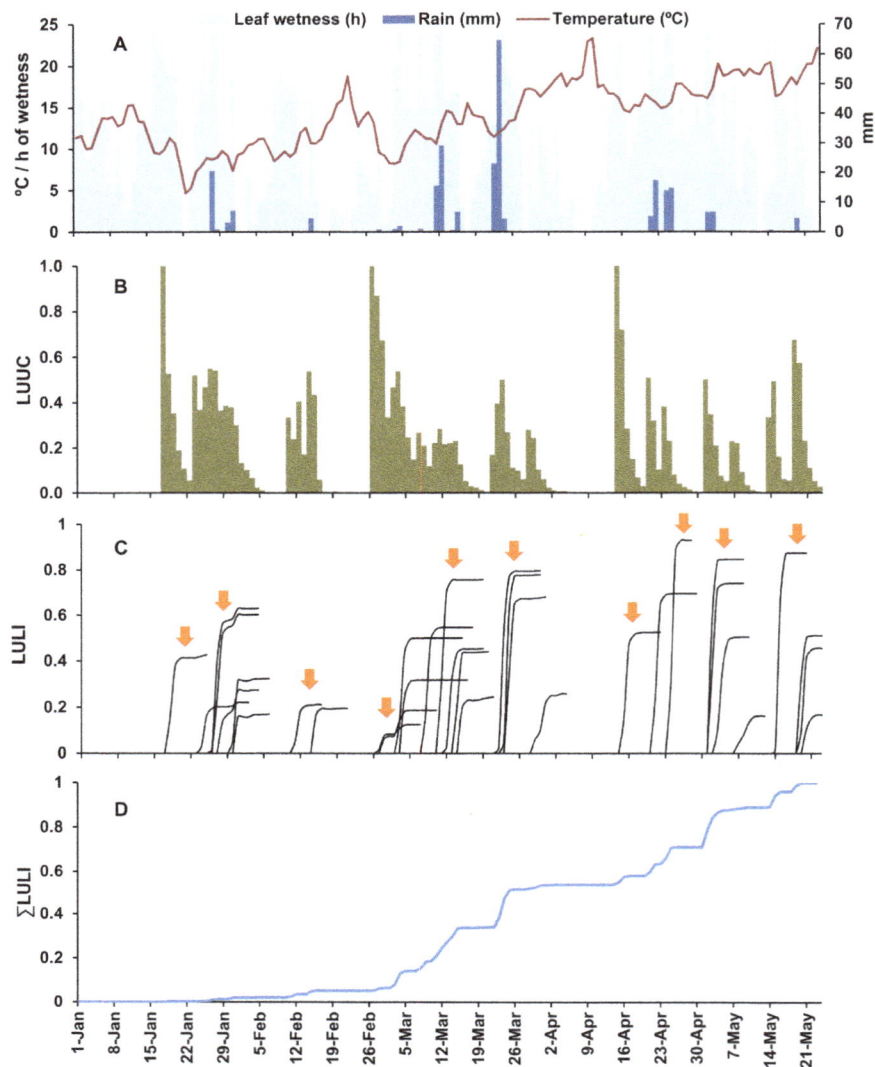

Figure 4. Weather data and model output in 2011. A: daily weather data; B: predicted frequency (%) of lesion units (LUs) with ungerminated conidia; C: predicted increase of LUs with latent infections (LULIs) for each infection period (arrows represent clusters of infection periods, clustering is based on an interval of at least 5 days between the beginning of two consecutive clusters); D: predicted seasonal dynamics of the cumulative values of LULI (ΣLULI).

coefficient, which is the ratio of the mean square error to the variance in the observed data, subtracted from unity (when the error is zero, NS = 1, and the provides a perfect fit); the W index of agreement which is the ratio between mean square error and total potential error (W = 1 represents a perfect fit); model efficiency (EF) which is a dimensionless coefficient that takes into account both the index of disagreement and the variance of the observed values (when EF increases toward 1, the fit increases); and the coefficient of residual mass (CRM) which is a measure of the tendency of the to overestimate or underestimate the observed values (a negative CRM indicates a tendency of the model toward overestimation).

For data set (iii), a one-way analysis of variance (ANOVA) was performed to determine whether the numbers of outbreaks predicted by the model in each category of loquat scab severity defined by the expert (i.e., low, medium, or high) were significantly different from one another.

Results of Model Validation

Predicted vs. observed disease incidence in orchards

In 2011 between 1 January (fruit set) and 23 May (harvest), 257.6 mm of rain fell, distributed in three main periods: the last week of January, the second week of March (with 64.8 mm of rain in 1 day), and the last 2 weeks of April (with daily temperature > 15°C) (Fig. 4A). According to the model, a total of 33 infection periods were triggered by these rain events, and the first was on 17 January (Fig. 4B and 4C). In the analysis of this model output, infection periods were clustered in "infection clusters" based on an interval of a minimum of 5 days elapsed between the beginning of two consecutive infection clusters (i.e., the protection provided by a copper-based fungicide application as described in [65]); therefore, there were 10 infection clusters in the considered period. ΣLULI began to increase from mid-January to mid-February (with three infection clusters), but three infection clusters in March resulted in a substantial increase in ΣLULI to 0.5;

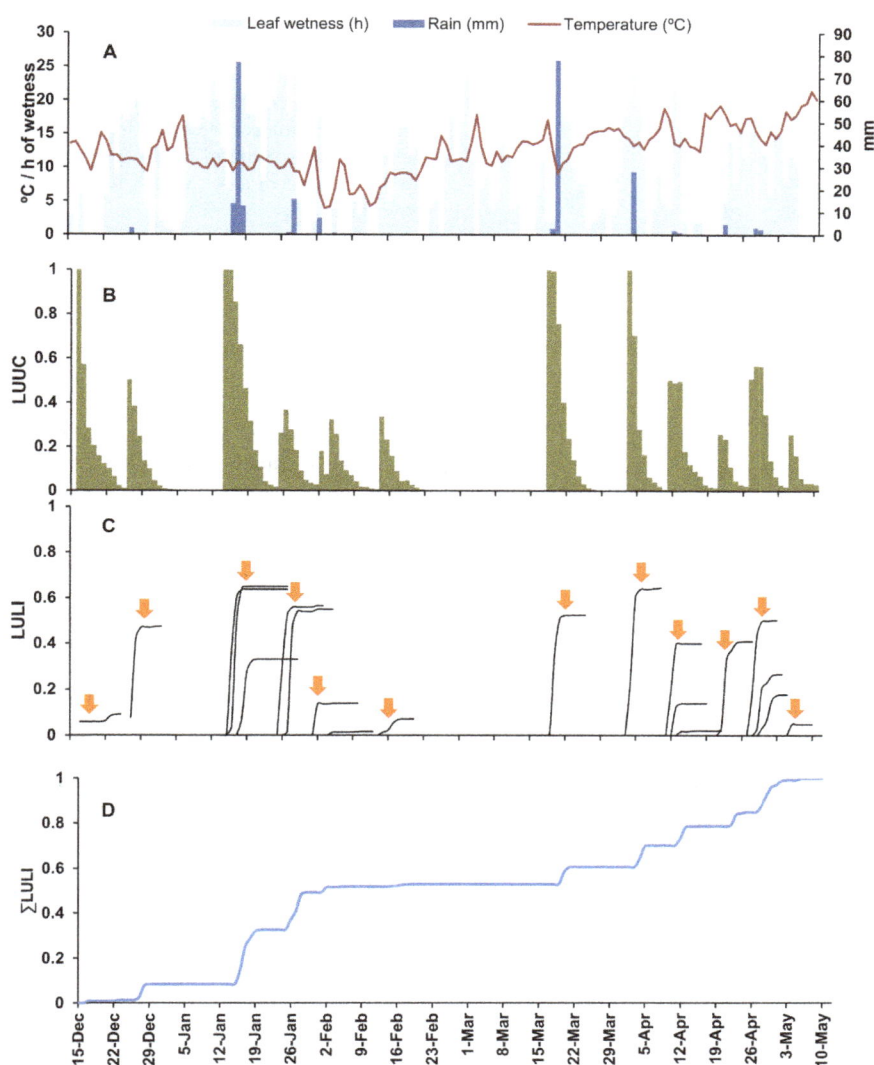

Figure 5. Weather data and model output in 2012. A: daily weather data; B: predicted frequency (%) of lesion units (LUs) with ungerminated conidia; C: predicted increase of LUs with latent infections (LULIs) for each infection period (arrows represent clusters of infection periods, clustering is based on an interval of at least 5 days between the beginning of two consecutive clusters); D: predicted seasonal dynamics of the cumulative values of LULI (ΣLULI).

March had 12 infections periods, and the repeated and abundant rain events provided >18 h of wetness on most days (Fig. 4). From mid-April to the end of the considered period, a constant increase in ΣLULI was associated with abundant rain events and increasing temperature, which triggered four infection clusters (Fig. 4).

In 2012, although the total volume of rain that fell from 15 December to 10 May was similar (255.8 mm) to that in 2011, there were fewer rain events. The model predicted 20 infection periods that were grouped into 12 infection clusters (Fig. 5B and 5C). In 2012, rainy periods were separated by dry periods; from the end of January to mid-March, dry periods caused no substantial infection to develop (Fig. 5C). Therefore, there were two main periods of ΣLULI increase: the last half of January and from the end of March to May (Fig. 5D).

Goodness-of-fit of predicted (ΣLULI) versus observed data (loquat scab incidence) was greater when a fixed period of 21 days was considered for the latency. In this case, values of R^2, CCC, r,

Cb, NS, W, and EF were >0.95 (Table 2). However, when a latency period of 420 DD was considered the values of R^2 and CCC were <0.88, and values of model efficacy (NS) and model efficiency (EF) were 0.75 (Table 2) as a consequence of the high dispersion of residues in 2012 (Figure 6). The model slightly overestimated scab incidence when a latency of 21 days was used (CRM = −0.009) and underestimated scab incidence when DD were used (CRM = 0.182) (Table 2; Figure 6). For both latency options, the regression equations of predicted versus observed data had slopes and intercepts that were not significantly different from 1 and 0, respectively.

Predicted vs. observed disease incidence in single-exposure experiments

From 4 February to 15 April 2013, the model predicted 15 loquat scab infection periods but disease outbreaks were substantial (i.e., they resulted in a >10% increase in severity) in only two exposure periods. In these two cases, LULI values were >0.1;

Table 2. Statistics and indices used for evaluating the goodness-of-fit of loquat scab infection predicted by the model versus disease observed in field.

Data set[a]	A[b]	b	P(a=0)	P(b=1)	R^2	CCC	r	Cb	NS	W	EF	CRM
Data set 1 (latency = 21 days)	0.038	0.939	0.190	0.070	0.952	0.974(0.946-0.988)	0.975	0.999	0.951	0.987	0.951	-0.009
Data set 1 (latency = 420 DD)	-0.066	0.928	0.274	0.110	0.841	0.882(0.758-0.944)	0.921	0.960	0.753	0.939	0.752	0.182
Data set 2	0.043	0.965	0.02	0.1	0.984	0.986(0.942-0.996)	0.993	0.993	0.971	0.992	0.971	-0.247

[a]Data set 1 corresponds to comparison of daily accumulated LUVI predicted by the model versus observed data of loquat scab incidence in an orchard in southeastern Spain during 2 years (2011 and 2012). The model used a latency period of 21 days (first row) or 420 DD (second row). Data set 2 compares the increase of model output in weeks in which loquat shoots were exposed to splashing rain (triggering infection) with final disease severity in those shoots.
[b]a and b, parameters of the regression line of the predicted against observed values; P, probability level for the null hypotheses that a = 0 and b = 1; R^2, coefficient of determination of the regression line; CCC, concordance correlation coefficient; r, Pearson product-moment correlation coefficient; Cb, bias estimation factor; NS, model efficacy; W, index of agreement; EF, model efficiency; CRM, coefficient of residual mass.

when there were no or light outbreaks, *LULI* values were <0.06 (Figure 7). The goodness-of-fit of predicted versus observed for data set (ii) (Table 2) provided values >0.97 for R^2, CCC, r, Cb, NS, W, and EF. Although the slope was not significantly different from 1, the intercept was different from 0 at $P = 0.02$ (Table 2). The negative value of CRM indicated that the model somewhat overestimated disease, mainly when observed disease severity was low (Figure 7).

Expert assessment

The loquat scab epidemics that occurred in the eight seasons of data set (iii) were considered by the expert to be of low, medium, or high severity in two, three, and three seasons, respectively. The number of outbreaks predicted by the model ranged from 4 to 17 among the eight seasons; in average, 8.5±0.5 outbreaks were predicted for years with low value of loquat scab severity, 10±3 for year with medium value and 12±2.9 for years with high value. Although the average number of outbreaks predicted by the model increased as the expert assessment of disease severity increased, the number of predicted epidemics did not significantly differ among the severity categories ($P = 0.71$).

Discussion

In this work, a dynamic model was developed to predict infection of loquat fruits by conidia of *F. eriobotryae*. The model uses a mechanistic approach to describe the infection process [53,66,67]: the model splits the disease cycle of *F. eriobotryae* into different state variables, which change from one state to the following state based on rate variables or switches that depend on environmental conditions by means of mathematical equations. The mathematical equations were developed using published data on *F. eriobotryae* conidial dispersion patterns [7] and on *F. eriobotryae* growth, conidial germination, and infection under different environmental conditions [1,6]. In the absence of precise information, assumptions were made based on available knowledge.

Model validation showed that the model correctly predicted the occurrence of infection periods and the severity of any infection period, as demonstrated by the goodness-of-fit for the data collected on fruits exposed to single rainy periods. Because the purpose of the model is to be part of a warning system for loquat scab management, the ability to correctly predict infection periods is crucial. Accuracy of the model was also confirmed by the comparison of model output with expert assessment. Even though the numbers of predicted outbreaks did not differ among seasons that the expert had categorized as having low, medium, or high disease severity, the number of predicted outbreaks increased with increases in assessed disease severity.

For model validation, the latency period required for the appearance of scab was expressed as a fixed number of days or of degree-days (DD) based on results from Sánchez-Torres et al. [1]. Goodness-of-fit of model prediction was overall better using a fixed period of 21 days instead of 420 DD. In particular, the model underestimated the disease in the early season of 2012 when DD were used. The underestimation was probably caused by low temperatures in that period, which delayed DD accumulation. This result is questionable, because the physiological development of fungi is usually more closely related to DD than to calendar days [68,69]. In this work, DD was fixed based on the latency period observed in loquat plants kept at the optimal temperature for *F. eriobotryae* development, i.e., 21 days at 20°C [1]. Therefore, the DD value used in this study did not account for the non-linear response of *F. eriobotryae* growth to temperatures between 5 and

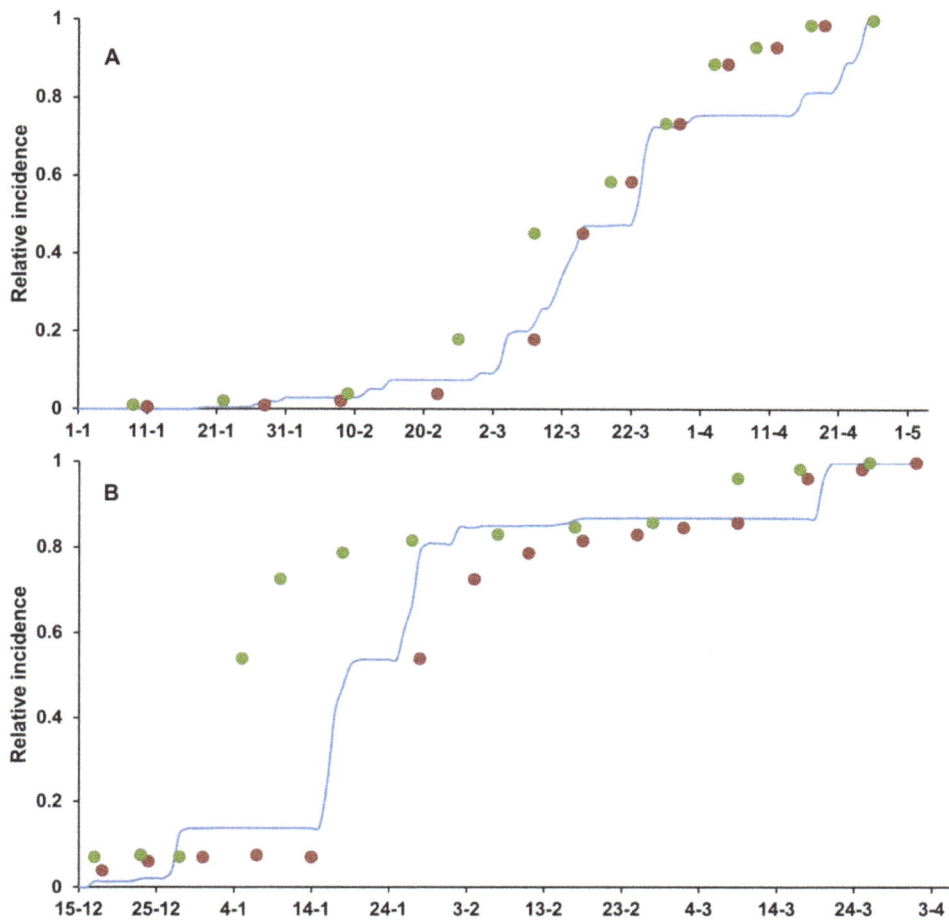

Figure 6. Comparison between model output and scab observed on loquat fruit in southeastern Spain. (A) data from 2011 and (B) data from 2012. Blue lines represent the rescaled infection predicted by the model as the seasonal summation of the lesion units with latent infections (ΣLULI). Points represent rescaled incidence of loquat fruit with scab observed in the orchards; rescaled incidence is shifted back by 21 days (red points) or 420 DD (base 0°C, green points) to account for the latency period, i.e., the time elapsed between infection and visible symptoms in the form of sporulating scab lesions.

30°C [6]. If a function for predicting the appearance of scab symptoms is needed in the model, such a function should be temperature dependent, as it is in models for *V. nashicola* [37] and *F. oleagineum* [40]. Salerno et al. [4] repeatedly exposed potted loquat plants under the canopies of affected trees for 3 days and then incubated these plants under a roof until the appearance of symptoms. Scab appeared in 11 to 26 days at temperatures ranging from 11.4 to 17°C (with a DD range of 157 to 340) and after >220 days at temperatures >20°C. Ptskialadze [5] found scab symptoms on both leaves and fruits 34 and 16 days after infection at 1–4°C and 21–25°C, respectively. Even though the calculation of latency can be improved, the model error in predicting disease onset due to a fix latency period may not reduce the ability of the model to correctly predict infection periods or reduce the value of the model for timing fungicides applications.

The model capitalized on recent research concerning loquat scab [1,6,7]. These studies have considered most of the components of the disease cycle, including dispersion of conidia, infection, incubation, and latency. Nevertheless, other components should be elucidated to improve our knowledge and thus to improve the model [66]. Currently, the model assumes that inoculum sources are always present in scab-affected loquat orchards and that viable *F. eriobotryae* inoculum is always present at fruit set (i.e., when the model begins operating) and beyond. Salerno et al. [4] found that lesions appear in autumn on leaves that were infected the previous spring, and Prota [3] found that the lesions appearing in autumn produce conidia for 5 to 6 months and that those viable conidia are present all year long. These observations were carried out in Sicily and Sardinia, respectively (i.e., under a Mediterranean climate); therefore, the model assumptions seem plausible. The assumptions that inoculum sources and viable conidia are always present in scab-affected loquat orchards are both precautionary because they can lead to over prediction of infection (which would occur if weather conditions were suitable for infection but no viable conidia were available) and thus to unnecessary applications of fungicides or other disease management measures. Because unnecessary fungicide applications entail costs for growers, consumers, and the environment [51], the model should be expanded to include the oversummering and availability of conidia.

With respect to oversummering, modeling the dormant stage of fungal pathogens is challenging [66], and the dormant stage has therefore been included in only a few models [70–74]. For this purpose, two key aspects must be addressed: (i) the inoculum dose (i.e., the quantity of inoculum that oversummers), which depends on the severity of the disease in each orchard at the end of the

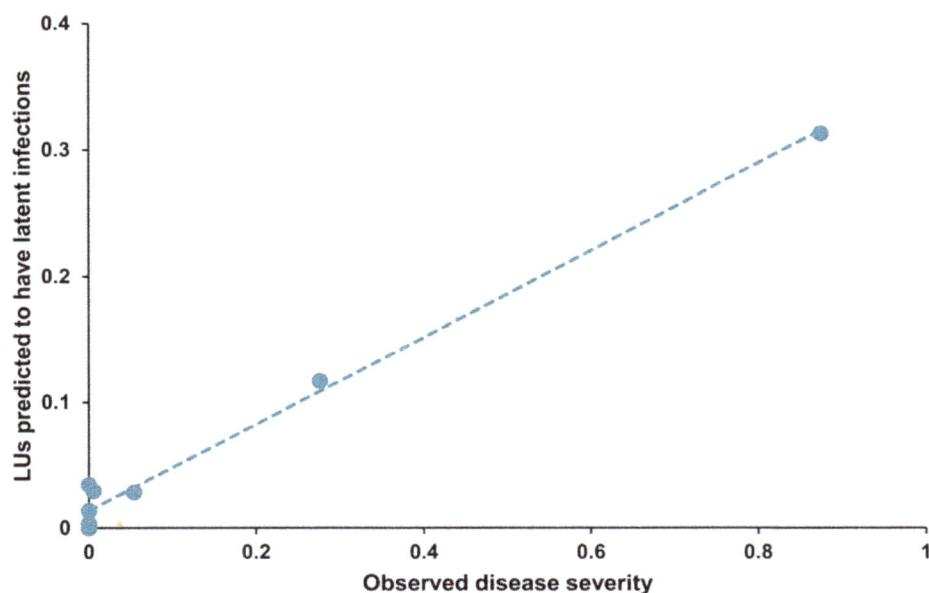

Figure 7. Comparison between model output and scab on loquat fruit in single-exposure experiments. Experiments were carried out in a loquat orchard in southeastern Spain in 2013. Observed data (X axis) are expressed as the rescaled disease severity in 11 groups of fruits that were exposed (for 7-day-long moving periods) to splashing rain in a severely affected orchard; model output (Y axis) is expressed as the summation of the lesion units with latent infections (ΣLULI) in the exposure period.

previous season; and (ii) the time when the primary inoculum begins to be available for infection. In other models, the inoculum dose was directly measured in the field [70,74] or broadly estimated as low/high disease pressure [72]. In our case, incorporation into the model of the specific farmer's assessment of the disease severity in the previous season may represent useful information regarding the potential primary inoculum dose. Modeling the sporulation patterns of *F. eriobotryae* may make it possible to estimate the available inoculum at each infection period. This estimation may consequently improve the ability of the model to predict the severity of each infection period. To account for the presence of inoculum in a model for *V. inaequalis*, Xu et al. [39] assumed a minimum interval of 7 hours between two successive infection processes to allow lesions to recover and sporulate, even though this approximation could introduce errors, because sporulation is highly dependent on temperature and RH [14].

Even without the above possible improvements, the present model can contribute to the practical control of loquat scab. The underutilization of disease predictive systems by farmers have been broadly discussed [46,47,51,75,76]. Rossi et al. [53] summarized the steps necessary for the practical implementation of a model as: (i) develop a computerized version of the model; (ii) create a network of agro-meteorological stations for collecting weather data; (iii) design a strategy for decision-making based on the model output; (iv) develop tools for supporting decision-making (e.g., decision support systems or disease warning systems); and (v) build user's confidence in the model by demonstrating the advantages of its use in comparison with the current options. Efforts devoted to the last three steps are crucial for the future applicability of the model [53] and requires a deep knowledge of the cultural context in which the model will be delivered, the farmers' perception of risk, and the current management of the disease [47].

In the main loquat cultivation areas of Spain, the regional plant protection services use the Mills-Laplante tables [77], which were

developed to control apple scab, to estimate the risk of infection by *F. eriobotryae* [78]. Researchers have indicated that the Mills-Laplante tables over-predict the number of infections for apple scab [37,79]. That the tables could over-predict the number of loquat scab infections has also been discussed, because the conidia of *F. eriobotryae* require longer times for leaf infection than those described by the Mills-Laplante tables for *V. inaequalis* and because the temperature range in which *F. eriobotryae* infection occurs is quite different [6,7]. Thus, the present model represents an improvement in loquat scab management, i.e., it should optimise scab management by helping loquat growers to schedule and probably to reduce fungicide applications.

The long-term existence of a warning system for loquat scab monitoring in Spain [80] may facilitate the implementation of the model developed in this area because i) extension agents and advisors are familiar with the use and interpretation of epidemiological models, and ii) loquat farmers are accustomed to considering the concept of "infection risk" when scheduling fungicide applications.

Because model building is "a never-ending story" [47,81], researchers will likely continue to improve the loquat scab model described here. As discussed in this manuscript, it will be necessary to define a relationship between model output and infection severity so as to identify appropiate thresholds for deciding when the treatments are needed [53,63].

Acknowledgments

We would like to thank E. Soler his contribution for model validation.

Author Contributions

Conceived and designed the experiments: EGD VR JA. Performed the experiments: EGD VR. Analyzed the data: EGD VR. Contributed reagents/materials/analysis tools: JA VR. Contributed to the writing of the manuscript: EGD JA VR.

References

1. Sánchez-Torres P, Hinarejos R, Tuset JJ (2009) Characterization and pathogenicity of *Fusicladium eriobotryae*, the fungal pathogen responsible for loquat scab. Plant Dis 93: 1151–1157.
2. Gladieux P, Caffier V, Devaux M, Le Cam B (2010) Host-specific differentiation among populations of *Venturia inaequalis* causing scab on apple, pyracantha and loquat. Fungal Genet Biol 47: 511–521.
3. Prota U (1960) Ricerche sulla «ticchiolatura del Nespolo del Giappone e sul suo agente (*Fusicladium eriobotryae* Cav.). I. Observazioni sull'epidemiologia della malattia e sui caratteri morfo-biologici del parassita in Sardegna. Stud di Sassari 8: 175–196.
4. Salerno M, Somma V, Rosciglione B (1971) Ricerche sull'epidemiologia della ticchiolatura del nespolo del Giappone. Tec Agric 23: 3–15.
5. Ptskialadze L (1968) The causal agent of loquat scab and its biological characteristics. Rev Appl Mycol 47: 268.
6. González-Domínguez E, Rossi V, Armengol J, García-Jiménez J (2013) Effect of environmental factors on mycelial growth and conidial germination of *Fusicladium eriobotryae*, and the infection of loquat leaves. Plant Dis 97: 1331–1338.
7. González-Domínguez E, Rossi V, Michereff SJ, García-Jiménez J, Armengol J (2014) Dispersal of conidia of *Fusicladium eriobotryae* and spatial patterns of scab in loquat orchards in Spain. Eur J Plant Pathol 139: 849–861.
8. González-Domínguez E, Martins RB, Del Ponte EM, Michereff SJ, García-Jiménez J, et al. (2014) Development and validation of a standard area diagram set to aid assessment of severity of loquat scab on fruit. Eur J Plant Pathol 139: 413–422.
9. Becker CM, Burr TJ (1994) Discontinuous wetting and survival of conidia of *Venturia inaequalis* on apple leaves. Phytopathology 84: 372–378.
10. Hartman JRR, Parisi L, Bautrais P (1999) Effect of leaf wetness duration, temperature, and conidial inoculum dose on apple scab infections. Plant Dis 83: 531–534.
11. Holb IJ, Heijne B, Withagen JCM, Jeger MJ (2004) Dispersal of *Venturia inaequalis* ascospores and disease gradients from a defined inoculum source. J Phytopathol 152: 639–646.
12. Rossi V, Giosue S, Bugiani R (2003) Influence of air temperature on the release of ascospores of *Venturia inaequalis*. J Phytopathol 151: 50–58.
13. Stensvand A, Gadoury DM, Amundsen T, Semb L, Seem RC (1997) Ascospore release and infection of apple leaves by conidia and ascospores of *Venturia inaequalis* at low temperatures. Phytopathology 87: 1046–1053.
14. Machardy WE (1996) Apple scab. Biology, epidemiology and management. St. Paul: APS Press. 545.
15. James J, Sutton TB (1982) Environmental factors influencing pseudothecial development and ascospore maturaion of *Venturia inaequalis*. Phytopathology 72: 1073–1080.
16. Boric B (1985) Influence of temperature on germability of spores of *Venturia inaequalis* (Cooke) Winter, and their viability as affected by age. Zast Bilja 36: 295–302.
17. Li B, Zhao H, Xu X-M (2003) Effects of temperature, relative humidity and duration of wetness period on germination and infection by conidia of the pear scab pathogen (*Venturia nashicola*). Plant Pathol 52: 546–552.
18. Li BH, Xu XM, Li JT, Li BD (2005) Effects of temperature and continuous and interrupted wetness on the infection of pear leaves by conidia of *Venturia nashicola*. Plant Dis 54: 357–363.
19. Umemoto S (1990) Dispersion of ascospores and conidia of causal fungus of japanese pear scab, *Venturia nashicola*. Ann Phytopathol Japan Soc 56: 468–473.
20. Rossi V, Salinari F, Pattori E, Giosuè S, Bugiani R (2009) Predicting the dynamics of ascospore maturation of *Venturia pirina* based on environmental factors. Phytopathology 99: 453–461.
21. Spotts RA, Cervantes A (1991) Effect of temperature and wetness on infection of pear by *Venturia pirina* and the relationship between preharvest inoculation and storage scab. Plant Dis 75: 1204–1207.
22. Spotts RA, Cervantes A, Cervantes LA (1994) Factors affecting maturation and release of ascospores of *Venturia pirina* in oregon. Phytopathology 84: 260–264.
23. Villalta O, Washington WS, Rimmington GM, Taylor PA (2000) Influence of spore dose and interrupted wet periods on the development of pear scab caused by *Venturia pirina* on pear (*Pyrus communis*) seedlings. Australas Plant Pathol 29: 255–262.
24. Villalta ON, Washington WS, Rimmington GM, Taylor PA (2000) Effects of temperature and leaf wetness duration on infection of pear leaves by *Venturia pirina*. Aust J Agric Res 51: 97–106.
25. Lan Z, Scherm H (2003) Moisture sources in relation to conidial dissemination and infection by *Cladosporium carpophilum* within peach canopies. Phytopathology 93: 1581–1586.
26. Lawrence E, Zehr E (1982) Enviromental effects on the development and dissemination of *Cladosporium carpophilum* on peach. Phytopathology 72: 773–776.
27. Gottwald TR, Bertrand PF (1982) Patterns of diurnal and seasonal airborne spore concentration of *Fusicladium effusum* and its impact on a pecan scab epidemic. Phytopathology 72: 330–335.
28. Gottwald TR (1985) Influence of temperature, leaf wetness period, leaf age, and spore concentration on infection of pecan leaves by conidia of *Cladosporium caryigenum*. Phytopathology 75: 190–194.
29. Latham AJ (1982) Effect of some weather factors and *Fusicladium effusum* conidium dispersal on pecan scab occurrence. Phytopathology 72: 1339–1345.
30. De Marzo L, Frisullo S, Lops F, Rossi V (1993) Possible dissemination of *Spilocaea oleagina* conidia by insects (*Ectopsocus briggsi*). EPPO Bull 23: 389–391.
31. Lops F, Frisullo S, Rossi V (1993) Studies on the spread of the olive scab pathogen, *Spilocaea oleagina*. EPPO Bull 23: 385–387.
32. Obanor FO, Walter M, Jones EE, Jaspers MV (2008) Effect of temperature, relative humidity, leaf wetness and leaf age on *Spilocaea oleagina* conidium germination on olive leaves. Eur J Plant Pathol 120: 211–222.
33. Obanor FO, Walter M, Jones EE, Jaspers MV (2010) Effects of temperature, inoculum concentration, leaf age, and continuous and interrupted wetness on infection of olive plants by *Spilocaea oleagina*. Plant Pathol 60: 190–199.
34. Viruega JR, Moral J, Roca LF, Navarro N, Trapero A (2013) *Spilocaea oleagina* in olive groves of southern Spain: survival, inoculum production, and dispersal. Plant Dis 97: 1549–1556.
35. Viruega JR, Roca LF, Moral J, Trapero A (2011) Factors affecting infection and disease development on olive leaves inoculated with *Fusicladium oleagineum*. Plant Dis 95: 1139–1146.
36. Eikemo H, Gadoury DM, Spotts RA, Villalta O, Creemers P, et al. (2011) Evaluation of six models to estimate ascospore maturation in *Venturia pyrina*. Plant Dis 95: 279–284.
37. Li B, Yang J, Dong S, Li B, Xu X (2007) A dynamic model forecasting infection of pear leaves by conidia of *Venturia nashicola* and its evaluation in unsprayed orchards. Eur J Plant Pathol 118: 227–238.
38. Rossi V, Bugiani R (2007) A-scab (Apple-scab), a simulation model for estimating risk of *Venturia inaequalis* primary infections. EPPO Bull 37: 300–308.
39. Xu X, Butt DJ, Santen VAN (1995) A dynamic model simulating infection of apple leaves by *Venturia inaequalis*. Plant Pathol 44: 865–876.
40. Roubal C, Regis S, Nicot PC (2013) Field models for the prediction of leaf infection and latent period of *Fusicladium oleagineum* on olive based on rain, temperature and relative humidity. Plant Pathol 62: 657–666.
41. Payne AF, Smith DL (2012) Development and evaluation of two pecan scab prediction models. Plant Dis 96: 1358–1364.
42. Trapman M, Jansonius PJ (2008) Disease management in organic apple orchards is more than applying the right product at the correct time. Ecofruit-13th International Conference on Cultivation Technique and Phytopathological Problems in Organic Fruit-Growing: Proceedings to the Conference from 18th February to 20th February 2008 at Weinsberg/Germany. 16–22.
43. Holb IJ, Jong PF, Heijne B (2003) Efficacy and phytotoxicity of lime sulphur in organic apple production. Ann Appl Biol 142: 225–233.
44. Jamar L, Cavelier M, Lateur M (2010) Primary scab control using a "during-infection" spray timing and the effect on fruit quality and yield in organic apple production. 14: 423–439.
45. Giosuè S, Bugiani R, Caffi T, Pradolesi GF, Melandri M, et al. (2010) Used of the A-scab model for rational control of apple scab. IOBC WPRS Bull 54: 345–349.
46. Rossi V, Caffi T, Salinari F (2012) Helping farmers face the increasing complexity of decision-making for crop protection. Phytopathol Mediterr 51: 457–479.
47. Gent DH, Mahaffee WF, McRoberts N, Pfender WF (2013) The use and role of predictive systems in disease management. Annu Rev Phytopathol 51: 267–289.
48. Boller EEF, Avilla J, Gendrier JP, Jörg E, Malavolta C (1998) Integrated Production in Europe: 20 years after the declaration of Ovronnaz. IOBC Bull 21: 1–34.
49. Alavanja MCR, Hoppin JA, Kamel F (2004) Health effects of chronic pesticide exposure: cancer and neurotoxicity. Annu Rev Public Health 25: 155–197.
50. Brent KJ, Hollomon DW (2007) Fungicide resistance in crop pathogens: How can it be managed? FRAC Monog 2. Fungicide Resistance Action Committee.
51. Shtienberg D (2013) Will decision-support systems be widely used for the management of plant diseases? Annu Rev Phytopathol 51: 1–16.
52. Leffelaar P (1993) On Systems Analysis and Simulation of Ecological Processes. Kluwer. London.
53. Rossi V, Giosuè S, Caffi T (2010) Modelling plant diseases for decision making in crop protection. In: Oerke E-C, Gerhards R, Menz G, Sikora RA, editors. Precision Crop Protection-the Challenge and Use of Heterogeneity.
54. Hui C (2006) Carrying capacity, population equilibrium, and environment's maximal load. Ecol Modell 192: 317–320.
55. Townsend C, Begon M, Harper J (2008) Essentials of ecology. John Wiley and Sons. New York. 510.
56. Zadoks J, Schein R (1979) Epidemiology and plant disease management. Oxford University Press, New York. 427.
57. Bennett JC, Diggle A, Evans F, Renton M (2012) Assessing eradication strategies for rain-splashed and wind-dispersed crop diseases. Pest Manag Sci 69: 955–963.
58. Caffi T, Rossi V, Bugiani R, Spanna F, Flamini L, et al. (2009) A model predicting primary infections of *Plasmopara viticola* in different grapevine-growing areas of Italy. J Plant Pathol 91: 535–548.

59. Ghanbarnia K, Dilantha Fernando WG, Crow G (2009) Developing rainfall- and temperature-based models to describe infection of canola under field conditions caused by pycnidiospores of *Leptosphaeria maculans*. Phytopathology 99: 879–886.

60. Gilligan CA, van den Bosch F (2008) Epidemiological models for invasion and persistence of pathogens. Annu Rev Phytopathol 46: 385–418.

61. Buck AL (1981) New equations for computing vapor pressure and enhancement factor. J Appl Meteorol 20: 1527–1532.

62. Teng P (1981) Validation of computer models of plant disease epidemics: a review of philosophy and methodology. J Plant Dis Prot 88: 49–63.

63. Madden L V, Hughes G, van den Bosch F (2007) The study of plant disease epidemics. APS press. St. Paul. 421.

64. Nash J, Sutcliffe J (1970) River flow forecasting through conceptual models part I. J Hidrol 10: 282–290.

65. González-Domínguez E, Rodríguez-Reina J, García-Jiménez J, Armengol J (2014) Evaluation of fungicides to control loquat scab caused by *Fusicladium eriobotryae*. Plant Heal Prog Accepted.

66. De Wolf ED, Isard SA (2007) Disease cycle approach to plant disease prediction. Annu Rev Phytopathol 45: 203–220.

67. Krause RA, Massie LB (1975) Predictive systems: modern approaches to disease control. Annu Rev Phytopathol 13: 31–47.

68. Fourie P, Schutte T, Serfontein S, Swart F (2013) Modeling the effect of temperature and wetness on *Guignardia pseudothecium* maturation and ascospore release in citrus orchards. Phytopathology 103: 281–292.

69. Gadoury DM, Machardy WE (1982) A model to estimate the maturity of ascospores of *Venturia inaequalis*. Phytopathology 72: 901–904.

70. Holtslag QA, Remphrey WR, Fernando WGD, Ash GHB (2004) The development of a dynamic disease- forecasting model to control *Entomosporium mespili* on Amelanchier alnifolia. Can J Plant Pathol 313: 304–313.

71. Legler SEE, Caffi T, Rossi V (2013) A Model for the development of Erysiphe necator chasmothecia in vineyards. Plant Pathol. DOI:10.1111/ppa.12145.

72. Luo Y, Michailides TJ (2001) Risk analysis for latent infection of prune by *Monilinia fructicola* in California. Phytopathology 91: 1197–1208.

73. Rossi V, Caffi T, Giosuè S, Girometta B, Bugiani R, et al. (2005) Elaboration and validation of a dynamic model for primary infections of *Plasmopara viticola* in North Italy. Riv Ital di Agrometeorol 13: 7–13.

74. Gadoury D, Machardy WE (1986) Forecasting ascospore dose of *Venturia inaequalis* in commercial apple orchards. Phytopathology 76: 112–118.

75. Gent DH, De Wolf E, Pethybridge SJ (2011) Perceptions of risk, risk aversion, and barriers to adoption of decision support systems and integrated pest management: an introduction. Phytopathology 101: 640–643.

76. Schut M, Rodenburg J, Klerkx L, van Ast A, Bastiaans L (2014) Systems approaches to innovation in crop protection. A systematic literature review. Crop Prot 56: 98–108.

77. Mills W, Laplante A (1954) Diseases and insect in the orchard. Cornell Ext Bull 711.

78. GVA (2013) Octubre-Noviembre 2013. Butlletí d'avisos 13.

79. Machardy WE, Gadoury DM (1989) A revisions of Mills's criteria for predicting apple scab infection periods. Phytopathology 79: 304–310.

80. González-Domínguez E, Armengol J, García-Jiménez J, Soler E (2013) El moteado del níspero en la Marina Baixa. Phytoma 247: 50–52.

81. Teng P (1981) Validation of computer models of plant disease epidemics: a review of philosophy and methodology. J Plant Dis Prot 88: 49–63.

10

Rapid Identification of Antifungal Compounds against *Exserohilum rostratum* Using High Throughput Drug Repurposing Screens

Wei Sun[1,9], Yoon-Dong Park[2,9], Janyce A. Sugui[2], Annette Fothergill[3], Noel Southall[1], Paul Shinn[1], John C. McKew[1], Kyung J. Kwon-Chung[2], Wei Zheng[1]*, Peter R. Williamson[2,4]*

1 National Center for Advancing Translational Sciences, National Institutes of Health, Bethesda, Maryland, United States of America, 2 Laboratory of Clinical Infectious Diseases, National Institute of Allergy and Infectious Diseases, National Institutes of Health, Bethesda, Maryland, United States of America, 3 University of Texas Health Science Center, San Antonio, Texas, United States of America, 4 Section of Infectious Diseases, Immunology and International Medicine, University of Illinois College of Medicine, Chicago, Illinois, United States of America

Abstract

A recent large outbreak of fungal infections by *Exserohilum rostratum* from contaminated compounding solutions has highlighted the need to rapidly screen available pharmaceuticals that could be useful in therapy. The present study utilized two newly-developed high throughput assays to screen approved drugs and pharmaceutically active compounds for identification of potential antifungal agents. Several known drugs were found that have potent effects against *E. rostratum* including the triazole antifungal posaconazole. Posaconazole is likely to be effective against infections involving septic joints and may provide an alternative for refractory central nervous system infections. The anti-*E. rostratum* activities of several other drugs including bithionol (an anti-parasitic drug), tacrolimus (an immunosuppressive agent) and floxuridine (an antimetabolite) were also identified from the drug repurposing screens. In addition, activities of other potential antifungal agents against *E. rostratum* were excluded, which may avoid unnecessary therapeutic trials and reveals the limited therapeutic alternatives for this outbreak. In summary, this study has demonstrated that drug repurposing screens can be quickly conducted within a useful time-frame. This would allow clinical implementation of identified alternative therapeutics and should be considered as part of the initial public health response to new outbreaks or rapidly-emerging microbial pathogens.

Editor: Ping Wang, Research Institute for Children and the Louisiana State University Health Sciences Center, United States of America

Funding: This work was supported by the Intramural Research Programs of the National Center for Advancing Translational Sciences and National Institute of Allergy and Infectious Diseases, National Institutes of Health. The funders had no role in study design, data collection and analysis, decision to publish, or preparation of the manuscript.

Competing Interests: The authors have declared that no competing interests exist.

* E-mail: williamsonpr@mail.nih.gov (PRW); wzheng@mail.nih.gov (WZ)

9 These authors contributed equally to this work.

Introduction

Unusual or highly antibiotic resistant organisms may subject large numbers of individuals to unexpected infectious diseases due to greater globalization that brings more widespread distribution networks and potential threats such as bioterrorism. Limited therapeutic options or failures in conventional therapy during these outbreaks can be encountered because of either intolerable drug toxicities or lack of efficacious drugs. Recently, a large outbreak of fungal infections has been caused by the widespread distribution of contaminated preservative-free methylprednisolone acetate prepared by a single compounding pharmacy [1,2,3,4]. It has currently resulted in 741 infections with 55 deaths [5]. *Exserohilum rostratum*, a dermatiaceous fungus commonly found on plants, in soil and in households, has been identified as one of the predominant pathogens in the current multistate outbreak of fungal meningitis and other infections associated with contaminated steroid injections. Although it rarely causes infections in healthy people, infections of skin and corneal tissues as well as

more disseminated infections in immunocompromised populations have been reported [6,7,8].

E. rostratum is sensitive to amphotericin B, a commonly used antifungal agent, but the severe and potentially lethal side-effects of this drug have limited its use in certain patients. While traditional antibiotic susceptibility testing has provided initial recommendations of using amphotericin B for treatment, the advanced age (median 69) of the patient population in this outbreak has limited the therapeutic efficacy in many patients, mainly due to drug toxicity. There are few alternative drugs that are known for the treatment of infections caused by *E. rostratum*. The conventional process for drug discovery and development cannot accommodate such a sudden outbreak as it requires 10–12 years in average to develop a drug. Recently, a drug repurposing screen with approved drugs and pharmacologically active agents [9] has emerged as an alternative approach to rapidly identify new therapeutic indications that have been successfully applied to several diseases [10,11,12,13,14,15]. This unique approach may also help to quickly identify alternative therapeutics for the

Table 1. Assay protocol for the measurements of fungicidal compounds.

Step	Parameter	Value	Description
1	PBS/Medium	2.5 μl/well	PBS or RPMI Medium
2	Compound	0.023 μl/well	Compound in DMSO solution
3	Hyphae/Conidia	2.5 μl/well	Hyphae in PBS or Conidia in RPMI Medium
4	Incubation	24 hr	37°C, 5% CO$_2$
5	Detection reagent	4 μl/well	ATP content assay reagent
6	Incubation	15 min	Room temperature
7	Plate reading	Luminescence mode	ViewLux plate reader

Note: PBS was dispensed in the hyphae assay following by transferring of compound and addition of hyphae cell suspension. RPMI Medium, compound and conidia cells in RPMI Medium were sequentially added in the conidia assays.

treatment of infections caused by outbreaks such as that by *E. rostratum*.

Here we report the development and optimization of two assays using *E. rostratum* hyphae and conidia in an ATP content assay format for high throughput screening. Both assays were screened in parallel against two known compound libraries including 4096 approved drugs and 1280 compounds with pharmacologically known activities. Within seven weeks, the activities of 20 known antifungals, 8 other anti-infectious agents and 10 other drugs against *E. rostratum* were identified from the screens. While some of these drugs may be considered as alternative therapeutics to treat *E. rostratum* infections, others could serve as tools for identification of new molecular targets for future drug development.

Materials and Methods

Materials

Amphotericin B (catalog # A9528) was purchased from Sigma-Aldrich (St. Louis, MO). The ATP content kit (ATPlite, catalog No. 6016941) was purchased from PerkinElmer (Waltham, MA). PBS (Catalog No. 10010049) was purchased from Life technologies. The 1536-well white sterile tissue culture treated polystyrene plates (Catalog No. 789092-F) were purchased from Greiner Bio-One (Monroe, NC).

Preparation of *Exserohilum rostratum* conidia and hyphal fragments

Conidia and hyphae of *E. rostratum* were obtained as described by Richard et al. [16], with the following modifications. Briefly, conidia were harvested from Potato Dextrose Agar (PDA) cultured media with 0.05% Tween 80, and the conidial suspension was filtered using a Cell Strainer (100 μm, BD Falcon REF 352340). After centrifugation at 700g for 10 min, the suspension was decanted and conidia were resuspended at 1×10^5 per ml in RPMI and counted in a hemocytometer. Hyphae were harvested from yeast extract peptone dextrose (YPD) culture media with 0.05% Tween 80. Hyphal fragments were sized by vortexing 15 sec twice with 0.4 mm glassbeads, and the hyphae suspension was filtered by cheese cloth twice. Microscopy was used to determine the size of hyphal fragments, which ranged between 10–50 μm. To normalize concentrations of hyphal fragments for batch to batch consistency, carbohydrate analysis was performed by a phenol-sulfuric acid method as previously described [17]. The final stock concentration of hyphae was adjusted to 1.0 (OD$_{490}$) per 100 μl.

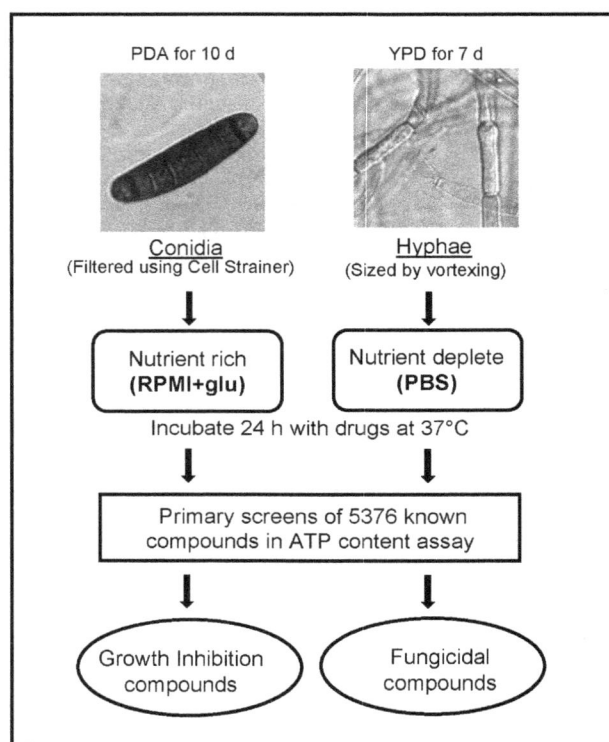

Figure 1. Scheme of hyphae and conidia assay methodology. Conidia harvested from colonies grown on potato dextrose agar (PDA) were incubated in nutrient-rich RPMI media with compounds to identify inhibitors of *E. rostratum* growth. Alternatively, hyphal fragments, obtained by growth on Yeast Extract Peptone Dextrose agar (YPD) and size-reduced by sonication, were incubated with compounds under nutrient-deprived conditions to identify fungicidal compounds.

Mammalian cell culture

Human neuroblastoma SH-SY5Y cell line (Catalog No. CRL-2266) was purchased from ATCC (Manassas, VA). SH-SY5Y cell line was cultured in 175-cm^2 tissue culture flasks (Costar, Cambridge, MA) with 30 ml of growth medium at 37°C in a 5% CO$_2$ humidified atmosphere. Growth medium was made with Dulbecco's Modified Eagle Medium: Nutrient Mixture F-12 with 10% fetal bovine serum (FBS). Growth medium was replaced every other day and cells were passed at 75% confluence.

A. Cell viability

B. S/B ratio

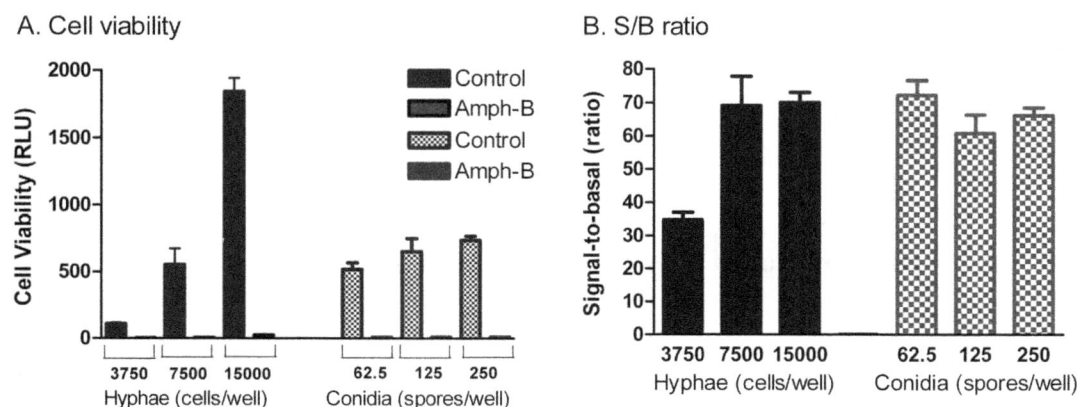

Figure 2. *E. rostratum* cell density optimization in a 1536-well plate assay format. (A) The viability of hyphal fragments was assayed at 3750, 7500 and 15000 fragments/well suspended in PBS after incubation with 46 µM amphotericin B (Amph-B) or DMSO (solvent control) for 24 h at 37°C using the ATP content assay. Similarly, the viability of conidia (spores) was assayed at 62.5, 125 and 250 cells/well in nutrition rich RPMI medium. **(B)** Signal-to-basal ratios of the hyphal fragment and conidia assays in different cell densities. The total signal and basal signal were defined by the cells treated with DMSO and 46 µM Amph-B, respectively. (RLU: relative luminescence units).

ATP content assay

The ATPLite assay kit, consisting of luciferase/luciferin and detergent, was used to quantitate cellular ATP levels as a marker for cell viability (Table 1). For assay development, hyphal fragments or conidia in PBS were plated in a seeding density of 3,750, 7,500 and 15,000 cellular particles/well with a final volume of 5 µl/well using the Multidrop-Combi dispenser in white 1536-well plates. Conidia or hyphae fragments were incubated for 24 h at 37°C in a 5% CO_2 humidified atmosphere. ATPLite detection reagents were added at 4 µl/well. The assay plates were incubated at room temperature for 15 minutes before reading luminescence intensity on a ViewLux plate reader (Perkin Elmer, Waltham, MA). Conidia suspended in RPMI medium were inoculated at concentrations of 62.5, 125 and 250 conidia/well with a final volume of 5 µl/well in white 1536-well plates. The conidia assay was developed and optimized in the same way as the hyphal assay.

Compound library and liquid handling instrument

The library of pharmacologically active compounds (LOPAC) consists of 1280 small molecules with characterized biological activities. The LOPAC library (Sigma-Aldrich) has been extensively used for HTS assay validations [10,18]. The NIH Chemical Genomics Center Pharmaceutical (NPC) collection was constructed in-house through a combined effort of compound purchasing and custom synthesis [9] and recently expanded to 4096 compounds. In this drug libarary, 52% of which are drugs approved for human or animal use by the United States Food and Drug Administration (FDA), 22% are drugs approved in Europe, Canada, or Japan, and the remaining 25% are compounds that have entered clinical trials or are research compounds commonly used in biomedical research. Compounds from both libraries were obtained as powder samples and dissolved in DMSO as 10 mM stock solutions, except several hundred from the NPC library that were prepared as 4.47 mM stock solutions due to solubility limitations. 2.5 µl/well PBS was dispensed into 1536-well plates using a Multidrop-Combi dispenser. Compound in DMSO solution was transferred in a volume of 23 nl/well using a NX-TR pintool station (WAKO Scientific Solutions, San Diego, CA). This additional PBS step for compound addition was designed to prevent the potential fungal contamination of compound libraries. The hyphal fragments in PBS were then added at 2.5 µl/well for a final seeding density of 15,000 hyphal fragments/well using the Multidrop-Combi dispenser. The final compound concentration was 46 µM in the primary screen. The assay plates were incubated

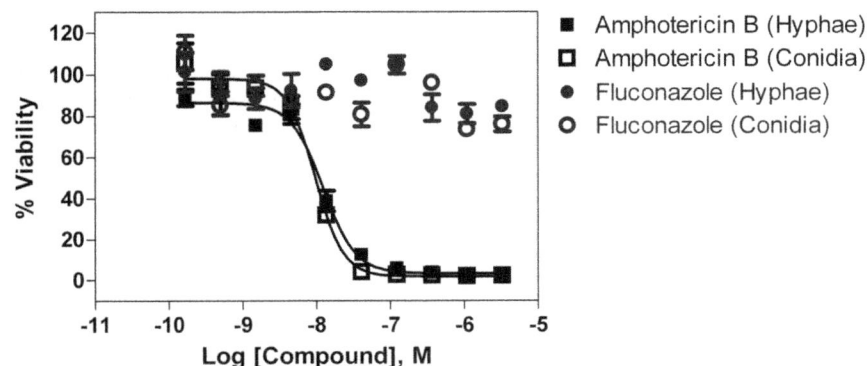

Figure 3. Concentration-response curves of Amphotericin B and fluconazole against *E. rostratum* in hyphal fragment and conidia assays. The IC$_{50}$ values of amphotericin B in the hyphae assay with 15000 fragments/well and conidia assay with 250 cells/well were 12.4 nM (9.93 to 15.6 nM, 95% confidence intervals) and 9.71 nM (8.24 to 11.4 nM), respectively. Fluconazole showed no activity in both assays.

Primary Screens in hyphae and conidia assays
(1280 compounds with known activities + 4096 compounds in an approved drug library)

↓

Compound selection for confirmation
(Hyphae: 87 hits + 26 antifungal drugs; Conidia: 86 hits + 26 antifungal drugs)

↓

Confirmation & Counter-Screen

↓

Hyphae Final Hits (22 compounds)   1   21   16   Conidia Final Hits (37 compounds)

↓

Colony Formation Assay to further confirm compound activities

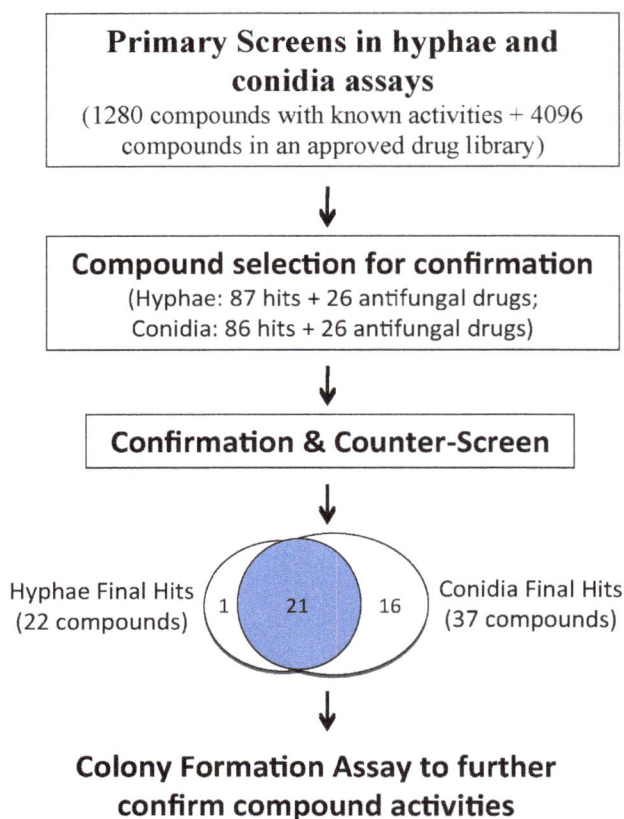

Figure 4. Flowchart of *E. rostratum* viability screens and compound confirmation in hyphal fragment (left) and conidia (right) assays. The primary screens of the LOPAC and approved drug libraries were carried out in the hyphal fragment and conidia assays. A group of 87 hits from the hyphae screen and a group of 86 hits from the conidia screen were selected for confirmation in the same assays along with an additional 26 antifungal drugs. A counter-screen using a mammalian SH-SY5Y cell line as a general cytotoxicity assay was also performed in parallel with the confirmation assays to determine the selectivity of these compounds against *E. rostratum*. The anti-*E. rostratum* activities of 22 compounds were confirmed in the hyphae assay and 37 compounds were confirmed in the conidia assay, with 22 of them being active in both assays. Eight selected compounds from the confirmed compounds were further tested in the traditional colony formation experiment to confirm their anti-*E. rostratum* activities.

for 24 h at 30°C before the detection of cellular viability with the ATP content assay. Screen of conidia was performed in a similar way except that PBS was replaced with RPMI medium and conidia were plated for a final seeding density of 250 conidia/well.

Colony formation
10 µl of hyphal suspension (processed as described above) was transferred to 1.7 mL microfuge tubes, each containing 28 nM amphotericin B, 242 nM posaconazole, or 10% DMSO in 1 ml PBS. Tubes containing hyphae and drug were incubated at 37°C in a shaking incubator for 24 h, then 200 µl of the hyphae drug suspension was spread on Yeast Extract Peptone Dextrose (YPD) plates for 24 h at 30°C and the number of colony forming units were scored. The experiments were performed in triplicate.

Data analysis
The primary screen data and curve fitting were analyzed using software developed internally at the NIH Chemical Genomics Center (NCGC) [19]. Total signals for cellular viability (100% cellular viability) were calculated from the wells with DMSO solvent and the complete cell killing (0% cellular viability) was calculated from 46 µM amphotericin B treated wells. IC_{50} values of compound confirmation data were calculated using the Prism software (Graphpad Software, Inc. San Diego, CA). All values were expressed as the mean +/− SD (n≥3).

Results

Development of fungicidal assays for hyphal fragments and conidia
In standard fungal growth inhibition assays, conidial suspensions are typically used to measure compound activities on the inhibition of fungal growth in a nutrient-rich RPMI medium [20]. Thus, the growth inhibition assay in RPMI medium was adapted for high-throughput screening to identify antifungal compounds (Fig. 1). However, fungal infections may also occur within nutrient-deprived environments such as in the central nervous system or abscess cavities [21]. In addition, many clinical data have indicated that the fungicidal activity of antifungal drugs, exemplified by the polyene amphotericin B, may be more important to clinical success for the treatments of fungal infections [22]. Thus, in addition to the standard growth inhibition assay using conidia, we also developed a hyphal fungicidal assay using nutrient-depleted medium for this drug repurposing screen (Fig. 1). In this study, both hyphae and conidia assays were used for a comprehensive evaluation of therapeutic potentials of antifungal drugs since the current outbreak of infections has been reported in both the CNS and within other closed spaces such as peripheral joints [3].

We used a commercially available ATP content assay kit to measure the viability of hyphal fragments or conidia after treatment with compounds. The bioluminescence signals are generated by the interaction of ATP in lysates of live cellular elements with luciferase and its substrate luciferin added from the assay kit. This assay was suitable for use in 1536-well plates because of its high sensitivity and simple assay procedure. The hyphal fragment density was first optimized using 3750, 7500 and 15000 fragments/well in 1536-well plates with amphotericin B as a positive control. The total ATP content signals in control group (without compound treatment) increased with the increase in hyphal densities while the signals in the group treated with amphotericin B for 24 hours reduced dramatically, indicating almost complete killing of hyphae (Fig. 2a). The signal-to-basal (S/B) ratio increased from 35 fold to 70 fold when the cellular densities rose from 3500 to 7500–15000 cellular equivalents/well (Fig. 2b). The hyphal fragment cellular density of 15000/well was then selected for further experiments. The conidia density was also optimized in a similar manner. The signals of viable conidia increased with increased cellular densities. We selected 250 conidia/well for subsequent experiments. The IC_{50} values of amphotericin B were similar in both our 1536-well plate assays (Fig. 3); 12.4 nM in the hyphal assay and 9.41 nM in the conidia assay that were 5 to 6 fold more potent than 60 nM determined from the microdilution method for filamentous fungi [23]. This discrepancy might be due to differences in assay formats as one measures the ATP contents in viable cellular elements and the microdilution method counts viable cell density directly. Fluconazole, another antifungal drug without activity against molds such as *E. rostratum*, showed no activity at concentrations up to 46 µM

Table 2. Confirmed active compounds against *E. rostratum*.

Compound Name	Hyphae		Conidia		SH-SY5Y		Reported drug activity	Reported mechanism of action
	IC$_{50}$ (μM)	% Max. Resp.	IC$_{50}$ (μM)	% Max. Resp.	% Max. Resp. Resp.	% Max. Resp.		
Known antifungals:								
Amphotericin B	0.014	84	0.012	104	inact.	inact.	Antifungal	Ergosterol synthesis
Posaconazole	0.121	78	0.097	106	inact.	inact.	Triazole antifungal	Ergosterol synthesis
Lanoconazole	0.203	65	0.38	103	inact.	inact.	Topical antifungal & antiparasite	Ergosterol synthesis
Azoxystrobin	0.198	81	0.304	83	inact.	inact.	Fungicide in agriculture	Respiration
Enilconazole	0.312	73	1.72	102	inact.	inact.	Fungicide in agriculture	Ergosterol synthesis
Captan	1.17	82	inact.	inact.	18.6	100	Phthalimide fungicide	Respiration
Itraconazole	2.19	80	1.24	100	inact.	inact.	Triazole antifungal	Ergosterol synthesis
Voriconazole	2.6	94	2.9	101	inact.	inact.	Triazole antifungal	Ergosterol synthesis
Ketoconazole	2.38	74	3.47	101	inact.	inact.	Broad spectrum antifungal	Ergosterol synthesis
Myriocin	2.75	81	13.6	92	inact.	inact.	Antifungal and immunosuppressive	Serine palmitoyltransferase
Amorolfine	4.51	81	7.86	102	inact.	inact.	Morpholine antifungal	Ergosterol synthesis
Pyrrolnitrin	5.59	80	0.409	101	inact.	inact.	Antifungal antibiotic	Respiration
Triticonazole	5.82	74	6.95	100	inact.	inact.	Fungicide	Ergosterol synthesis
Tioconazole	5.97	100	8.55	101	inact.	inact.	Antifungal medication	Ergosterol synthesis
Terconazole	6.12	100	9.06	101	inact.	inact.	Antifungal medication	Ergosterol synthesis
Luliconazole	6.4	82	4.39	100	inact.	inact.	Topical antifungal	Ergosterol synthesis
Thimerosal	inact.	inact.	0.139	102	2.63	96	Topic antiseptic and antifungal	Organomercury
Phenylmercuric acetate	inact.	inact.	0.146	99	0.417	96	Fungicide	Organomercury
Clioquinol	inact.	inact.	0.975	103	37.2	60	Antifungal and antiprotozoal	Not clear
Hexachlorophene	inact.	inact.	3.97	109	inact.	inact.	Disinfectant and fungicide	Respiration
Fenticlor	inact.	inact.	5.72	107	inact.	inact.	Topical antibacterial and antifungal	Not clear
Anti-infectious agents:								
Pentamidine isethionate	0.468	82	1.11	108	inact.	inact.	Antiprotozoal	No clear
Dequalinium dichloride	0.673	93	2.54	109	inact.	inact.	Topical bacteriostat	Not clear
Broxyquinoline	4.51	96	1.6	98	58.9	100	Anti-infective	Not clear
Calcimycin	inact.	inact.	0.735	51	3.95	94	Ionophorous, polyether antibiotic	Ionophore
Phenylmercuric borate	inact.	inact.	0.149	103	0.417	99	Topical antiseptic and disinfectant	Organomercury
Bithionol	inact.	inact.	1.87	100	inact.	inact.	Halogenated anti-infective	NF-κB inhibitor
Bleomycin sulfate	inact.	inact.	1.24	98	inact.	inact.	Glycopeptide antibiotic	DNA synthesis
Broquinaldol	inact.	inact.	7.75	101	inact.	inact.	Antimicrobial	Not clear
Other drugs:								
Tacrolimus	0.032	83	0.024	101	inact.	inact.	Immunosuppressive	Calcineurin phosphatase

Table 2. Cont.

Compound Name	Hyphae IC$_{50}$ (µM)	Hyphae % Max. Resp.	Conidia IC$_{50}$ (µM)	Conidia % Max. Resp.	SH-SY5Y IC$_{50}$ (µM)	SH-SY5Y % Max. Resp.	Reported drug activity	Reported mechanism of action
Floxuridine	0.846	77	0.288	29	inact.	inact.	Antineoplastic antimetabolite	DNA synthesis
Oxaliplatin	0.935	76	4.95	95	inact.	inact.	Antineoplastic	DNA synthesis
Cerivastatin sodium	1.61	79	0.472	101	inact.	inact.	Hydroxymethylglutaryl-CoA Reductase Inhibitor	HMG-CoA reductase
Tetradonium bromide	inact.	inact.	2.44	103	inact.	inact.	Detergents and surface-active	Not clear
Sulfadiazine silver	inact.	inact.	1.3	98	26.3	87	Topical antibacterial	Not clear
Tribromosalan	inact.	inact.	1.93	99	inact.	inact.	N/A	NF-κB inhibitor
Iodoquinol	inact.	inact.	2.75	99	26.3	58	Intestinal antiseptic	Not clear
Tyrphostin AG 879	inact.	inact.	6.16	92	inact.	inact.	Inhibitors of nerve growth factor (NGF) TrkA	TRK receptor
Tyrphostin A9	inact.	inact.	1.98	100	18.6	61	Receptor tyrosine kinase inhibitor	CRAC channels

Note: % Max. Resp.: % maximal response (46 µM Amphotericin B is considered as 100%); Inact. in the conidia assay or SH-SY5Y cytotoxicity assay: no significant activity at the highest compound concentration (46 µM); Inact. in the hyphae assay - no significant activity in the primary screen.

in either the hyphae or conidial assay format (Fig. 3). The results indicate that both hyphae and conidia assays in the ATP content assay format are suitable for compound screening to identify anti-*E. rostratum* agents.

Compound library screening and hit confirmation

Two small known compound libraries including the LOPAC (1280 compounds) and approved drugs (4096 compounds) were screened in parallel using the above optimized assays for hyphae and conidia in 1536-well plates (Fig. 4). A total of 87 primary hits were identified from the hyphal assay and 86 compounds were found from the conidia screen. These hits were retested in the same assays for their concentration response curves along with 26 additional known antifungal drugs (Table S1). A mammalian SH-SY5Y cell line was used in parallel as a counter-screen with the same ATP content assay kit to assess the specificity of these compounds for *E. rostratum*. As a result, 22 compounds in the hyphal assay and 37 compounds in the conidial assay showed activity against *E. rostratum* with IC$_{50}$ values under 10 µM and with greater than 10-fold selectivity over the SH-SY5Y cells (Table 2). Among these confirmed compounds, 21 of them were active in both hyphal and conidia assays along with one compound active in the hyphal assay only and 16 compounds selective to the conidia assay.

Several triazole antifungal drugs were among the confirmed compounds, including the most potent one – posaconazole (hyphae IC$_{50}$ = 121 nM, conidia IC$_{50}$ = 97 nM), as well as lanoconazole (hyphae IC$_{50}$ = 203 nM, conidia IC$_{50}$ = 380 nM), voriconazole (hyphae IC$_{50}$ = 2.64 µM, conidia IC$_{50}$ = 2.91 µM), itraconazole (hyphae IC$_{50}$ = 2.19 µM, conidia IC$_{50}$ = 1.24 µM), and luliconazole (hyphae IC$_{50}$ = 6.40 µM, conidia IC$_{50}$ = 4.39 µM) (Fig. 5, Table 2). The IC$_{50}$ values were similar to these (posaconazole: 0.25 µg/ml or 360 nM; voriconazole: 2 µg/ml or 5.7 µM) determined using the microdilution method (Clinical and Laboratory Standards Institute) for filamentous fungi M38-A2 for this isolate [23].

In addition to known antifungals, several other known drugs were identified as potential anti-*E. rostratum* agents. Bithionol, an old anti-parasitic drug with central nervous system penetration, was active with an IC$_{50}$ of 1.87 µM in the conidia assay (Table 2). In addition, a number of known drugs with immunosuppressive or anti-cancer properties were also found having anti-*E. rostratum* acitivies, though they may not be suitable for primary therapy as they could suppress the immune response of patients. However, these drugs may provide prophylactic activity against *E. rostratum* infection if the patient was on these therapies for other reasons. For example, tacrolimus (Fig. 5D), an immunosuppressive agent used after organ transplantation, exhibited potent activity in both hyphae (IC$_{50}$ = 32 nM) and conidia (IC$_{50}$ = 24 nM) assays. Another known drug with anti-*E. rostratum* activity was floxuridine (Table 2), an antimetabolite used to treat a variety of tumors including brain tumors [24]. Its IC$_{50}$ value in the hyphal assay was 846 nM. Floxuridine crosses the blood brain barrier with brain concentrations above 600 nM 8 hours after an intravenous (iv) administration of 20 mg/kg of the drug [25]. Among the 38 confirmed hits (Table 2), many may not be suitable for direct use to treat infections with *E. rostratum*. Diverse molecular targets and mechanism of action (Fig. 6) have been reported for the 8 'anti-infective agents' and 10 'other agents' listed in Table 2. Taken together, we have identified several non-antifungal drugs that exhibited fungicidal activity against *E. rostratum*. Since most are approved drugs, they may be used in either single use or in a combination with amphotericin B for the treatment or concomitant prophylaxis of infections caused by *E. rostratum*.

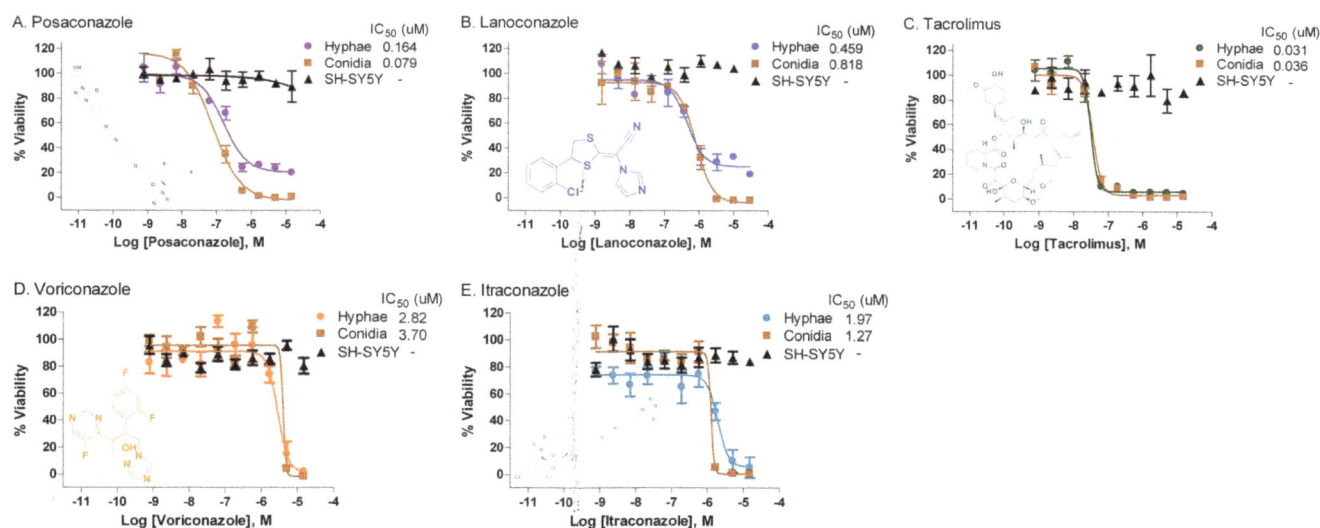

Figure 5. Representative confirmed compounds against *E. rostratum*. The antifungal activities of five compounds including (A) posaconazole, (B) lanoconazole, (C) tacrolimus, (D) voriconazole, and (E) itraconazole were compared in the hyphae and conidia assays with their cytotoxicities in SH-SY5Y mammalian cells. All of these five compounds exhibited similar antifungal activities in both hyphae and conidia assays with no significant cytotoxicity in the mammalian cells at the concentration up to 46 μM.

Confirmation of fungicidal activity in colony formation assays

To further validate the fungicidal activities of compounds identified from our new assays, we measured viability by colony forming units of *E. rostratum* hyphal fragments in the presence of amphotericin B, posaconazole, lanoconazole, voriconazole, itraconazole, luliconazole, tacrolimus and floxuridine (Fig. 7). After treatment with these compounds at 37°C for 24 h, all the compounds at concentrations of 2 times their respective IC_{50} values significantly reduced hyphae-derived colony forming units (CFU) compared to that in the DMSO control (p<0.01). Together, the results further demonstrate the activities of these compounds against *E. rostratum*.

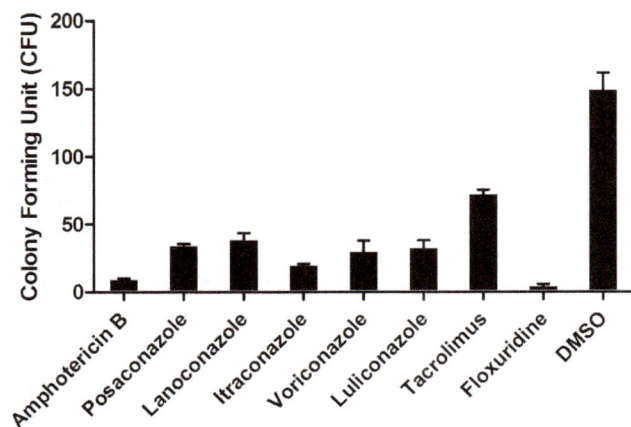

Figure 6. Distribution of molecular targets of 18 confirmed compoundss. The antifungal activities against *E. rostratum* were identified for 8 other anti-infectious agents and 10 other drugs (for details, see Table 2). Their molecular targets and mechanism of action are known although their antifungal activities were not previously reported. These reported molecular targets may implicate potential new directions for anti-*E. rostratum* drug development.

Discussion

The *E. rostratum* infection outbreak thus far has resulted in approximately 741 cases with a large number of CNS infections and 55 deaths. Advanced age of the patient population has limited therapeutic applications in some patients and many have responded suboptimally, requiring changes in therapy. The present pilot screen successfully identified several known drugs as alternative therapeutic choices from a collection of known drugs and pharmaceutically active compounds within a seven-week period. The original goal of such a screen was to identify a spectrum of antifungal agents such as posaconazole that could be used immediately to treat patients with *E. rostratum* infection. We also found a group of novel agents such as bithionol that could be developed further in animal models, preclinical and clinical studies in the event the outbreak continued or drug resistance was encountered. The results from this drug repurposing screen were also helpful in excluding agents that are unlikely to be useful in clinical therapy. For example, albendazole has shown activity against other molds such as *Aspergillus* [26] but is not active against *E. rostratum* in our experiments. Therefore, this study demonstrates that such a drug repurposing screen can be conducted within a useful time-frame and should be considered as part of the initial public health response to new outbreaks or emerging microbial pathogens.

To improve assessment of the *in vitro* activity of antifungal agents against hyphae, quantitative colorimetric assays using 2,3-bis(2-methoxy-4-nitro-5-[(sulphenylamino)carbonyl]-2H-tetrazolium-hydroxide (XTT) and a fluorometric assay using AlarmaBlue have been developed [27]. However, the lower signal-to-basal ratio in either assay limits their applications in high-throughput screening. An alternative cell viability assay that measures cellular ATP levels (termed ATP content assay) with a luminescence readout has been successfully applied in high throughput screening to identify lead compounds for a verity of targets [10,28,29,30,31]. ATP is the primary energy storage in all cells and can be used as an important marker for the functional integrity of live cells. The ATP content assay has been used to determine compound cytotoxicity in

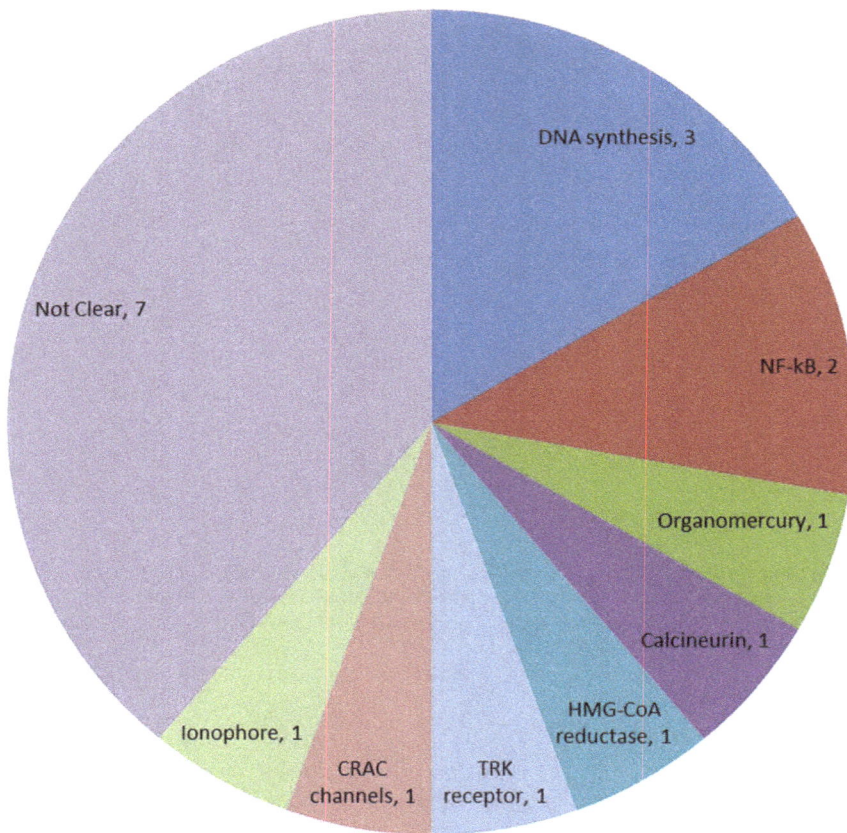

Figure 7. Validation of anti-*E. rostratum* activities of 8 confirmed compounds in colony forming assay. Hyphal fragment suspensions were treated with either 28 nM amphotericin B, 242 nM posaconazole, 406 nM lanoconazole, 4.38 μM itraconazole, 5.2 μM voriconazole, 12.8 μM luliconazole, 64 nM tacrolimus, 1.692 μM floxuridine or 1% DMSO (a solvent control) in PBS. Mixtures were cultured at 37°C in a shaking incubator for 24 h, then the hyphal suspensions were inoculated on Yeast Extract Peptone Dextrose (YPD) plates that were incubated for 1 day. The number of colonies on each YPD plate was counted. Results from three independent experiments were averaged; error bars show standard deviation.

mammalian cells and more recently *Thielavia subthermophila* fungal growth [32,33,34]. This assay is a homogenous assay with 'one step' reagent addition in which no cell wash step is required. A luminescence signal is generated upon the release of cellular ATP that is detectable following a 10-minute incubation. A unique feature of the ATP content assay is that the signal-to-basal ratio is characteristically much higher in comparison to the AlamarBlue assay, partly due to low innate luminescence background generated by cells, reagents or plates. Our results have demonstrated that the new assay in the 1536-well plate format can measure the sensitivities of thousands of known drugs against *E. rostratum* in a short period of time that is not possible for conventional antibiotic susceptibility testing (AST). This new high throughput method may also be extended to other fungi or bacteria for identification of useful drugs from all known approved compounds.

Our data indicate that two potent triazole antifungal agents including posaconazole and lanoconazole (under development) may be useful as alternative agents for the treatment of infections with *E. rostratum*. The high potency and fungicidal activity of posaconazole to *E. rostratum* may suggest a role in the treatment of non-CNS infections such as septic joints that have also featured predominantly in the current outbreak and may allow prolonged oral therapy. Posaconazole has been reported to penetrate the blood brain barrier and may thus be an effective alternative for fungal infections within the central nervous system, though it does

not reach reliable CNS levels compared to azoles such as voriconazole, [20].

Bithionol, an old anti-parasitic drug, is not previously known for its antifungal activity. The serum concentration of this drug has been found to be between 0.08 and 0.18 mg/l during administration and has been used successfully in the treatment of CNS *Paragonimus* infections [21,22]. Thus, it might be a good candidate for the further clinical trials to treat CNS infections of *E. rostratum*.

Although tacrolimus, an immunosuppressive agent, showed potent activity against *E. rostratum* in this study, it would not be useful for treatment of infections with *E. rostratum* because the immunosuppression may counteract its fungicidal activity. The mechanism of action of tacrolimus has been reported through inhibition of calcineurin that, in mammalian cells, prevents transport of the nuclear factor of activated T cells (NF-ATc) to the nucleus and reduces protective IL-2 production [23]. Tacrolimus is a lipophilic agent that may have several neurotoxic effects, especially on lipid-rich white matter which is reversible after dose reduction [24,25]. However, the potent effect of tacrolimus against *E. rostratum* suggests that anti-calcineurin compounds without these immunosuppressive properties may be of benefit against pigmented molds. This indicates that calcineurin might be a new target for drug development to treat infections with *E. rostratum*.

Floxuridine, an antimetabolite cancer drug, exhibited high nanomolar activity against *E. rostratum*. However, its cytotoxicity would likely prevent its use in treatment of infection with *E. rostratum*. Thus, the structure of this compound requires modification to reduce cytotoxicity and to enhance potency against *E. rostratum* before being applied as a therapeutic drug.

While many of the confirmed hits may not be suitable for treatment of fungal infections, their fungicidal activities may suggest new targets for future drug development after their potencies are optimized. For example, azoxystrobin, captan, pyrrolnitrin and hexachlorophene inhibit respiration [35,36,37,38]; myriocin inhibits serine palmitoyltransferase [39,40]; and thimerosal and phenylmercuric acetate are toxic organomercuries [41,42]; while clioquinol and fenticlor have unclear mechanisms of action. Bleomycin sulfate, floxuridine and oxaliplatin inhibit DNA synthesis [43,44,45], which suggests that a selective fungal DNA synthesis inhibitor might be developed as a new antifungal agent against molds such as *E. rostratum*. Tribromosalan and bithionol are nuclear factor-kappa B (NF-κB) inhibitors [11]. Phenylmercuric borate is an organomercury [46,47]. Tacrolimus binds to the immunophilin FKBP12 to inhibit calcineurin [48]. Cerivastatin sodium is an inhibitor of hydroxymethylglutaryl-coenzyme A (HMG-CoA) reductase [49]. Tyrphostin AG 879 is an inhibitor of tropomyosin-receptor-kinase (TRK) receptor [50]. Tyrphostin A9 is an inhibitor of calcium release-activated calcium (CRAC) channels [51]. Calcimycin is an ionophore [52]. The MOA & molecular targets of the other 6 compounds are yet to be elucidated. Thus, the data indicate that a rich collection of new drug targets can be identified from such a drug repurposing screen.

In conclusion, we have developed two high throughput assays using conidia and hyphae of *E. rostratum* in an ATP content assay format. A screen of 4096 compounds from an approved drug library and a pharmaceutically active compound collection using this assay led to identification and confirmation of a group of potential antifungal agents. While several triazole antifungal agents including posaconazole and lanoconazole could be used as alternative therapies for the treatment of infections with *E. rostratum*, the other confirmed compounds may implicate new drug targets for further drug development to improve their selectivity against molds such as *E. rostratum*. We believe that implementation of such a drug repurposing screen within a 7 week window provides a useful paradigm. It meets the challenges of future public health threats by enabling the rapid identification of potential therapeutic agents for infected patients and for identification of new drug development targets.

Author Contributions

Conceived and designed the experiments: PRW WZ. Performed the experiments: WS YP AF. Analyzed the data: WS YP AF. Contributed reagents/materials/analysis tools: JAS NS PS JCM KJKC. Wrote the paper: WS YP.

References

1. Kainer MA, Reagan DR, Nguyen DB, Wiese AD, Wise ME, et al. (2012) Fungal infections associated with contaminated methylprednisolone in Tennessee. The New England journal of medicine 367: 2194–2203.
2. Kauffman CA, Pappas PG, Patterson TF (2012) Fungal Infections Associated with Contaminated Methylprednisolone Injections - Preliminary Report. The New England journal of medicine.
3. Smith RM, Schaefer MK, Kainer MA, Wise M, Finks J, et al. (2012) Fungal Infections Associated with Contaminated Methylprednisolone Injections - Preliminary Report. The New England journal of medicine.
4. Kerkering TM, Grifasi ML, Baffoe-Bonnie AW, Bansal E, Garner DC, et al. (2013) Early Clinical Observations in Prospectively Followed Patients With Fungal Meningitis Related to Contaminated Epidural Steroid Injections. Annals of internal medicine 158: 154.
5. Finks J, Collins J, Miller C, et al. (2013) Spinal and paraspinal infections associated with contaminated methylprednisolone acetate injections - michigan, 2012–2013. MMWR Morbidity and mortality weekly report 62: 377–381.
6. Adler A, Yaniv I, Samra Z, Yacobovich J, Fisher S, et al. (2006) Exserohilum: an emerging human pathogen. European journal of clinical microbiology & infectious diseases : official publication of the European Society of Clinical Microbiology 25: 247–253.
7. Lin SC, Sun PL, Ju YM, Chan YJ (2009) Cutaneous phaeohyphomycosis caused by Exserohilum rostratum in a patient with cutaneous T-cell lymphoma. International journal of dermatology 48: 295–298.
8. Juhas E, Reyes-Mugica M, Michaels MG, Grunwaldt LJ, Gehris RP (2012) Exserohilum Infection in an Immunocompromised Neonate. Pediatric dermatology.
9. Huang R, Southall N, Wang Y, Yasgar A, Shinn P, et al. (2011) The NCGC pharmaceutical collection: a comprehensive resource of clinically approved drugs enabling repurposing and chemical genomics. Science translational medicine 3: 80ps16.
10. Chen CZ, Kulakova L, Southall N, Marugan JJ, Galkin A, et al. (2011) High-throughput Giardia lamblia viability assay using bioluminescent ATP content measurements. Antimicrobial agents and chemotherapy 55: 667–675.
11. Miller SC, Huang R, Sakamuru S, Shukla SJ, Attene-Ramos MS, et al. (2010) Identification of known drugs that act as inhibitors of NF-kappaB signaling and their mechanism of action. Biochemical pharmacology 79: 1272–1280.
12. Shen M, Zhang Y, Saba N, Austin CP, Wiestner A, et al. (2013) Identification of Therapeutic Candidates for Chronic Lymphocytic Leukemia from a Library of Approved Drugs. Manuscript in submission.
13. Xia M, Huang R, Sakamuru S, Alcorta D, Cho M, et al. (2013) Identification of repurposed small molecule drugs for chordoma therapy. Cancer Biology and Therapy In press.
14. Xu M, Liu K, Swaroop M, Porter FD, Sidhu R, et al. (2012) delta-Tocopherol reduces lipid accumulation in Niemann-Pick type C1 and Wolman cholesterol storage disorders. The Journal of biological chemistry 287: 39349–39360.
15. Yuan H, Myers S, Wang J, Zhou D, Woo JA, et al. (2012) Use of reprogrammed cells to identify therapy for respiratory papillomatosis. The New England journal of medicine 367: 1220–1227.
16. Richard JL, Cysewski SJ, Fichtner RE (1971) Harvest and survival of aspergillus fumigatus Fresenius spores. Mycopathologia et mycologia applicata 43: 165–168.
17. Masuko T, Minami A, Iwasaki N, Majima T, Nishimura S, et al. (2005) Carbohydrate analysis by a phenol-sulfuric acid method in microplate format. Analytical biochemistry 339: 69–72.
18. Lea WA, Xi J, Jadhav A, Lu L, Austin CP, et al. (2009) A high-throughput approach for identification of novel general anesthetics. PloS one 4: e7150.
19. Wang Y, Jadhav A, Southal N, Huang R, Nguyen DT (2010) A grid algorithm for high throughput fitting of dose-response curve data. Current chemical genomics 4: 57–66.
20. CLSI (2012) Reference method for Broth Dilution Antifungal Susceptibility Testing of Yeasts; Fourth Informational Supplement. CLSI document M27-S4 Wayne: Clinical and Laboratory Standards Institute (CLSI).
21. Diamond RD, Bennett JE (1974) Prognostic Factors in Cryptococcal Meningitis - Study in 111 Cases. Annals of internal medicine 80: 176–181.
22. Lewis JS, Graybill JR (2008) Fungicidal versus fungistatic: what"s in a word? Expert opinion on pharmacotherapy 9: 927–935.
23. National Committee for Clinical Laboratory Standards (2008) Reference Method for Broth Dilution Susceptibility Testing of Filamentous Fungi: Approved Standard-Second Edition. Wayne, PA: Clinical and Laboratory Standards Institute.
24. Nakagawa H, Yamada M, Fukushima M, Shimizu K, Ikenaka K (1998) In vivo study on intrathecal use of 5-fluoro-2 '-deoxysuridine (FdUrd) in meningeal dissemination of malignant tumor. Neurological Surgery 26: 873–879.
25. Wang JX, Sun X, Zhang ZR (2002) Enhanced brain targeting by synthesis of 3 ',5 '-dioctanoyl-5-fluoro-2 '-deoxyuridine and incorporation into solid lipid nanoparticles. European Journal of Pharmaceutics and Biopharmaceutics 54: 285–290.
26. Berthet N, Faure O, Bakri A, Ambroise-Thomas P, Grillot R, et al. (2003) In vitro susceptibility of Aspergillus spp. clinical isolates to albendazole. J Antimicrob Chemother 51: 1419–1422.
27. Rabjohns JLA, Park YD, Dehdashti J, Zheng W, Williamson PR (2013) A High Throughput Screening Assay for Fungicidal Compounds against Cryptococcus neoformans. Journal of Bimolecular Screen In submission.
28. Cho SH, Warit S, Wan B, Hwang CH, Pauli GF, et al. (2007) Low-oxygen-recovery assay for high-throughput screening of compounds against nonreplicat-

ing Mycobacterium tuberculosis. Antimicrobial agents and chemotherapy 51: 1380–1385.

29. Coconnier-Polter MH, Lievin-Le Moal V, Servin AL (2005) A Lactobacillus acidophilus strain of human gastrointestinal microbiota origin elicits killing of enterovirulent Salmonella enterica Serovar Typhimurium by triggering lethal bacterial membrane damage. Applied and environmental microbiology 71: 6115–6120.

30. Hu G, Hacham M, Waterman SR, Panepinto J, Shin S, et al. (2008) PI3K signaling of autophagy is required for starvation tolerance and virulenceof Cryptococcus neoformans. The Journal of clinical investigation 118: 1186–1197.

31. Kim K, Pollard JM, Norris AJ, McDonald JT, Sun Y, et al. (2009) High-throughput screening identifies two classes of antibiotics as radioprotectors: tetracyclines and fluoroquinolones. Clinical cancer research : an official journal of the American Association for Cancer Research 15: 7238–7245.

32. Kusari S, Zuhlke S, Kosuth J, Cellarova E, Spiteller M (2009) Light-independent metabolomics of endophytic Thielavia subthermophila provides insight into microbial hypericin biosynthesis. Journal of natural products 72: 1825–1835.

33. Kangas L, Gronroos M, Nieminen AL (1984) Bioluminescence of cellular ATP: a new method for evaluating cytotoxic agents in vitro. Medical biology 62: 338–343.

34. Cree IA, Andreotti PE (1997) Measurement of cytotoxicity by ATP-based luminescence assay in primary cell cultures and cell lines. Toxicology in vitro : an international journal published in association with BIBRA 11: 553–556.

35. Esser L, Quinn B, Li YF, Zhang MQ, Elberry M, et al. (2004) Crystallographic studies of quinol oxidation site inhibitors: A modified classification of inhibitors for the cytochrome bc(1) complex. Journal of Molecular Biology 341: 281–302.

36. Rousk J, Demoling LA, Baath E (2009) Contrasting Short-Term Antibiotic Effects on Respiration and Bacterial Growth Compromises the Validity of the Selective Respiratory Inhibition Technique to Distinguish Fungi and Bacteria. Microbial Ecology 58: 75–85.

37. Tripathi RK, Gottlieb D (1969) Mechanism of action of the antifungal antibiotic pyrrolnitrin. Journal of bacteriology 100: 310–318.

38. Frederick JJ, Corner TR, Gerhardt P (1974) Antimicrobial actions of hexachlorophene: inhibition of respiration in Bacillus megaterium. Antimicrobial agents and chemotherapy 6: 712–721.

39. Miyake Y, Kozutsumi Y, Nakamura S, Fujita T, Kawasaki T (1995) Serine Palmitoyltransferase Is the Primary Target of a Sphingosine-Like Immunosuppressant, Isp-1/Myriocin. Biochemical and Biophysical Research Communications 211: 396–403.

40. Yamaji-Hasegawa A, Takahashi A, Tetsuka Y, Senoh Y, Kobayashi T (2005) Fungal metabolite sulfamisterin suppresses sphingolipid synthesis through inhibition of serine palmitoyltransferase. Biochemistry 44: 268–277.

41. Melnick JG, Yurkerwich K, Buccella D, Sattler W, Parkin G (2008) Molecular structures of thimerosal (Merthiolate) and other arylthiolate mercury alkyl compounds. Inorganic Chemistry 47: 6421–6426.

42. Geier J, Lessmann H, Uter W, Schnuch A (2005) Patch testing with phenylmercuric acetate. Contact Dermatitis 53: 117–118.

43. Hecht SM (2000) Bleomycin: New perspectives on the mechanism of action. Journal of natural products 63: 158–168.

44. Murakami Y, Kazuno H, Emura T, Tsujimoto H, Suzuki N, et al. (2000) Different mechanisms of acquired resistance to fluorinated pyrimidines in human colorectal cancer cells. International Journal of Oncology 17: 277–283.

45. Graham J, Muhsin M, Kirkpatrick P (2004) Oxaliplatin. Nature Reviews Drug Discovery 3: 11–12.

46. Cortat M (1978) Studies on the mode of action of phenylmercuric borate on Escherichia coli. II. Biochemical localization and inhibition of some metabolic activities. Zentralblatt fur Bakteriologie, Parasitenkunde, Infektionskrankheiten und Hygiene Erste Abteilung Originale Reihe B: Hygiene, praventive Medizin 166: 528–539.

47. Cortat M (1978) Studies on the mode of action of phenylmercuric borate on Escherichia coli. I. Structural localization and kinetics of incorporation. Zentralblatt fur Bakteriologie, Parasitenkunde, Infektionskrankheiten und Hygiene Erste Abteilung Originale Reihe B: Hygiene, praventive Medizin 166: 517–527.

48. Liu J, Farmer JD, Lane WS, Friedman J, Weissman I, et al. (1991) Calcineurin Is a Common Target of Cyclophilin-Cyclosporine-a and Fkbp-Fk506 Complexes. Cell 66: 807–815.

49. Ganne F, Vasse M, Beaudeux JL, Peynet J, Francois A, et al. (2000) Cerivastatin, an inhibitor of HMG-CoA reductase, inhibits urokinase/urokinase-receptor expression and MMP-9 secretion by peripheral blood monocytes - A possible protective mechanism against atherothrombosis. Thrombosis and Haemostasis 84: 680–688.

50. Ohmichi M, Pang L, Ribon V, Gazit A, Levitzki A, et al. (1993) The Tyrosine Kinase Inhibitor Tyrphostin Blocks the Cellular Actions of Nerve Growth-Factor. Biochemistry 32: 4650–4658.

51. Denys A, Aires V, Hichami A, Khan NA (2004) Thapsigargin-stimulated MAP kinase phosphorylation via CRAC channels and PLD activation: inhibitory action of docosahexaenoic acid. Febs Letters 564: 177–182.

52. Abbott BJ, Fukuda DS, Dorman DE, Occolowitz JL, Debono M, et al. (1979) Microbial Transformation of A23187, a Divalent-Cation Ionophore Antibiotic. Antimicrobial agents and chemotherapy 16: 808–812.

11

Histatin 5 Resistance of *Candida glabrata* Can Be Reversed by Insertion of *Candida albicans* Polyamine Transporter-Encoding Genes *DUR3* and *DUR31*

Swetha Tati, Woong Sik Jang¤, Rui Li, Rohitashw Kumar, Sumant Puri, Mira Edgerton*

Department of Oral Biology, University at Buffalo, Buffalo, New York, United States of America

Abstract

Candida albicans and *Candida glabrata* are predominant fungi associated with oral candidiasis. Histatin 5 (Hst 5) is a small cationic human salivary peptide with high fungicidal activity against *C. albicans*, however many strains of *C. glabrata* are resistant. Since Hst 5 requires fungal binding to cell wall components prior to intracellular translocation, reduced Hst 5 binding to *C. glabrata* may be the reason for its insensitivity. *C. glabrata* has higher surface levels of β-1,3-glucans as compared with *C. albicans*; however these differences did not account for reduced Hst 5 uptake and killing in *C. glabrata*. Similarly, the biofilm matrix of *C. glabrata* contained significantly higher levels of β-1,3-glucans compared with *C. albicans*, but it did not reduce the percentage of Hst 5 positive fungal cells in the biofilm. Hst 5 enters *C. albicans* cell through polyamine transporters Dur3p and Dur31p that are uncharacterized in *C. glabrata*. *C. glabrata* strains expressing CaDur3 and CaDur31 had two-fold higher killing and uptake of Hst 5. Thus, neither *C. glabrata* cell surface or biofilm matrix β-1,3-glucan levels affected Hst 5 toxicity; rather the crucial rate limiting step is reduced uptake that can be overcome by expression of *C. albicans* Dur proteins in *C. glabrata*.

Editor: Martine Bassilana, Université de Nice-CNRS, France

Funding: This work was supported by grant R01DE010641 (ME) from the National Institute of Dental and Craniofacial Research, National Institutes of Health, USA. The funders had no role in study design, data collection and analysis, decision to publish, or preparation of the manuscript. No additional external funding was received for this study.

Competing Interests: The authors have declared that no competing interests exist.

* E-mail: edgerto@buffalo.edu

¤ Current address: Department of General Education and Teaching Profession Hoseo University, Asan, Chungnam, South Korea

Introduction

Candida albicans and *Candida glabrata* rank as the first and second most prevalent fungi, respectively, that cause oral and systemic candidiasis in the United States [1,2]. *C. glabrata* previously was considered to be a relatively non-pathogenic fungus of the normal flora in healthy humans, and was not initially associated with serious infections. However it is now known that *C. glabrata* can rapidly disseminate throughout the body; and infection with this species is associated with a high mortality rate. Moreover *C. glabrata* is of added concern because of its propensity to develop resistance to commonly used antifungal drugs such as fluconazole [3].

Histatins are basic histidine-rich proteins secreted in human parotid and submandibular-sublingual saliva in humans and higher primates [4]. Histatin 5 (Hst 5) is a proteolytic cleavage product of the larger Histatin 3 family member [5,6]. Among Histatins, Hst 5 has the most potent fungicidal activity against pathogenic fungi including *C. albicans* and other medically important Candida species such as *Candida kefyr*, *Candida krusei*, and *Candida parapsilosis* (MIC_{50} 10–20 μg/ml), as well as *Cryptococcus neoformans* and *Aspergillus fumigatus* (MIC_{50} 5–6 μg/ml) [4,7–9]. However, many strains of *C. glabrata* have been shown to be significantly more resistant to Hst 5 as well as other Hst family members for reasons that are unknown [10]. Some *C. glabrata* strains (ATCC 90030, 2001 and 64677) are completely insensitive

to Hst 5 even at high concentrations (IC50>225 μg/ml) [10]. *C. glabrata* planktonic cells and biofilms exhibited reduced susceptibility to Hst 5 compared with *C. albicans* [11].

Azole drug resistance in *C. glabrata* is very well studied and is often due to enhanced drug efflux through over-expression of ATP-binding cassette transporter genes *CgCDR1* and *CgCDR2* [12,13]. However, almost nothing is known about the mechanism underlying the variable strain resistance of *C. glabrata* to histatins. In azole resistant *C. glabrata* clinical isolates, gain of function mutations in the transcription factor *CgPdr1* resulted in intrinsically higher expression of the drug transporter gene *CgCDR1* as well as up-regulation of *PUP1* that encodes a mitochondrial protein [14,15]. These gain of function mutations in *CgPdr1* also supported enhanced virulence of *C. glabrata* in animal models of systemic infection [15]. Similarly, an azole resistant *C. glabrata* petite mutant (respiration incompetent), selected *in vivo* under azole therapy, had increased virulence that correlated with increased expression of genes involved in cell wall biogenesis and remodeling [16]. *C. glabrata* biofilms grown in the presence of antifungal drugs Caspofungin, Amphotericin B, Nystatin, and Ketoconazole resulted in adaptation and drug resistance via differential metabolic activity [17]. However neither respiratory (mitochondrial) deficiency or deletion of *C. glabrata* multidrug efflux transporter genes *CgCDR1* and *CgCDR2* affected cell susceptibility to Hst 5 [18], showing that the mechanism of azole and Hst 5 resistance in *C. glabrata* is fundamentally different.

Histatin 5 fungicidal activity in *C. albicans* is a distinctive multistep mechanism requiring binding to Candida cell wall, followed by translocation to intracellular compartments. Lethality of Hst 5 is caused by non-lytic release of intracellular ions and small nucleotides, followed by induction of reactive oxygen species and osmotic stress [19,20]. Two critical events for Hst 5 antifungal activity are its ability to bind to the fungal cell wall and sequential transportation into the cytosol. Among various *C. albicans* cell surface polysaccharides, we identified laminarins (beta-glucans) as primary surface binding moieties for Hst 5 [21], followed by Ssa1 and Ssa2 binding proteins within the cell wall [22,23]. Like *C. albicans*, β,(1–3)-D-glucans are major carbohydrate components of the outer cell wall of *C. glabrata* [24]. These cell surface moieties are recognition sites for the host immune system [25] and potential binding sites for antifungal drugs or peptides. The Candida biofilm matrix is also primarily comprised of β-1,3-glucans that sequester antifungal drugs and contribute to fluconazole resistance in the cells of the biofilm [26–28]. Therefore, it is possible that differences in cell surface or biofilm matrix glucans between *C. albicans* and *C. glabrata* may alter initial Hst 5 binding to the fungal cells and/or biofilm matrix components of these two species.

We and others found that Hst 5 fungicidal activity requires energy dependent translocation to the cytosol, so that cells treated with azide or cold do not take up Hst 5 and do not suffer consequential toxicity [20,21,29]. Recently, we identified *C. albicans* spermidine transporters Dur3 and Dur31 as major conduits for intracellular translocation of Hst 5 [30] as Hst 5 is potentially recognized as a polyamine analogue due to its small size and cationic charge. Deletion of *C. albicans DUR3* and *DUR31* resulted in loss of Hst 5 uptake and reduced fungicidal activity [30], and *DUR31* knock-out mutants were more susceptible to killing by human neutrophils and were less virulent *in vivo* [31]. Although *C. glabrata* and *C. albicans* belong to the same genus, *C. glabrata* is more phylogenetically related to *Saccharomyces cerevisiae* than *C. albicans*. For example, *C. albicans* has six Dur transporter family members, whereas *C. glabrata* and *S. cerevisiae* have only two (*DUR3*: CAGL0108613; and *DUR31*: CAGL0K03157) and one (*DUR3* YHL016C) spermidine transporter homologs, respectively. Neither of the *C. glabrata* Dur proteins has been characterized in terms of its polyamine substrate specificity or ability to take up Hst 5 as a spermidine analogue. We report here that although Hst 5 has lower binding to *C. glabrata*, this was not a result of increased β-1,3-glucan levels at the cell surface or within the biofilm matrix. Instead, fungicidal activity of Hst 5 was significantly increased upon expression of *C. albicans* spermidine transporters *DUR3* and *DUR31* in *C. glabrata*, showing that a major reason for *C. glabrata* resistance to Hst 5 is due to its poor uptake.

Results

C. glabrata strains are resistant to Hst 5 compared with *C. albicans*

Fungicidal activity of Hst 5 against three strains of *C. glabrata* (*Cg 931010*, *Cg 90030*, and *Cg 90032*) was significantly lower than *C. albicans CAI4* (Figure 1 A). Candidacidal activity was concentration dependent and resulted in 65% cell death upon incubation with 31 μM Hst 5 in *C. albicans CAI4*. In contrast, only 22% killing with 31 μM Hst 5 was observed even in the most susceptible *C. glabrata* strain *Cg 931010* (*Cg10*), while *Cg 90030* (*Cg30*) and *Cg 90032* (*Cg32*) had maximal killing of only 8 and 10% respectively (Figure 1 A). Higher concentrations of Hst 5 (>31 μM) did not result in higher killing among any of the strains of *C. glabrata* (data not

shown). Since Hst 5 candidacidal activity requires both cell wall binding and intracellular translocation of the peptide [21], we examined *C. glabrata* cells to determine whether either process was defective. Cell wall binding and cytosolic concentrations of Hst 5 were quantified by western blotting directly from cell wall and cytosolic preparations after exposing *CAI4*, *Cg10*, *Cg30*, and *Cg32* cells to 31 μM biotin labeled Hst 5 (B-Hst 5) for 30 min at 37°C. *Cg10* had a 30% reduction in cell wall binding while its cytosolic levels were reduced by more than 50% compared with *C. albicans* (Figure 1 B). Among *C. glabrata* strains, *Cg10* had more cell wall associated and cytosolic levels of Hst 5 when compared to *Cg30* and *Cg32* (Figure 1 B). Next, cells were exposed to FITC labeled Hst 5 (F-Hst 5) for 30 minutes and examined by confocal microscopy for both cell envelope association of Hst 5 as well as total cellular uptake of F-Hst 5 (Figure 1 C). After 30 min, all *C. albicans* cells showed conspicuous surface binding of F-Hst 5 and nearly 95% of cells examined showed intracellular uptake of Hst 5. In contrast, although all *C. glabrata* strains were able to bind Hst 5, only 20–25% of *Cg10* cells and 5–10% of *Cg30* and *Cg32* cells contained translocated cytosolic F-Hst 5 (as determined from multiple microscopic fields).

C. glabrata strains show reduced binding and intracellular uptake of Hst 5

We next quantitatively compared Hst 5 binding and translocation among *C. glabrata* strains. To differentiate between cell surface bindings and total cellular uptake of Hst 5, we performed a time course experiment of cells exposed to F-Hst 5 using a Fluorescently Activated Cell Sorter (FACScan). Baseline cell surface binding of F-Hst 5 (15 μM) was measured in *C. albicans* and *C. glabrata* cells that had been incubated on ice for one hour since these cells do not translocate Hst 5 due to suspension of energy generation that is needed for transport (Figure 2). In line with previous results with B-Hst (Figure 1 B) Hst 5 binding to cold treated cells was reduced by about 30% in *C. glabrata Cg10* (Mean Fluorescence Intensity, MFI = 11) compared with *C. albicans* (MFI = 15) (Figure 2 A). Among the *C. glabrata* strains, *Cg10* had significantly higher surface binding of F-Hst 5 than for *Cg30* (MFI = 8) or *Cg32* (MFI = 3). In contrast, *C. albicans* cells cultured in warm media at 37°C showed rapid intracellular accumulation of F-Hst 5 over 30 min. The level of total cellular Hst 5 increased at 5 min (Figure 2 B, solid gray line) to MFI = 21 and reached a maximum of MFI = 62 at 30 min (Figure 2 B, black broken line). In contrast, uptake of F-Hst 5 by *C. glabrata Cg10* cells increased only slightly at 5 min (MFI = 13) and by 30 min reached a maximum of only MFI = 15 (Figure 2 C). Thus, there was a significant reduction in both binding and uptake of Hst 5 in all *C. glabrata* strains compared with *C. albicans* and either may potentially account for differences in susceptibility to Hst 5 killing.

C. glabrata has higher surface exposed glucans that do not influence Hst 5 toxicity

Since Candidal cell surface β-1,3-glucans are important binding moieties for Hst 5 in *C. albicans*, we examined their role in binding Hst 5 in *C. glabrata*. Unexpectedly, we found that *C. glabrata* strains had significantly higher (6–7 fold) surface content of β-1,3-glucan when compared to *C. albicans*, although there were no statistically significant differences among the *C. glabrata* strains (Figure 3 A). Pre-treatment of *C. albicans* cells (performed at 4°C to block Hst 5 uptake) with β-1,3-glucan antibody reduced F-Hst 5 surface binding by 35% (Figure 3 B). We expected that antibody blocking of cell surface β-1,3-glucan in *C. glabrata* cells would result in greater reduction of Hst 5 binding due to its higher surface glucan

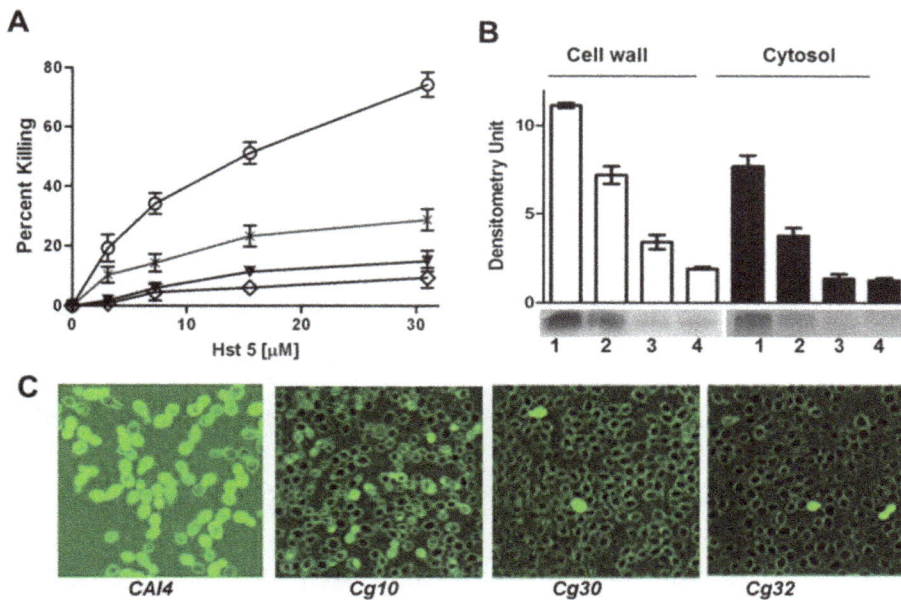

Figure 1. Hst 5 has lower toxicity and intracellular uptake in *C. glabrata* compared to *C. albicans*. A. Candidacidal assays were performed on *C. albicans* CAI4, *C. glabrata* Cg10, Cg30, and Cg32 strains. Killing activity of Hst 5 was significantly higher in *C. albicans* CAI4 (O) than in the *C. glabrata* strains tested. *C. glabrata* Cg10 (*) was significantly more sensitive to Hst 5 than either Cg30 (△) or Cg32 (◇) (n = 4). **B.** Hst 5 binding (cell wall) and uptake (cytosol) assays were performed by incubating CAI4, Cg10, Cg30, and Cg32 strains with biotin-labeled Hst 5 (B-Hst 5) at 37°C for 30 min (31.5 µM). *C. albicans* CAI4 (1) showed the highest amount of cell wall (white bars) and cytosolic (black bars) Hst 5 compared to *C. glabrata* Cg10 (2), Cg30 (3), and Cg32 (4) strains (n = 3). Among *C. glabrata* strains, Cg10 showed higher amounts of cell wall bound and cytosolic Hst 5 than the other two strains. **C.** Translocation of Hst 5 was visualized using time-lapse confocal microscopy with FITC Hst 5 (31.5 µM) for 30 min, and the percentage of Hst 5 positive cells was quantified in at least three independent fields from three independent experiments. *C. albicans* CAI4 had the largest number of Hst 5 containing cells (95%) compared with *C. glabrata* Cg10 (20–25%) and Cg30 and Cg32 (5–10%).

content. Indeed, β-1,3-glucan antibody pre-treatment inhibited Hst 5 binding to *C. glabrata* Cg10 cells by 80% (Figure 3 B). Hst 5 binding to *C. glabrata* Cg30 and Cg32 cells was reduced by 75% and 40%, respectively, by similar pre-treatment. Overall, Hst 5 surface binding was reduced to a similar level (MFI = 3) in all *C. glabrata* strains following β-1,3-glucan antibody pre-treatment, compared with reduction to MFI = 10 in *C. albicans*. To determine whether Hst 5 binding to β-1,3-glucans affect Hst 5 mediated toxicity, candidacidal assays were done with *C. albicans* and *C. glabrata* cells pre-incubated with β-1,3-glucan antibody (Figure 3 C). Pretreatment of *C. albicans* cells showed a 40% reduction in Hst 5 mediated killing at 31 µM (Figure 3 C), similar to the percentage reduction in its binding. However, killing of *C. glabrata* strains by Hst 5 was reduced by only 20–25% upon pre-treatment (P<0.05) (Figure 3 C), far less than its percentage reduction in binding. Thus, while Hst 5 binding of *C. albicans* surface β-1,3-glucan is closely linked to its toxicity, Hst 5 killing in *C. glabrata* requires much lower levels of surface accessible β-1,3-glucan.

C. glabrata biofilms have higher matrix density of Hst 5 but a lower percentage of Hst 5 containing cells

Glucans are a major component of the extracellular biofilm matrix [26,32]. We hypothesized that higher surface glucan content of *C. glabrata*, in comparison to *C. albicans*, might result in differences in biofilm matrix composition among these two Candida species. Therefore, we quantified β-1,3-glucan content of 24 h biofilm matrix of CAI4 and Cg30 strains. *C. glabrata* biofilm matrix consisted of 104±4 ng β-1,3-glucan/mg dry weight, in agreement with other studies [27,28], compared with the significantly lower (P<0.05) matrix β-1,3-glucan content

(93±3 ng/mg dry weight) of *C. albicans* biofilm matrix. To determine whether higher β-1,3-glucan levels in the biofilm matrix of *C. glabrata* might bind more Hst 5 and reduce its ability to disseminate to fungal cells within the biofilm, we examined the relative concentration of Hst 5 in biofilm matrices. FITC-Hst 5 applied to the surface of 24 h biofilms readily diffused into the biofilm matrix formed by both *C. albicans* and *C. glabrata* and became concentrated within the bottom regions of the matrix within 30 min (Figure 4 A). Hst 5 associated with the matrix formed by *C. glabrata* was significantly higher (P<0.05), at all depths analyzed (2–16 µm), compared with *C. albicans* biofilms and was an average of 32% higher in *C. glabrata* matrices (Figure 4 A). However, this elevation of matrix associated Hst 5 (32% increase compared to *C. albicans*) in *C. glabrata* biofilms was three fold higher than the corresponding difference in β-1,3-glucan content (11% increase compared to *C. albicans*), pointing to additional matrix components other than glucans that could potentially bind Hst 5.

We next examined the relative proportions of Hst 5 labeled cells within each region of the biofilm with the expectation that regions with high matrix associated Hst 5 would have less peptide available for diffusion and thus have reduced cell associated Hst 5. Surprisingly, the bottom regions (2.5 µm) of the biofilm of *C. glabrata* with the highest matrix density of Hst 5 also had the highest percentage (20%) of Hst 5-labeled *C. glabrata* cells (Figure 4 B). In upper layers (15 µm), the percentage of Hst 5 labeled *C. glabrata* cells was reduced proportionally with the descending matrix gradient of Hst 5 so that the lowest percentage of Hst 5 labeled cells and lowest Hst 5 matrix density were at the uppermost (top) regions of the biofilm. Thus Hst 5 sequestration within the biofilm matrix does not limit its availability for uptake into fungal cells.

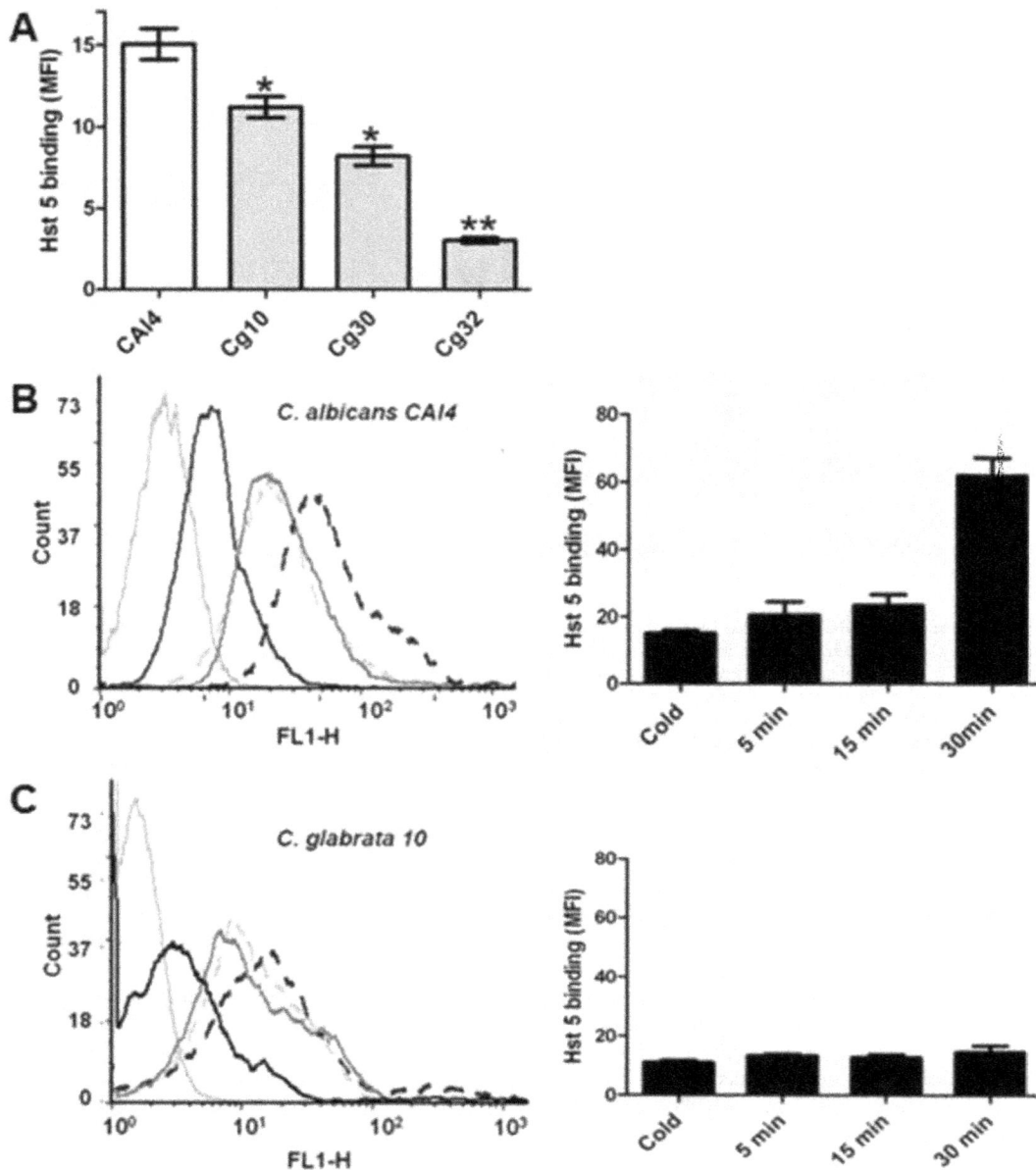

Figure 2. *C. glabrata* **has reduced cell surface binding and uptake of Hst 5. A.** FITC-Hst 5 (15 μM) was incubated with Candida strains *CAI4*, *Cg10*, *Cg30* and *Cg32* for 15 min and quantified using flow cytometry. *C. glabrata* strains (gray bars) *Cg10*, *Cg30* and *Cg32* had significantly (* $P \leq 0.05$; ** $P \leq 0.001$) less cell surface bound Hst 5 than *CAI4* (white bars) (n = 4). FACS analysis of Hst 5 binding and uptake was performed using *CAI4* (**B**) and *Cg10* (**C**) strains. Cells not exposed to Hst 5 were used as controls (light gray solid lines -). Cells were pre-incubated on ice (cold) before incubation with F-Hst 5 for 15 min to block energy dependent uptake of Hst 5 and to quantify cell wall bound Hst 5 (black solid lines -), then warmed cells were treated with F-Hst 5 for 5 min (dark gray solid lines -), 15 minutes (light gray broken lines ---), and 30 minutes (black broken lines ---). *CAI4* had significantly higher cell wall bound Hst 5 than *Cg10*. No significant differences were observed between cell wall bound Hst 5 and translocated Hst 5 in *C. glabrata* strain *Cg10* (n = 3). The bar graphs represent mean fluorescence intensities of F-Hst 5 *CAI4* (**B**) and *Cg10* (**C**).

C. albicans biofilms had a different distribution of cells containing Hst 5 that did not follow the concentration of Hst 5 within the matrix, in contrast to *C. glabrata* biofilms. The highest percentage of Hst 5 positive *C. albicans* cells (33%) was found in the middle regions (8 μm) of the biofilm, while bottom regions with the highest density of matrix Hst 5 and top regions with the lowest density of Hst 5 had equivalent percentages of Hst 5 containing cells (≈25%) (Figure 4 B). In comparing the total biofilms, *C. albicans* biofilms had more Hst 5 containing cells (28%) than *C. glabrata* biofilms (17%). Interestingly, the relative proportion of Hst

5 containing cells (*C. albicans* to *C. glabrata*) in biofilms was nearly equal to that measured with planktonic cells (Figure 2 A), suggesting that biofilm phase cells take up Hst 5 to the same degree as planktonic cells. Thus, although Hst 5 readily diffused and bound to biofilm matrices, its presence here did not reduce its ability to bind to and enter biofilm cells. Furthermore, differences in the proportions of matrix β-1,3-glucan did not account for differences in the amount of total bound Hst 5 either within the matrix or with biofilm cells.

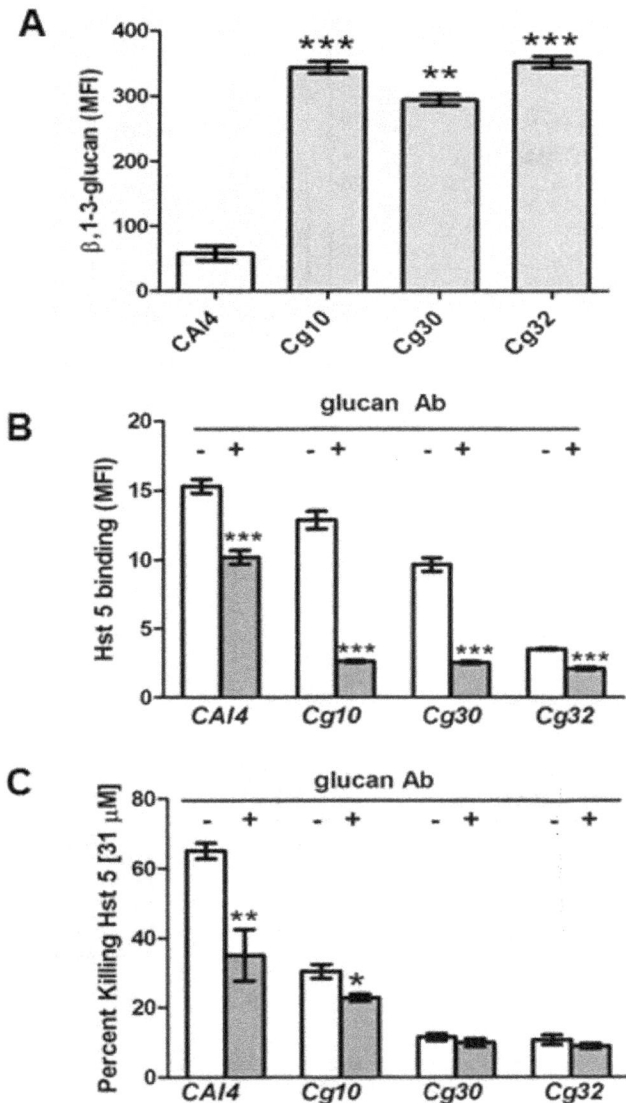

Figure 3. Cell surface β-1,3-glucans differentially influence Hst 5 binding and killing. A. Candida cell surface exposed β-1,3-glucans were quantified using flow cytometry. *C. albicans* CAI4 had significantly lower levels (four to six fold) of surface exposed β-1,3-glucans compared to the *C. glabrata* strains tested (n = 3). Candida cells were pre-incubated with β-1,3-glucan antibody to block Hst 5 cell surface binding components and then treated with FITC-Hst 5 or Hst 5 for binding (**B**) and candidacidal assays (**C**). CAI4, Cg10, Cg30 and Cg32 all had a significant (*** P<0.0001) decrease in Hst 5 binding following pre-incubation with β-1,3-glucan Ab (n = 3) (**B**), although only CAI4 (60%) and Cg10 (22%) had a significant decrease in Hst 5 killing following blocking with β-1,3-glucan Ab (n = 4) (**C**). Statistical analysis of differences was calculated by Student's t-test.

C. glabrata cells that express *C. albicans* DUR transporters have increased Hst 5 translocation and killing

Next, we investigated the role of altered Hst 5 uptake mechanisms between *C. albicans* and *C. glabrata* as a probable reason for their differential susceptibilities to Hst 5. *C. albicans* contains six *DUR* transporter family members that are responsible for polyamine (spermidine, spermine, and putrescine) uptake. We found that *DUR3* (Orf 19.781) and *DUR31* (Orf 19.6656) genes

encode polyamine transporters that facilitate Hst 5 uptake and that Hst 5 killing was significantly decreased in Δdur3, Δdur31, and Δdur3/Δdur31 strains [30]. We performed BLASTp analysis of *C. albicans DUR3* and *DUR31* translated protein sequences with the *C. glabrata* genome and found that the highest sequence similarities are with gene products of *CAGL0K03157g* and *CAGL0I08613g* that have not been characterized in *C. glabrata*. *C. glabrata* CAGL0K03157g product has 54% identity and 71% similarity with *C. albicans DUR3* (orf19.781); and *C. glabrata* CAGL0I08613g product has 51% identity and 69% similarity with *C. albicans DUR31* (orf19.6656). However, *C. glabrata* CAGL0K03157g (Ca-DUR3 homologue) gene product has even higher homology with *S. cerevisiae DUR3*, having 65% identity and 76% similarity. The *in silico* predicted function of these *C. glabrata* proteins is transmembrane transport; however their specific functions remain uncharacterized. To determine whether spermidine uptake levels were comparable to *C. albicans*, spermidine uptake rates were measured in Cg10, Cg30, Cg32, and CAI4 strains as we have previously described [30]. There were no significant differences in rates of spermidine uptake among the three strains of *C. glabrata* (0.163 ± 0.015 nanomoles/10^6 cells/min) compared to *C. albicans* (0.160 ± 0.014 nanomoles/10^6 cells/min); thus showing polyamine transporter activity is equivalent between *C. albicans* and *C. glabrata* strains in respect to spermidine. However, the significantly lower uptake of Hst 5 suggested that *C. glabrata* polyamine transporter proteins might have differences in substrate specificities resulting in lower Hst 5 uptake capacity.

We hypothesized that introduction of *C. albicans* Dur3 and Dur31 transporters into *C. glabrata* may increase the uptake of Hst 5. To this end, we constructed *C. glabrata* strains expressing either CaDUR3 or CaDUR31. Since *C. albicans DUR31* has no CUG codons and *DUR3* has only 3 CUG codons (one in transmembrane domain 2 (TMD), one in the external loop between TMD1 and TMD2, and one in the internal loop between TMD7 and TMD8) we did not expect differences in translation of these codons by *C. glabrata* would affect protein function. Indeed, we found that *C. glabrata* expressing either CaDUR3 or CaDUR31 both had similar amounts of Hst 5 uptake and sensitivity to Hst 5 (see data below), suggesting that these three CUG codons in *DUR3* are likely in non-conserved regions. Due to lack of nutrient auxotrophies, Cg10, Cg30, and Cg32 could not be used for making insertional *DUR* mutants. Instead, *C. glabrata* BG14 strain [33] was used for insertion of CaDUR3 and CaDUR31 since it is auxotrophic for the selection marker uridine and we found that this strain was similar to Cg10 in Hst 5 binding as well as its low sensitivity to Hst 5 as shown below.

Expression levels of CaDUR3 and CaDUR31 in *C. glabrata* expressing *C. albicans DUR* genes were examined by RT-PCR (Figure 5). Both *C. albicans DUR3* and *DUR31* genes were expressed in *C. glabrata* and we found no evidence of amplification of other *DUR* related genes or *C. glabrata* homologues (empty vector), highlighting the low similarity between *C. albicans* and *C. glabrata DUR* genes. However, the expression level of both CaDUR3 and CaDUR31 genes in *C. glabrata* was about half of that found in native *C. albicans* CAI4. Thus, we expected that the phenotype of the *C. glabrata* mutant expressing CaDUR3 and CaDUR31 genes would be attenuated with respect to Dur functions compared to *C. albicans*. To test these functions, we examined both fungicidal activity and Hst 5 uptake in *C. glabrata* mutants.

A significant increase (p<0.05) in Hst 5 mediated killing was observed in *C. glabrata* mutants expressing both *C. albicans DUR3* or *DUR31* (Figure 6 A). Killing of Cg BG14-CaDUR3 was increased to 43% and Cg BG14-CaDUR31 to 55% compared with only 25% in

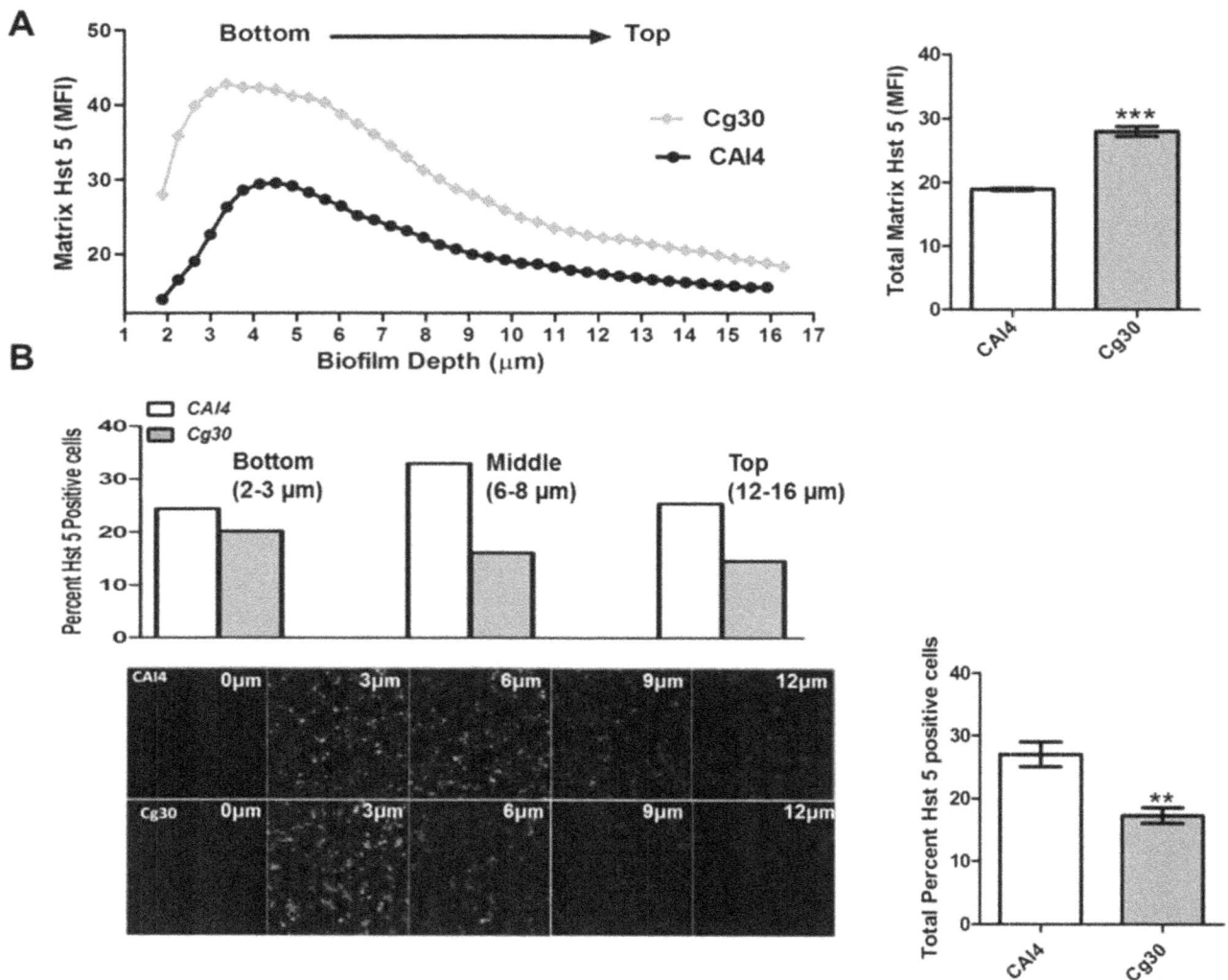

Figure 4. Hst 5 binding to Candida biofilm matrix components. Confocal microscopy was used to analyze FITC-labeled Hst 5 (F-Hst 5) binding to *CAI4* and *Cg30* biofilms. **A.** The biofilm matrix content of F-Hst 5 was quantified in areas without cells at 5 different locations and mean fluorescent intensity (MFI) was calculated across all Z-stacks (44). The matrix density of F-Hst 5 was highest at lower biofilm depths (3–5 μm) although the *Cg30* biofilm matrix (■) uniformly had higher Hst 5 content than *CAI4* (●) matrix. The total matrix content of F-Hst 5 was significantly higher (P<0.05) for *Cg30* than for *CAI4* (right). **B.** F-Hst 5 labeled Candida cells were counted at three different depths of the biofilm for *CAI4* and *Cg30* from the entire plane section (left); and percentage of F-Hst 5 labeled to unlabeled cells was calculated at Bottom (2–3 μm), Middle (6–8 μm), and Top (12–16 μm) sections. *Cg30* had fewer F-Hst 5 labeled cells at all biofilm depths (right), although the highest percentage of F-Hst 5 labeled *Cg30* was also found at Bottom depths with highest Hst 5 matrix density. The total percentage of F-Hst 5 labeled *Cg30* cells was significantly less (P<0.001) than F-Hst 5 labeled *CAI4* cells.

cells not expressing *C. albicans DUR* genes following treatment with Hst 5 [60 μM]. These strains had a 50 percent gain of killing function that was found at all concentrations of Hst 5 examined and was specific to cells carrying *C. albicans DUR3* and *DUR31*, as the empty vector control strain had no change in Hst 5 sensitivity (Figure 6 A). To determine if the increase in sensitivity of these strains to Hst 5 was due to an alteration in binding or uptake, we examined cells by FACScan. No difference in Hst 5 surface binding was observed among cold treated *C. glabrata* cells (Figure 6 B, white bars), showing that the expression of *C. albicans* Dur transporters does not alter *C. glabrata* cell wall composition in terms of Hst 5 binding. However, both *Cg BG14-CaDUR3* and *Cg BG14-CaDUR31* cells had significantly more cell associated Hst 5 when treated at 37°C when compared with cells containing only an empty vector (Figure 6 B, grey bars). These *C. glabrata* strains

expressing *C. albicans DUR* genes showed a time dependent increase in Hst 5 uptake from MFI = 9 to MFI = 17.5 at 30 min, both of which were significantly higher than the empty vector control (MFI = 9 at 30 min). To verify that *C. glabrata* cells expressing *CaDUR3* and *CaDUR31* had a higher proportion of Hst 5 translocation, we examined these strains using confocal microscopy (Figure 6 B). Both *Cg BG14-CaDUR3* and *Cg BG14-CaDUR31* had multiple cells per field that contained intracellular F-Hst 5 at 15 min, while none of the *Cg BG14-Empty vector* cells had any uptake of Hst 5 at this time. By 30 min, only a few *Cg BG14-Empty vector* cells contained F-Hst 5, while double to triple the number of *Cg BG14-CaDUR3* and *Cg BG14-CaDUR31* were positive for F-Hst 5 (Figure 6 B). Thus, total intracellular uptake of Hst 5 as measured both by FACScan and confocal microscopy in *Cg BG14-CaDUR3* and *Cg BG14-CaDUR31* very closely matched

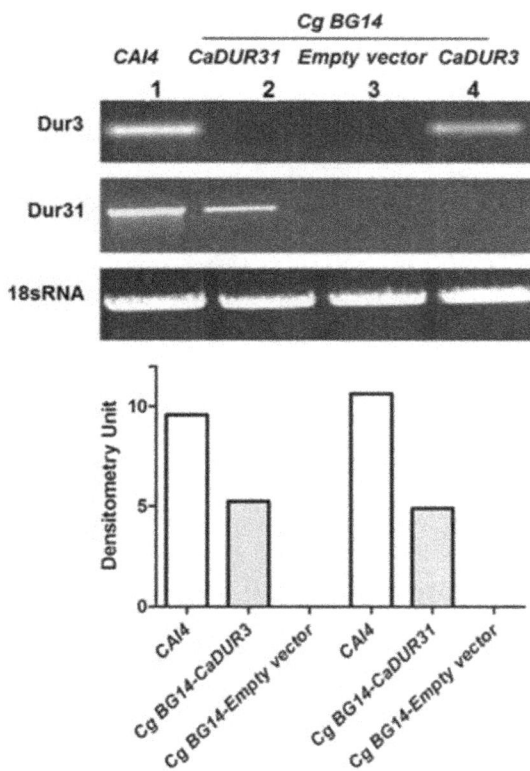

Figure 5. Expression of *C. albicans DUR3* and *DUR31* transporters in *C. glabrata*. *C. albicans DUR3* and *DUR31* were expressed in *C. glabrata BG14* using the pGRB2.2 vector. *BG14* cells expressing an empty vector, Ca*DUR3*, and Ca*DUR31*, respectively, along with CAI4 were used to perform Reverse Transcription PCR to determine expression levels of *DUR3* and *DUR31* genes using ribosomal *18S* as a control. Both *C. albicans DUR3* and *DUR31* genes were expressed specifically in *C. glabrata*, although their expression levels were only half of that found in their native site in CAI4 (n = 3). Densitometric quantification is shown in the bottom panel for ease of visualization.

the increased killing by Hst 5, clearly showing that Dur3 and Dur31 mediated uptake is required for toxicity. To determine whether β-1,3-glucan dependent toxicity was restored by increased uptake as a result of placement of *C. albicans* Dur proteins, we performed β-1,3-glucan blocking experiments. Indeed, *C. glabrata* expressing *C. albicans DUR* genes showed a significant (p<0.05) decrease in percent killing by Hst 5 at (60 µM) (Figure 6 C), showing that Hst 5 uptake is the limiting process for its toxicity in *C. glabrata* cells.

Discussion

Cell surface carbohydrates play an important role in protection and maintenance of fungal cells while serving as the point of contact with the host environment. Since we previously identified surface β-1,3-glucans as important for Hst 5 binding in *C. albicans* [21], we expected that this carbohydrate would also have a function in Hst 5 interactions with *C. glabrata*. However, despite higher surface levels of surface β-1,3-glucan in *C. glabrata* (Figure 3), we observed that blocking cell surface exposed polysaccharides in *C. glabrata 10* reduced binding of Hst 5 by more than 60%, although its killing was only reduced by 20% (Figure 3). This anomaly was conceivably based on our observations that uptake of Hst 5 was extremely reduced in *C. glabrata* strains (Figures 1 and 2)

and therefore reduction in binding to surface β-1,3-glucan does not further impact Hst 5 mediated toxicity. Indeed, expression of *C. albicans* Dur3 and Dur31 transporters in *C. glabrata* not only increased Hst 5 uptake and toxicity, but also restored β-1,3-glucan dependent binding for toxicity (Figure 6). However, our data do not explain why Hst 5 cell wall binding to some *C. glabrata* strains (*Cg30* and *Cg32*, Figure 3) was poor even though these strains had equivalent surface content of β-1,3-glucan when compared to *C. albicans*. Like *C. albicans*, *C. glabrata* cell walls contain Pir (Proteins with internal repeats) proteins that are covalently linked to β-1,3-glucan by mild-alkali-sensitive linkages [34] and are likely a major source of surface exposed β-1,3-glucan molecules. However additional cell wall proteins also contain Pir repeats connecting them to β-1,3-glucans, thus illustrating the "mosaic-like nature of the external protein coat" [35]. We speculate that variations in the surface distribution of these "mosaics" might account for the low binding activity of Hst 5 to certain *C. glabrata* strains. It is also possible that other *C. glabrata* cell wall proteins may mediate critical binding; however we found no differences in Hst 5 binding to purified cell wall preparations that were treated with detergents and denaturing agents to remove non-covalently linked cell wall proteins (our unpublished data). Other carbohydrate or lipid cell wall components remain to be examined as possible binding sites for Hst 5 in *C. glabrata*.

Among the known virulence factors of *C. glabrata*, biofilm formation is well studied and has become increasingly recognized as an important clinical problem [36–38]. The significantly higher levels of cell surface β-1,3-glucan in *C. glabrata* compared to *C. albicans* suggested that the secreted glucans in its biofilm matrix may contribute to its protection from Hst 5 similar to the role of β-1,3-glucan in fluconazole sequestration within *C. albicans* biofilms [26]. Indeed, we found significantly elevated matrix associated Hst 5 in *C. glabrata* biofilms at all depths of the biofilm compared with *C. albicans* biofilms (Figure 4). However, this did not result in sequestration of Hst 5 as regions with the highest percentage of Hst 5 labeled *C. glabrata* cells were in regions with highest matrix density of Hst 5. From this data, we propose that the biofilm matrix serves as a locally sequestered reservoir of Hst 5 that subsequently diffuses to cells throughout the biofilm without loss of antifungal activity. Indeed, 30 min after Hst 5 application to the biofilm surface, the highest concentration of Hst 5 was found in the matrix and cells within the middle and lowest regions of the biofilm (Figures 4 A and 4 B). This is in contrast to other antifungal compounds such as flucytosine, fluconazole, amphotericin B, and voriconazole that have poor diffusion through fungal biofilms [39], thus bolstering potential therapeutic use of Hst 5 against fungal biofilms.

In *C. albicans* Hst 5 initially binds to cell wall carbohydrates, then translocates to the cytoplasm through polyamine transporters, specifically Dur3 and Dur31 transporters. Unlike other pore forming antimicrobial compounds such as bactenecins, defensins, magainins, and tachyplesins, Hst 5 cannot insert into membranes due to its weak amphipathic nature. Like polyamines, Hst 5 is highly polar, hydrophilic and cationic. Based on biophysical studies [6], Hst 5 is unstructured in aqueous environments and this structural flexibility may be the cause of its ability to be transported through polyamine transporters in *C. albicans*. In most *S. cerevisiae* strains, Hst 5 is not transported into the cytosol nor is it fungicidal, suggesting that like *C. glabrata*, its transporters do not carry Hst 5 as a substrate. However, insertion of either *C. albicans DUR3* or *DUR31* in *C. glabrata* increased the uptake of Hst 5 and fungicidal activity by more than 40% (Figure 6 A), thus underscoring the crucial role of polyamine transporters for Hst 5 uptake. It is likely that the differential uptake of Hst 5 between the

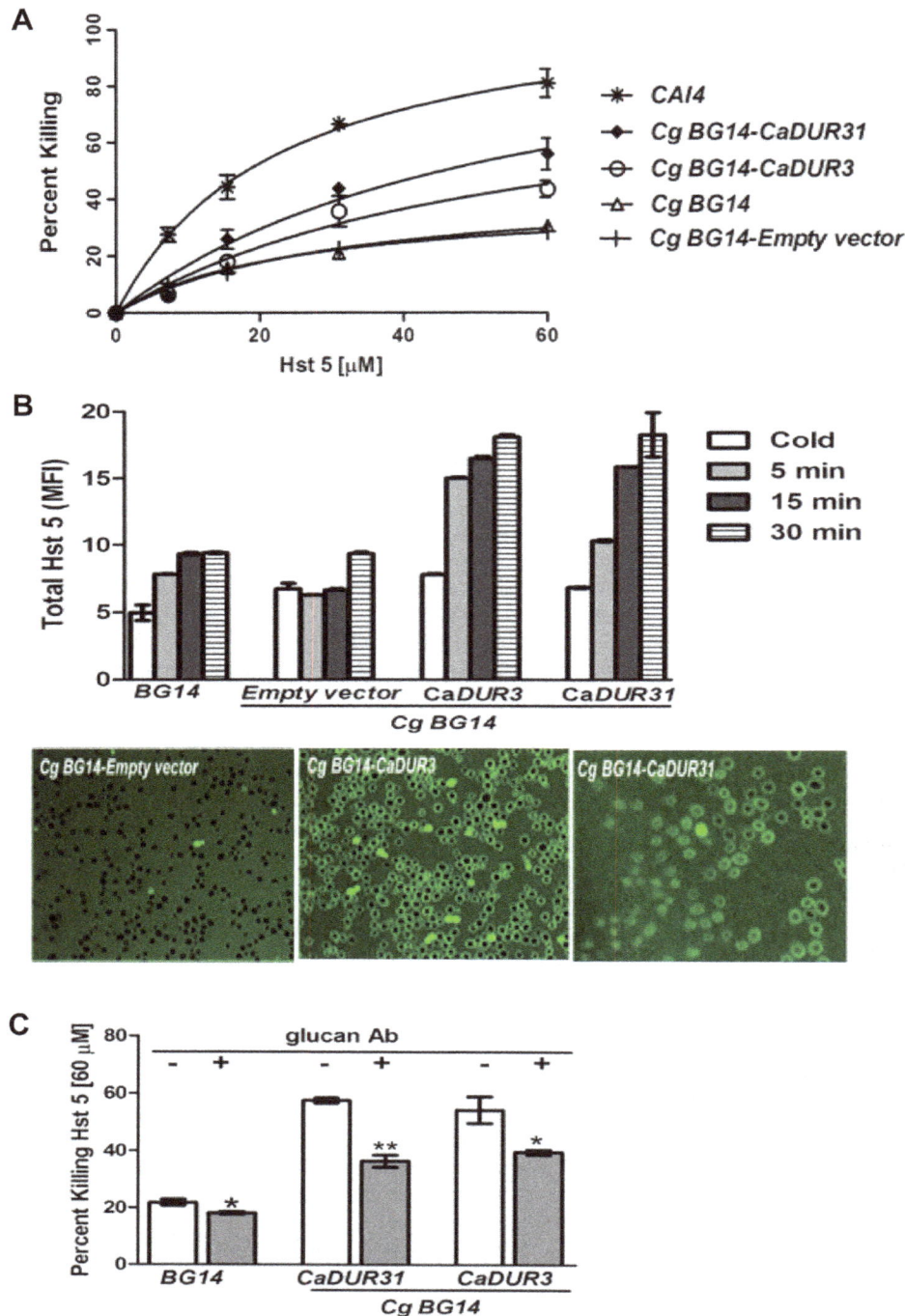

Figure 6. Hst 5 sensitivity and uptake in *C. glabrata* strains expressing *C. albicans* DUR transporters is increased. A. Candidacidal assays were performed on *C. albicans* WT strain *CAI4* (*), WT strain of *C. glabrata BG14* (△), and *C. glabrata* expressing *C. albicans* DUR genes *Cg BG14-CaDUR3* (○) and *Cg BG14-CaDUR31* (◆); and *Cg BG14-Empty vector* (I). *C. glabrata* expressing *C. albicans* DUR genes *Cg BG14-CaDUR3* and *Cg BG14-CaDUR31* showed a 40%–50% percent increase in the sensitivity to Hst 5 (n = 3). **B.** Hst 5 binding and uptake were quantified by flow cytometry on *Cg BG14-CaDUR3*, *Cg BG14-CaDUR31*, and *Cg BG14-Empty vector* (control) strains. Cold treated cells were used for quantification of Hst 5 surface binding (white bars); warmed cells were assessed for cellular uptake of Hst 5 (gray bars). A significant increase (P<0.001) in the uptake of Hst 5 was observed in *C. glabrata* expressing *C. albicans* DUR genes compared to the parental strain (BG14) and *Cg BG14-Empty vector* (control). No significant difference was observed in the binding of Hst 5 in *C. glabrata* expressing *C. albicans* DUR genes (white bars). *Cg BG14-CaDUR3*, *Cg BG14-CaDUR31*, and *Cg BG14-Empty vector* (control) strains were treated with FITC-Hst 5 and the translocation of this peptide was observed using confocal microscopy (**B**, lower). *C. glabrata* expressing *CaDUR3 and CaDUR31* showed increased Hst 5 uptake compared with the control strain (*Cg BG14-Empty vector*). Images are shown after 30 min incubation with Hst 5 (n = 3). **C.** Candidacidal assays were performed on *C. glabrata* expressing *C. albicans* DUR genes and *BG14* strains with (+) and without (−) preincubation with blocking antibodies to β-1,3-glucan. Both *Cg BG14-CaDUR3* and *Cg BG14-CaDUR31* strains showed a significant reduction in Hst 5 killing following pre-incubation with β-1,3-glucan Ab (n = 3).

C. glabrata strains examined here is due to structural differences in Dur3 transporters that are reflected in their varying ability to utilize Hst 5 as a transported substrate. Alternatively, strain differences in Hst 5 uptake might be due to differing cellular requirements for polyamines related to intracellular polyamine stores or alternative biosynthesis. For example, the C. glabrata enzyme spermidine synthase SPE'3P: CAGL0D01408g is differentially expressed in azole resistant C. glabrata strains [40,41]. More information is needed to identify and map the polyamine biosynthesis pathway in C. glabrata. This information will open the possibility for treatment of C. glabrata with polyamine biosynthesis inhibitors that increase the activity of native polyamine transporters and/or upregulate their expression levels, thereby resulting in higher uptake of Hst 5. Hst 5 could be used in combination with spermidine synthase inhibitors and/or be coupled with spermidine to improve its efficacy by increasing its uptake in C. glabrata.

We identify here for the first time that the basis for differential resistance of C. glabrata to salivary Hst 5 is due to its low uptake, and is not a result of reduced binding to the cell surface despite differences in surface carbohydrate content. Hst 5 uptake and fungicidal activity were substantially increased by expression of C. albicans DUR3 or DUR31 polyamine transporters, stressing the importance of this uptake mechanism for Hst 5 activity. This insight provides a basis for design of Hst 5 peptides that have improved intracellular uptake in fungal cells, such as Hst 5-polyamine conjugates.

Materials and Methods

Strains and Media

Candida strains used in this study are summarized in Table 1. Candida albicans CAI4, Candida glabrata Cg 931010 (Cg10), Cg 90030 (Cg30), Cg 90032 (Cg32), and BG14 were used as WT strains. Cg BG14-Empty vector, Cg BG14-CaDUR3, Cg BG14-CaDUR31 strains were created in this study and used as C. glabrata wild type strain expressing C. albicans DUR genes. Overnight cultures were grown in yeast extract/peptone/dextrose (YPD; 1% yeast extract, 2% bacto peptone, 2% glucose, Difco, Detroit, MI) at 30°C to an OD600 of 2.0. C. glabrata wild type strains expressing C. albicans DUR genes were grown in Yeast Nitrogen Base media (YNB; Difco, Detroit, MI) without uridine and supplemented with 2% glucose.

Insertion of C. albicans DUR3 and DUR31 genes in C. glabrata BG14. C. glabrata strains Cg BG14-Empty vector, Cg BG14-CaDUR3, and Cg BG14-CaDUR31 were created using an episomal plasmid

pGRB2.2 by inserting C. albicans DUR3 (EcoRI-SalI site) and DUR31 (BamHI-SalI site) using uridine and neomycin as selection markers. Candida albicans DUR3 and DUR31 were amplified using gene specific primers with CAI4 genomic DNA as a template. CaDUR3: forward primer: GAATTC ATG GCT GAT TCA TAT GTC CA, reverse primer: GTCGAC CTA TCC TTT CTT CTC CTC TAA TGC AT. CaDUR31, forward primer: GGATCC ATG GCA CAA CTA TCA TCA CAG G, reverse primer: GTCGAC TTA GAC CAC CTT TTT AGT ATC TGA TTC. PCR amplified DNA was ran on 1.2% agarose gel, purified, ligated to linearized T-cloning vector pGEM®-T Easy Vector System I (Promega Inc) and transformed into DH5α cells that were spread onto X-gal/IPTG plates for blue white screening. T-cloning vector DNA was digested with specific restriction enzymes (EcoRI, SalI to CaDUR3 and BamHI, SalI to CaDUR31) to isolate CaDUR3 and CaDUR31 fragments. Plasmid pGRB2.2 (which is an URA3 CEN/ARS plasmid, using the S. cerevisiae PGK1 promoter and the C. glabrata HIS3 3′ untranslated region) vector DNA was digested using specific restriction enzymes to ligate CaDUR3 and CaDUR31. Subsequently, this insert was cloned into pGRB2.2 at the same sites (EcoRI, SalI to CaDUR3 and BamHI, SalI to CaDUR31) to yield plasmids pGRB2.2-CaDUR3 and pGRB2.2-CaDUR31. The resulting pGRB2.2 plasmid DNA with insert was linearized and transformed into C. glabrata BG14 strain. C. glabrata transformation was performed as previously described [42]. Briefly, overnight cultures of BG14 cells were re-suspended in fresh YPD to OD600 = 0.4 and grown 3–4 h at 37°C with shaking to reach OD600 = 1.0. Cells were harvested, washed twice with water; cell pellet was re-suspended in Tris EDTA buffer and collected by centrifugation. Cells were re-suspended in 0.15 M Lithium acetate (LiOAC), 1 mm EDTA, and 10 mm Tris (pH 7.5), and incubated at 30°C for 1 h. Cells were harvested and re-suspended in 0.15 M LiOAc. The cell suspension was transformed with plasmid DNA using pGRB2.2 with the CaDUR3 insert; pGRB2.2 with the CaDUR31 insert; or PGRB2.2 DNA, along with 20 μg of denatured salmon sperm DNA and was incubated at 37°C for 30 min. Polyethylene glycol 4000 (52.5%) with 0.15 M LiOAC was mixed with cells and incubated for 45 min at 42°C. This mixture was spread onto the selection media (YNB without uridine and with gentamicin (50 ng/ml)) and grown at 37°C.

Determination of expression levels of CaDUR3 & CaDUR31

For RT-PCR analysis, RNA was extracted from the CAI4, BG14, and C. glabrata insertional mutant strains using the RNeasy

Table 1. Strains used in this study.

Candida Strains	Genotype	Reference
C. albicans CAI4	Δura3::imm434/Δura3::imm434	[44]
C. glabrata BG2	Wild type	[45]
C. glabrata BG14	ura3Δ::Tn903	[46]
C. glabrata Cg BG14-Empty vector	ura3Δ::Tn903 G418R+pGRB2.2G418R	This study
C. glabrata Cg BG14-CaDUR3	ura3Δ::Tn903 G418R+pGRB2.2 CaDUR3	This study
C. glabrata Cg BG14-CaDUR31	ura3Δ::Tn903 G418R+pGRB2.2 CaDUR31	This study
C. glabrata Cg 931010	Wild type	[47]
C. glabrata Cg 90030	Wild type	ATCC
C. glabrata Cg 90032	Wild type	ATCC

Mini-Kit (Qiagen). cDNA was synthesized using 1 µg of total RNA and oligo (dT) primers and Moloney murine leukemia virus reverse transcriptase (Retroscript, Ambion, Austin,TX). Using 1 µl of synthesized cDNA, PCR was performed using GoTaq® Hot Start polymerase (Promega Corp., WI). PCR was performed using: 5'-GACCAATGACTGCTGCTGAA-3' and 5'-GCCAGTTTTGACGTTTGGAT-3' for DUR3; 5'-GAT-CATCTGTGCTGCTGGAA-3' and 5'-AGCAGCTGAAGC-CAATGT-3' for DUR31; 5'-CGATGGAAGTTTGAGG-CAATA-3' and 5'-CTCTCGGCCAAGGCTTATACT-3' for 18S RNA. Amplified PCR products were separated with 1.2% agarose gel and visualized by ethidium bromide staining.

Candidacidal Assays of Hst 5

Candidacidal assays were performed using microdilution plate assays [21]. Briefly, single colonies of *C. albicans CAI4*, *C. glabrata Cg10*, *Cg30*, *Cg32*, and *BG14* strains were inoculated in YPD media; *Cg BG14-CaDUR3*, *Cg BG14-CaDUR31*, and *Cg BG14-Empty vector* were inoculated in YNB media without uridine and grown overnight ($A_{600} = 1.6–1.8$). Cells were washed twice with 10 mM sodium phosphate buffer (NaPB) (pH 7.2) and cells (1×10^6) were incubated at 30°C for 30 min with different concentrations of Hst 5. Aliquots of 500 cells were spread onto YPD (WT strains) or YNB - uridine (*C. glabrata* expressing *C. albicans DUR* genes) agar plates and incubated for 48 h to visualize surviving colonies. Blocking experiments were performed using cells pre-incubated with β-1,3-glucan monoclonal antibody (Biosupplies) at room temperature for 30 min. All killing assays were performed in triplicate and repeated at least thrice. Percent cell killing was calculated as 1−(number of colonies from suspensions with Hst 5/numbers of colonies from control suspensions)×100.

Time Lapse Confocal Microscopy for Hst 5 Binding and Uptake

CAI4, *Cg10*, *Cg30*, *Cg32*, *BG14*, *Cg BG14-CaDUR3*, *Cg BG14-CaDUR31*, and *Cg BG14-Empty* vector were treated with fluorescein isothiocyanate FITC-labeled Hst 5 (F-Hst 5, synthesized by Genemed Synthesis, Inc.) to observe relative binding and uptake of salivary Hst 5 as described previously [21]. FITC alone does not bind *C. albicans* or *C. glabrata* cells. Cells grown overnight ($A_{600} = 0.8–1.0$) were diluted to obtain 10^6 cells/ml in NaPB. Chambered cover glass slides (Lab-TekII) were coated with concanavalin A (100 µg/ml) for 30 min and washed twice with water. Cells (1×10^6) were fixed on concanavalin A-coated slides for 30 min at room temperature. The plates were then washed twice with 10 mM NaPB followed by addition of 31 µM FITC-Hst 5. Images were captured using a Zeiss LSM 510 Meta Confocal Microscope and Plan Apochromat 63/1.4 (oil) objective. The average fluorescence intensity was calculated using ImageJ software. Confocal images of cells were compared to determine the relative binding and uptake of Hst 5. Hst 5 binding to *Candida* biofilms was analyzed using confocal microscopy on biofilms of *CAI4* and *Cg30* strains. Biofilms were formed by addition of 500 µl cells (cultured overnight in YPD at 28°C, then washed and diluted to $OD_{600} = 1.0$ in PBS) to each well of culture dishes (MatTek, MA) and incubated at 37°C for 3 h. Non-adherent cells were removed by gentle washing and 500 µl media was added to each well. The dishes were then incubated at 37°C for 24 h to allow biofilm formation. F-Hst 5 (31.5 µM) was added to each well. Biofilms were measured using a series of horizontal (x–y) optical sections with a thickness of 0.38 µm taken throughout the full length of the biofilm. Z-stack images and thickness measurements of biofilms were obtained using AxioVision 4.4 software (Carl

Zeiss LSM Micro Imaging). Mean Fluorescence Intensities (MFI) of Hst 5 were measured from the biofilm matrix that did not contain cells from five different areas across each Z- stack (44 stacks) from top to bottom of the biofilm using Image J software. The total number of F-Hst 5 labeled cells was quantified in *CAI4* and *Cg30* biofilms manually from three independent stacks originating at the Bottom (2–3 µm), Middle (6–8 µm) or Top (12–16 µm) regions of the biofilm. Values were plotted using Graphpad Prism 5 software.

Hst 5 Cell Wall Binding and Cytoplasmic Transport Assays

Hst 5 cell wall binding assays were performed as we have described previously [43]. Briefly, early log phase cells (1×10^7) of *CAI4*, *Cg10*, *Cg30*, and *Cg32* strains were washed with 10 mM NaPB and suspended in 1 ml of NaPB containing biotin-labeled Hst 5 (B-Hst 5) to a final concentration of 31 µM and incubated at 37°C for 30 min. The cells were washed with 10 mM NaPB to remove non-adherence Hst 5. Cell wall bound B-Hst 5 was measured by extracting cell wall components using ammonium carbonate buffer (pH 8.0) containing 1% (vol/vol) β-mercapto-ethanol (β-ME). Cells were then washed twice with 10 mM NaPB; and the cell pellet was incubated in 1 volume of cold lysis buffer supplemented with protease inhibitors (10 mM NaPB, 1 mM phenylmethylsulfonyl fluoride, 1 mM EDTA, 1 µg/ml aprotinin, 1 µg/ml pepstatin A, 1 µg/ml leupeptin, and 1 µg/ml benzamidine) and processed using a FastPrep homogenizer at 4°C. Cell wall extracts and cytosolic proteins were normalized to total protein content using a BCA assay (Pierce). *Candida* cell wall proteins and cytosolic proteins (10 µg) were subjected to SDS-PAGE, transferred to polyvinylidene difluoride (PVDF) membranes, and visualized with streptavidin conjugated with horse-radish peroxidase (Pierce). Data was analyzed with Quantity One software (version 4.2).

Flow Cytometry for Hst 5 binding and Uptake

CAI4, *Cg 931010*, *Cg 90030*, *Cg 90032*, *BG14*, *Cg BG14-CaDUR3*, *Cg BG14-CaDUR31*, and *Cg BG14-Empty* vector were treated with F-Hst 5 to observe relative binding and uptake of Hst 5. Cells grown overnight were diluted with fresh media to $A_{600} = 0.4$ and grown till they reached an OD of $A_{600} = 0.8–1.0$; then were diluted to obtain 10^6 cells/ml in NaPB. Cells were incubated with F-Hst 5 (15 µM) in 10 mM NaPB buffer at 37°C for 15 min in the dark with shaking and washed twice with PBS. For analyzing Hst 5 binding, cells were pre-incubated on ice for one hour prior to treating with F-Hst 5 (15 min on ice), then washed twice with ice cold PBS. For uptake assays, cells were incubated with F-Hst 5 at 37°C for 5, 15, or 30 min before washing. Cells were then re-suspended in 500 µl PBS and flow cytometry analysis was performed with FACSCalibur flow cytometry and Cellquest Pro Software (BD-Biosciences) with 10,000 cells collected and analyzed. Data analysis was performed using FCS Express 4 Flow Cytometry software (De Novo Software). For quantification of cell surface exposed glucan, cells (3×10^6) were incubated with anti-β-1,3-glucan monoclonal antibody (Biosupplies) at room temperature for 30 min, followed by incubation with Alexa-Fluor 647 conjugated secondary antibody (Cell Signaling Technology) for 30 min on ice and washed twice with the cold PBS. Cells were re-suspended in 500 µl PBS and flow cytometric analyses were performed with as described above.

Statistical analysis

Statistical analyses were performed using GraphPad Prism version 5.0 (GraphPad Software, San Diego, CA, USA) using unpaired Student's t-tests. Differences of P<0.05 were considered significant.

Acknowledgments

We thank Wade J. Sigurdson, and Raymond J. Kelleher from the Confocal Microscopy and Flow Cytometry Core Facility, University at Buffalo, for assistance with microscopy, FACScan, and helpful discussions. *Candida glabrata* strains *BG2*, *BG14* and plasmid pGRB2.2 were kindly provided by Dr. Brendan P. Cormack (Johns Hopkins University, Baltimore, MD). We thank all laboratory members for helpful discussions.

Author Contributions

Conceived and designed the experiments: ST WSJ ME. Performed the experiments: ST WSJ RL. Analyzed the data: ST WSJ RL RK ME. Contributed reagents/materials/analysis tools: ME. Wrote the paper: ST WSJ RL SP ME.

References

1. Pfaller MA, Diekema DJ (2007) Epidemiology of invasive Candidiasis: a persistent public health problem. Clin Microbiol Rev 20: 133–163.
2. Wisplinghoff H, Bischoff T, Tallent SM, Seifert H, Wenzel RP, et al. (2004) Nosocomial bloodstream infections in US Hospitals: analysis of 24,179 cases from a prospective Nationwide surveillance study. Clin Infect Dis 39: 309–317.
3. Fidel PL, Vazquez JA, Sobel JD (1999) *Candida glabrata*: Review of epidemiology, pathogenesis, and clinical disease with comparison to *C. albicans*. Clin Microbiol Rev 12: 80–96.
4. Oppenheim FG, Xu T, McMillian FM, Levitz SM, Diamond RD, et al. (1988) Histatins, a novel family of histidine-rich proteins in human parotid secretion. Isolation, characterization, primary structure, and fungistatic effects on *Candida albicans*. J Biol Chem 263: 7472–7477.
5. Raj PA, Edgerton M, Levine MJ (1990) Salivary histatin 5: dependence of sequence, chain length, and helical conformation for candidacidal activity. J Biol Chem 265: 3898–3905.
6. Raj PA, Marcus E, Sukumaran DK (1998) Structure of human salivary histatin 5 in aqueous and nonaqueous solutions. Biopolymers 45: 51–67.
7. Xu T, Levitz SM, Diamond RD, Oppenheim FG (1991) Anticandidal activity of major human salivary histatins. Infect Immun 59: 2549–2554.
8. Tsai H, Bobek LA (1997) Human salivary histatin-5 exerts potent fungicidal activity against *Cryptococcus neoformans*. Biochim Biophys Acta 1336: 367–369.
9. Helmerhorst EJ, Reijnders IM, van't Hof W, Simoons-Smit I, Veerman EC, et al. (1999) Amphotericin B and fluconazole-resistant *Candida* spp., *Aspergillus fumigatus*, and other newly emerging pathogenic fungi are susceptible to basic antifungal peptides. Antimicrob Agents Chemother 43: 702–704.
10. Helmerhorst EJ, Venuleo C, Beri A, Oppenheim FG (2005) *Candida glabrata* is unusual with respect to its resistance to cationic antifungal proteins. Yeast 22: 705–714.
11. Konopka K, Dorocka-Bobkowska B, Gebremedhin S, Düzgüneş N (2010) Susceptibility of *Candida* biofilms to histatin 5 and fluconazole. Antonie van Leeuwenhoek 97: 413–417.
12. Sanglard D, Ischer F, Calabrese D, Majcherczyk PA, Bille J (1999) The ATP Binding Cassette transporter gene *CgCDR1* from *Candida glabrata* is involved in the resistance of clinical isolates to azole antifungal agents. Antimicrob Agents Chemother 43: 2753–2765.
13. Sanglard D, Ischer F, Bille J (2001) Role of ATP-Binding-Cassette transporter genes in high-frequency acquisition of resistance to azole antifungals in *Candida glabrata*. Antimicrob Agents Chemother 45: 1174–1183.
14. Ferrari SlN, Sanguinetti M, Torelli R, Posteraro B, Sanglard D (2011) Contribution of *CgPDR1* regulated genes in enhanced virulence of azole-resistant *Candida glabrata*. PLoS ONE 6: e17589.
15. Ferrari SlN, Ischer Fo, Calabrese D, Posteraro B, Sanguinetti M, et al. (2009) Gain of function mutations in *CgPDR1* of *Candida glabrata* not only mediate antifungal resistance but also enhance virulence. PLoS Pathog 5: e1000268.
16. Ferrari SlN, Sanguinetti M, De Bernardis F, Torelli R, Posteraro B, et al. (2011) Loss of mitochondrial functions associated with azole resistance in *Candida glabrata* results in enhanced virulence in mice. Antimicrob Agents Chemother 55: 1852–1860.
17. Seneviratne CJ, Wang Y, Jin L, Abiko Y, Samaranayake LP (2010) Proteomics of drug resistance in *Candida glabrata* biofilms. Proteomics 10: 1444–1454.
18. Helmerhorst EJ, Venuleo C, Sanglard D, Oppenheim FG (2006) Roles of cellular respiration, *CgCDR1*, and *CgCDR2* in *Candida glabrata* resistance to Histatin 5. Antimicrob Agents Chemother 50: 1100–1103.
19. Helmerhorst EJ, Troxler RF, Oppenheim FG (2001) The human salivary peptide histatin 5 exerts its antifungal activity through the formation of reactive oxygen species. Proc Natl Acad Sci 98: 14637–14642.
20. Koshlukova SE, Lloyd TL, Araujo MWB, Edgerton M (1999) Salivary histatin 5 induces non-lytic release of ATP from *Candida albicans* leading to cell death. J Biol Chem 274: 18872–18879.
21. Jang WS, Bajwa JS, Sun JN, Edgerton M (2010) Salivary histatin 5 internalization by translocation, but not endocytosis, is required for fungicidal activity in *Candida albicans*. Mol Microbiol 77: 354–370.
22. Sun JN, Li W, Jang WS, Nayyar N, Sutton MD, et al. (2008) Uptake of the antifungal cationic peptide histatin 5 by *Candida albicans* Ssa2p requires binding to non-conventional sites within the ATPase domain. Mol Microbiol 70: 1246–1260.
23. Li XS, Sun JN, Okamoto-Shibayama K, Edgerton M (2006) *Candida albicans* cell wall Ssa proteins bind and facilitate import of salivary histatin 5 required for toxicity. J Biol Chem 281: 22453–22463.
24. Lowman DW, West LJ, Bearden DW, Wempe MF, Power TD, et al. (2011) New insights into the structure of $(1\rightarrow3,1\rightarrow6)$-β-D-Glucan side chains in the *Candida glabrata* cell wall. PLoS ONE 6: e27614.
25. Keppler-Ross S, Douglas L, Konopka JB, Dean N (2010) Recognition of yeast by murine macrophages requires mannan but not glucan. Eukaryot Cell 9: 1776–1787.
26. Taff HT, Nett JE, Zarnowski R, Ross KM, Sanchez H, et al. (2012) A *Candida* biofilm-induced pathway for matrix glucan delivery: implications for drug resistance. PLoS Pathog 8: e1002848.
27. Nett J, Lincoln L, Marchillo K, Massey R, Holoyda K, et al. (2007) Putative role of β-1,3 glucans in *Candida albicans* biofilm resistance. Antimicrob Agents Chemother 51: 510–520.
28. Al-Fattani MA, Douglas LJ (2006) Biofilm matrix of *Candida albicans* and *Candida tropicalis*: chemical composition and role in drug resistance. J Med Microbiol 55: 999–1008.
29. Mochon AB, Liu H (2008) The antimicrobial peptide Histatin-5 causes a spatially restricted disruption on the *Candida albicans* surface, allowing rapid entry of the peptide into the cytoplasm. PLoS Pathog 4: e1000190.
30. Kumar R, Chadha S, Saraswat D, Bajwa JS, Li RA, et al. (2011) Histatin 5 uptake by *Candida albicans* utilizes polyamine transporters Dur3 and Dur31 proteins. J Biol Chem 286: 43748–43758.
31. Mayer FoL, Wilson D, Jacobsen ID, Miramón P, Große K, et al. (2012) The novel *Candida albicans* transporter Dur31 is a multi-stage pathogenicity factor. PLoS Pathog 8: e1002592.
32. Baillie GS, Douglas LJ (2000) Matrix polymers of *Candida* biofilms and their possible role in biofilm resistance to antifungal agents. J Antimicrob Chemother 46: 397–403.
33. Ma B, Pan S-J, Domergue R, Rigby T, Whiteway M, et al. (2009) High-affinity transporters for NAD+precursors in *Candida glabrata* are regulated by Hst1 and induced in response to niacin limitation. Mol Cell Biol 29: 4067–4079.
34. de Groot PWJ, Kraneveld EA, Yin QY, Dekker HL, Groß U, et al. (2008) The cell wall of the human pathogen *Candida glabrata*: differential incorporation of novel adhesin-like wall proteins. Eukaryot Cell 7: 1951–1964.
35. Sorgo AG, Heilmann CJ, Dekker HL, Brul S, de Koster CG, et al. (2010) Mass spectrometric analysis of the secretome of *Candida albicans*. Yeast 27: 661–672.
36. Jain N, Kohli R, Cook E, Gialanella P, Chang T, et al. (2007) Biofilm formation by and antifungal susceptibility of *Candida* isolates from urine. Appl Environ Microbiol 73: 1697–1703.
37. Lewis RE, Kontoyiannis DP, Darouiche RO, Raad II, Prince RA (2002) Antifungal activity of amphotericin b, fluconazole, and voriconazole in an in vitro model of Candida catheter-related bloodstream infection. Antimicrob Agents Chemother 46: 3499–3505.
38. Thein ZM, Samaranayake YH, Samaranayake LP (2007) In vitro biofilm formation of *Candida albicans* and non-albicans *Candida* species under dynamic and anaerobic conditions. Arch Oral Biol 52: 761–767.
39. Al-Fattani MA, Douglas LJ (2004) Penetration of *Candida* biofilms by antifungal agents. Antimicrob Agents Chemother 48: 3291–3297.
40. Loureiro y Penha CV, Kubitschek PHB, Larcher G, Perales J, Rodriguez LI, et al. (2010) Proteomic analysis of cytosolic proteins associated with petite mutations in *Candida glabrata*. Braz J Med Biol Res 43: 1203–1214.
41. Rogers PD, Vermitsky JP, Edlind TD, Hilliard GM (2006) Proteomic analysis of experimentally induced azole resistance in *Candida glabrata*. J Antimicrob Chemother 58: 434–438.
42. Ito H, Fukuda Y, Murata K, Kimura A (1983) Transformation of intact yeast cells treated with alkali cations. J Bacteriol 153: 163–168.
43. Jang WS, Li XS, Sun JN, Edgerton M (2008) The P-113 fragment of histatin 5 requires a specific peptide sequence for intracellular translocation in *Candida albicans*, which is independent of cell wall binding. Antimicrob Agents Chemother 52: 497–504.
44. Fonzi WA, Irwin MY (1993) Isogenic strain construction and gene mapping in *Candida albicans*. Genetics 134: 717–728.
45. Fidel PL, Cutright Jr LJ, Tait L, Sobel JD, (1996) A murine model of *Candida glabrata* vaginitis. J Infect Dis 173: 425–431.
46. Cormack BP, Falkow S (1999) Efficient homologous and illegitimate recombination in the opportunistic yeast pathogen *Candida glabrata*. Genetics 151: 979–987
47. Joly S, Maze C, McCray PB Jr, Guthmiller JM (2004) Human beta-defensins 2 and 3 demonstrate strain-selective activity against oral microorganisms. J Clin Microbiol 42:1024–1029

12

Non-Hodgkin Lymphoma Risk and Insecticide, Fungicide and Fumigant Use in the Agricultural Health Study

Michael C. R. Alavanja[1]*, **Jonathan N. Hofmann**[1], **Charles F. Lynch**[2], **Cynthia J. Hines**[3], **Kathryn H. Barry**[1], **Joseph Barker**[4], **Dennis W. Buckman**[4], **Kent Thomas**[5], **Dale P. Sandler**[6], **Jane A. Hoppin**[6], **Stella Koutros**[1], **Gabriella Andreotti**[1], **Jay H. Lubin**[1], **Aaron Blair**[1], **Laura E. Beane Freeman**[1]

1 Division of Cancer Epidemiology and Genetics, National Cancer Institute, Rockville, Maryland, United States of America, 2 College of Public Health, University of Iowa, Iowa City, Iowa, United States of America, 3 National Institute for Occupational Safety and Health, Cincinnati, Ohio, United States of America, 4 IMS, Inc, Calverton, Maryland, United States of America, 5 National Exposure Research Laboratory, U.S. Environmental Protection Agency, Research Triangle Park, North Carolina, United States of America, 6 Epidemiology Branch, National Institute for Environmental Health Sciences, Research Triangle Park, North Carolina, United States of America

Abstract

Farming and pesticide use have previously been linked to non-Hodgkin lymphoma (NHL), chronic lymphocytic leukemia (CLL) and multiple myeloma (MM). We evaluated agricultural use of specific insecticides, fungicides, and fumigants and risk of NHL and NHL-subtypes (including CLL and MM) in a U.S.-based prospective cohort of farmers and commercial pesticide applicators. A total of 523 cases occurred among 54,306 pesticide applicators from enrollment (1993–97) through December 31, 2011 in Iowa, and December 31, 2010 in North Carolina. Information on pesticide use, other agricultural exposures and other factors was obtained from questionnaires at enrollment and at follow-up approximately five years later (1999–2005). Information from questionnaires, monitoring, and the literature were used to create lifetime-days and intensity-weighted lifetime days of pesticide use, taking into account exposure-modifying factors. Poisson and polytomous models were used to calculate relative risks (RR) and 95% confidence intervals (CI) to evaluate associations between 26 pesticides and NHL and five NHL-subtypes, while adjusting for potential confounding factors. For total NHL, statistically significant positive exposure-response trends were seen with lindane and DDT. Terbufos was associated with total NHL in ever/never comparisons only. In subtype analyses, terbufos and DDT were associated with small cell lymphoma/chronic lymphocytic leukemia/marginal cell lymphoma, lindane and diazinon with follicular lymphoma, and permethrin with MM. However, tests of homogeneity did not show significant differences in exposure-response among NHL-subtypes for any pesticide. Because 26 pesticides were evaluated for their association with NHL and its subtypes, some chance finding could have occurred. Our results showed pesticides from different chemical and functional classes were associated with an excess risk of NHL and NHL subtypes, but not all members of any single class of pesticides were associated with an elevated risk of NHL or NHL subtypes. These findings are among the first to suggest links between DDT, lindane, permethrin, diazinon and terbufos with NHL subtypes.

Editor: Suminori Akiba, Kagoshima University Graduate School of Medical and Dental Sciences, Japan

Funding: This work was supported by the Intramural Research Program of the NIH, National Cancer Institute, Division of Cancer Epidemiology and Genetics (Z01CP010119) and The National Institutes of Environmental Health Sciences (Z01ES049030). The funders had no role in study design, data collection and analysis, decision to publish, or preparation of the manuscript. IMS, Inc, provided support in the form of salaries for authors Joseph Barker and Denis W Buckman, but did not have any additional role in the study design, data collection and analysis, decision to publish, or preparation of the manuscript. The specific roles of these authors are articulated in the 'author contributions' section.

Competing Interests: The authors have the following interests. Joseph Barker and Dennis W. Buckman are employed by IMS, Inc. There are no patents, products in development or marketed products to declare.

* Email: alavanjm@mail.nih.gov

Introduction

Since the 1970s, epidemiologic studies of non-Hodgkin lymphoma (NHL) and multiple myeloma (MM) have shown increased risk among farmers and associations with the type of farming practiced [1–6]. While farmers are exposed to many agents that may be carcinogenic [7]; there has been a particular focus on pesticides. Studies from around the world have suggested increased risk of NHL or MM [8,9] and other NHL subtypes [10] in relation to the use of specific pesticides in different functional classes (i.e., insecticides, fungicides, fumigants and herbicides). A

meta-analysis of 13 case-control studies published between 1993–2005 observed an overall significant meta-odds ratio (OR) between occupational exposure to pesticides and NHL (OR = 1.35; 95% CI: 1.2–1.5) [11]. This risk was greater among individuals with more than 10 years of exposure (OR = 1.65; 95% CI: 1.08–1.95) [11], but the meta-analysis lacked details about the use of specific pesticides and other risk factors [11]. Although the International Agency for Research on Cancer (IARC) has classified "Occupational exposures in spraying and application of non-arsenical insecticides" as "probably carcinogenic to humans", the human evidence for the 17 individual pesticides evaluated in this monograph was determined to be inadequate for nine and there were no epidemiological studies for eight pesticides [12]. Since then, more studies have focused on cancer risk from specific pesticides, although the information is still relatively limited for many cancer-pesticide combinations [8,9].

To help fill the current information gap we evaluated the relationships between the use of specific insecticides, fungicides and fumigants and NHL in the Agricultural Health Study (AHS), a prospective cohort of licensed private (i.e., mostly farmer) and commercial pesticide applicators. Because the etiology of NHL and its B and T cell subtypes may differ by cell type[13], we also evaluated risk by subtype while controlling for potential confounding factors suggested from the literature [13], and the AHS data.

Novelty and Impact

These findings on occupationally exposed pesticide applicators with high quality exposure information are among the first to suggest links between DDT, lindane, permethrin, diazinon and terbufos and specific NHL subtypes in a prospective cohort study.

Materials and Methods

Study Population

The AHS is a prospective cohort study of 52,394 licensed private pesticide applicators (mostly farmers) in Iowa and North Carolina and 4,916 licensed commercial applicators in Iowa (individuals paid to apply pesticides to farms, homes, lawns, etc.), and 32,346 spouses of private applicators. Only applicators are included in this analysis. The cohort has been previously described in detail [14,15] and study questionnaires are available on the AHS website (www.aghealth.nih.gov). Briefly, individuals seeking licenses to apply restricted use pesticides were enrolled in the study from December 1993 through December 1997 (82% of the target population enrolled). At enrollment, subjects did not sign a written informed consent form. However, the cover letter of the questionnaire booklet informed subjects of the voluntary nature of participation, the ability to not answer any question, and it provided an assurance of confidentiality (including a Privacy Act Notification statement). The letter also included a written summary of the purpose of research, time involved, benefits of research, and a contact for questions about the research. The cover letter to the take-home questionnaire included all of the above and also informed the participant that they had the right to withdraw at any time. Finally, subjects were specifically informed that their contact information (including Social Security Number) would be used to search health and vital records in the future. The participants provided consent by completing and returning the questionnaire booklet. These documents and procedures were approved in 1993 by all relevant institutional review boards (i.e., National Cancer Institute Special Studies Institutional Review Board, Westat Institutional Review Board, and the University of Iowa Institutional Review Board-01).

Excluded from this analysis were study participants who had a history of any cancer at the time of enrollment (n = 1094), individuals who sought pesticide registration in Iowa or North Carolina but did not live in these states at the time of registration (n = 341) and were thus outside the catchment area of these cancer registries and individuals that were missing information on potential confounders (i.e., race or total herbicides application days [n = 1,569]). This resulted in an analysis sample of 54,306. We obtained cancer incidence information by regular linkage to the population-based cancer registry files in Iowa and North

Carolina. In addition, we linked cohort members to state mortality registries of Iowa and North Carolina and the nation-wide National Death Index to determine vital status, and to the nation-wide address records of the Internal Revenue Service, state-wide motor vehicle registration files, and pesticide license registries of state agricultural departments to determine residence in Iowa or North Carolina. The current analysis included all incident primary NHL, as well as CLL and MM (which are now classified as NHL) [13] (n = 523) diagnosed from enrollment (1993–1997) through December 31, 2010 in North Carolina and from enrollment (1993–1997) through December 31, 2011 in Iowa, the last date of complete cancer incidence reports in each state. We ended follow-up and person-year accumulation at the date of diagnosis of any cancer, death, movement out of state, or December 31, 2010 in North Carolina and December 31, 2011 in Iowa, whichever was earlier.

Tumor Characteristics

Information on tumor characteristics was obtained from state cancer registries. We followed the definition of NHL and six subtypes of NHL used by the Surveillance Epidemiology and End Results (SEER) coding scheme [16] which was based on the Pathology Working Group of the International Lymphoma Epidemiology Consortium (ICD-O-3 InterLymph modification) classification (Table S1 in File S1, [17], i.e., 1. Small B-cell lymphocytic lymphomas (SLL)/chronic B-cell lymphocytic lymphomas (CLL)/mantle-cell lymphomas (MCL); 2. Diffuse large B-cell lymphomas; 3. Follicular lymphomas; 4. 'Other B-cell lymphomas' consisting of a diverse set of B-cell lymphomas; 5. Multiple myeloma; and 6. T-cell NHL and undefined cell type). There were too few T-cell NHL cases available for analysis [n = 19] so this cell type was not included in the subtype analysis. The ICD-O-3 original definition (used in many earlier studies of pesticides and cancer) of NHL [18] was also evaluated in relation to pesticide exposure to allow a clearer comparison of our results with previous studies.

Exposure Assessment

Initial information on lifetime use of 50 specific pesticides (Table S2 in File S1), including 22 insecticides, 6 fungicides and 4 fumigants was obtained from two self-administered questionnaires [14,15] completed during cohort enrollment (Phase 1). All 57,310 applicators completed the first enrollment questionnaire, which inquired about ever/never use of 50 pesticides, as well as duration (years) and frequency (average days/year) of use for a subset of 22 pesticides including 9 insecticides, 2 fungicides and 1 fumigant. In addition, 25,291 (44%) of the applicators returned the second (take-home) questionnaire, which inquired about duration and frequency of use for the remaining 28 pesticides, including 13 insecticides, 4 fungicides and 3 fumigants.

A follow-up questionnaire, which ascertained pesticide use since enrollment, was administered approximately 5 years after enrollment (1999–2005, Phase 2) and completed by 36,342 (63%) of the original participants. The full text of the questionnaires is available at www.aghealth.nih.gov. For participants who did not complete the Phase 2 questionnaire (20,968 applicators, 37%), a data-driven multiple imputation procedure which used logistic regression and stratified sampling [19] was employed to impute use of specific pesticides in Phase 2. Information on pesticide use from Phase 1, Phase 2 and imputation for Phase 2 was used to construct three cumulative exposure metrics: (i) lifetime days of pesticide use (i.e., the product of years of use of a specific pesticide and the number of days used per year); (ii) intensity-weighted lifetime days of use (i.e., the product of lifetime days of use and a measure of exposure intensity) and (iii) ever/never use data for each pesticide. Intensity was derived from an exposure-algorithm, which was based on exposure measurements from the literature and individual information on pesticide use and practices (e.g., whether or not they mixed pesticides, application method, whether or not they repaired equipment and use of personal protective equipment) obtained from questionnaires completed by study participants [20].

Statistical Analyses

We divided follow-up time into 2-year intervals to accumulate person-time and update time-varying factors, such as attained age and pesticide use. We fit Poisson models to estimate rate ratios (RRs) and 95% confidence intervals (95% CI) to evaluate the effects of pesticide use on rates of overall NHL and the five NHL subtypes.

We evaluated pesticides with 15 or more exposed cases of total NHL, thereby excluding aluminum phosphide, carbon tetrachloride/carbon disulfide, ethylene dibromide, trichlorfon, and ziram leaving 26 insecticides, fungicides and fumigants for analysis (permethrin for animal use and crop use were combined into one category, all insecticides, fungicides and fumigants are listed in Table S2 in File S1). For each pesticide, we evaluated ever vs. never exposure, as well as tertiles of exposure which were created based on the distribution of all NHL exposed cases and compared to those unexposed. In the NHL subtype analysis and in circumstances where multiple pesticides were included in the model we categorized exposure for each pesticide into unexposed (i.e., never users) and two exposed groups (i.e., low and high) separated at the median exposure level. The number of exposed cases included in the ever/never analysis and in the trend analysis can differ because of the lack of information necessary to construct quantitative exposure metrics for some individuals.

Several lifestyle and demographic factors associated with NHL in the AHS cohort or previously suggested as possible confounders in the NHL literature[13] were evaluated as potential confounders in this analysis. These included: age at enrollment, gender, race, state, license type, education, autoimmune diseases, family history of lymphoma in first-degree relatives, body mass index, height, cigarette smoking history, alcohol consumption per week and several occupational exposures[1-13] including number of livestock, cattle, poultry, whether they raised poultry, hogs or sheep, whether they provided veterinary services to their animals, number of acres planted, welding, diesel engine use, number of years lived on the farm, total days of any pesticide use, and total days of herbicide use. However, since most of these variables did not change the risk estimates for specific pesticides, we present results adjusted for age, race, state and total days of herbicide use, which impacted risk estimates by more than 10% for some subtypes. We also performed analyses adjusting for specific insecticides, fungicides and fumigants shown to be associated with NHL or a specific NHL subtype in the current analysis. Tests for trend used the median value of each exposure category. All tests were two-sided and conducted at $\alpha = 0.05$ level. Analysis by NHL subtype was limited to insecticides, fungicides, and fumigants with 6 or more exposed cases.

We also fit polytomous logit models, where the dependent variable was a five-level variable (i.e., five NHL subtypes) and a baseline level (i.e., no NHL) to estimate exposure-response odds ratios (ORs) and 95% confidence intervals (CIs) for each subtypes of NHL. We then used polytomous logit models to estimate exposure-response trend while adjusting for age, state, race and total days of herbicide use, as in the Poisson models, and tested homogeneity among the 5 NHL subtypes.

Poisson models were fit using the GENMOD procedure and polytomous logit models were fit using the LOGISTIC procedure of the SAS 9.2 statistical software package (SAS Institute, Cary, NC). Summary estimates of NHL and NHL subtype risks for both Poisson models and polytomous logit models incorporated imputed data and were calculated along with standard error estimates, confidence intervals, and p-values, using multiple imputation methods implemented in the MIANALYZE procedure of SAS 9.2.

We also evaluated the impact of the additional pesticide exposure information imputed for Phase 2 on risk estimates. We compared risk estimates for those who completed both the phase 1 enrollment and take-home questionnaires and the phase 2 questionnaires (n = 17,545) with risk estimates obtained from the combined completed questionnaire data plus the imputed phase 2 data (n = 54,306). We also explored the effect of lagging exposure data 5 years because recent exposures may not have had time to have an impact on cancer development. For comparison to previous studies, we also assessed the exposure-response association for NHL using the original ICD-O-3 definition of NHL [18] and the new definition [16] in Table S3 in File S1. Unless otherwise specified, reported results show un-lagged exposure information from both Phase 1 and Phase 2 including Phase 2 imputed data for lifetime exposure-days and intensity-weighted lifetime days of use and NHL defined by the InterLymph modification of ICD-O-3 [17]. Data were obtained from AHS data release versions P1REL201005.00 (for Phase 1) and P2REL201007.00 (for Phase 2).

Results

The 54,306 applicators in this analysis contributed 803,140 person-years of follow-up from enrollment through December 31, 2010 in North Carolina and December 31, 2011 in Iowa (Table 1). During this period, there were 523 incident cases of NHL, including 148 SLL/CLL/MCL, 117 diffuse large B-cell lymphomas, 67 follicular lymphomas, 53 'other B-cell lymphomas' (consisting of a diverse set of B-cell lymphomas) and 97 cases of MM. Another 41 cases consisting of T-cell lymphomas (n = 19) and non-Hodgkin lymphoma of unknown lineage (n = 22) were excluded from cell type-specific analyses because of small numbers of cases with identified cell types. Between enrollment and the end of follow-up, 6,195 individuals were diagnosed with an incident cancer other than NHL, 4,619 died without a record of cancer in the registry data, and 1,248 cohort members left the state and could not be followed-up for cancer. Person-years of follow-up accumulated for all of these study participants after enrollment until they were censored for the incident cancer, death or moving out of the state (data not shown). The risk of NHL increased significantly and monotonically with age in the AHS cohort in this analysis (p = 0.001) and age-adjusted risks were significant for state and NHL overall and race for multiple myeloma (data not shown). Total days of herbicide use had a small but significant effect on the risk of some NHL subtypes, but not on NHL overall. No other demographic or occupational factors showed evidence of confounding so they were not included in the final models.

In Table 2 we present ever/never results for 26 insecticides, fungicides and fumigants by total NHL and by NHL subtype adjusted for age, race, state and herbicide use (total life-time days). Terbufos was the only pesticide associated with an increased risk of total NHL in the ever/never use analysis (RR = 1.2 [1.0–1.5]), although the trend for increasing use and risk of total NHL was not significant (p trend = 0.43) (Table 3). In contrast, there were a few chemicals that were not associated with ever/never use, but

Table 1. Baseline characteristics of AHS study participants in the NHL incidence analysis[1,2].

Variables	All NHL cases (%)	Cohort Person-years.
Age at Enrollment		
<45	84 (16.1)	426,288
45–49	51 (9.8)	101,018
50–54	75 (14.3)	84,998
55–59	90 17.2)	74,440
60–64	78 (14.9)	56,978
65–69	79 (15.1)	35,071
≥70	66 (12.6)	24,347
Race		
White	509 (97.3)	787,799
Black	14 (2.7)	15,341
State		
IA	332 (63.5)	537,252
NC	191 (36.5)	265,888
Lifetime Total Herbicide Exposure Days		
0–146 days	170 (32.5)	251,401
147–543 days	169 (32.3)	273,107
544–2453 days	184 (35.2)	278,632

[1]During the period from enrollment (1993–1997) to December 31, 2010 in NC and December 31, 2011 in Iowa.
[2]Individuals with missing ever/never exposure information or missing confounding variable information were not included in the table.

did show evidence of an exposure-response association. Lindane was the only pesticide that showed a statistically significant increasing trend in risk for NHL with both exposure metrics, for lifetime-days of lindane use the RR were = 1.0 (ref), 1.2 (0.7–1.9), 1.0 (0.6–1.7), 2.5 (1.4–4.4); p trend = 0.004 and intensity-weighted lifetime-days of use the: RR were: = 1.0 (ref), 1.3 (0.8–2.2), 1.1 (0.7–1.8), 1.8 (1.0–3.2); p trend = 0.04. DDT showed a significant trend for NHL risk with life-time days of use RR = 1.0 (ref), 1.3 (0.9–1.8), 1.1 (0.7–1.7), 1.7 (1.1–2.6); p trend = 0.02, while the intensity weighted lifetime days of use of DDT was of borderline significance: RR = 1.0 (ref), 1.2 (0.8–1.8), 1.1 (0.8–1.7), 1.6 (1.0–2.3); p trend = 0.06. The number of lifetime days of use of lindane and DDT was weakly correlated (coefficient of determination = 0.04), and the pattern of NHL risk showed little change when both were included in the model. The results for lindane adjusted for DDT were, RR = 1.0 (ref), 1.2 (0.7–2.0), 1.0 (0.5–1.8), 1.6 (0.9–3.3); p trend = 0.07 and the results for DDT adjusted for lindane were, RR = 1.0 (ref), 1.3 (0.9–2.0), 0.9 (0.6–1.6), 1.6 (0.9–2.6); p trend = 0.08).

We also evaluated pesticides by NHL sub-type. In the ever/never analyses (Table 2), permethrin was significantly associated with multiple myeloma, RR = 2.2 (1.4–3.5) and also demonstrated an exposure-response trend (RR = 1.0 (ref), 1.4 (0.8–2.7), 3.1 (1.5–6.2); p trend = 0.002) (Table 4). Similarly, there was an elevated risk of SLL/CLL/MCL with terbufos in ever/never analyses RR = 1.4 (0.97–2.0) and an exposure response trend (RR = 1.0 (ref), 1.3 (0.8–2.0), 1.6 (1.0–2.5); p trend = 0.05). For follicular lymphoma, lindane showed an elevated but non-significant association for ever use, RR = 1.7 (0.96–3.2) and a significant exposure-response association (RR = 1.0 (ref), 4.9 (1.9–12.6), 3.6 (1.4–9.5); p trend = 0.04). There were also two chemicals with evidence of exposure-response that were not associated with specific subtypes in the ever/never analyses: DDT (Dichlorodiphenyltrichloroethane) with SLL/CLL/MCL (RR = 1.0 (ref), 1.0

(0.5–1.8), 2.6 (1.3–4.8; p trend = 0.04); and diazinon with follicular lymphoma (RR = 1.0 (ref), 2.2 (0.9–5.4), 3.8 (1.2–11.4); p trend = 0.02) (Table 4).

The pattern of increased CLL/SLL/MCL risk with increased use of DDT and terbufos remained after both insecticides were placed in our model concurrently. CLL/SLL/MCL risk increased with DDT use (RR = 1.0 (ref), 0.9 (0.5–4.7); 2.4 (1.1–4.7); p trend = 0.04), and a pattern of increased CLL/SLL/MCL risk was also observed with terbufos use (RR = 1.0 (ref), 1.1 (0.6–2.1), 1.7 (0.9–3.3) p trend = 0.07), although the trend was not significant for terbufos. Similarly, the pattern of increased follicular lymphoma risk with lindane use and diazinon use remained after both insecticides were placed in our model concurrently. Follicular lymphoma risk increased with diazinon use (RR = 1.0 (ref), 4.1 (1.5–11.1); 2.5 (0.9–7.2); p trend = 0.09), and a similarly, pattern of increased follicular lymphoma risk was observed with lindane use (RR = 1.0 (ref), 1.6 (0.6–4.1), 2.6 (0.8–8.3) p trend = 0.09), although neither remained statistically significant (Table 4).

Three chemicals showed elevated risks in ever/never analyses for certain subtypes, with no apparent pattern in exposure-response analyses: metalaxyl and chlordane with SLL/CLL/MCL, RR = 1.6 (1.0–2.5) and RR = 1.4 (0.97–2.0) respectively, and methyl bromide with diffuse large B-cell lymphoma RR = 1.9 (1.1–3.3). Although there was evidence of association by subtype, and polytomous logit models indicated homogeneity across subtypes for lindane (p = 0.54), DDT (p = 0.44) and any other pesticide evaluated in this study (e.g., permethrin (p = 0.10), diazinon (p = 0.09), terbufos (p = 0.63), (last column in Table 4).

There was no evidence of confounding of the total NHL associations with either lindane or DDT. We also calculated RR for those who completed both the phase 1 enrollment and take-home questionnaires and the phase 2 questionnaire (n = 17,545) and found no meaningful difference in the RR that also included imputed exposures, although there was an increase in precision of

Table 2. Pesticides exposure (ever/never) and adjusted Relative Risk of total NHL and NHL Subtype[1].

Insecticide

Pesticide (chemical-functional class)	Total NHL Cases[2]		SLL/CLL/MCL Cases[2]		Diffuse Large B-Cell Cases[2]		Follicular B-Cell Cases[2]		Other B-cell Cases[2]		Multiple Myeloma Cases[2]	
	Ever/Never Exposed	RR[3,4] (95% CI)	Ever/Never Exposed	RR[3,4] (95% CI)	Ever/Never Exposed	RR[3,4] (95% CI)	Ever/Never Exposed	RR[3,4] (95% CI)	Ever/Never Exposed	RR[3,4] (95% CI)	Ever/Never Exposed	RR[3,4] (95% CI)
Aldicarb (carbamate-insecticide)	47/435	1 (0.7-1.4)	14/124	1.1 (0.6-1.8)	8/98	0.7 (0.4-1.5)	6/54	0.9 (0.3-2.2)	7/41	1.6 (0.7-3.5)	10/82	1.2 (0.6-2.2)
Carbofuran (carbamate-insecticide)	147/317	1.1 (0.9-1.3)	48/86	1.2 (0.8-1.8)	26/78	0.8 (0.5-1.3)	18/39	1 (0.5-1.7)	13/31	0.8 (0.4-1.6)	31/56	1.3 (0.8-2.1)
Carbaryl (carbamate-insecticide)	272/225	1 (0.8-1.2)	75/66	1 (0.7-1.5)	58/53	0.8 (0.5-1.3)	37/24	0.8 (0.5-1.3)	24/28	0.9 (0.5-1.6)	58/34	0.9 (0.6-1.4)
Chlorpyrifos (organophosphate-insecticide)	210/300	1 (0.8-1.2)	62/84	1 (0.7-1.4)	44/70	0.9 (0.6-1.4)	32/33	1.3 (0.8-2.2)	21/31	0.8 (0.5-1.5)	36/58	1 (0.6-1.5)
Coumaphos (organophos-phate-insecticide)	46/411	1.1 (0.8-1.5)	15/120	1.2 (0.7-2.1)	10/93	1 (0.6-1.4)	8/48	1.6 (0.8-3.5)	5/40	xxx	7/78	1 (0.1-2.1)
DDVP (dimethyl phosphate-insecticide)	55/407	1.1 (0.8-1.5)	13/124	0.8 (0.5-1.5)	10/93	1 (0.5-2.1)	8/48	1.3 (0.6-2.7)	6/39	1	12/73	1.7 (0.9-3.2)
Diazinon (organophosphorous-insecticide)	144/342	1 (0.8-1.3)	46/93	1.3 (0.9-1.9)	30/78	0.9 (0.6-1.4)	22/38	1.3 (0.7-2.3)	12/37	0.8 (0.4-1.6)	27/64	1
Fonofos (organophosphorous-insecticide)	115/349	1.1 (0.9-1.4)	35/100	1.1 (0.7-1.6)	25/81	1.2 (0.7-1.9)	13/45	0.9 (0.5-1.7)	15/30	1.3 (0.7-2.5)	19/66	1.3 (0.8-2.3)
Malathion (organophosphorous-insecticide)	332/163	0.9 (0.8-1.1)	99/43	1 (0.7-1.4)	72/37	0.9 (0.6-1.4)	46/14	1.3 (0.7-2.4)	30/21	0.6 (0.3-1.0)	61/32	0.9 (0.6-1.5)
Parathion (ethyl or methyl) (organophosphorous insecticide)	69/411	1.1 (0.8-1.4)	20/117	1 (0.7-1.4)	14/91	1	10/48	1.1 (0.8-1.5)	7/44	1.1 (0.7-1.5)	14/77	1
Permethrin (animal and crop applications) (pyrethroid insecticide)	112/363	1.1 (0.8-1.3)	32/106	1 (0.6-1.5)	18/81	0.7 (0.4-1.2)	18/81	1.1 (0.6-2.0)	9/14	0.8 (0.4-1.6)	20/72	**2.2 (1.4-3.5)**
Phorate (organophosphorous-insecticide)	160/325	1 (0.8-1.2)	53/87	1.1 (0.8-1.6)	31/76	0.9 (0.5-1.3)	20/40	0.9 (0.5-1.6)	19/31	0.9 (0.5-1.6)	26/64	1
Terbufos (organophosphorous-insecticide)	201/267	**1.2 (1.0-1.5)**	64/72	1.4 (0.99-2.1)	42/63	1.1 (0.7-1.7)	31/26	1.2 (0.7-2.1)	26/19	1.8 (0.94-3.2)	32/59	1.2 (0.7-1.9)
Chlorinated Insecticides												
Aldrin (chlorinated insecticide)	116/364	0.9 (0.7-1.1)	53/99	0.9 (0.6-1.4)	15/91	0.8 (0.4-1.6)	13/45	0.8 (0.4-1.6)	12/37	0.6 (0.3-1.3)	29/62	1.5 (0.9-2.5)
Chlordane (chlorinated insecticide)	136/344	1 (0.8-1.3)	49/90	1.4 (0.99-2.1)	20/86	0.6 (0.4-1.0)	18/41	1.2 (0.7-2.1)	13/36	1	31/60	1.2 (0.8-1.9)
DDT	182/300	1	59/79	1.2	34/73	0.8	18/41	0.9	20/31	1.1	40/50	1.1

Table 2. Cont.

Insecticide

Pesticide (chemical-functional class)	Total NHL Cases[2] Ever/Never Exposed	RR[3,4] (95% CI)	SLL/CLL/MCL Cases[2] Ever/Never Exposed	RR[3,4] (95% CI)	Diffuse Large B-Cell Cases[2] Ever/Never Exposed	RR[3,4] (95% CI)	Follicular B-Cell Cases[2] Ever/Never Exposed	RR[3,4] (95% CI)	Other B-cell Cases[2] Ever/Never Exposed	RR[3,4] (95% CI)	Multiple Myeloma Cases[2] Ever/Never Exposed	RR[3,4] (95% CI)
(chlorinated insecticide)		(0.8–1.3)		(0.8–1.8)		(0.5–1.3)		(0.5–1.6)		(0.6–2.1)		(0.7–1.8)
Dieldrin (chlorinated insecticide)	35/442	0.9 (0.6–1.2)	5/130	xxx	4/101	xxx	4/54	xxx	7/42	1	10/81	0.9 (0.5–1.4)
Heptachlor (chlorinated insecticide)	90/384	1 (0.7–1.2)	33/104	1.1 (0.7–3.0)	10/95	1.1	9/48	1.1	13/36	0.9	17/72	1.1 (0.6–2.0)
Lindane (chlorinated insecticide)	85/396	1 (0.8–1.2)	27/113	1.2 (0.6–1.5)	12/95	0.6 (0.3–1.1)	16/41	1.7 (0.96–3.2)	9/40	0.7 (0.4–1.2)	13/73	1.1 (0.5–2.0)
Toxaphene (chlorinated insecticide)	79/397	1 (0.7–1.2)	21/116	0.9 (0.5–1.5)	14/90	0.8 (0.4–1.4)	9/47	1	10/40	1.1 (0.6–2.0)	19/73	1.1 (0.6–1.9)
Fungicides												
Benomyl (carbamate fungicide)	54/428	1.1 (0.8–1.5)	18/123	1.2 (0.7–2.0)	12/95	1.1 (0.6–1.9)	4/51	xxx	4/51	xxx	11/80	1.1 (0.6–2.0)
Captan (phthalimide fungicide)	60/406	1.1 (0.8–1.4)	18/118	1.1 (0.6–1.8)	12/91	0.9 (0.5–1.8)	5/51	xxx	6/39	1.1 (0.5–2.7)	12/76	1.2 (0.6–2.2)
Chloro-thalonil (poly-chlorinated aromatic thalonitrile fungicide)	35/474	0.8 (0.5–1.2)	9/135	0.9 (0.4–1.9)	6/107	0.5 (0.2–1.3)	5/60	xxx	2/50	xxx	11/84	1.2 (0.6–2.3)
Maneb/Mancozeb (dithiocarbamate fungicide)	44/437	0.9 (0.7–1.3)	13/127	1.1 (0.6–2.1)	12/95	1.1 (0.6–2.1)	4/60	xxx	5/49	xxx	10/79	0.8 (0.4–1.7)
Metalaxyl (acylalanine fungicide)	108/381	1 (0.8–1.3)	34/106	**1.6 (1.0–2.5)**	27/82	1.1 (0.6–1.8)	10/48	0.7 (0.4–1.4)	10/40	0.9 (0.4–1.7)	21/71	0.8 (0.4–1.3)
Fumigant												
Methyl bromide (methyl halide fumigant)	85/425	1.1 (0.9–1.5)	18/126	0.9 (0.5–1.7)	28/86	**1.9 (1.1–3.3)**	7/58	0.6 (0.2–1.4)	8/44	2.2 (0.9–5.7)	19/76	1 (0.6–1.8)

[1] During the period from enrollment (1993–1997) to December 31, 2010 in NC and December 31, 2011 in Iowa.
[2] Numbers of cases by NHL subtype do not sum to total number of NHL cases (n = 523) due to missing data.
[3] Adjusted RR: age (<45, 45–49, 50–54, 55–59, 60–64, 65–69, ≥70), State (NC vs. IA), Race (White vs. Black), AHS herbicides (tertiles of total herbicide use-days). Statistically significant RR and 95% confidence limits are bolded.
[4] RR was not calculated if the number of exposed cases in a pesticide-NHL subtype cell was <6 and the missing RR was marked with an XXX. Statistically significant RRs and 95% confidence limits are bolded.

Table 3. Pesticide exposure (lifetime-days & intensity weighted life-time days) and adjusted risks of total NHL incidence[1].

Insecticides

Pesticide (chemical-functional class) [days of lifetime exposure for each category]	NHL Cases[2]	Non-Cases[2]	RR[3,4] (95% CI) by Total Days of Exposure	NHL Cases[2,]	Non-Cases	RR[3,4] (95% CI) Intensity-weighted days of exposure
Aldicarb (carbamate-insecticide)						
None	238	21557	1.0 (ref)	238	21557	1.0 (ref)
Low [≤8.75]	7	633	1.1 (0.5–2.3)	6	383	1.3 (0.6–3.3))
Medium [>8.75–25.5]	5	522	0.9 (0.3–2.5)	6	853	0.9 (0.4–1.9)
High [>25.5–224.75]	5	1266	0.5 (0.2–1.3)	5	1183	0.5 (0.2–1.3)
			P trend = 0.23			P trend = 0.22
Carbofuran (carbamate-insecticide)						
None	317	36296	1.0 (ref)	317	36296	1.0 (ref)
Low [≤8.75]	63	4775	1.2 (0.9–1.6)	46	3695	1.2 (0.9–1.6)
Medium [>8.75–38.75]	32	3648	0.8 (0.6–1.2)	46	4590	1.0 (0.7–1.3)
High [>38.75–767.25]	44	4370	0.97 (0.7–1.4)	45	4477	1.0 (0.7–1.4)
			P trend = 0.69			P trend = 0.74
Carbaryl (carbamate-insecticide)						
None	128	12864	1.0 (ref)	128	12864	1.0 (ref)
Low [≤8.75]	54	4128	1.1 (0.7–1.6)	46	3962	1.0 (0.7–1.5)
Medium [8.75–56]	43	5096	0.9 (0.6–1.2)	45	4433	0.9 (0.7–1.5)
High [>56–737.5]	39	3281	1.0 (0.7–1.6)	44	4029	1.0 (0.6–1.5)
			P trend = 0.87			P trend = 0.94
Chlorpyrifos (organophosphate-insecticide)						
None	300	30393	1.0 (ref)	300	30393	1.0 (ref)
Low [≤8.75]	71	6493	1.1 (0.9–1.5)	61	6383	1.1 (0.8–1.4)
Medium [>8.75–44]	65	6892	1.1 (0.8–1.4)	60	7549	0.9 (0.7–1.2)
High [>44–767.25]	67	9380	0.8 (0.6–1.1)	60	7044	1.0 (0.7–1.3)
			P trend = 0.11			P trend = 0.85
Coumaphos (organophosphate-insecticide)						
None	411	44846	1.0 (ref)	411	44846	1.0 (ref)
Low [<8.75]	16	1510	1.0 (0.6–1.7)	15	1132	1.3 (0.8–2.1)
Medium [>8.75–38.75]	14	1076	1.2 (0.7–2.1)	14	1452	1.0 (0.6–1.6)
High [>38.75–1627.5]	13	1175	1.2 (0.7–2.0)	14	1170	1.2 (0.7–2.1)
			P for trend = 0.50			P trend = 0.48
DDVP (dimethyl phosphate-insecticide)						
None	407	44551	1.0 (ref)	407	44551	1.0 (ref)
Low [≤8.75]	19	1342	1.4 (0.9–2.1)	18	1281	1.4 (0.9–2.3)
Medium [>8.75–87.5]	17	1519	1.2 (0.7–1.9)	18	1633	1.1 (0.7–1.8)
High [>87.5–2677.5]	17	1893	0.9 (0.6–1.5)	17	1824	1.0 (0.6–1.6)
			P trend = 0.78			P trend = 0.83
Diazinon (organophosphorous-insecticide)						
None	187	17943	1.0 (ref)	187	17943	1.0 (ref)
Low [≤8.75]	28	2506	1.1 (0.7–1.6)	23	2047	1.1 (0.7–1.8)
Medium [>8.75–25]	19	1515	1.0 (0.6–1.8)	24	2246	0.9 (0.5–1.5)
High [>25–457.25]	23	1990	1.2 (0.7–1.9)	22	1708	1.3 (0.8–2.1)
			P trend = 0.52			P trend = 0.33

Table 3. Cont.

Insecticides

Pesticide (chemical-functional class) [days of lifetime exposure for each category]	NHL Cases[2]	Non-Cases[2]	RR[3,4] (95% CI) by Total Days of Exposure	NHL Cases[2,]	Non-Cases	RR[3,4] (95% CI) Intensity-weighted days of exposure
Fonofos (organophosphorous-insecticide)						
None	349	39570	1.0 (ref)	349	39570	1.0 (ref)
Low [≤20]	47	3812	1.3 (0.96–1.8)	37	2906	1.4 (0.97–1.9)
Medium [>20–50.75]	28	2819	1.1 (0.7–1.6)	38	3487	1.1 (0.8–1.6)
High [>50.75–369.75]	37	3385	1.1 (0.7–1.5)	36	3606	1.0 (0.7–1.4)
			P trend = 0.83			P trend = 0.87
Malathion (organophosphorous-insecticide)						
None	90	8368	1.0 (ref)	90	8368	1.0 (ref)
Low [≤8.75]	75	7284	0.97 (0.7–1.3)	60	5535	1.0 (0.7–1.4)
Medium [>8.75–38.75]	47	5779	0.7 (0.5–1.1)	59	6899	0.8 (0.6–1.1)
High [>38.75–737.5]	57	5037	0.9 (0.6–1.3)	59	5588	0.9 (0.6–1.2)
			P trend = 0.63			P trend = 0.46
Parathion (ethyl or methyl) (organophosphorous insecticide)						
None	228	21457	1.0 (ref)	228	21457	1.0 (ref)
Low [≤8.75]	9	693	1.0 (0.5–2.0)	7	612	0.9 (0.4–2.0)
Medium [>8.75–24.5]	6	351	1.4 (0.6–3.2)	8	462	1.4 (0.7–2.9)
High [>.24.5–1237.5]	6	652	0.8 (0.3–1.8)	6	621	0.8 (0.4–1.9)
			P trend = 0.64			P trend = 0.74
Permethrin (animal and crop applications) (pyrethroid insecticide)						
None	371	37496	1.0 (ref)	371	37496	1.0 (ref)
Low [≤8.75]	38	4315	1.1 (0.8–1.5)	33	4263	0.9 (0.6–1.3)
Medium [>8.75–50.75]	31	4611	0.8 (0.5–1.2)	33	4200	1.0 (0.7–1.4)
High [>50.75–1262.25]	33	4121	1.2 (0.8–1.7)	32	4553	1.0 (0.7–1.5)
			P trend = 0.54			P trend = 0.99
Phorate (organophosphorous-insecticide)						
None	171	16834	1.0 (ref)	171	16834	1.0 (ref)
Low [≤8.75]	27	2621	0.8 (0.5–1.2)	26	2320	0.9 (0.6–1.4)
Medium [8.75–24.5]	33	1819	1.4 (0.96–2.1)	27	1951	1.1 (0.7–1.7)
High [>24.5–224.75]	18	2246	0.6 (0.4–1.1)	25	2409	0.8 (0.5–1.3)
			P trend = 0.25			P trend = 0.44
Terbufos (organophosphorous-insecticide)						
None	267	31076	1.0 (ref)	267	31076	1.0 (ref)
Low [≤24.5]	82	8410	1.2 (0.9–1.5)	64	6895	1.1 (0.9–1.5)
Medium [>24.5–56]	54	3925	1.6 (1.2–2.1)	64	4642	1.6 (1.2–2.2)
High [>56–1627.5]	57	6080	1.1 (0.8–1.5)	63	6842	1.1 (0.8–1.5)
			P trend = 0.43			P trend = 0.44
Chlorinated Insecticides						
Aldrin (chlorinated insecticide)						
None	193	19743	1.0 (ref)	193	19743	1.0 (ref)
Low [≤8.75]	27	1613	0.9 (0.6–1.4)	20	1212	0.9 (0.6–1.4)
Medium [>8.75–24.5]	16	1002	0.8 (0.5–1.3)	20	1279	0.8 (0.5–1.3)

Table 3. Cont.

Insecticides						
Pesticide (chemical-functional class) [days of lifetime exposure for each category]	NHL Cases[2]	Non-Cases[2]	RR[3,4] (95% CI) by Total Days of Exposure	NHL Cases[2,]	Non-Cases	RR[3,4] (95% CI) Intensity-weighted days of exposure
High [>24.5–457.25]	17	903	0.9 (0.5–1.5)	19	1026	0.9 (0.6–1.5)
			P trend = 0.58			P trend = 0.74
Chlordane (chlorinated insecticide)						
None	179	19115	1.0 (ref)	179	19115	1.0 (ref)
Low [≤8.75]	47	2687	1.3 (0.97–1.9)	23	1303	1.4 (0.9–2.2)
Medium[5]	0	0	xxx	24	1747	1.0 (0.6–1.5)
High [>8.75–1600]	23	1450	1.1 (0.7–1.7)	22	1085	1.4 (0.9–2.2)
			P trend = 0.43			P trend = 0.16
DDT (chlorinated insecticide)						
None	152	18543	1.0 (ref)	152	18543	1.0 (ref)
Low [≤8.75]	43	2121	1.3 (0.9–1.8)	33	1601	1.2 (0.8–1.8)
Medium [>8.75–56]	28	1598	1.1 (0.7–1.7)	32	1760	1.1 (0.8–1.7)
High [>56–1627.5]	27	953	1.7 (1.1–2.6)	32	1305	1.6 (1.0–2.3)
			P trend = 0.02			P trend = 0.06
Dieldrin (chlorinated insecticide)						
None	235	22510	1.0 (ref)	235	22510	1.0 (ref)
Low [≤8.75]	7	472	0.7 (0.3–1.5)	6	363	0.8 (0.4–1.8)
Medium [>8.75–24.5]	8	154	2.3 (1.1–4.7)	5	106	2.2 (0.9–5.3)
High [>24.5–224.75]	2	140	0.7 (0.2–2.9)	5	298	0.8 (0.3–2.0)
			P trend = 0.47			P trend = 0.84
Heptachlor (chlorinated insecticide)						
None	205	20844	1.0 (ref)	205	20844	1.0 (ref)
Low [≤8.75]	21	1261	1.0 (0.6–1.6)	15	1110	0.8 (0.5–1.4)
Medium [>8.75–24.5]	18	679	1.5 (0.9–2.4)	16	425	2.0 (1.2–3.4)
High [>24.5–457.25]	7	600	0.7 (0.3–1.4)	14	1001	0.8 (0.5–1.4)
			P trend = 0.82			P trend = 0.88
Lindane (chlorinated insecticide)						
None	205	20375	1.0 (ref)	205	20375	1.0 (ref)
Low [≤8.75]	18	1285	1.2 (0.7–1.9)	15	976	1.3 (0.8–2.2)
Medium [>8.75–56]	13	1103	1.0 (0.6–1.7)	16	1205	1.1 (0.7–1.8)
High [>56–457.25]	14	467	2.5 (1.4–4.4)	14	673	1.8 (1.0–3.2)
			P trend = 0.004			**P trend = 0.04**
Toxaphene (chlorinated insecticide)						
None	214	20911	1.0 (ref)	214	20911	1.0 (ref)
Low [≤8.75]	14	1198	0.8 (0.5–1.4)	11	630	1.3 (0.7–2.3)
Medium [>8.75–24.5]	13	564	1.5 (0.9–2.7)	12	931	0.9 (0.5–1.6)
High [>24.5–457.25]	6	686	0.6 (0.3–1.4)	10	886	0.8 (0.4–1.5)
			P trend = 0.50			P trend = 0.38
Fungicides						
Benomyl (carbamate fungicide)						
None	219	21425	1.0 (ref)	219	21425	1.0 (ref)
Low [≤12.25]	14	896	1.7 (0.9–2.9)	9	432	2.2 (1.1–4.3)
Medium [>12.25–24.5]	4	214	2.4 (0.9–6.6)	10	732	1.7 (0.9–3.2)

Table 3. Cont.

Insecticides						
Pesticide (chemical-functional class) [days of lifetime exposure for each category]	NHL Cases[2]	Non-Cases[2]	RR[3,4] (95% CI) by Total Days of Exposure	NHL Cases[2,]	Non-Cases	RR[3,4] (95% CI) Intensity-weighted days of exposure
High [>24.5–457.25]	8	834	1.0 (0.5–2.1)	7	779	0.9 (0.4–2.0)
			P trend = 0.93			P trend = 0.75
Captan (phthalimide fungicide)						
None	407	43433	1.0 (ref)	407	43433	1.0 (ref)
Low [≤0.25]	15	2334	0.8 (0.5–1.4)	15	2108	0.9 (0.6–1.5)
Medium [>0.25–12.25]	16	1004	1.5 (0.8–2.6)	15	1171	1.2 (0.7–2.2)
High [>12.25–875]	14	1823	0.8 (0.5–1.5)	14	1805	0.8 (0.5–1.5)
			P trend = 0.69			P trend = 0.52
Chlorothalonil (polychlorinated aromatic thalonitrile fungicide)						
None	474	48442	1.0 (ref)	474	48442	1.0 (ref)
Low [≤12.25]	13	1509	0.9 (0.5–1.6)	10	1800	0.6 (0.3–1.2)
Medium [>12.25–64]	9	1492	0.8 (0.4–1.6)	11	1501	0.9 (0.5–1.7)
High [>64–395.25]	9	1678	0.6 (0.3–1.3)	9	1362	0.8 (0.4–1.6)
			P trend = 0.16			PP trend = 0.52
Maneb/Mancozeb (dithiocarbamate fungicide)						
None	228	21512	1.0 (ref)	228	21512	1.0 (ref)
Low [≤7]	8	400	1.9 (0.9–3.9)	8	486	1.6 (0.8–3.3)
Medium [>7–103.25]	9	990	0.9 (0.4–1.7)	9	680	1.3 (0.6–2.6)
High [>103.25–737.5]	7	454	1.4 (0.6–2.9)	7	677	0.9 (0.4–1.9)
			P trend = 0.49			P trend = 0.78
Metalaxyl (acylalanine fungicide)						
None	209	18833	1.0 (ref)	209	18833	1.0 (ref)
Low [≤6]	16	1439	1.0 (0.6–1.8)	15	1079	1.3 (0.8–2.2)
Medium [>6–28]	15	2182	0.7 (0.4–1.3)	15	2203	0.8 (0.4–1.3)
High [>28–224.75]	13	1566	1.1 (0.6–2.1)	14	1893	0.9 (0.5–1.6)
			P trend = 0.76			P trend = 0.63
Fumigant						
Methyl bromide (methyl halide fumigant)						
None	425	45265	1.0 (ref)	425	45265	1.0 (ref)
Low [≤8]	37	2060	2.0 (1.4–2.9)	26	1680	1.8 (1.2–2.7)
Medium [>8–28]	24	3011	0.9 (0.6–1.4)	25	2501	1.1 (0.7–1.8)
High [>28–387.5]	17	2768	0.6 (0.4–1.0)	25	3571	0.8 (0.5–1.2)
			P trend = 0.04			P trend = 0.10

[1]During the period from enrollment (1993–1997) to December 31, 2010 in NC and December 31, 2011 in Iowa.
[2]Numbers of cases in columns do not sum to total number of NHL cases (n = 523) due to missing data. In the enrollment questionnaire, lifetime-days & intensity weighted life-time days of pesticide use was obtained for the insecticides: carbofuran, chlorpyrifos, coumaphos, DDVP, fonofos, permethrin and terbufos; the fungicides: captan, chlothalonil and the fumigant: methyl bromide. In the take home questionnaire lifetime-days & intensity weighted life-time days of pesticide use were obtained for the insecticides: aldicarb, carbaryl, diazinon, malathion, parathion, and phorate, the chlorinated insecticides: aldrin, chlordane, DDT, dieldrin, heptachlor, lindane, and toxaphene, the fungicides: benomyl, maneb/mancozeb and metalaxyl, therefore, numbers of NHL cases can vary among pesticides listed in the table.
[3]Adjusted RR: age (<45, 45–49, 50–54, 55–59, 60–64, 65–69, ≥70), State (NC vs. IA), Race (White vs. Black), AHS herbicides (tertiles of total herbicide use-days). Statistically significant P trends are bolded.
[4]Permethrin for animal use and crop use were combined into one category.
[5]The distribution of life-time days of chlordane exposure was clumped into two exposed groups those who with, ≤8.75 life-time days of exposure and those with >8.75 life-time days of exposure.

Table 4. Pesticide exposure (Lifetime-Days of Exposure) and adjusted risks for NHL Subtypes.

Insecticides

	SLL, CLL, MCL		Diffuse Large B-cell		Follicular B-cell		Other B-cell types		Multiple Myeloma		
	RR[3,4] (95% CI)	N[2]	RR[3,4] (95% CI)	N[2]	RR[3,4] (95% CI)	N[2]	RR[3,4] (95% CI)	N[2]	RR[3,4] (95% CI)	N[2]	NHL subtype Homo-geneity Test (p-value)
Carbaryl											
None	1.0 (ref)	42	1.0 (ref)	29	1.0 (ref)	11	1.0 (ref)	14	1.0 (ref)	22	
Low	1.1 (0.6–2.2)	19	0.8 (0.4–1.6)	17	1.6 (0.6–3.9)	10	1.8 (0.7–4.3)	10	0.7 (0.3–1.4)	14	
High	0.6 (0.3–1.3)	15	1.3 (0.6–2.8)	15	2.8 (1.0–7.4)	10	0.4 (0.1–1.5)	3	1.1 (0.7–1.8)	13	
	p trend = 0.16		p trend = 0.33		p trend = 0.06		p trend = 0.63		p trend = 0.98		0.19
Carbofuran											
None	1.0 (ref)	87	1.0 (ref)	78	1.0 (ref)	39	1.0 (ref)	33	1.0 (ref)	56	
Low	1.1 (0.7–1.8)	28	0.9 (0.5–1.7)	13	1.3 (0.7–2.4)	15	0.8 (0.4–1.8)	8	1.9 ((1.1–3.3)	16	
High	1.5 (0.9–2.5)	19	0.8 (0.5–1.3)	13	0.4 (0.1–1.4)	3	0.7 (0.2–2.0)	4	0.9 (0.4–1.6)	12	
	p trend = 0.16		p trend = 0.37		p trend = 0.31		p trend = 0.46		p trend = 0.57		0.52
Chlorpyrifos											
None	1.0 (ref)	84	1.0 (ref)	70	1.0 (ref)	33	1.0 (ref)	31	1 (ref)	58	
Low	1.2 (0.8–1.8)	31	0.9 (0.6–1.5)	22	1.6 (0.9–2.9)	20	1.2 (0.6–2.2)	14	1.0 (0.6–1.8)	17	
High	0.9 (0.6–1.3)	30	1.1 (0.6–1.7)	22	1.0 (0.5–2.1)	11	0.5 (0.2–1.3)	7	0.7 (0.4–1.3)	14	
	p trend = 0.45		p trend = 0.80		p trend = 0.94		p trend = 0.13		p trend = 0.27		0.90
Coumaphos											
None	1.0 (ref)	120	1.0 (ref)	92	1.0 (ref)	48	1.0 (ref)	40	1.0 (ref)	78	
Low	1.1 (0.5–2.2)	8	0.7 (0.3–1.9)	4	2.1 (0.7–5.8)	4	xxx-	4	0.7 (0.2–2.2)	3	
High	1.5 (0.6–3.4)	6	1.6 (0.6–4.5)	4	1.4 (0.5–4.0)	4	xxx-	1	1.2 (0.4–4.0)	3	
	p trend = 0.35		p trend = 0.42		p trend = 0.47		p trend = xxx		p trend = 0.84		0.63
Diazinon											
None	1.0 (ref)	53	1.0 (ref)	40	1.0 (ref)	15	1.0 (ref)	20	1.0 (ref)	41	
Low	1.4 (0.7–2.7)	14	1.5 (0.7–3.2)	9	2.2 (0.9–5.4)	8	xxx	3	0.4 (0.1–1.2)	4	
High	1.9 (0.98–3.6)	12	1.1 (0.5–2.4)	8	3.8 (1.2–11.4)	7	xxx	2	0.5 (0.2–1.7)	3	
	p trend = 0.06		p trend = 0.72		**p trend = 0.02**		p trend = xxx		p trend = 0.35		0.09
DDVP											
None	1.0 (ref)	124	1.0 (ref)	93	1.0 (ref)	48	1.0 (ref)	39	1.0 (ref)	73	
Low	0.8 (0.4–1.9)	6	1.1 (0.4–2.7)	5	1.5 (0.6–3.9)	5	1.1 (0.4–3.7)	3	2.7 (1.2–5.8)	7	
High	0.7 (0.3–1.7)	6	0.9 (0.4–2.3)	5	1.0 (0.3–3.4)	3	0.9 (0.3–3.1)	3	1.0 (0.3–2.7)	4	
	p trend = 0.49		p trend = 0.87		p trend = 0.90		p trend = 0.91		p trend = 0.81		0.96
Fonofos											
None	1.0 (ref)	100	1.0 (ref)	81	1.0 (ref)	45	1.0 (ref)	30	1.0 (ref)	66	
Low	1.2 (0.7–2.0)	20	1.2 (0.7–2.2)	13	1.5 (0.8–3.0)	11	1.4 (0.6–3.1)	8	1.2 (0.6–2.5)	9	
High	1.0 (0.6–1.8)	15	1.2 (0.6–2.3)	11	0.3 (0.1–1.2)	2	1.1 (0.4–2.7)	6	1.4 (0.7–3.0)	9	
	p trend = 0.96		p trend = 0.65		p trend = 0.19		p trend = 0.84		p trend = 0.33		0.35
Malathion											
None	1.0 (ref)	27	1.0 (ref)	20	1.0 (ref)	6	1.0 (ref)	11	1.0 (ref)	17	
Low	0.7 (0.4–1.3)	29	0.96 (0.5–1.8)	23	1.0 (0.4–2.9)	12	1.0 (0.5–2.4)	11	1.0 (0.5–2.1)	18	
High	1.0 (0.6–1.8)	22	1.0 (0.5–2.0)	20	1.6 (0.6–4.4)	11	0.3 (0.1–0.8)	6	1.0 (0.5–2.0)	17	
Ever/Never	1.0 (0.7–1.4)		0.9 (0.6–1.4)		1.3 (0.7–2.4)		0.6 (0.3–1.0)		0.9 (0.6–1.5)		
	p trend = 0.65		p trend = 0.88		p trend = 0.25		p trend = 0.17		p trend = 0.86		0.33
Permethrin											

Table 4. Cont.

Insecticides	SLL, CLL, MCL		Diffuse Large B-cell		Follicular B-cell		Other B-cell types		Multiple Myeloma		NHL subtype
	RR[3,4] (95% CI)	N[2]	RR[3,4] (95% CI)	N[2]	RR[3,4] (95% CI)	N[2]	RR[3,4] (95% CI)	N[2]	RR[3,4] (95% CI)	N[2]	Homo-geneity Test (p-value)
None	1.0 (ref)	108	1.0 (ref)	89	1.0 (ref)	41	1.0 (ref)	38	1.0 (ref)	64	
Low	1.1 (0.6–2.0)	15	0.6 (0.3–1.2)	8	1.3 (0.6–2.7)	8	0.9 (0.3–2.7)	5	1.4 (0.8–2.7)	13	
High	0.8 (0.5–1.5)	15	1.0 (0.5–2.1)	8	1.0 (0.5–2.4)	8	0.5 (0.2–1.7)	4	3.1 (1.5–6.2)	12	
	p trend = 0.53		p trend = 0.99		p trend = 0.88		p trend = 0.28		**p trend = 0.002**		0.10
Phorate											
None	1.0 (ref)	48	1.0 (ref)	37	1.0 (ref)	20	1.0 (ref)	16	1.0 (ref)	36	
Low	1.0 (0.6–1.9)	14	1.4 (0.7–2.7)	15	1.1 (0.4–3.0)	5	0.9 (0.3–2.2)	6	0.7 (0.3–1.8)	6	
High	0.8 (0.4–1.6)	11	0.7 (0.3–2.1)	4	0.8 (0.3–2.2)	5	1.1 (0.4–3.5)	4	0.8 (0.3–2.4)	4	
	p trend = 0.51		p trend = 0.80		p trend = 0.67		p trend = 0.91		p trend = 0.73		0.77
Terbufos											
None	1.0 (ref)	72	1.0 (ref)	63	1.0 (ref)	31	1.0 (ref)	19	1.0 (ref)	59	
Low	1.3 (0.8–2.0)	32	1.2 (0.8–1.9)	29	1.6 (0.9–3.1)	15	1.8 (0.9–3.6)	17	1.1 (0.6–1.9)	12	
High	1.6 (1.0–2.5)	31	1.0 (0.5–2.0)	12	0.8 (0.4–1.7)	10	1.6 (0.7–3.9)	8	1.3 (0.7–2.7)	5	
	p trend = 0.05		p trend = 0.90		p trend = 0.48		p trend = 0.29		p trend = 0.42		0.63
Chlorinated Insecticides											
Aldrin											
None	1.0 (ref)	53	1.0 (ref)	46	1.0 (ref)	22	1.0 (ref)	20	1.0 (ref)	34	
Low	1.0 (0.5–2.0)	11	xxx	2	1.2 (0.4–3.8)	4	0.4 (0.1–1.5)	3	2.1 (0.9–4.7)	8	
High	1.0 (0.5–2.0)	10	xxx	3	0.8 (0.3–2.5)	4	1.1 (0.3–3.9)	3	1.2 (0.5–3.2)	6	
	p trend = 0.70		p trend = xxx		p trend = 0.21		p trend = 0.67		p trend = 0.40		0.98
Chlordane											
None	1.0 (ref)	48	1.0 (ref)	42	1.0 (ref)	20	1.0 (ref)	21	1.0 (ref)	32	
Low	1.8 (1.0–3.1)	16	1.0 (0.5–2.2)	8	1.7 (0.7–4.3)	6	xxx	2	1.7 (0.9–3.3)	13	
High	1.5 (0.7–3.3)	8	1.4 (0.6–3.3)	7	1.3 (0.4–4.6)	3	xxx	2	0.7 (0.2–2.2)	3	
	p trend = 0.34		p trend = 0.69		p trend = 0.70		p trend = xxx		p trend = 0.57		0.85
DDT											
None	1.0 (ref)	42	1.0 (ref)	34	1.0 (ref)	17	1.0 (ref)	16	1.0 (ref)	28	
Low	1.0 (0.5–1.8)	16	1.6 (0.4–3.1)	2	3.3 (1.4–8.1)	9	0.4 (0.3–2.5))	5	1.2 (0.6–2.6)	10	
High	2.6 (1.3–4.8)	15	1.4 (0.6–3.5)	3	1.1 (0.3–3.6)	4	2.1 (0.7–6.5)	5	0.8 (0.4–1.8)	9	
	p trend = 0.04		P trend = 0.17		p trend = 0.80		p trend = 0.64		p trend = 0.37		0.44
Heptachlor											
None	1.0 (ref)	58	1.0 (ref)	47	1.0 (ref)	24	1.0 (ref)	21	1.0 (ref)	40	
Low	1.1 (0.5–2.3)	9	xxx	3	xxx	2	xxx	3	1.3 (0.4–3.8)	4	
High	1.4 (0.7–3.0)	9	xxx	1	xxx	1	xxx	2	1.2 (0.4–3.6)	4	
	p trend = 0.16		p trend = xxx		p trend = xxx		p trend = xxx		p trend = 0.91		0.68
Lindane											
None	1.0 (ref)	57	1.0 (ref)	49	1.0 (ref)	16	1.0 (ref)	21	1.0 (ref)	43	
Low	1.2 (0.6–2.5)	10	0.6 (0.2–1.7)	4	4.9 (1.9–12.6)	6	xxx	2	xxx	3	
High	2.6 (1.2–5.6)	9	2.0 (0.6–6.5)	3	3.6 (1.4–9.5)	6	xxx	1	xxx	2	
	p trend = 0.13		p trend = 0.96		**p trend = 0.04**		p trend = xxx		p trend = xxx		0.54
Toxaphene											
None	1.0 (ref)	68	1.0 (ref)	47	1 (ref)	23	1.0 (ref)	22	1.0 (ref)	40	

Table 4. Cont.

Insecticides

	SLL, CLL, MCL RR[3,4] (95% CI)	N[2]	Diffuse Large B-cell RR[3,4] (95% CI)	N[2]	Follicular B-cell RR[3,4] (95% CI)	N[2]	Other B-cell types RR[3,4] (95% CI)	N[2]	Multiple Myeloma RR[3,4] (95% CI)	N[2]	NHL subtype Homo-geneity Test (p-value)
Low	0.9 (0.4–2.3)	5	1.3 (0.5–3.3)	5	xxx	2	xxx	3	0.7 (0.2–2.0)	4	
High	0.4 (0.1–1.6)	2	0.9 (0.3–3.0)	3	xxx	2	xxx	2	0.7 (0.2–2.9)	2	
	p trend = 0.08		p trend = 0.77		p trend = xxx		p trend = xxx		p trend = 0.64		0.34
Fungicides											
Captan											
None	1.0 (ref)	118	1.0 (ref)	91	1.0 (ref)	52	1.0 (ref)	39	1.0 (ref)	76	
Low	0.9 (0.4–1.9)	7	1.1 (0.5–2.4)	7	xxx	2	xxx	3	1.4 (0.5–3.4)	5	
High	1.1 (0.5–2.6)	7	0.7 (0.1–3.1)	4	xxx	1	xxx	2	1.2 (0.5–2.9)	5	
	p trend = 0.78		p trend = 0.58		p trend = xxx		p trend = xxx		p trend = 0.75		0.92
Chlorothalonil											
None	1.0 (ref)	135	1.0 (ref)	107	1.0 (ref)	60	1.0 (ref)	50	1.0 (ref)	84	
Low	0.9 (0.4–2.3)	5	1.1 (0.4–3.1)	4	xxx	3	−xxx	1	1.1 (0.4–2.8)	5	
High	1.1 (0.4–3.3)	4	0.3 (0.1–1.2)	2	xxx	2	−xxx	1	0.7 (0.6–2.3)	3	
	p trend = 0.83		p trend = 0.09		p trend = xxx		p trend = xxx		p trend = 0.56		0.76
Metalaxyl											
None	1.0 (ref)	60	1.0 (ref)	45	1.0 (ref)	25	1.0 (ref)	23	1.0 (ref)	39	
Low	2.8 (1.4–5.8)	9	1.1 (0.4–2.6)	7	xxx	3	−xxx	2	0.4 (0.1–1.1)	4	
High	1.1 (0.4–2.8)	6	1.0 (0.4–2.7)	5	xxx	2	−xxx	1	1.1 (0.4–3.2)	4	
	p trend = 0.99		p trend = 0.97		p trend = xxx		p trend = xxx		p trend = 0.87		0.92
Maneb/ Mancozeb											
None	1.0 (ref)	69	1.0 (ref)	49	1.0 (ref)	25	1.0 (ref)	26	1.0 (ref)	41	
Low	2.1 (0.7–6.0)	4	4.0 (1.4–11.6)	4	xxx	2	−xxx	0	1.0 (0.4–2.5)	5	
High	1.2 (0.3–4.0)	3	0.9 (0.3–3.1)	3	−xxx	1	−xxx	0	2.2 (0.5–9.5)	2	
	p trend = 0.84		p trend = 0.74		p trend = xxx		p trend = xxx		p trend = 0.28		0.82
Fumigant											
Methyl Bromide											
None	1.0 (ref)	126	1.0 (ref)	86	1.0 (ref)	58	1.0 (ref)	44	1.0 (ref)	76	
Low	1.1 (0.5–2.2)	9	4.0 (2.2–7.4)	15	1.4 (0.5–4.2)	4	3.6 (1.3–9.8)	5	1.0 (0.5–2.1)	8	
High	0.8 (0.4–1.8)	8	1.0 (0.5–2.1)	11	0.3 (0.1–1.1)	3	1.3 (0.3–5.0)	3	0.8 (0.4–1.8)	8	
	p trend = 0.58		p trend = 0.67		p trend = 0.08		p trend = 0.56		p trend = 0.63		0.59

[1]During the period from enrollment (1993–1997) to December 31, 2010 in NC and December 31, 2011 in Iowa.

[2]Numbers of cases in columns do not sum to total number of NHL cases (n = 523) due to missing data. Ever/never use of all 26 pesticides (table 3) do not always match with exposure-response data in table 4 because of missing data to calculate lifetime-days of use.

[3]Adjusted for age (<45, 45–49, 50–54, 55–59, 60–64, 65–69, ≥70), State (NC vs. IA), Race (White vs. Black), AHS herbicides (in tertiles of total herbicide use-days). Significant RR and 95% confidence limits are bolded.

[4]RR was not calculated if the number of exposed cases for any NHL subtype was <6 and these cells are marked XXX. Four pesticides included in Table 2 (i.e., aldicarb, benomyl, dieldrin and parathion) were not included in Table 4 because no NHL subtype included≥6 cases of a specific cell types with lifetime-days of exposure.

risk estimates (i.e., narrower confidence intervals) when we included phase 2 imputed data (n = 54,306) (data not shown). Lagging exposures by five years did not meaningfully change the association between lindane or DDT and total NHL (data not shown). The significant exposure-response trends linking use of a particular pesticide to NHL and certain NHL subtypes did not always correspond to a significant excess risk among those who ever used the same pesticide. For chemicals for which the detailed information was only asked about in the take-home questionnaire, we evaluated potential differences between the ever/never analyses based on the enrolment questionnaire and data from the same sub-set of participants who completed the exposure-response in the take-home questionnaire and found no meaningful differences in the results. We also evaluated the impact of using an updated definition of NHL; when using the original ICD-O-3 definition of NHL[19], lifetime-days of lindane use remained significantly associated with NHL risk (RR = 1.0 (ref), 1.3 (0.7–2.6), 1.2 (0.6–2.8), 2.7 (1.3–5.4), p trend = 0.006). The trend between total NHL and lifetime-days of DDT, however, was less clear and not statistically significant (RR = 1.0 (ref) 1.3 (0.9–1.8), 1.1 (0.5–2.1), 1.4 (0.8–2.6), p trend = 0.32) [Table S3 in File S1]. Carbaryl and diazinon showed non-significant trends with the older definition of NHL, but not with the newer definition used here.

Discussion

A significant exposure–response trend for total NHL was observed with increasing lifetime-days of use for two organochlorine insecticides, lindane and DDT, although RRs from ever/never comparisons were not elevated. On the other hand, terbufos use showed a significant excess risk with total NHL in ever vs. never exposed analysis, but displayed no clear exposure-response trend. Several pesticides showed significant exposure-response trends with specific NHL subtypes however, when polytomous models were used to test the difference in parametric estimates of trend among the five NHL subtypes, there was no evidence of heterogeneity in the sub-types for specific chemicals. The subtype relationships that looked particularly interesting were DDT and terbufos with the SLL/CLL/MCL subtype, lindane and diazinon with the follicular subtype, and permethrin with MM. These pesticide-NHL links should be evaluated in future studies.

Lindane (gamma-hexachlorocyclohexane) is a chlorinated hydrocarbon insecticide. Production of lindane was terminated in the United States in 1976, but imported lindane was used to treat scabies and lice infestation and for agricultural seed treatment [21] until its registration was cancelled in 2009 [22], the same year production was banned worldwide [23]. In our study, 3,410 people reporting ever using lindane (6%) prior to enrollment, 433 reported use at the phase 2 questionnaire (1%), indicating that use had dropped substantially. Oral administration of lindane has increased the incidence of liver tumors in mice and less clearly, thyroid tumors in rats [24]. Lindane produces free radicals and oxidative stress (reactive oxygen species [ROS]) [25] and has been linked with chromosomal aberrations in human peripheral lymphocytes in vitro [26].

Lindane has been linked with NHL in previous epidemiologic studies. A significant association between lindane use and NHL was observed in a pooled analysis of three population-based case-control studies conducted in the Midwestern US, with stronger relative risks observed for greater duration and intensity of use [27]. NHL was also associated with lindane use in a Canadian case-control study [28]. Lindane was significantly associated with NHL risk in an earlier report from the AHS [29]. We are not aware of any previous study that assessed the association between a NHL subtype and lindane use. The exposure-response pattern with total NHL and the follicular lymphoma subtype indicates a need for further evaluation of lindane and NHL.

DDT is an organochlorine insecticide that was used with great success to control malaria and typhus during and after World War II [29] and was widely used for crop and livestock pest control in the United States from the mid-1940s to the 1960s [30]. Its registration for crop use was cancelled in the US in 1972 [30] and banned worldwide for agricultural use in 2009, but continues to be used for disease vector control in some parts of the world [23]. In our study, 12,471 participants (23%) reported ever using DDT

prior to enrollment; 12%, 8.7% and 2.3% responding to the take-home questionnaire reported their first use occurred prior to the 1960s, during the 1960s, and during the 1970s, respectively. The National Toxicology Program classifies DDT as "reasonably anticipated to be a human carcinogen" [31] and IARC classifies DDT as a "possible human carcinogen (2B)" [12], both classifications were based on experimental studies in which excess liver tumors were observed in two rodent species. Epidemiology data on the carcinogenic risk of DDT is inconsistent. NHL was not associated with use of DDT in a pooled analysis of three case-control studies in the U.S. where information on exposure was obtained from farmers by questionnaire [32]. There also was no association between the use of DDT and NHL in our study when we used an earlier definition of NHL [18], suggesting some of the inconsistency may be due to disease definition. In the large Epilymph study, no meaningful links between DDT and the risk of NHL, or diffuse large B cell lymphoma were observed, and only limited support was found for a link to CLL [33], although a case-control study of farmers in Italy suggested increased risk of NHL and CLL with DDT exposure [34]. NHL was not associated with serum levels of DDT in a prospective cohort study from the U.S. [35], but NHL was associated with the DDT-metabolite p, p'-DDE, as well as chlordane and heptachlor-related compounds (oxychlordane, heptachlor epoxide) and dieldrin, in a study with exposure measured in human adipose tissue samples [36]. In a Danish cohort, a higher risk of NHL was associated with higher prediagnostic adipose levels of DDT, cis-nonachlor, and oxychlordane [37]. In a Canadian study, analytes from six insecticides/insecticide metabolites (beta-hexachlorocyclohexane, p, p'-dichloro-DDE, hexachlorobenzene (HCB), mirex, oxychlordane and transnonachlor) were linked with a significant increased risk with NHL [38]. However, in an analysis of plasma samples from a case-control study in France, Germany and Spain, the risk of NHL did not increase with plasma levels of hexachlorobenzene, beta-hexachlorobenzene or DDE [39]. In this analysis, NHL was significantly associated with reported use of DDT, but not with the other organochlorine insecticides studied (i.e., aldrin, chlordane, dieldrin, heptachlor, toxaphene). Our findings add further support for an association between DDT and total NHL and our results on SLL/CLL/MCL are novel and should be further explored.

Permethrin is a broad-spectrum synthetic pyrethroid pesticide widely used in agriculture and in home and garden use as an insecticide and acaricide, as an insect repellant, and as a treatment to eradicate parasites such as head lice or mites responsible for scabies [40]. This synthetic pyrethroid was first registered for use in the United States in 1979 [40]. The U.S. Environmental Protection Agency classified permethrin as "likely to be carcinogenic to humans" largely based on the observed increase incidence of benign lung tumors in female mice, liver tumors in rats and liver tumors in male and female mice [41]. Permethrin was not associated with NHL overall in our study, nor in pooled case-control studies of NHL from the U.S (the NHL definition in use at the time of the study did not include MM) [42]. In our analysis, however, the risk of MM increased significantly with lifetime-days of exposure to permethrin, as had been noted in an earlier analysis of AHS data [43]. We are unaware of other studies that have found this association.

Terbufos is an organophosphate insecticide and nematicide first registered in 1974 [44]. The EPA classifies terbufos as Group E, i.e., "Evidence of Non-Carcinogenicity for Humans" [44]. We found some evidence for an association between terbufos use and NHL, particularly for the SLL/CLL/MCL subtype. NHL was not associated with terbufos in the pooled case-control studies from the

U.S. [42] but there was a non-significant association between terbufos and small cell lymphocytic lymphoma [10].

Diazinon is an organophosphate insecticide registered for a variety of uses on plants and animals in agriculture [45]. It was commonly used in household insecticide products until the EPA phased out all residential product registrations for diazinon in December 2004 [45.46]. In an earlier evaluation of diazinon in the AHS, a significant exposure-response association was observed for leukemia risk with lifetime exposure-days [47]. While there was no link between diazinon and NHL overall in this analysis, there was a statistically significant exposure-response association between diazinon and the follicular lymphoma subtype and an association with the SLL/CLL/MCL subtype that was not statistically significant. Diazinon was previously associated with NHL in pooled case-control studies from the U.S. and particularly with SLL [10].

Several other insecticides, fungicides and fumigants cited in recent reviews of the pesticide-cancer literature suggested etiological associations with total NHL [8,9], these include: oxychlordane, trans-nonachlor, and cis-nonachlor which are metabolites of chlordane; and dieldrin and toxaphene among NHL cases with t(14,18) translocations. We did not find a significant association between chlordane and total NHL nor with any NHL subtype, but we did not have information about chlordane metabolites to make a more direct comparison. Similarly we did not observe a significant association between dieldrin nor toxaphene and total NHL nor with any NHL subtypes. Mirex (1,3-cyclopentadiene), an insecticide, and hexachlorobenzene, a fungicide, were also associated with NHL risk [8,9] but we did not examine these compounds in the AHS.

This study has a number of strengths. It is a large population of farmers and commercial pesticide applicators who can provide reliable information regarding their pesticide use history [48]. Information on pesticide use and application practices was obtained prior to onset of cancer. An algorithm that incorporated several exposure determinants which predicted urinary pesticide levels was used to develop an intensity-weighted exposure metric in our study [20]. Exposure was ascertained prior to diagnosis of disease, which should eliminate the possibility of case-response bias [14]. Because of the detailed information available on pesticide use, we were able to assess the impact for the use of multiple pesticides. For example, we evaluated total pesticide use-days, and specific pesticides found to be associated with NHL or its subtypes in the AHS. We found no meaningful change in the associations with DDT, lindane, permethrin, diazinon and terbufos from such adjustments. Information on many potential NHL risk factors was available and could be controlled in the analysis.

Most epidemiological investigations of NHL prior to 2007 [17] did not include CLL and MM as part of the definition. These two subtypes made up 37% (193/523) of the NHL cases in this analysis. This is a strength of our study in that the definition of NHL used here is based on the most recent classification system [16,17] and will be relevant for comparisons with future studies. On the other hand, the inclusion of MM and CLL in the recent definition of NHL makes comparisons of our findings with earlier literature challenging, because the NHL subtypes may have different etiologies. For example, DDT was not significantly associated with NHL using the older definition, but was significantly associated with the NHL using the most recent definition of NHL because of its association with the SLL/CLL/MCL subtype (Table S1 in File S1). On the other hand, carbaryl and diazinon were associated with the old definition of NHL (although non-significantly) but not with the new definition. Lindane, however, was associated with both definitions of NHL.

Lindane was significantly associated with the follicular lymphoma subtype and this subtype was included in the older and newer definition of NHL. No other pesticides were significantly associated with NHL under the old definition (Table S3 in File S1).

Although this is a large prospective study, limitations should be acknowledged. A small number of cases exposed to some specific pesticides could lead to false positive or negative findings. We also had reduced statistical power to evaluate some pesticides for total days of use and intensity-weighted days of use because some participants did not complete the phase one take-home questionnaire and the tests of homogeneity between specific pesticides and specific NHL subtypes were underpowered. Some chance associations could occur because of multiple testing, i.e., a number of pesticides, several NHL subtypes, and more than one exposure metric. Despite the generally high quality of the information on pesticide use provided by AHS participants [48,50], misclassification of pesticide exposures can occur and can have a sizeable impact on estimates of relative risk, which in a prospective cohort design would tend to produce false negative results [49].

Conclusion

Our results showed pesticides from different chemical and functional classes were associated with an excess risk of NHL and NHL subtypes, but not all members of any single class of pesticides were associated with an elevated risk of NHL or NHL subtypes, nor were all chemicals of a class included on our questionnaire. Significant pesticide associations were between total NHL and reported use of lindane and DDT. Links between DDT and terbufos and SLL/CLL/MCL, lindane and diazinon and follicular lymphoma, and permethrin and MM, although based on relatively small numbers of exposed cases, deserve further evaluation. The epidemiologic literature on NHL and these pesticides is inconsistent and although the findings from this large, prospective cohort add important information, additional studies that focus on NHL and its subtypes and specific pesticides are needed. The findings from this large, prospective cohort add important new information regarding the involvement of pesticides in the development of NHL. It provides additional information regarding specific pesticides and NHL overall and some new leads regarding possible links with NHL subtypes that deserve evaluation in future studies.

Supporting Information

File S1   This file contains Table S1, Table S2, and Table S3. Table S1, Frequency of NHL in Agricultural Health Study applicators using New (Interlymph hierarchical classification of lymphoid neoplasms) and Older Definitions (ICD-O-3). Table S2, Pesticides included in the Agricultural Health Study questionnaires by Chemical/Functional Class. Table S3, Pesticide exposure (lifetime-days) and adjusted risks of total NHL incidence (Older definition [ICD-O-3]).

Acknowledgments

Disclaimer: The findings and conclusions in this report are those of the author(s) and do not necessarily represent the views of the National Institute for Occupational Safety and Health. The United States Environmental Protection Agency through its Office of Research and Development partially funded and collaborated in the research described here under Contracts 68-D99–011 and 68-D99–012, and through Interagency Agreement DW-75–93912801–0. It has been subjected to Agency review and approved for publication.

This work was supported by the Intramural Research Program of the NIH, National Cancer Institute, Division of Cancer Epidemiology and

Genetics (Z01CP010119) and the National Institutes of Environmental Health Science (Z01ES049030). The authors have no conflicts of interest in connection with this manuscript.

Ms. Marsha Dunn and Ms. Kate Torres, (employed by Westat, Inc. Rockville, Maryland) are gratefully acknowledged for study coordination. The ongoing participation of the Agricultural Health Study participants is indispensable and sincerely appreciated.

Author Contributions

Conceived and designed the experiments: MCA DPS AB. Performed the experiments: MCA CFL KT CJH. Analyzed the data: MCA JNH CFL CJH KHB JB DWB KT DPS JAH SK GA JHL AB LEB. Contributed reagents/materials/analysis tools: MCA JB DWB CFL. Wrote the paper: MCA LEBF JNH CFL CJH KT AB DWB JHL. Designed the software: JB DWB.

References

1. Milham S (1971) Leukemia and multiple myeloma in farmers. Am J Epidemiol 94: 507–510.
2. Cantor KP (1982) Farming and mortality from non-Hodgkin's lymphoma: a case-control study. Int J Cancer 29: 239–247.
3. Blair A, Malker H, Cantor KP, Burmeister L, Wiklund K (1985) Cancer among farmers: A review. Scand J Work Environ Health 11: 397–407.
4. Pearce NE, Smith AH, Fischer DO (1985) Malignant lymphoma and multiple myeloma linked with agricultural occupation in a New Zealand cancer registration-base-study. Am J Epidemiol 121: 235–237.
5. Baris D, Silverman DT, Brown LM, Swanson GM, Hayes RB, et al. (2004) Occupation, pesticide exposure and risk of multiple myeloma. Scand J Work Environ Health 30(3): 215–222.
6. Beane Freeman LE, DeRoos AJ, Koutros S, Blair A, Ward MH, et al. (2012) Poultry and livestock exposure and cancer risk among farmers in the agricultural health study. Cancer Causes Control 23: 663–670.
7. Cordes DH, Rea DF (1991) Farming: A Hazardous Occupation. In: Health Hazards of Farming. Occupational Medicine: State of the Art Reviews. Vol6(3). Hanley & Belfus, Inc., Philadelphia, PA.
8. Alavanja M, Bonner M (2012) Occupational pesticide exposure and cancer risk. A Review. J. Toxicol Environ Health B Critic Review. 1594: 238–263.
9. Alavanja MCR, Ross MK, Bonner MR (2013) Increased cancer burden among pesticide applicators and others due to pesticide exposure. CA, Cancer J Clin; 63(2): 120–142.
10. Waddell BL, Zahm SH, Baris D, Weisenburger DD, Holmes F, et al. (2001). Agricultural use of organophosphate pesticides and the risk of non-Hodgkin's lymphoma among male farmers (United States). Cancer Causes and Control 12: 509–517.
11. Merhi M, Raynal H, Cahuzac E, Vinson F, Cravedi JP, et al. (2007) Occupational exposure to pesticides and risk of hematopoietic cancers: meta-analysis of case-control studies. Cancer Causes Control. 18: 1209–1226.
12. IARC (1991) International Agency for Research on Cancer (IARC). Occupational Exposures in Insecticide applications and some pesticides. Lyon, France: IARC, 1991. Monographs on the Evaluation of Carcinogenic Risk to Humans, volume 53.
13. Morton LM, Slager SL, Cerhan JR, Wang SS, Vajdic CM, et al. (2014) Etiologic Heterogeneity Among NHL Subtypes: The InterLymph NHL Subtypes Project. J Natl Cancer Institute 48: 130–144.
14. Alavanja MCR, Sandler DP, McMaster SB, Zahm SH, McDonnell CJ, et al. (1996) The Agricultural Health Study. Environ Health Perspect 104: 362–369.
15. Alavanja MC, Samanic C, Dosemeci M, Lubin J, Tarone R, et al. (2003) Use of agricultural pesticides and prostate cancer risk in the Agricultural Health Study Cohort. Am J Epidemiol 157(9): 800–814.
16. SEER Program, National Cancer Institute. Available: http://seer.cancer.gov/lymphomarecode Accessed September 15, 2013.
17. Morton LM, Turner JJ, Cerhan JR, Linet MS, Treseler PA, et al. (2007) Proposed classifciation of lymphoid neoplasms for epidemiologic research from the Pathology Working Group of the International Lymphoma Epidemiology Consortium (InterLymph). Blood 110(2): 695–708.
18. Percy C, Fritz A, Ries L (2001) Conversion of neoplasms by topography and morphology from the International Classification of Disease for Oncology, second edition, to International Classification of Diseases for Oncology, 3rd ed. Cancer Statistics Branch, DCCPS, SEER Program, National Cancer Institute; 2001.
19. Heltshe SL, Lubin JH, Koutros S, Coble JB, Ji B-T, et al. (2012) Using multiple imputation to assign pesticide use for non-respondents in the follow-up questionnaire in the Agricultural Health Study J. Exp Sci Environ Epidemiol 22(4): 409–416.
20. Coble J, Thomas KW, Hines CJ, Hoppin JA, Dosemeci M, et al. (2011) An updated algorithm for estimation of pesticide exposure intensity in the Agricultural Health Study. Int J Environ Res Public Health. 8(12): 4608–4622.
21. ATSDR (2005) Agency for Toxic Substances and Disease Registry. Toxicological profile for Alpha-, Beta-, Gamma- and Delta- Hexachlorcyclohexane, August, 2005. Available: http://www.atsdr.cdc.gov/Toxprofiles/tp43.pdf. Accessed 2013 Sep 15.
22. EPA (2006a) US Environmental Protection Agency (2006). Lindane; Cancellation Order December 13, 2006. Federal Register/Vol 71, number 239, page 74905.
23. Stockholm Convention Report (2009) Report of the conference of the Parties of the Stockholm Convention on Persistent Organic Pollutants on the work of its fourth meeting. Convention on Persistent Organic Pollutants. Fourth Meeting, Geneva, 4–8 May 2009. Available: http://chm.pops.int/Portals/0/Repository/COP4/UNEP-POPS-COP.4-38.English.pdf.
24. IARC (1987) International Agency for Research on Cancer (IARC). Overall evaluation of carcinogenicity: an updating of IARC monographs volume 1 to 42. Lyon, France: IARC, 1987. Monographs on the Evaluation of Carcinogenic Risk to Humans, Supplement 7.
25. Piskac-Collier AL, Smith MA (2009) Lindane-induced generation of reactive oxygen species and depletion of glutathione do not result in necrosis in renal distal tube cells. J Toxicol and Environ Health, Part A. 72: 1160–1167.
26. Rupa DS, Reddy PP, Reddi OS (1989). Genotoxic effect of benzene hexachloride in cultured human lymphocytes. Hum Genet 83: 271–273.
27. Blair A, Cantor KP, Zahm SH (1998) Non-Hodgkin's lymphoma and agricultural use of the insecticide lindane. Am J Ind Med 33: 82–87.
28. McDuffie HH, Pahwa P, Mclaughlin JR, Spinelli JJ, Fincham S, et al. (2001) Non-Hodgkin's lymphoma and specific pesticide exposures in Men: Cross-Canada Study of Pesticides and Health. Cancer, Epidemiology, Biomarkers & Prevention 10: 1155–1163.
29. Purdue MP, Hoppin JA, Blair A, Dosemeci M, Alavanja MCR (2007) Occupational exposure to organochlorine insecticides and cancer incidence in the Agricultural Health Study. Int J Cancer. 120(3): 642–649.
30. EPA (2012) US Environmental Protection Agency 2012. DDT-A Brief History and Status. Available: http://www.epa.gov/pesticides/factsheets/chemicals/ddt-brief-history-status.htm. Accessed 2013 Sep 15.
31. NTP (2011) National Toxicology Program, Report on Carcinogen- Twelfth Edition. Available: http://ntp.niehs.nih.gov/go/roc12. Accessed 2013 Sep 15.
32. Baris D, Zahm SH, Cantor KP, Blair A (1998). Agricultural use of DDT and the risk of non-Hodgkin's lymphoma: pooled analysis of three case-control studies in the United States, Occup Environ Med 55: 522–527.
33. Cocco P, Satta G, Dubois S, Pilli C, Pillieri M, et al. (2013) Lymphoma risk and occupational exposure to pesticides: results of the Epilymph study. Occupational Environ Med 70(2): 91–98.
34. Nanni O, Amadori D, Lugaresi C, Falcini F, Scarpi E, et al. (1996) Chronic lymphocytic leukaemia and non-Hodgkin's lymphomas by histological type in farming-animal breeding workers: a population case-control study based on a priori exposure matrics Occup Environ Med 53(10): 652–657.
35. Rothman N, Cantor KP, Blair A, Bush D, Brock JW, et al. (1997) A nested case-control study of non-Hodgkin lymphoma and serum organochlorine residues. The Lancet 350: 240–244.
36. Quintana PJE, Delfino RJ, Korrick S, Ziogas A, Kutz FW, et al. (2004) Adipose tissue levels of organochlorine pesticides and chlorinated biphenyls and the risk of non-Hodgkin's lymphoma. Environ Health Perspect 112: 854–861.
37. Brauner EV, Sorensen MA, Gaudreau E, LeBlanc A, Erikson KT, et al. (2012) A prospective study of organochlorines in adipose tissue and risk of non-Hodgkin lymphoma. Environ Health Perspect. 120(1): 105–111.
38. Spinelli JJ, Ng CH, Weber JP, Connors JM, Gascoyne RD, et al. (2007) Organochlorines and risk of non-Hodgkin lymphoma. Int J Cancer. 121(12): 2767–2775.
39. Cocco P, Brennan P, Ibba A, de Sanjose Llongueras S (2008) Plasma polychlorobiphenyl and organochlorine pesticide level and risk of major lymphoma subtypes. Occup Environ Med 65: 132–140.
40. EPA 2006(b). U.S. Environmental Protection Agency. 2006. Re-registration Eligibility Decision for Permethrin: Available: http://www.epa.gov/oppsrrd1/REDs/permethrin_red.pdf. Accessed 2013 Sep 15.
41. NPIC (2012). National Pesticide Information Center. Chemicals Evaluated for Carcinogenic Potential. Office of Pesticide Programs. U.S. Environmental Protection Agency. November 2012. Available: http://npic.orst.edu/chemicals_evaluated.pdf. Accessed 2013 Sep 15.
42. De Roos AJ, Zahm SH, Cantor KP, Weisenburger DD, Holmes FF, t al. (2003) Integrative assessment of multiple pesticides as risk factors for non-Hodgkin's lymphoma among men Occup Environ Med 60: E11.
43. Rusiecki JA, Patel R, Koutros S, Beane Freeman LE, Landgren O, et al. (2009) Cancer incidence among pesticide applicators exposed to permethrin in the Agricultural Health Study. Environ Health Perspect 117: 582–586.
44. EPA 2006(b). U.S. Environmental Protection Agency. 2006. Re-registration Eligibility Decision for Terbufos: Available: http://www.epa.gov/pesticides/reregistration/REDs/terbufos_red.pdf. Accessed 2013 Nov 18.
45. Environmental Protection Agency, 2004. Interim registration eligibility decision: diazinon. Available: http://www.epa.gov/pesticides/reregistration/REDs/diazinon_red.pdf. Accessed 2013 Nov 18.
46. Donalson D, Kieley T, Grube A (2002) Pesticide industry sales and usage: 1998 and 1999 market estimates. Washington, DC.: US Environmental Protection Agency, 2002 (EPA-733-R-02–001).

47. Beane Freeman LE, Bonner MR, Blair A, Hoppin JA, Sandler DP, et al. (2005). Cancer incidence among male pesticide applicators in the Agricultural Health Study cohort exposed to diazinon. Am J Epidemiol. 162: 1070–1079.

48. Blair A. Tarone R, Sandler D, Lynch CF, Rowland A, et al. (2002) Reliability of reporting on life-style and agricultural factors by a sample of participants in the Agricultural Health Study from Iowa. Epidemiology 13(1): 94–99.

49. Blair A, Thomas HT, Coble J, Sandler DP, Hines CJ, et al. (2011). Impact of pesticide exposure on misclassification on estimates of relative risks in the Agricultural Health Study. Occup Environ Med 68: 537–541.

50. Thomas KW, Dosemeci M, Coble JB, Hoppin JA, Sheldon LS, et al. (2010) Assessment of a pesticide exposure intensity algorithm in the Agricultural Health Study. J Expo Sci Environ Epidemiol 20(6): 559–569.

13

Fungicide Effects on Fungal Community Composition in the Wheat Phyllosphere

Ida Karlsson[1]*, **Hanna Friberg**[2], **Christian Steinberg**[3], **Paula Persson**[1]

1 Dept. of Crop Production Ecology, Swedish University of Agricultural Sciences (SLU), Uppsala, Sweden, **2** Dept. of Forest Mycology and Plant Pathology, SLU, Uppsala, Sweden, **3** INRA, UMR 1347 Agroécologie, Pole IPM, Dijon, France

Abstract

The fungicides used to control diseases in cereal production can have adverse effects on non-target fungi, with possible consequences for plant health and productivity. This study examined fungicide effects on fungal communities on winter wheat leaves in two areas of Sweden. High-throughput 454 sequencing of the fungal ITS2 region yielded 235 operational taxonomic units (OTUs) at the species level from the 18 fields studied. It was found that commonly used fungicides had moderate but significant effect on fungal community composition in the wheat phyllosphere. The relative abundance of several saprotrophs was altered by fungicide use, while the effect on common wheat pathogens was mixed. The fungal community on wheat leaves consisted mainly of basidiomycete yeasts, saprotrophic ascomycetes and plant pathogens. A core set of six fungal OTUs representing saprotrophic species was identified. These were present across all fields, although overall the difference in OTU richness was large between the two areas studied.

Editor: Newton C. M. Gomes, University of Aveiro, Portugal

Funding: The study was funded by The Swedish Research Council for Environment, Agricultural Sciences and Spatial Planning (http://www.formas.se/en/), registration number 2010-1945. The funder had no role in study design, data collection and analysis, decision to publish, or preparation of the manuscript.

Competing Interests: The authors have declared that no competing interests exist.

* Email: ida.karlsson@slu.se

Introduction

The phyllosphere, defined as the total above-ground parts of plants, provides a habitat for many microorganisms [1]. Phyllosphere microorganisms, including fungi, have been shown to perform important ecological functions and can be both beneficial and harmful to their host plant [2]. In agricultural crops, some phyllosphere fungi are important pathogens, while others have antagonistic properties [3] or can influence the physiology of the plant [4]. Understanding the influence of agricultural practices on phyllosphere fungal communities is important in order to create the best conditions for crop development.

Wheat is one of the most important crops worldwide and the wheat-associated fungal community was one of the first phyllosphere communities to be studied [5]. The wheat phyllosphere has been found to contain many basidiomycete yeasts such as *Cryptococcus* spp., *Sporobolomyces roseus* and filamentous saprotrophs, e.g. *Cladosporium* spp., *Alternaria* spp., *Epicoccum* spp., and plant pathogens [5–8]. Fungi can be present both as epiphytes and endophytes on wheat leaves. This is reflected in the different sets of fungi retrieved when washed leaf pieces are cultured compared with leaf wash liquid [9]. The main components of the fungal wheat leaf community differ in studies conducted at different sites and at different times and the mechanisms that lie behind the dynamics of fungal communities in the phyllosphere of agricultural crops are not well understood.

Plant pathogens are an important and well-studied group of wheat-associated microorganisms. Important fungal wheat leaf diseases world-wide include different types of rusts (*Puccinia* spp.), powdery mildew (*Blumeria graminis*) and leaf blotch diseases such as septoria tritici blotch (*Mycosphaerella graminicola* (*Zymoseptoria tritici*)). Septoria tritici blotch has been one of the most serious diseases of European wheat since the early 1980 s, causing up to 50% yield losses [10].

Foliar fungicides are routinely used in conventional agriculture to control fungal diseases. However, besides the desired effect on fungal pathogens, non-target fungi are also subjected to the fungicide treatment. It is important to understand the effect of fungicides on non-target fungi given the antagonistic potential of some phyllosphere fungi and interactions between different pathogens [1,11]. Applying a fungicide to control one pathogen might even increase the problems with another, as has been shown for *Fusarium* spp. causing fusarium head blight in cereals [12,13]. It has been hypothesised that fungicides suppress saprotrophic fungi that otherwise would act as competitors against *Fusarium* [13]. On the other hand, phyllosphere saprotrophs have been shown to accelerate leaf senescence, which could explain some of the yield increase after fungicide treatment not explained by attack of pathogens [9,14]. More knowledge on the effect of fungicides on phyllosphere fungal communities is important in order to optimise fungicide application strategies.

Fungicides have different modes of action and can be both broad range or target a specific group of fungi [15] and the fungicide type and use vary for different crops. Previous studies examining fungicide effects on non-target fungi in the wheat phyllosphere using culture-dependent methods have shown that fungicides with different modes of action have differing effects on individual fungal taxa [7–9,16–18]. Some of the biases of culture-dependent methods can be overcome using DNA-based methods. Recently, high-throughput sequencing technologies have revolutionized the study of microbial diversity in the phyllosphere. Consequently, knowledge on bacterial phyllosphere communities on agricultural crops is growing, but less is known about fungi [19]. So far, fungicide effects on fungal communities in the phyllosphere has only been investigated to a limited extent using DNA-based fingerprinting methods [20,21] and high-throughput sequencing [22], but none of these studies focused on cereals.

The aims of this study were: 1) to identify the fungal community in the wheat phyllosphere using 454 high-throughput sequencing, 2) to study the effect of fungicides on fungal community composition in the wheat phyllosphere, and 3) to study differences between phyllosphere fungal communities in two areas characterised by different climate conditions and agricultural management regimes. Fungicide-treated and non-fungicide treated leaves were sampled from winter wheat fields in two areas in Sweden and fungal community composition on the leaves was analysed by amplification and 454-sequencing of the fungal ITS2 region of the ribosomal DNA.

Materials and Methods

Ethics statement

Permission from the farmers was obtained through the Plant Protection Centres of the Swedish Board of Agriculture in Skara (for the Northern area) and Alnarp (for the Southern area) respectively. The study did not involve any protected or endangered species.

Sampling and plant material

Sampling of wheat fields was carried out in two important agricultural production areas of Sweden, a Northern sampling area located in the region of Västergötland and a Southern sampling area in the Skåne region (Fig. 1). The Southern area is characterised by a milder and drier climate. The two areas also differ in agricultural management, for example in terms of cropping sequence [23], the choice of wheat variety and fungicides are used more frequently in the Southern area [24]. The average winter wheat yield is about 2000 kg/ha higher in the Southern area [23]. At the time of sampling, fields in the Northern area had reached anthesis, while in the Southern area the developmental stage ranged from anthesis to the early dough ripening stage (Table 1).

Wheat (*Triticum aestivum*) leaves were sampled in pest surveillance plots, disease control and variety trials placed in conventionally managed farmers' fields during summer 2011. The pest surveillance plots are used for monitoring the incidence of pests and diseases, so fungicides or insecticides are not applied within these plots. Leaf samples representing seven different winter wheat varieties were collected from a total of 18 fields (Table 1).

All fields had received 1–3 fungicide treatments containing one or several of the following active ingredients: azoxystrobin, bixafen, cyprodinil, difenoconazole, fenpropimorph, metrafenone, picoxystrobin, prochloraz, propiconazole, prothioconazole and pyraclostrobin (see Tables S1 and S2 for further details). In fields with pest surveillance plots, fungicide application outside the plots was managed by the farmer. In field trials, fungicide application was carried out by field trial management staff.

The leaf below the flag leaf was sampled from 10 randomly chosen plants in each plot. For pest surveillance plots, plants were sampled from the fungicide-treated crop outside the plot and in the non-fungicide treated surveillance plot itself. In field trials,

plants were sampled from fungicide-treated plots and from untreated control plots. However, there was no untreated plot available from one field in the Southern area. Leaves from each plot were pooled into one sample. In total, 42 samples (Table 1) and 420 leaves were used for the study. Gloves were used when picking leaves to avoid cross-contamination between fungicide-treated and untreated leaves, and between fields. The leaves were collected in clean plastic bags and stored overnight in the refrigerator and then transferred to −20°C until DNA extraction.

DNA extraction, PCR amplification and sequencing

In order to capture both endophytic and epiphytic fungi, the whole leaf tissue was used for DNA extraction. Leaves were split into halves, with the middle vein left on the half used for extraction. The 10 halved leaves were cut into smaller pieces and placed in a plastic bag. The samples were frozen in liquid nitrogen, homogenised with a pestle and 100 mg of each sample were transferred to another bag (Bioreba AG, Switzerland). The DNA was then extracted using the DNeasy Plant Mini kit (QIAGEN AB, Sweden) according to the manufacturer's instructions, except for the lysis buffer, for which a larger volume was used (530 µl). The DNeasy kit was used with the QiaCube (QIAGEN AB, Sweden) with the standard plant cells and tissues protocol.

The ITS2 region was amplified on a 2720 Thermal Cycler (Life Technologies, CA, USA) using the forward primer fITS7 (GTGARTCATCGAATCTTTG; [25] and the reverse primer ITS4 (TCCTCCGCTTATTGATATGC; [26]. The length of the ITS2 is variable among fungi, ranging between ~122 and 245 bp [25]. The ITS4 primer was tagged with an 8 bp barcode. PCR was run in 50-µl reactions with 0.8 ng/µl template, 200 µM of each nucleotide, 2.75 mM $MgCl_2$, forward primer at 500 nM, tagged primer at 300 nM and 0.02 U/µl polymerase (DreamTaq Green, Thermo Scientific, MA, USA) in PCR buffer. PCR conditions were 5 min at 94°C, 30–32 cycles of [30 s at 94°C, 30 s at 57°C, 30 s at 72°C] and 7 min at 72°C. The number of cycles was adapted for each sample to give weak to moderately strong bands on the agarose gel with approximately the same strength for all samples to avoid oversaturation and distortion of the PCR pool. To determine the number of cycles necessary for each sample, test runs were conducted with non-barcoded primers starting at 25 PCR cycles before samples were run with the barcoded primers.

PCR products were cleaned using AMPure (Beckman Coulter, CA, USA) according to the manufacturer's instructions. DNA concentration was measured on a NanoDrop 1000 spectrophotometer (Thermo Scientific, MA, USA) and the samples were pooled in equimolar amounts. The sample pool was freeze-dried and sent to LGC Genomics (Germany) for adaptor ligation and sequencing on 1/16th of a plate on a GS FLX Titanium sequencer (Roche, Switzerland). Demultiplexed raw sequence data were deposited in the Sequence Read Archive (http://www.ncbi.nlm.nih.gov/sra) under the accession number SRP042192.

Bioinformatics and taxonomic assignment

The raw sequence data were analysed using the SCATA pipeline (http://scata.mykopat.slu.se; [27]). Sequences were screened for tags and primer sequences, allowing for one mismatch for the primers in addition to degenerate bases. Sequences shorter than 200 bp and those with a mean quality score lower than 20 and containing bases with a score lower than 10 were discarded. The option "extract high quality region" was used.

Sequences were clustered into operational taxonomic units (OTUs) at a clustering level that was chosen to approximate species level. The sequences passing the quality control were clustered at 1.5% dissimilarity cut-off in SCATA, using single

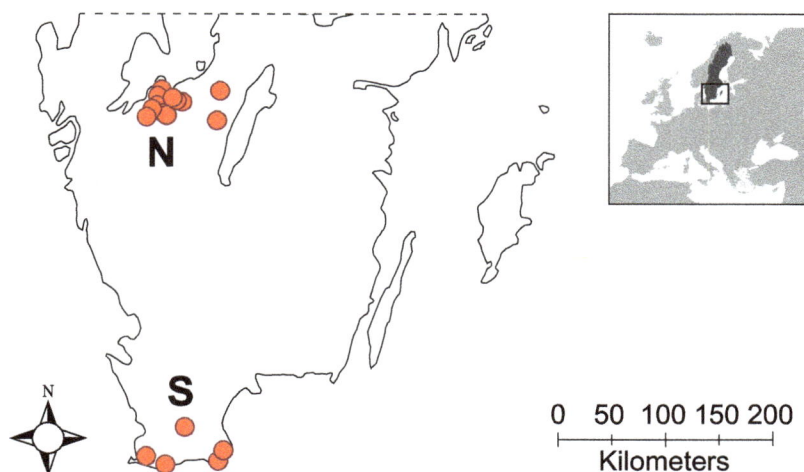

Figure 1. Wheat leaves were sampled in two important agricultural production areas of Sweden. Dots represent position of individual fields within in the two sampling areas. The Northern sampling area (N) is located in the Västergötland region and the Southern area (S) in the Skåne region.

linkage clustering with default settings (85% alignment, collapse homopolymers to 3 bp, usearch as cluster engine, miss match penalty 1, gap open penalty 0, gap extension 1 and end gap weight 0). The level of intraspecific variation in the *ITS* sequence is variable within fungi [28], and thus using a single cut-off level will not perfectly reflect biological species. However, we found 1.5% dissimilarity to be the most appropriate level in this dataset, as higher cut-off levels would group some basidiomycete species into the same OTU.

Singletons in the full dataset were removed, as many of them were considered to represent sequencing errors [29]. In addition, singletons in each sample were removed in an effort to limit the effects of tag switching [30].

We focused on taxonomically assigning the OTUs represented by at least 10 sequences globally in the dataset (67 OTUs). Some of these could be taxonomically assigned in SCATA by including reference sequences from isolates from the Fungal Biodiversity Center CBS (http://www.cbs.knaw.nl/) and from the UNITE database including 'species hypotheses' accessions (version 6

Table 1. Distribution of wheat leaf samples between two geographical areas, fields and treatments, including weather data.

	Northern area	Southern area
Sampling date	20-June	27-June
No. of fields	13	5
No. of samples	26	16
No. of control samples	13	5
No. of fungicide-treated samples	13	11
Wheat developmental stage (DC)	61–69	61–83 nd
	06-June—20-June	13-June—27-June
Mean temp. (°C)	14.7	14.3
Mean rel. humidity (%)	81.6	82.6
Acc. rainfall (mm)	123.4	37.6
	13-June—20-June	20-June—27-June
Mean temp. (°C)	13.4	14.5
Mean rel. humidity (%)	80.5	81.4
Acc. rainfall (mm)	47.8	17.2
	19-June	26-June
Mean temp. (°C)	12.4	14.5
Mean rel. humidity (%)	93.5	78.2
Acc. rainfall (mm)	23.4	0.0

Mean temperature, mean relative humidity and accumulated rainfall[1] are given for two weeks, one week and the day before sampling.
DC = developmental stage according to the Zadoks scale, nd = not determined for all fields.
1 Weather data from the Lantmet weather stations in Skara and Anderslöv respectively.

09.02.2014; [31]) in the clustering. Kõljalg *et al.* [31] introduced the term 'species hypothesis' (SH) in order to facilitate the communication of fungal taxa discovered when clustering DNA sequences at different similarity cut-offs. Stable accession codes for all species hypotheses in the UNITE database have now been introduced [31], thus facilitating comparisons among sequence-based studies of fungi.

OTUs that did not cluster with any reference sequence in SCATA were blasted against GenBank (http://www.ncbi.nlm.nih.gov/genbank) in order to find suitable reference sequences. The most abundant sequence in each OTU was used for this purpose. Taxonomic assignment was then performed with the help of neighbour-joining trees (Figs. S1 and S2). First, the OTUs were divided into basidiomycetes and ascomycetes. Then, a multiple alignment for each phylum was generated in the stand-alone version of MAFFT (v7.058b; [32]) using the G-IN-Si option. The alignments were cut so that all sequences had the same length. Subsequently, neighbour-joining trees were constructed with the ProtDist/FastDist+BioNJ option on the phylogeny.fr web service (http://www.phylogeny.fr; [33]) using 1000 bootstraps. The OTUs were assigned to the finest possible taxonomic level.

Statistical analyses

First, the relationship between sequencing depth and OTU richness was analysed using rarefaction curves for each area and treatment, generated with Analytical Rarefaction 1.3 (Steven M. Holland 2003, http://strata.uga.edu/software/anRareReadme.html) in steps of 1000 specimens. The relationship between the number of samples and OTU richness was analysed with species accumulation curves. The distribution of treated and untreated samples was uneven in the two sampling areas. Therefore, when more than two samples were taken per field, one fungicide-treated and one untreated sample were randomly chosen from each field for inclusion in the analysis. Species accumulation curves were generated using the *specaccum* function with the random method in the 'Vegan' package (version 2.0–10; [34]) in R (version 3.0.2)

As the number of sequences per sample was unequal, the dataset was rarefied to 197 sequences per sample, which was the size of the smallest sample. The rarefaction was performed using the *rrarefy* function in 'Vegan' (version 2.0–10; [34]). The rarefaction was repeated 1000 times on the sample-by-OTU table and the mean was taken over the 1000 matrices and used for subsequent analysis.

Second, we tested the effect of fungicide treatment and geographical area on OTU richness and community evenness (Pielou's evenness index [35]) using linear mixed models (LMM). We used the *lmer* function in the 'lme4' R package [36]. A model including treatment, geographical area and their interaction, with field and the interaction between field and treatment as random factors was fitted to both OTU richness and evenness. Significance tests were performed with a Kenward-Roger modification for performing F-tests, the *KRmodcomp* function in the 'pbkrtest' package [37]. The LMM analyses were performed both on the full dataset and on a smaller dataset excluding two fields in the Southern area where the control samples were dominated by one single OTU, namely *Puccinia striiformis*.

Third, non-metric multidimensional scaling (NMDS) was used to explore the fungal community composition using the function *metaMDS* in the 'Vegan' package [34] in R. The NMDS was performed using Bray-Curtis dissimilarities with square root transformation and Wisconsin double standardisation. Subsequently, 95% confidence areas were fitted to the ordination using the *ordiellipse* function.

Fourth, we tested the effect of fungicide treatment and geographical area on both community composition and individual OTUs using generalised linear models (GLMs). First, a model was fitted in order to test the effect of geographical area, treatment and their interaction. The interaction was not significant and a model including field as a block factor and treatment as a fixed factor could be fitted to test the effect of the fungicide treatment in both areas. For these analyses, a GLM was fitted to each OTU using the *manyglm* function in the 'mvabund' package in R (version 3.8.4; [38]) using a negative binomial probability distribution. The rarefied sample-by-OTU table was input as the response variable. Next, the models were tested using the function *anova* in 'mvabund', providing both a multivariate test for the whole community and univariate tests for each OTU. The score test was used and the cor.type argument set to shrink to allow for correlated response among OTUs. P-values were adjusted for multiple testing. The analysis was performed both at the level of species and order.

For the NMDS, LMM and the GLM analyses, only the 67 taxonomically assigned OTUs (those represented by at least 10 sequences globally) were included, since they represented the majority of the sequences in the dataset.

Results and Discussion

Sequence data quality

In all, 56% of the 454 reads from the pool of 420 wheat leaves passed the SCATA quality filtering. Singletons made up 1.7% of the sequences in the filtered dataset (471 global singletons and 324 per-sample singletons) and were removed. The removal of per-sample singletons resulted in a loss of 30 OTUs represented by 2–5 sequences each. Non-fungal sequences were also removed and these constituted 3.5% of the dataset, mostly wheat sequences. Per sample, 0–45% non-fungal sequences were removed. The quality-controlled dataset contained 44 245 sequences in 42 samples. The number of sequences per sample ranged between 197–2978, with a mean of 1053 sequences per sample.

Taxonomic composition and richness of the fungal community of wheat leaves

The fungal community composition in the wheat phyllosphere was characterised using 454 high-throughput sequencing. We found 235 fungal OTUs in the pool of 420 wheat leaves when clustering at 1.5% dissimilarity level. The rarefaction curves approached saturation (Fig. 2a) for all conditions, while the species accumulation curves did not (Fig. 2b).

We taxonomically assigned the OTUs containing more than 10 reads in the total dataset (67 OTUs, Table S3). These OTUs accounted for 90–100% of the sequences in the samples and none of the species in the tail were present in more than 10% of the samples. Overall, 45% of the OTUs were identified to species level and the highest taxonomic level of identification was at the order level.

The fungal community in the present study consisted of almost equal proportions of ascomycetes (54%) and basidiomycetes (46%). The most common orders in the dataset were Sporidiobolales, Tremellales, Capnodiales and Pleosporales (Fig. 3).

We identified a 'core' fungal community of six OTUs that was found across all treatments and sites. In fact, these six most abundant OTUs were found in all samples except two, including all fields. Of these six, five were identified as basidiomycete yeasts (OTU_0_*Sporobolomyces_roseus*, OTU_3_*Dioszegia_fristingensis*, OTU_4_*Cryptococcus_tephrensis*, OTU_12_*Cryptococcus*_sp and OTU_9_*Dioszegia_hungarica*), and one as the ascomycete

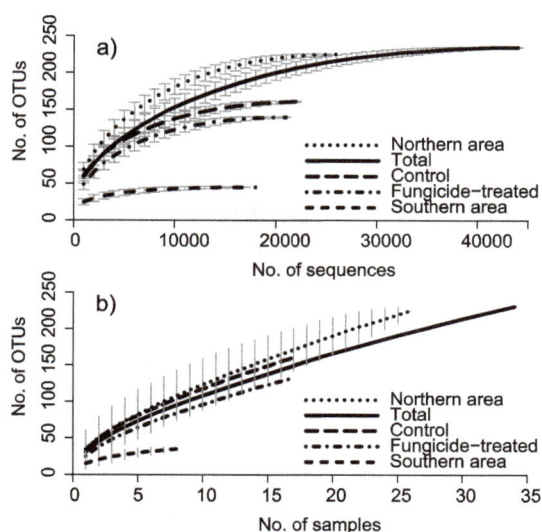

Figure 2. Rarefaction and species accumulation curves. a) Rarefaction curves presenting the relationship between sequencing depth and species richness in operational taxonomic units (OTUs). Error bars indicate 95% confidence intervals. b) Sample-based species accumulation curves. When more than two samples were taken per field, one fungicide-treated and one untreated sample was randomly chosen from each field for inclusion in the analysis. Error bars indicate 95% confidence intervals, only shown for Northern and Southern area respectively.

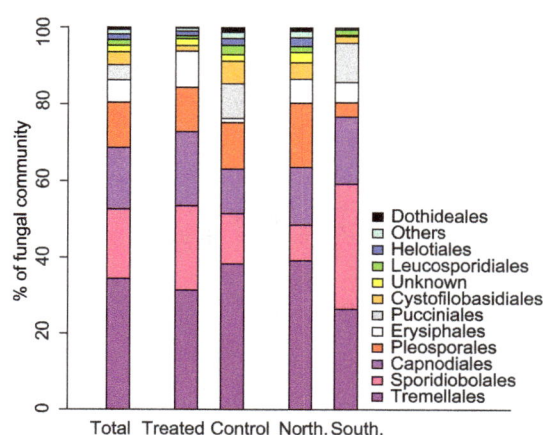

Figure 3. Fungal community composition on wheat leaves at the order level. Community composition is presented for the total dataset, fungicide-treated, control and samples from the Northern and Southern area respectively. Orders with low abundance have been merged to the group 'Others' to improve visual representation.

OTU_5_*Cladosporium*_spp. The phyllosphere fungal community of wheat has often been described as consisting of 'pink' yeasts (*Sporobolomyces* and *Rhodotorula* producing carotenoid pigments), 'white' yeasts (*Cryptococcus*) and ascomycete saprotrophs such as *Cladosporium* and *Alternaria* [17,18]. We were able to confirm the presence of these fungal taxa previously identified by culture-dependent methods (Table 2). Comparison of DNA-based and culture-based studies may give misleading results, since culture-dependent studies typically group morphologically similar species into one category. Thus, comparing the fungal community composition at the genus level might overestimate the similarity to previous studies of the phyllosphere of wheat. Using high-throughput sequencing, we were able to describe the fungal community in more detail. Blixt et al. (2010) [6] identified16 fungal species from wheat leaves in Sweden using cloning and sequencing, 13 of these were found among the 235 OTUs in our study.

The number of ITS copies has been reported to vary by an order of magnitude among different fungal species [39,40] and within the same fungal species [41]. This will to some extent bias quantitative comparisons among different taxa. Therefore, we focused on comparing the relative abundance of each OTU in fungicide-treated and untreated samples.

Fungicide effects on fungal community composition

The application of fungicides had a significant effect on fungal community composition on wheat leaves (Fig. 4 and Table 3). The total OTU richness was lower for the fungicide-treated sample pool (Fig. 2). There was also a tendency for a lower mean OTU richness per ten leaves in the fungicide-treated samples (19.4±1.8 SE) than in the control samples (24.3±2.1 SE), but the difference was not significant (p>0.05) (Fig. 5a, Table 4). There was no interaction between fungicide treatment and geographical area for neither community composition (Table 3) nor OTU richness (p>

0.05) (Table 4). When samples from fields infected with *Puccinia striiformis* were included in the analysis, the same pattern was observed (Fig. S3a, Table S4).

Fungicide treatment affected community evenness negatively (p<0.01), and there was no interaction with geographical area (p>0.05) (Fig. 5b, Table 4). However, the pattern was different when samples from fields infected with *P. striiformis* were included. When these samples were included, there was a significant interaction between area and fungicide treatment (p<0.05) and evenness tended to be higher in fungicide-treated samples than in control samples in the Southern area (Fig. S3b, Table S4). *P. striiformis* dominated the control samples when present in a field, and consequently the evenness in these samples was very low. This dominance might have been further amplified due to the pathogenic activity of *P. striiformis*, physically changing the leaf surface and thus possibly making it less suitable to other fungi, or by a high amount of *P. striiformis* biomass masking the less abundant community members since samples were pooled in equimolar amounts [42].

The community composition at the order level was significantly different for fungicide-treated and untreated samples (Fig. 3). The proportion of Leucosporidiales (p<0.05) and Dothideales (p<0.05) was lower in fungicide-treated samples than in the control samples. This was reflected at the species level, where univariate tests showed that the relative abundance of three OTUs: OTU_6_*Dioszegia*_sp (p<0.05), OTU_28_*Aureobasidium_pullulans*_a (p<0.05) and OTU_25_*Leucosporidium_golubevii* (p<0.05), was lower in fungicide-treated leaves than control leaves. OTU_6_*Dioszegia*_sp was similar to the *ITS* sequences of both *D. crocea* and *D. aurantiaca*. These species have been isolated from both the phyllosphere [43] and the rhizosphere of different plants [44,45]. *Leucosporidium golubevii* is a yeast discovered in freshwater [46], and has been reported from the phyllosphere of balsam poplar [47]. In addition, the relative abundance of OTU_16_*Phaeosphaeria_juncophila* (p<0.01) was higher in fungicide-treated leaves (Fig. 6). *Phaeosphaeria juncophila* was first isolated from the rush *Juncus articulatus*, but little is known about its ecology.

The fungicide sensitivity of phyllosphere fungi has mostly been investigated with fungicides that are no longer used or have been prohibited in Sweden, except for some sterol biosynthesis inhibitors (SBI). SBI fungicides have been shown to have no or

Table 2. Taxonomic and functional assignment of the 30 most abundant operational taxonomic units (OTUs).

OTU ID	No. of reads	Taxonomic assignment	UNITE SH-accesssion	SH %[2]	Phylum	Class	Order	Putative functional assignment [64,65]
OTU_0	8190	Sporobolomyces roseus	SH196706.06FU	98.5	Basidiomycota	Microbotryomycetes	Sporidiobolales	yeast
OTU_3	5475	Dioszegia fristingensis	SH196962.06FU	98.5	Basidiomycota	Tremellomycetes	Tremellales	yeast
OTU_5	4760	Cladosporium spp	-	-	Ascomycota	Dothideomycetes	Capnodiales	saprotroph, pathogen
OTU_4	3510	Cryptococcus tephrensis	SH198056.06FU	98.5	Basidiomycota	Tremellomycetes	Tremellales	yeast
OTU_12	2848	Cryptococcus spp	SH154124.06FU	97.5	Basidiomycota	Tremellomycetes	Tremellales	yeast
OTU_9	2330	Dioszegia hungarica	SH196961.06FU	98.5	Basidiomycota	Tremellomycetes	Tremellales	yeast
OTU_1	2301	Puccinia striiformis	SH205903.06FU	98.5	Basidiomycota	Pucciniomycetes	Pucciniales	pathogen
OTU_14	2223	Mycosphaerella graminicola	SH044710.06FU	98	Ascomycota	Dothideomycetes	Capnodiales	pathogen
OTU_13	2012	Blumeria graminis	SH195226.06FU	98.5	Ascomycota	Leotiomycetes	Erysiphales	pathogen
OTU_7	1268	Udeniomyces pannonicus	SH217650.06FU	98.5	Basidiomycota	Tremellomycetes	Cystofilobasidiales	yeast
OTU_6	1136	Dioszegia spp	SH196959.06FU	98.5	Basidiomycota	Tremellomycetes	Tremellales	yeast
OTU_24	929	Phaeosphaeriaceae sp1a	-	-	Ascomycota	Dothideomycetes	Pleosporales	-
OTU_27	836	Ascochyta sp	SH233950.06FU	98.5	Ascomycota	Dothideomycetes	Pleosporales	saprotroph, pathogen [57]
OTU_16	560	Phaeosphaeria juncophila	SH227803.06FU	98.5	Ascomycota	Dothideomycetes	Pleosporales	saprotroph
OTU_18	555	Ascochyta skagwayensis	-	-	Ascomycota	Dothideomycetes	Pleosporales	saprotroph, pathogen
OTU_26	492	Leucosporidiella fragaria	SH212317.06FU	98.5	Basidiomycota	Microbotryomycetes	Leucosporidiales	yeast
OTU_22	459	Alternaria malorum	-	-	Ascomycota	Dothideomycetes	Pleosporales	pathogen, saprotroph [66]
OTU_19	424	Phaeosphaeriaceae sp5a	SH227813.06FU	98.5	Ascomycota	Dothideomycetes	Pleosporales	-
OTU_47	327	Phaeosphaeria nodorum	SH206989.06FU	98.5	Ascomycota	Dothideomycetes	Pleosporales	pathogen
OTU_34	297	Helotiales sp1a	-	-	Ascomycota	Leotiomycetes	Helotiales	-
OTU_28	249	Aureobasidium pullulans a	-	-	Ascomycota	Dothideomycetes	Dothideales	saprotroph, antagonist [52]
OTU_25	179	Leucosporidium golubevii	SH212315.06FU	98.5	Basidiomycota	Microbotryomycetes	Leucosporidiales	yeast
OTU_50	171	Cystofilobasidiales sp a	-	-	Basidiomycota	Tremellomycetes	Cystofilobasidiales	-
OTU_32	154	Cryptococcus wieringae	SH216503.06FU	98.5	Basidiomycota	Tremellomycetes	Filobasidiales	yeast
OTU_35	152	Pleosporales sp1	-	-	Ascomycota	Dothideomycetes	Pleosporales	-
OTU_43	145	Helotiales sp1b	-	-	Ascomycota	Leotiomycetes	Helotiales	-
OTU_44	124	Bullera globospora	SH197114.06FU	98.5	Basidiomycota	Tremellomycetes	Tremellales	yeast
OTU_20	123	Monographella spp	SH216927.06FU	98.5	Ascomycota	Sordariomycetes	Xylariales	pathogen
OTU_33	114	Phaeosphaeriaceae sp5b	-	-	Ascomycota	Dothideomycetes	Pleosporales	-
OTU_70	111	Leptospora rubella	-	-	Ascomycota	Dothideomycetes	Incertae sedis	saprotroph [67]

Species hypothesis accession codes in the UNITE[1] database version are indicated when available. Functional assignment as 'pathogen' is only used for taxa known to be pathogenic on wheat.
[1]Version 6 (02.09.2014) http://unite.ut.ee/, 2 Similarity cut-off for clustering in UNITE.

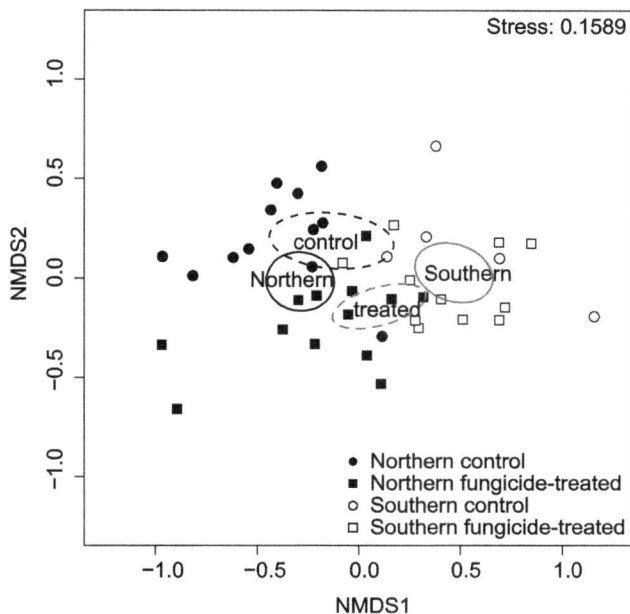

Figure 4. Non-metric multidimensional scaling (NMDS) of phyllosphere fungal communities of wheat. Ordination of samples with fitted environmental variables. Ellipses represent 95% confidence areas for Southern area (solid grey line), Northern area (solid black line), fungicide-treated (dashed grey line) and control (dashed black line) groups respectively. The NMDS was performed on the mean of 1000 rarefied datasets.

a moderate effect on the 'pink' and 'white' phyllosphere yeasts, with a somewhat stronger effect on 'white' yeasts [16]. *Aureobasidium pullulans* is reported to be sensitive to propiconazole [16] and prochloraz [17,18] and these two fungicides and other SBI fungicides have been used in almost all the fields in this study (Table S1). The ascomycete *A. pullulans* is one of the most common inhabitants of the phyllosphere of many crops [48] and is also present in many other habitats [49]. *Aureobasidium pullulans* is known to be antagonistic towards necrotrophic pathogens, e.g. grey mould in strawberries [50] and powdery mildew in durum wheat [51]. The mechanism of the antagonism is hypothesised to be competition for nutrients [52]. Some strains of *A. pullulans* also produce a type of antibiotic called aureobasidins [53]. The antagonistic potential of biological control agents is frequently strain-specific, depending on the mechanism of antagonism. From

the sequence data, we cannot make inferences about the antagonistic capacity of an OTU. Hence, it is unknown whether the reduction in the relative abundance of *A. pullulans* in fungicide-treated leaves has an impact on the antagonistic capacity of the fungal community.

Several OTUs were identified as common wheat pathogens in the dataset: OTU_14_*Mycosphaerella_graminicola*, OTU_13_*Blumeria_graminis*, OTU_1_*Puccinia_striiformis*, OTU_47_*Phaeosphaeria_nodorum* (*Parastagonospora nodorum*), OTU_20_*Monographella_spp* and OTU_224_*Pyrenophora_tritici-repentis*. Surprisingly, there was no significant effect of fungicide treatment on the relative abundance of any of these OTUs. On the contrary, there was a tendency for higher variability in the relative abundance of *M. graminicola* and *B. graminis* in treated samples (Fig. 6), and the share of *B. graminis* (the only member of Erysiphales) was larger in treated samples (Fig. 3). On the other hand, *P. striiformis* tended to dominate the fungal community in untreated samples and was nearly absent from the fungicide-treated samples from the same fields, only being present in two fields in the Southern area (Fig. 3). Fungicide resistance in common pathogens is an increasing problem and could be an explanation for the high variability in the relative abundance of the pathogens observed here. Resistance to strobilurines in *M. graminicola* and even more so in *B. graminis* is widespread in the Nordic and Baltic countries. In addition, resistance to demethylation inhibitor fungicides is increasing in both pathogens [10]. 454 sequencing is a semi-quantitative method only allowing quantification of the relative abundance of different OTUs [25]. Thus, we were unable to determine whether the fungicide treatment had an effect on absolute abundance of the pathogens.

In soil, DNA from dead fungal mycelia has been shown to degrade rapidly [54], but data on the rate of DNA degradation in the phyllosphere are lacking. It is possible that DNA from fungi killed by the fungicides might have been detected by the PCR. The difference in community composition between fungicide-treated and untreated samples may therefore be underestimated. However, the large difference in the relative abundance of *P. striiformis* between fungicide-treated and untreated samples does not support this hypothesis.

Fungicide use was significantly correlated with changes in the relative abundance of certain fungal taxa (Fig. 6). Several hypotheses can be put forward to explain the cause of these negative or positive correlations. First, a difference in fungicide sensitivity [20] can cause some taxa to decrease relative to others in the community. Secondly, a specific taxon may be affected by the fungicides indirectly through changes in the abundance of

Table 3. Effect of environmental and management factors on fungal community composition in the wheat phyllosphere.

Multivariate test (anova.manyglm)

Data	Factor	Res. Df	score	Pr(>score)
Abund	Intecept	41		
Abund	Area	40	78.29	0.001 ***
Abund	Fungicide treatment	39	67.38	0.001 ***
Abund	Area x fungicide treatment	38	22.53	0.546
Abund	Intercept	41		
Abund	Field	24	690.8	0.001 ***
Abund	Fungicide treatment	23	125	0.001 ***

Significance codes: *** = 0.001, Abund = 67 most abundant OTUs.

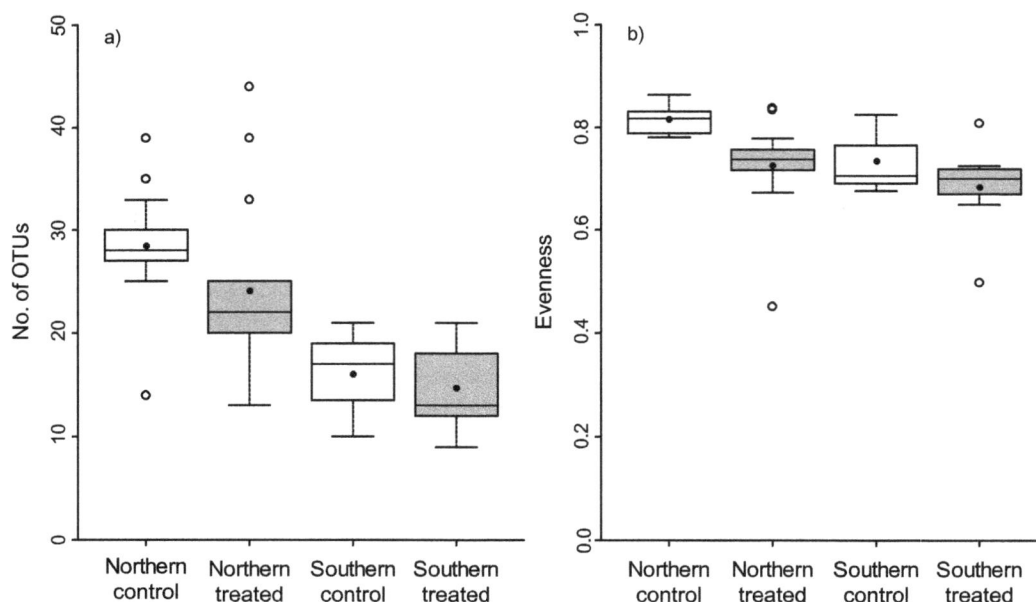

Figure 5. Richness of operational taxonomic units (OTUs) and community evenness. Boxplots with interquartile ranges of a) OTU richness and b) community evenness grouped by treatment (fungicide-treated and control samples) and geographical area. Horizontal lines represent medians and dots mean values. Samples from fields infected with yellow rust (*Puccinia striiformis*) in the Southern area (fields 15 and 16, Table S1) have been removed.

Table 4. Summary of the linear mixed model analysis of OTU richness and evenness.

OTU Richness

Random effects	Variance	Standard deviation	
Field	16.0	4.00	
Field x fungicide treatment	0.00	0.00	
Residual	36.7	6.06	

Fixed effects	Estimate	Standard error	t-value
Intercept	28.5	2.01	14.1
Fungicide treatment (treated)	−4.39	2.38	−1.84
Area (Southern)	−12.3	4.82	−2.55
Fungicide treatment x Area	2.84	5.06	0.56

Evenness

Random effects	Variance	Standard deviation	
Field	0.000	0.000	
Field x fungicide treatment	0.002	0.040	
Residual	0.004	0.065	

Fixed effects	Estimate	Standard error	t-value
Intercept	0.816	0.021	38.6
Fungicide treatment (treated)	−0.089	0.030	−2.99
Area (Southern)	−0.073	0.052	−1.40
Fungicide treatment x Area	0.025	0.066	0.380

Samples from fields infected with yellow rust (*Puccinia striiformis*) (fields 15 and 16, Table S1) were excluded.

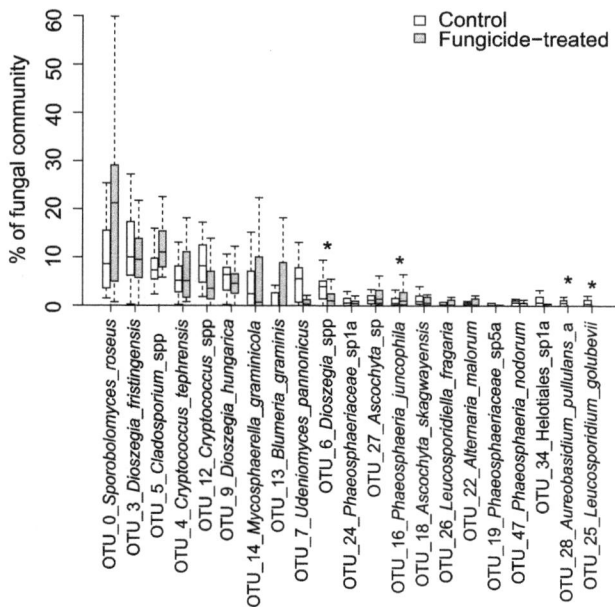

Figure 6. Distribution of community abundance for the most abundant OTUs grouped by treatment. Boxplots with interquartile ranges showing the relative abundances of the 21 most abundant operational taxonomic units (OTUs) in the dataset grouped by treatment. Outliers are not shown, OTU_1_*Puccinia_striiformis* is therefore excluded. Significant (p<0.05) differences are marked with an asterisk.

other, potentially competing, members of the community. It is also possible that taxonomic groups are closely interconnected by unknown functional interactions leading to pairwise co-occurrence or decrease in response to another underlying factor, such as fungicide-induced changes in plant physiology [55].

Spatial variation and biogeographical patterns

There were significant differences between fungal communities on wheat leaves sampled from two different areas in Sweden (Fig. 4, Table 3). The two areas were chosen as they differed in terms of climate conditions and agricultural management. The mean OTU richness per ten leaves was significantly lower (p< 0.05) in the Southern area (13.8±1.1 SE) than in the Northern (26.3±1.6 SE) (Fig. 5a, Table 4) as well as the total OTU richness in the sample pool (Fig. 2), although the Southern area was only represented by five fields. There were more fungicide-treated samples from the Southern area but the difference in overall OTU richness persisted also when comparing the same number of fungicide-treated and untreated samples in the two areas (Fig. 2b). The community evenness tended to be lower in the Southern area (Fig. 5b, Table 4), but there was no significant difference (p>0.05) when samples dominated with *P. striiformis* had been removed (Fig. S3b, Table S4).

The variation in community composition among fields was high, as field was a significant factor in the GLM analysis (Table 3). In addition, most of the OTUs (155 out of 235) only occurred in one sample in the dataset. For OTU richness, the variable field explained one third of the random variation, while for evenness, field did not explain any of the random variation (Table 4).

At the order level, Sporidiobolales had a significantly higher relative abundance in the Southern area (p<0.001), while Pleosporales (p<0.01), Helotiales (p<0.05) and the unassigned

group of OTUs (p<0.05) were relatively more abundant in the Northern area (Fig. 3). There were many OTUs in the *Phaeosphariaceae* and in the Pleosporales that did not match any known species. These may represent undescribed fungal species, but could also reflect intragenomic variation, although this phenomenon does not seem to be wide-spread in fungi [56]. At the species level, OTU_16_*Phaeosphaeria_juncophila* (p<0.05) and OTU_18_*Ascochyta_skagwayensis* (p<0.01) were relatively more abundant in the Northern than in the Southern area. Fungi in the genus *Ascochyta* can be weak pathogens on cereals or have a saprotrophic lifestyle [57].

OTU_0_*Sporobolomyces_roseus* (p<0.001) was relatively more abundant in the Southern area and it was the largest community member in that area, while OTU_3_*Dioszegia_fristingensis* was the most abundant species in the Northern area (Fig. 7). Both of these species produce pigments and ballistospores, two characters considered to be a sign of adaptation to the phyllosphere [48]. Several studies have reported *Sporobolomyces roseus* as very common on wheat leaves [5,7]. In contrast, Blixt et al. [6] only found a small proportion of this species in their study, but they selectively collected leaves diseased with *Phaeospharia nodorum*. *Dioszegia fristingensis* was described relatively recently in Germany [58], and has been reported from China [59]. It has been suggested that a group of *Dioszegia*, including *D. fristingensis*, is restricted to colder climates [58].

Climate is an important factor shaping phyllosphere communities. The Northern area in the present study had received more precipitation than the Southern area, and the relative humidity was higher on the day before sampling (Table 1). This could be a possible explanation for the higher fungal species richness observed in the Northern area. Levetin and Dorsey [60] found that rainfall was the most important factor for leaf surface fungi,

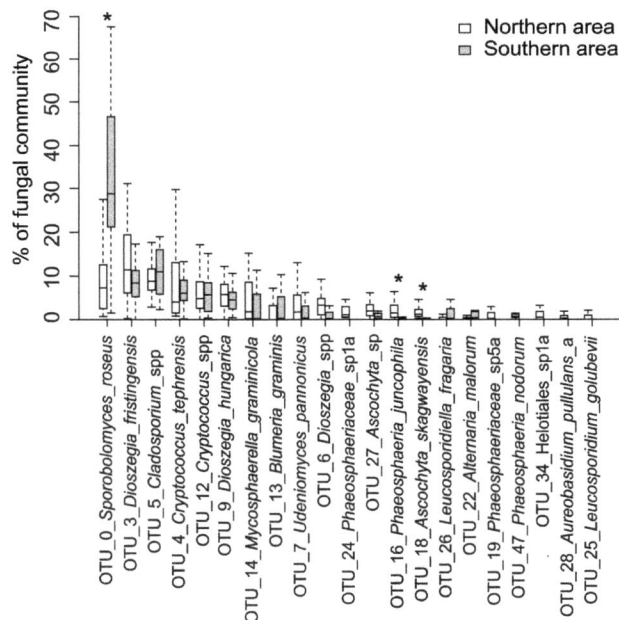

Figure 7. Distribution of community abundance for the most abundant OTUs grouped by geographical area. Boxplots with interquartile ranges showing the relative abundances of the 21 most abundant operational taxonomic units (OTUs) in the dataset grouped by geographical area. Outliers are not shown, OTU_1_*Puccinia_striiformis* is therefore excluded. Significant differences (p<0.05) are marked with an asterisk.

with the number of yeasts and *Phoma* spp. correlating positively with the amount of rainfall in their study. In our study, some *Cryptococcus* and *Dioszegia* yeasts were more abundant in the Northern area, while the opposite was true for *Sporobolomyces roseus* (Fig. 7).

The atmosphere is also an important source of phyllosphere microorganisms [61]. The local air spora is one factor influencing phyllosphere community composition in different areas. Levetin and Dorsey [60] found an overlap between fungi found in the phyllosphere and the air spora. Similarly, we identified species that are commonly found in air, e.g. *Aureobasidium pullulans* and *Cladosporium* spp. [60,62,63]. The differences we found between areas and fields indicate that local conditions were important for fungal community composition and richness in the wheat phyllosphere in this study.

Conclusions

Fungicide-use was associated with moderate but significant changes in fungal community composition on wheat leaves. Community evenness was negatively correlated with fungicide use. Fungicides had no effect on OTU richness on a per-plant basis, but there were fewer OTUs in the fungicide-treated sample pool. On the species level, the relative abundance of several saprotrophs was significantly affected in fungicide-treated samples. However, it is unclear whether the saprotrophic species that persist on treated leaves are capable of resisting and/or degrading the fungicides used, or what role they play in the control of pathogens and disease suppression. Interestingly, there was no significant difference in the relative abundance of common wheat pathogens, although *P. striiformis* tended to dominate the community in control samples when present. Further research is necessary to identify the mechanisms behind fungicide-fungi interactions in the phyllosphere of agricultural crops. Identification of the interactions between pathogenic and saprotrophic phyllosphere fungi and management practices has the potential to guide the development of sustainable disease control strategies.

Supporting Information

Figure S1   Neighbour-joining tree of the most abundant ascomycete ITS2 sequences in the dataset. The most abundant sequence in each operational taxonomic unit (OTU_x) is included together with publicly available reference sequences and selected environmental sequences. OTUs marked with an asterisk were taxonomically assigned in SCATA. Species hypothesis accession codes in the UNITE database are given when available. Dotted lines represent branches with bootstrap values lower than 70%. *Sporobolomyces roseus* is included as an outgroup.

Figure S2   Neighbour-joining tree of the most abundant basidiomycete ITS2 sequences in the dataset. The most

abundant sequence in each operational taxonomic unit (OTU_x) is included together with publicly available reference sequences and selected environmental sequences. OTUs marked with an asterisk were taxonomically assigned in SCATA. Species hypothesis accession codes in the UNITE database are given when available. Dotted lines represent branches with bootstrap values lower than 70%. *Rhizopus oryzae* and *Rhizopus microsporus* are included to form an outgroup.

Figure S3   Richness of operational taxonomic units (OTUs) and community evenness in the full dataset. Boxplots with interquartile ranges of a) OTU richness and b) community evenness grouped by treatment (fungicide-treated and control samples) and geographical area. Horizontal lines represent medians and dots mean values. Also samples from fields infected with yellow rust (*Puccinia striiformis*) in the Southern area (fields 15 and 16, Table S1) were included. F-tests with Kenward-Roger approximation showed a significant effect of geographical area on OTU richness ($p<0.001$) and of geographical area ($p<0.01$) and the interaction between treatment and area ($p<0.05$) on community evenness.

Table S1   Wheat variety, fungicide, dose and application date for wheat leaf samples collected.

Table S2   Active ingredients in fungicides used in the sampled wheat fields.

Table S3   Taxonomic assignment and sequence data for the 67 most abundant operational taxonomic units (OTUs) in the dataset.

Table S4   Summary of the linear mixed model analysis of OTU richness and evenness including all samples.

Acknowledgments

We are grateful to the Plant Protection Centres of the Swedish Board of Agriculture, Cecilia Lerenius and Tove Nilsson (Skara) and Gunilla Berg (Alnarp), for help with field sampling and data collection. We would also like to thank Maria Jonsson for technical assistance and Mikael Brandström-Durling for demultiplexing of sequence data.

Author Contributions

Conceived and designed the experiments: PP HF CS IK. Performed the experiments: IK. Analyzed the data: IK. Contributed to the writing of the manuscript: PP HF CS IK.

References

1. Newton AC, Gravouil C, Fountaine JM (2010) Managing the ecology of foliar pathogens: ecological tolerance in crops. Ann Appl Biol 157: 343–359. doi:10.1111/j.1744-7348.2010.00437.x.

2. Andrews JH, Harris RF (2000) The ecology and biogeography of microorganisms of plant surfaces. Annu Rev Phytopathol 38: 145–180. doi:10.1146/annurev.phyto.38.1.145.

3. Blakeman JP, Fokkema NJ (1982) Potential for biological control of plant diseases on the phylloplane. Annu Rev Phytopathol 20: 167–190. doi:10.1146/annurev.py.20.090182.001123.

4. Friesen ML, Porter SS, Stark SC, von Wettberg EJ, Sachs JL, et al. (2011) Microbially mediated plant functional traits. Annu Rev Ecol Evol Syst 42: 23–46. doi:10.1146/annurev-ecolsys-102710-145039.

5. Last FT (1955) Seasonal incidence of *Sporobolomyces* on cereal leaves. Trans Br Mycol Soc 38: 221–239. doi:10.1016/S0007-1536(55)80069-1.

6. Blixt E, Olson Å, Lindahl B, Djurle A, Yuen J (2010) Spatiotemporal variation in the fungal community associated with wheat leaves showing symptoms similar to stagonospora nodorum blotch. Eur J Plant Pathol 126: 373–386. doi:10.1007/s10658-009-9542-z.

7. Dickinson CH, Wallace B (1976) Effects of late applications of foliar fungicides on activity of micro-organisms on winter wheat flag leaves. Trans Br Mycol Soc 67: 103–112. doi:16/S0007-1536(76)80014-9.

8. Southwell RJ, Brown JF, Welsby SM (1999) Microbial interactions on the phylloplane of wheat and barley after applications of mancozeb and triadimefon. Australas Plant Pathol 28: 139–148. doi:10.1071/AP99024.

9. Dickinson CH (1973) Effects of ethirimol and zineb on phylloplane microflora of barley. Trans Br Mycol Soc 60: 423–431. doi:10.1016/S0007-1536(73)80027-0.

10. European Network for Durable Exploitation of crop protection strategies (ENDURE) (2008) Deliverable DR 1.2: Best control practices of diseases in winter wheat. Available: http://www.endure-network.eu/content/download/5610/43683/file/ENDURE_DR1.2.pdf.

11. Arnold AE, Mejía LC, Kyllo D, Rojas EI, Maynard Z, et al. (2003) Fungal endophytes limit pathogen damage in a tropical tree. Proc Natl Acad Sci 100: 15649–15654. doi:10.1073/pnas.2533483100.

12. Birzele B, Meier A, Hindorf H, Kramer J, Dehne H (2002) Epidemiology of *Fusarium* infection and deoxynivalenol content in winter wheat in the Rhineland, Germany. Eur J Plant Pathol 108: 667–673. doi:10.1023/A:1020632816441.

13. Henriksen B, Elen O (2005) Natural *Fusarium* grain infection level in wheat, barley and oat after early application of fungicides and herbicides. J Phytopathol 153: 214–220. doi:10.1111/j.1439-0434.2005.00955.x.

14. Bertelsen JR, de Neergaard E, Smedegaard-Petersen V (2001) Fungicidal effects of azoxystrobin and epoxiconazole on phyllosphere fungi, senescence and yield of winter wheat. Plant Pathol 50: 190–205. doi:10.1046/j.1365-3059.2001.00545.x.

15. Morton V, Staub T (2008) A Short History of Fungicides. APSnet Feature Artic. Available: https://www.apsnet.org/publications/apsnetfeatures/Pages/Fungicides.aspx. Accessed 9 September 2014.

16. Fokkema NJ (1988) Agrochemicals and the beneficial role of phyllosphere yeasts in disease control. Ecol Bull 39: 91–93.

17. Fokkema NJ, Nooij MP (1981) The effect of fungicides on the microbial balance in the phyllosphere. EPPO Bull 11: 303–310. doi:10.1111/j.1365-2338.1981.tb01937.x.

18. Magan N, Lacey J (1986) The phylloplane microflora of ripening wheat and effect of late fungicide applications. Ann Appl Biol 109: 117–128. doi:10.1111/j.1744-7348.1986.tb03190.x.

19. Rastogi G, Coaker GL, Leveau JHJ (2013) New insights into the structure and function of phyllosphere microbiota through high-throughput molecular approaches. FEMS Microbiol Lett 348: 1–10. doi:10.1111/1574-6968.12225.

20. Čadež N, Zupan J, Raspor P (2010) The effect of fungicides on yeast communities associated with grape berries. FEMS Yeast Res 10: 619–630. doi:10.1111/j.1567-1364.2010.00635.x.

21. Moulas C, Petsoulas C, Rousidou C, Perruchon C, Karas P, et al. (2013) Effects of systemic pesticides Imidacloprid and Metalaxyl on the phyllosphere of pepper plants. Biomed Res Int: 969750. doi:10.1155/2013/969750.

22. Pinto C, Pinho D, Sousa S, Pinheiro M, Egas C, et al. (2014) Unravelling the Diversity of Grapevine Microbiome. PLoS ONE 9: e85622. doi:10.1371/journal.pone.0085622.

23. Sverige, Statens jordbruksverk, Sverige, Statistiska centralbyran (2012) Jordbruksstatistisk arsbok 2012: med data om livsmedel = Yearbook of Agricultural statistics. 2012. Stockholm, Orebro, Publikationstjansten, SCB.

24. Statistiska centralbyran (2011) Växtskyddsmedel i jord- och trädgårdsbruket 2010 Användning i grödor = Plant protection products in agriculture and horticulture. Use in crops. Available: http://www.scb.se/statistik/mi/mi0502/2009i10/mi0502_2009i10_sm_mi31sm1101.pdf.

25. Ihrmark K, Bödeker ITM, Cruz-Martinez K, Friberg H, Kubartova A, et al. (2012) New primers to amplify the fungal ITS2 region – evaluation by 454-sequencing of artificial and natural communities. FEMS Microbiol Ecol 82: 666–677. doi:10.1111/j.1574-6941.2012.01437.x.

26. White TJ, Bruns SL, Taylor JW (1990) Amplification and direct sequencing of fungal ribosomal RNA genes for phylogenetics. In: Innis MA, Gefland D, Sninsky J, White TJ, editors. PCR Protocols: A Guide to Methods and Applications. San Diego, CA: Academic Press. pp. 315–322.

27. Durling MB, Clemmensen KE, Stenlid J, Lindahl B (2011) SCATA - An efficient bioinformatic pipeline for species identification and quantification after high-throughput sequencing of tagged amplicons (submitted). Available: http://scata.mykopat.slu.se/.

28. Nilsson RH, Kristiansson E, Ryberg M, Hallenberg N, Larsson K-H (2008) Intraspecific ITS variability in the kingdom Fungi as expressed in the international sequence databases and its implications for molecular species identification. Evol Bioinforma Online 4: 193–201.

29. Tedersoo L, Nilsson RH, Abarenkov K, Jairus T, Sadam A, et al. (2010) 454 Pyrosequencing and Sanger sequencing of tropical mycorrhizal fungi provide similar results but reveal substantial methodological biases. New Phytol 188: 291–301. doi:10.1111/j.1469-8137.2010.03373.x.

30. Carlsen T, Aas AB, Lindner D, Vrålstad T, Schumacher T, et al. (2012) Don't make a mista(g)ke: is tag switching an overlooked source of error in amplicon pyrosequencing studies? Fungal Ecol 5: 747–749. doi:10.1016/j.funeco.2012.06.003.

31. Kõljalg U, Nilsson RH, Abarenkov K, Tedersoo L, Taylor AFS, et al. (2013) Towards a unified paradigm for sequence-based identification of fungi. Mol Ecol 22: 5271–5277. doi:10.1111/mec.12481.

32. Katoh K, Misawa K, Kuma K, Miyata T (2002) MAFFT: a novel method for rapid multiple sequence alignment based on fast Fourier transform. Nucleic Acids Res 30: 3059–3066. doi:10.1093/nar/gkf436.

33. Dereeper A, Guignon V, Blanc G, Audic S, Buffet S, et al. (2008) Phylogeny.fr: robust phylogenetic analysis for the non-specialist. Nucleic Acids Res 36: W465–W469. doi:10.1093/nar/gkn180.

34. Oksanen J, Blanchet FG, Kindt R, Legendre P, Minchin PR, et al. (2013) vegan: Community Ecology Package. Available: http://CRAN.R-project.org/package=vegan.

35. Pielou EC (1966) The measurement of diversity in different types of biological collections. J Theor Biol 13: 131–144. doi:10.1016/0022-5193(66)90013-0.

36. Bates D, Mächler M, Bolker B, Walker S (2014) Fitting Linear Mixed-Effects Models using lme4. ArXiv14065823 Stat. Available: http://arxiv.org/abs/1406.5823. Accessed 16 September 2014.

37. Halekoh U, Højsgaard S (2014) A Kenward-Roger Approximation and Parametric Bootstrap Methods for Tests in Linear Mixed Models – The R Package pbkrtest. J Stat Softw 59. Available: http://www.jstatsoft.org/v59/i09.

38. Wang Y, Naumann U, Wright ST, Warton DI (2012) mvabund- an R package for model-based analysis of multivariate abundance data. Methods Ecol Evol 3: 471–474. doi:10.1111/j.2041-210X.2012.00190.x.

39. Maleszka R, Clark-Walker GD (1993) Yeasts have a four-fold variation in ribosomal DNA copy number. Yeast 9: 53–58. doi:10.1002/yea.320090107.

40. Rodland KD, Russell PJ (1982) Regulation of ribosomal RNA cistron number in a strain of *Neurospora crassa* with a duplication of the nucleolus organizer region. Biochim Biophys Acta BBA - Gene Struct Expr 697: 162–169. doi:10.1016/0167-4781(82)90072-0.

41. Liti G, Carter DM, Moses AM, Warringer J, Parts L, et al. (2009) Population genomics of domestic and wild yeasts. Nature 458: 337–341. doi:10.1038/nature07743.

42. Adams RI, Amend AS, Taylor JW, Bruns TD (2013) A unique signal distorts the perception of species richness and composition in high-throughput sequencing surveys of microbial communities: a case study of fungi in indoor dust. Microb Ecol 66: 735–741. doi:10.1007/s00248-013-0266-4.

43. Takashima M, Deak T, Nakase T (2001) Emendation of *Dioszegia* with redescription of *Dioszegia hungarica* and two new combinations, *Dioszegia aurantiaca* and *Dioszegia crocea*. J Gen Appl Microbiol 47: 75–84. doi:10.2323/jgam.47.75.

44. Renker C, Blanke V, Borstler B, Heinrichs J, Buscot F (2004) Diversity of *Cryptococcus* and *Dioszegia* yeasts (Basidiomycota) inhabiting arbuscular mycorrhizal roots or spores. Fems Yeast Res 4: 597–603. doi:10.1016/j.fe.syr.2004.01.001.

45. Smit E, Leeflang P, Glandorf B, van Elsas JD, Wernars K (1999) Analysis of fungal diversity in the wheat rhizosphere by sequencing of cloned PCR-amplified genes encoding 18S rRNA and temperature gradient gel electrophoresis. Appl Environ Microbiol 65: 2614–2621.

46. Sampaio JP, Gadanho M, Bauer R, Weiß M (2003) Taxonomic studies in the Microbotryomycetidae: *Leucosporidium golubevii* sp. nov., *Leucosporidiella* gen. nov. and the new orders Leucosporidiales and Sporidiobolales. Mycol Prog 2: 53–68. doi:10.1007/s11557-006-0044-5.

47. Balint M, Tiffin P, Hallstroem B, O'Hara RB, Olson MS, et al. (2013) Host genotype shapes the foliar fungal microbiome of balsam poplar (Populus balsamifera). Plos One 8: e53987. doi:10.1371/journal.pone.0053987.

48. Fonseca A, Inacio J (2006) Phylloplane yeasts. In: Rosa C, Peter G, editors. Biodiversity and ecophysiology of yeasts. The yeast handbook. Berlin Heidelberg: Springer. pp. 263–301.

49. Pitt JI, Hocking AD (2009) Fungi and food spoilage. US: Springer. 524 p.

50. Sylla J, Alsanius B, Krüger E, Reineke A, Bischoff-Schaefer M, et al. (2013) Introduction of *Aureobasidium pullulans* to the Phyllosphere of Organically Grown Strawberries with Focus on Its Establishment and Interactions with the Resident Microbiome. Agronomy 3: 704–731. doi:10.3390/agronomy3040704.

51. De Curtis F, De Cicco V, Lima G (2012) Efficacy of biocontrol yeasts combined with calcium silicate or sulphur for controlling durum wheat powdery mildew and increasing grain yield components. Field Crops Res 134: 36–46. doi:10.1016/j.fcr.2012.04.014.

52. Castoria R, De Curtis F, Lima G, Caputo L, Pacifico S, et al. (2001) *Aureobasidium pullulans* (LS-30) an antagonist of postharvest pathogens of fruits: study on its modes of action. Postharvest Biol Technol 22: 7–17. doi:10.1016/S0925-5214(00)00186-1.

53. Ikai K, Takesako K, Shiomi K, Moriguchi M, Umeda Y, et al. (1991) Structure of Aureobasidin-A. J Antibiot (Tokyo) 44: 925–933. doi:10.7164/antibiotics.44.925.

54. Herdina, Neate S, Jabaji-Hare S, Ophel-Keller K (2006) Persistence of DNA of *Gaeumannomyces graminis* var. *tritici* in soil as measured by a DNA-based assay. FEMS Microbiol Ecol 47. doi:10.1016/S0168-6496(03)00255-1.

55. Petit A-N, Fontaine F, Vatsa P, Clément C, Vaillant-Gaveau N (2012) Fungicide impacts on photosynthesis in crop plants. Photosynth Res 111: 315–326. doi:10.1007/s11120-012-9719-8.

56. Lindner DL, Carlsen T, Henrik Nilsson R, Davey M, Schumacher T, et al. (2013) Employing 454 amplicon pyrosequencing to reveal intragenomic divergence in the internal transcribed spacer rDNA region in fungi. Ecol Evol 3: 1751–1764. doi:10.1002/ece3.586.

57. Cromey MG, Butler RC, Mace MA, Cole ALJ (2004) Effects of the fungicides azoxystrobin and tebuconazole on *Didymella exitialis*, leaf senescence and grain yield in wheat. Crop Prot 23: 1019–1030. doi:10.1016/j.cropro.2004.03.002.

58. Inacio J, Portugal L, Spencer-Martins I, Fonseca A (2005) Phylloplane yeasts from Portugal: Seven novel anamorphic species in the Tremellales lineage of the Hymenomycetes (Basidiomycota) producing orange-coloured colonies. Fems Yeast Res 5: 1167–1183. doi:10.1016/j.femsyr.2005.05.007.

59. Wang Q-M, Jia J-H, Bai F-Y (2008) Diversity of basidiomycetous phylloplane yeasts belonging to the genus *Dioszegia* (Tremellales) and description of

Dioszegia athyri sp nov., *Dioszegia butyracea* sp nov and *Dioszegia xingshanensis* sp nov. Antonie Van Leeuwenhoek Int J Gen Mol Microbiol 93: 391–399. doi:10.1007/s10482-007-9216-9.

60. Levetin E, Dorsey K (2006) Contribution of leaf surface fungi to the air spora. Aerobiologia 22: 3–12. doi:10.1007/s10453-005-9012-9.

61. Kinkel LL (1997) Microbial population dynamics on leaves. Annu Rev Phytopathol 35: 327–347. doi:10.1146/annurev.phyto.35.1.327.

62. Fröhlich-Nowoisky J, Pickersgill DA, Després VR, Pöschl U (2009) High diversity of fungi in air particulate matter. Proc Natl Acad Sci 106: 12814–12819. doi:10.1073/pnas.0811003106.

63. Pitkäranta M, Meklin T, Hyvärinen A, Paulin L, Auvinen P, et al. (2008) Analysis of fungal flora in indoor dust by ribosomal DNA sequence analysis, quantitative PCR, and culture. Appl Environ Microbiol 74: 233–244. doi:10.1128/AEM.00692-07.

64. Agrios GN (2004) Plant pathology. 5th ed. Amsterdam; Boston: Elsevier Academic Press. 922 p.

65. Kurtzman CP, Fell JW, editors (2000) The yeasts: a taxonomic study. 4th ed. Amsterdam; New York: Elsevier. 1055 p.

66. Goetz J, Dugan FM (2006) *Alternaria malorum*: a mini-review with new records for hosts and pathogenicity. Pac Northwest Fungi 1.

67. Crous PW, Verkley GJM, Groenewald JZ (2006) Eucalyptus microfungi known from culture. 1. *Cladoriella* and *Fulvoflamma* genera nova, with notes on some other poorly known taxa. Stud Mycol 55: 53–63. doi:10.3114/sim.55.1.53.

14

Biofabricated Silver Nanoparticles Act as a Strong Fungicide against *Bipolaris sorokiniana* Causing Spot Blotch Disease in Wheat

Sandhya Mishra[1,9]**, Braj Raj Singh**[2,9]**, Akanksha Singh**[1]**, Chetan Keswani**[1]**, Alim H. Naqvi**[2]**, H. B. Singh**[1]*

1 Department of Mycology and Plant Pathology, Institute of Agricultural Sciences, Banaras Hindu University, Varanasi, India, **2** Centre of Excellence in Materials Science (Nanomaterials), Department of Applied Physics, Z. H. College of Engineering and Technology, Aligarh Muslim University, Aligarh, India

Abstract

The present study is focused on the extracellular synthesis of silver nanoparticles (AgNPs) using culture supernatant of an agriculturally important bacterium, *Serratia* sp. BHU-S4 and demonstrates its effective application for the management of spot blotch disease in wheat. The biosynthesis of AgNPs by *Serratia* sp. BHU-S4 (denoted as bsAgNPs) was monitored by UV–visible spectrum that showed the surface plasmon resonance (SPR) peak at 410 nm, an important characteristic of AgNPs. Furthermore, the structural, morphological, elemental, functional and thermal characterization of bsAgNPs was carried out using the X-ray diffraction (XRD), electron and atomic microscopies, energy dispersive X-ray (EDAX) spectrometer, FTIR spectroscopy and thermogravimetric analyzer (TGA), respectively. The bsAgNPs were spherical in shape with size range of ~10 to 20 nm. The XRD and EDAX analysis confirmed successful biosynthesis and crystalline nature of AgNPs. The bsAgNPs exhibited strong antifungal activity against *Bipolaris sorokiniana*, the spot blotch pathogen of wheat. Interestingly, 2, 4 and 10 μg/ml concentrations of bsAgNPs accounted for complete inhibition of conidial germination, whereas in the absence of bsAgNPs, conidial germination was 100%. A detached leaf bioassay revealed prominent conidial germination on wheat leaves infected with *B. sorokiniana* conidial suspension alone, while the germination of conidia was totally inhibited when the leaves were treated with bsAgNPs. The results were further authenticated under green house conditions, where application of bsAgNPs significantly reduced *B. sorokiniana* infection in wheat plants. Histochemical staining revealed a significant role of bsAgNPs treatment in inducing lignin deposition in vascular bundles. In summary, our findings represent the efficient application of bsAgNPs in plant disease management, indicating the exciting possibilities of nanofungicide employing agriculturally important bacteria.

Editor: Jae-Hyuk Yu, University of Wisconsin - Madison, United States of America

Funding: The authors have no support or funding to report.

Competing Interests: The authors have declared that no competing interests exist.

* E-mail: hbs1@rediffmail.com

9 These authors contributed equally to this work.

Introduction

Agricultural production is constantly suffering from plethora of threats instigated by various classes of plant pathogens leading to huge economic losses. Consequently, innovative technologies are being introduced in modern agriculture to minimise such losses. Among such technological innovations, nanotechnology is gathering noteworthy considerations due to its robust applications in agriculture [1,2]. In general, nanoparticles have attracted considerable attention in the fields of medicine, pharmaceuticals, cosmetics and electronics due to unique physical, chemical and biological properties [3]. Contemporarily, the development and application of biosynthesized nanoparticles has opened new avenues in agricultural research oriented to developing ecofriendly and effective means of controlling plant diseases. Most importantly, biological synthesis of silver nanoparticles (AgNPs) has offered a consistent, non toxic and ecofriendly approach for plant disease management due to their strong antimicrobial properties [4–6].

Antimicrobial property of AgNPs has been exploited to a great extent against a broad range of human pathogens [7–10].

However, the full potential of AgNPs is still largely unexplored for crop protection and presently there has been a growing interest to utilize their antimicrobial property for plant disease management. Earlier findings have advocated the toxicity of AgNPs on fungal hyphae and conidial development [11,12]. In recent years, *in vitro* assays conducted by Krishnaraj and colleagues [13] and Gopinath and Velusamy [14] showed the strong inhibitory effects of biosynthesized AgNPs against various fungal plant pathogens. In another study, AgNPs synthesized using cow milk, have also shown strong antifungal activity against a range of phytopathogens [15].

The above reports clearly provide substantial basis for exploring the possibility of an efficient application of AgNPs in controlling phytopathogens under *in vitro* conditions. However, a rigorous *in vivo* assessment is essential to authenticate their functional applications under field conditions. Furthermore, considering the significance of agriculturally important microbes, their use for synthesizing AgNPs with strong antimicrobial properties can certainly provide an alternate means for plant protection. Therefore, the prime objective of the present study is to use a plant growth promoting rhizobacterium (PGPR) for synthesis of AgNPs and to evaluate the biocontrol potential of bsAgNPs both under *in vitro* and *in vivo* conditions. In this regard, we report a newly isolated *Serratia* sp. BHU-S4 from agricultural soil that has the potential to synthesize AgNPs with strong antifungal activity against *Bipolaris sorokiniana*, the spot blotch pathogen of wheat. Spot blotch is considered as one of the most dreadful disease causing 20% yield loss in South Asia [16], while in India, yield loss has been estimated upto 22% [17]. The increasing threat of spot blotch pathogen leads to 80% disease severity and the losses could be as high as 100% under severe conditions [18,19]. Hence, the present study was initiated with an aim to determine the efficiency of biosynthesized AgNPs (denoted as bsAgNPs) in controlling infection of *B. sorokiniana* which causes spot blotch disease in wheat. Apart from *in vitro* assays, plant test under greenhouse condition was also conducted to ascertain the efficiency of bsAgNPs in controlling disease incidence.

Materials and Methods

Isolation and Identification of Bacterial Strain BHU-S4

Bacteria were isolated from soil samples collected from agricultural fields of the Institute of Agricultural Sciences, Banaras Hindu University campus, Varanasi (25°20′N latitude & 83°01′E longitude), Ramnagar (25°18′N latitude & 83°02′E longitude), Mirzapur (25°10′N latitude & 82°37′E longitude) and Chunar (25°07′N latitude & 82°54′E longitude), India. The soil samples were subjected to serial dilution and plated on Nutrient Agar (NA) medium. After incubation at 30°C for 24 h, bacterial colonies were subcultured and further purified on NA. On the basis of rapid reduction of $AgNO_3$ into AgNPs, the bacterial strain BHU-S4 was selected for further studies. The preliminary characterization of the bacterial strain BHU-S4 based on physiological and biochemical characteristics was carried out according to Bergey's Manual of Determinative Bacteriology [20]. Further authentic taxonomic characterization of bacterial strain BHU-S4 was done by 16S rDNA gene sequence homology and phylogenetic analysis [21,22].

The qualitative plant growth promoting activities of the strain BHU-S4 were also ascertained. The inorganic phosphate solubilizing ability was determined using NBRIP-BPB medium as described earlier [23]. The indole acetic acid (IAA) production was determined according to the method described by Brick et al. [24] and siderophore production was tested using Chromazural S (CAS) assay as described by Dwivedi et al. [25].

Extracellular Biosynthesis of AgNPs by *Serratia* sp. BHU-S4

The bacterial strain BHU-S4 was inoculated in Nutrient Broth (NB) medium and incubated at 30°C for 48 h to attain the early stationary phase. The culture supernatant was obtained by centrifugation at 10,000 rpm for 10 min and transferred to another sterile conical flask. The fresh stock of silver nitrate ($AgNO_3$) was prepared in sterile distilled water and added to the culture supernatant at the final concentration of 1 mM. The conical flask was incubated in dark and the synthesis of AgNPs was monitored by visual color change from yellow (original color of culture supernatant) to dark brown (color of culture supernatant after adding $AgNO_3$) [14,26]. The bsAgNPs were air dried in sterile conditions and obtained in the form of powder for further studies.

Characterization of bsAgNPs

The optical, structural, morphological, elemental, functional and thermal characterization of the bsAgNPs was carried out using the UV–visible spectrophotometery, X-ray diffraction (XRD), electron and atomic microscopies, energy dispersive X-ray (EDAX) spectrometer, Fourier transform infrared (FTIR) spectroscopy and thermogravimetric analyzer, respectively. In order to ascertain the optical characteristics of bsAgNPs in reaction solution, the absorption spectrum was recorded by UV–visible spectrophotometer (Perkin Elmer Life and Analytical Sciences, CT, USA) in the wavelength range of A_{200} to $A_{600 \text{ nm}}$ using quartz cuvette [27]. The structural characteristic of bsAgNPs powder sample was recorded on MiniFlex II benchtop XRD system (Rigaku Corporation, Tokyo, Japan) operating at 40 kV [22]. The scanning electron microscopy (SEM) was carried out using fine powder of the bsAgNPs on a carbon tape in JSM 6510LV scanning electron microscope (JEOL, Tokyo, Japan) at an accelerating voltage of 20 kV. The elemental analysis of bsAgNPs was done using Oxford Instruments INCAx-sight EDAX spectrometer equipped SEM [28]. The transmission electron microscopy (TEM) of bsAgNPs was carried out on JEOL 100/120 kV TEM (JEOL, Tokyo, Japan) with an accelerating voltage of 200 kV [29]. For atomic force microscopy, a thin film of bsAgNPs was prepared on the borosilicate glass slide to analyse the surface morphology. The prepared thin film was analysed on the atomic force microscope (AFM; Innova SPM, Veeco) in the tapping mode. The commercial etched silicon tips were used as scanning probes with typical resonance frequency of 300 Hz (RTESP, Veeco). The microscope was placed on a pneumatic anti-vibration desk, under a damping cover and the analysis was done using SPM Lab software [30,31]. The images of electron microscopies of EDAX and AFM were obtained and converted into an enhanced meta file format. For the functional characterization of the bsAgNPs, the powder was mixed with spectroscopic grade potassium bromide (KBr) in the ratio of 1:100 and spectra was recorded in the range of 400–4000 wavenumber (cm^{-1}) on Perkin Elmer FT-IR spectrometer Spectrum Two (Perkin Elmer Life and Analytical Sciences, CT, USA) in the diffuse reflectance mode at a resolution of 4 cm^{-1} in KBr pellets [32]. The functionalization and thermal stability of the bsAgNPs was investigated by thermogravimetric analysis (TGA) (Perkin Elmer Pyris 1 TGA Thermogravimetric Analyzer) at a heating rate of 10°C/min under nitrogen atmosphere [33].

Antagonistic Potential of bsAgNPs against *Bipolaris sorokiniana*

Cavity slide experiment. A cavity slide experiment was designed in triplicate to evaluate the antagonistic potential of bsAgNPs against *B. sorokiniana*. Conidial suspension of *B. sorokiniana* was prepared in sterile distilled water from 10 days old pure culture under aseptic conditions. The conidial suspension (40 μl) was mixed with different concentrations (2, 4 and 10 μg/ml) of bsAgNPs and filled in each cavity. The control set was maintained separately by filling the cavity with conidial suspension without bsAgNPs. Three replicates each for control and treated set were maintained. The cavity slides were incubated at 25±2°C in moist chambers (Petri dishes, 90 mm diameter, containing blotting papers soaked in sterile distilled water) for 24 h. After incubation period, slides were examined under microscope to evaluate inhibitory effect of bsAgNPs on conidia germination of *B. sorokiniana*.

Detached leaf assay. Wheat leaves were washed with sterile distilled water, surface sterilized by dipping into 1% sodium hypochlorite for 30 sec and further rinsed 5 times with sterile

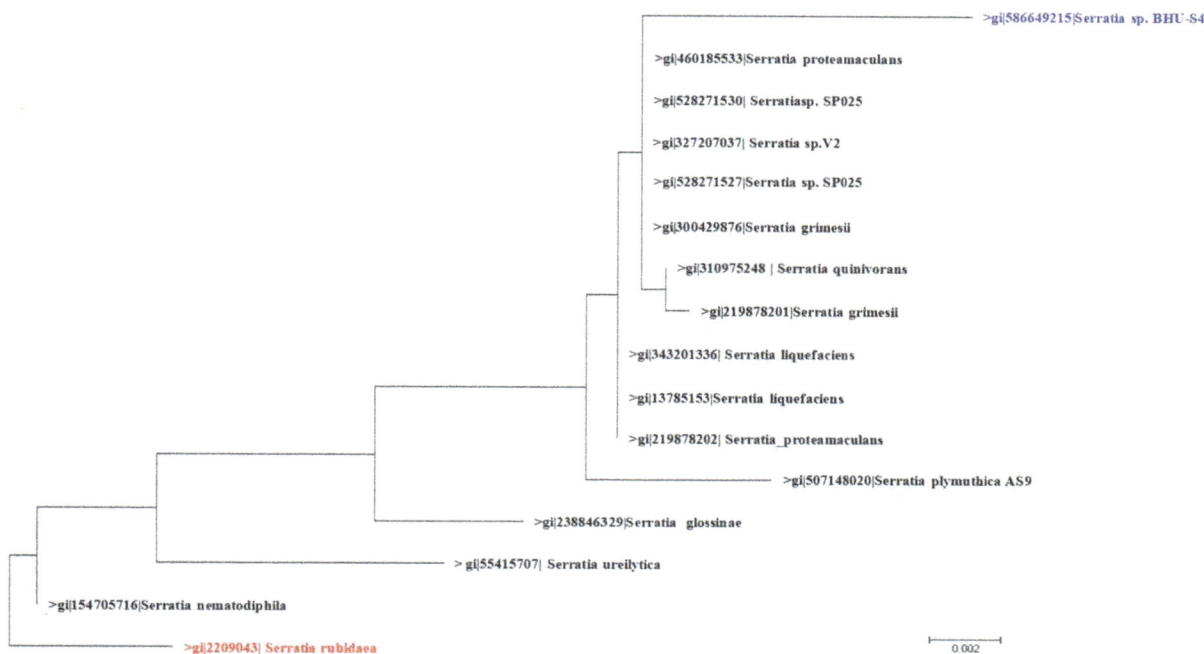

Figure 1. The phylogenetic tree of *Serratia* **sp. BHU-S4 based on 16S rDNA sequence.** The partial 16S rDNA sequence has been deposited in NCBI GenBank, nucleotide sequence database under the accession number KF863906. *Bar* 0.002 substitutions per site.

water under aseptic conditions. Leaves were placed in a petriplate containing 0.5% agar and inoculated (5 spots per leaf) by pipetting 10 μl droplets of *B. sorokiniana* conidial suspension mixed with 4 μg/ml of bsAgNPs. Leaves spotted with 10 μl droplets of *B. sorokiniana* conidial suspension served as positive control.

Greenhouse experiment. Plant test was performed under greenhouse conditions using wheat as a host plant. Wheat (*Triticum aestivum*) var. HUW-234 seeds were sown in pots (9 cm diameter) containing sterilized soil and irrigated with water to maintain 20% soil moisture. The plant test consisted of following treatments: control (C), pathogen challenged control (BC), pathogen challenged + bsAgNPs treated (B4). For each treatment, six pots were used and ten seeds were sown in each pot. After 15 days of growth, treatment of pathogen and bsAgNPs was given. *B. sorokiniana* conidial suspension (containing 10^9 conidia) was prepared in sterile distilled water. Conidial suspension was mixed with bsAgNPs suspension in 1:1 ratio and sprayed over the plants while plants sprayed with conidial suspension alone served as pathogen challenged control. Plants without conidial suspension and

bsAgNPs treatments served as healthy control. Pathogen challenged plants were covered with polybags for 48 h to maintain the humidity level favoring pathogen infection. Data on plant growth parameters was recorded after 30 days of sowing and subjected to one-way ANOVA followed by Waller-Duncan posthoc test at $p < 0.05$ using SPSS 16.0.

SDS-PAGE Analysis

The total protein from wheat leaf samples was isolated using the protocol described by Sarkar et al. [34]. The concentration of protein was measured by the Bradford method [35]. An amount of 100 μg protein was resolved in 12% polyacrylamide gel and further stained with CBB G-250 (BioRad).

Histochemical Staining

After harvesting the wheat plants, transverse sections of the stem were cut and mounted on a glass slide in aqueous solution of phloroglucinol in 20% HCl and observed under light microscope

Table 1. Physiological and biochemical characteristics of *Serratia* sp. BHU-S4.

Characteristics	*Serratia* sp. BHU-S4
Gram reaction	G−ve
Auxin production	+
Phosphate solubilization	+
Siderophore production	+
Catalase	+
Proteolytic activity	+
Cellulolytic activity	+

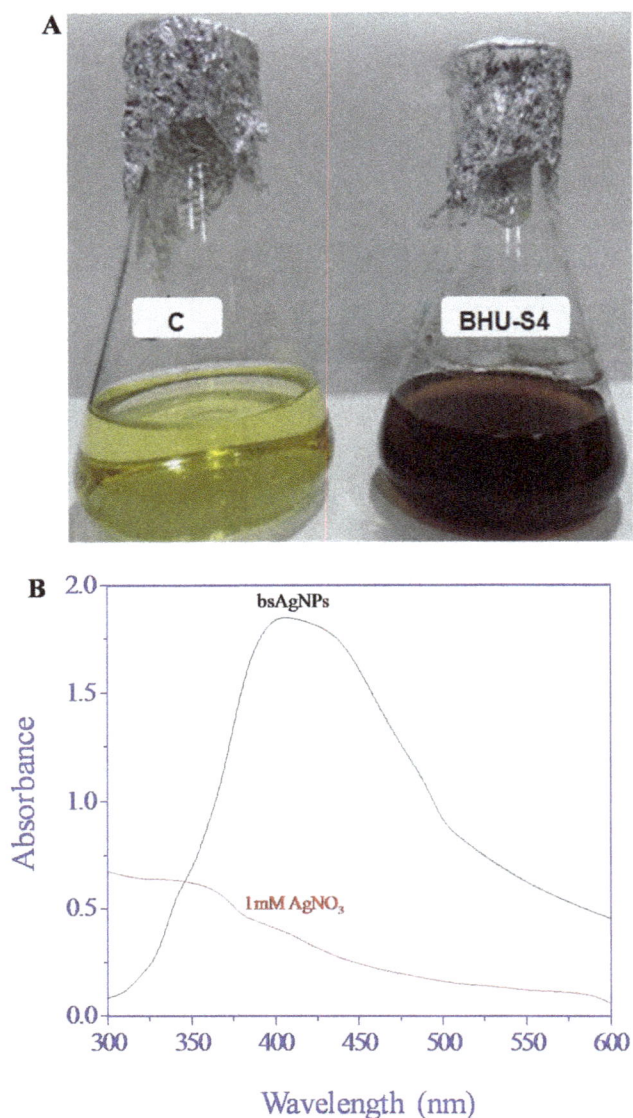

Figure 2. Biosynthesis of AgNPs by supernatant of *Serratia* sp. BHU-S4 (A). Culture supernatant without 1 mM AgNO₃ showed no color change (C) and after adding 1 mM AgNO₃ showed visual color change from yellow to dark brown. (B) UV-visible absorption spectrum of bsAgNPs and 1 mM aqueous solution of AgNO₃.

(Nikon DS-fi1, Japan) for lignin staining indicated by red-violet color [36].

Ethics Statement

The locations for the collection of soil samples are very common agricultural fields, so it did not need specific permission. It is also confirmed that these fields did not involve any endangered or protected species. We used sampling procedures that did not harm the plant diversity of the locations.

Results and Discussion

In the present study, the bacterial strain BHU-S4 was isolated from agricultural soil with the aim of exploiting its AgNPs synthesizing potential for agricultural purposes. Studies related to

its plant growth promoting characters revealed its inorganic phosphate solubilization, IAA and siderophore production activities (Table 1). This bacterial strain was identified as a member of *Serratia* species by 16S rDNA gene sequencing, which has been deposited in NCBI GenBank (Accession Number: KF863906). The information obtained by the BLAST program and phylogenetic analysis indicated a close genetic relatedness of strain BHU-S4 with *Serratia* sp. and was therefore designated as *Serratia* sp. BHU-S4 (Figure 1). The species of genus *Serratia* are well known plant growth promoting rhizobacteria with biocontrol potential against various phytopathogens [37–39]. Subsequently, extracellular AgNPs synthesis by *Serratia* sp. BHU-S4 was performed and monitored by UV–visible spectrum that showed a strong and broad surface plasmon resonance (SPR) peak at 410 nm, which is a characteristic of AgNPs (Figure 2). However, this characteristic peak was absent in the aqueous solution of AgNO₃. It is well known that due to Mie scattering, AgNPs exhibit absorption at the wavelength range of 390 to 420 nm [27,40]. As evident from previous reports, the presence of single SPR peak indicates spherical shape of AgNPs which was further confirmed by XRD and electron microscopy [14,41].

Characterization of Biosynthesized AgNPs

The crystalline nature of bsAgNPs was characterized by XRD with Cu Kα radiation ($\lambda = 0.15418$ nm). The data revealed that the well resolved Bragg reflections were obtained at $2\theta = 38.4°$, $44.5°$, $64.6°$ and $76.9°$ which correspond to the crystal lattice planes [111], [200], [220] and [311] of face centered cubic (*fcc*) structures of silver (JCPDS files No. 03-0921), respectively (Figure 3A). The unassigned Bragg reflections at $2\theta = 28.1°$ and $32.9°$ marked with green circles are probably due to the crystallization of bioorganic phase that occurs on the surface of AgNPs. The XRD data indicated the successful synthesis of bsAgNPs in this study. The average crystallite size (*d*) of synthesized bsAgNPs was calculated following the Debye-Scherrer formula:

$$D = k\lambda/\beta\cos\theta$$

where, $k = 0.9$ is the shape factor, λ is the X-ray wavelength of Cu Kα radiation (1.54 Å), θ is the Bragg diffraction angle, and β is the full width at half maximum (FWHM) of the (111) plane diffraction peak. The average crystallite size was estimated to be 18.9 nm. The nanostructure of bsAgNPs was analyzed by SEM, TEM and AFM. The Figure 3B shows the scanning electron micrograph recorded from powder of bsAgNPs deposited on a carbon tape. The micrograph clearly demonstrated the aggregation of bsAgNPs into microparticles in powdered form. However, the TEM image showed a noticeable spherical morphology of bsAgNPs with a size range of ~10–20 nm (Figure 3C), which was found to be consistent with the results obtained from XRD [42,43]. TEM characterization revealed that bsAgNPs were smooth, well dispersed and crystalline in nature. However, agglomeration might be due to the presence of biological macromolecules on the surface of bsAgNPs. The AFM images also showed the irregular spherical surface morphology of bsAgNPs in the size range of ~8 to 22 nm (Figure 3D and 3E), which is in accordance with the results obtained by XRD, SEM and TEM. The irregular spherical surface morphology of AgNPs observed in this study might be due to the biological macromolecules like proteins and enzymes present on the bsAgNPs surface. The elemental analysis using EDAX indicated the presence of mainly silver (Ag) in the bsAgNPs (Figure 4A), which reconfirmed that bsAgNPs were indeed metallic and crystalline in nature. Further, the occurrence

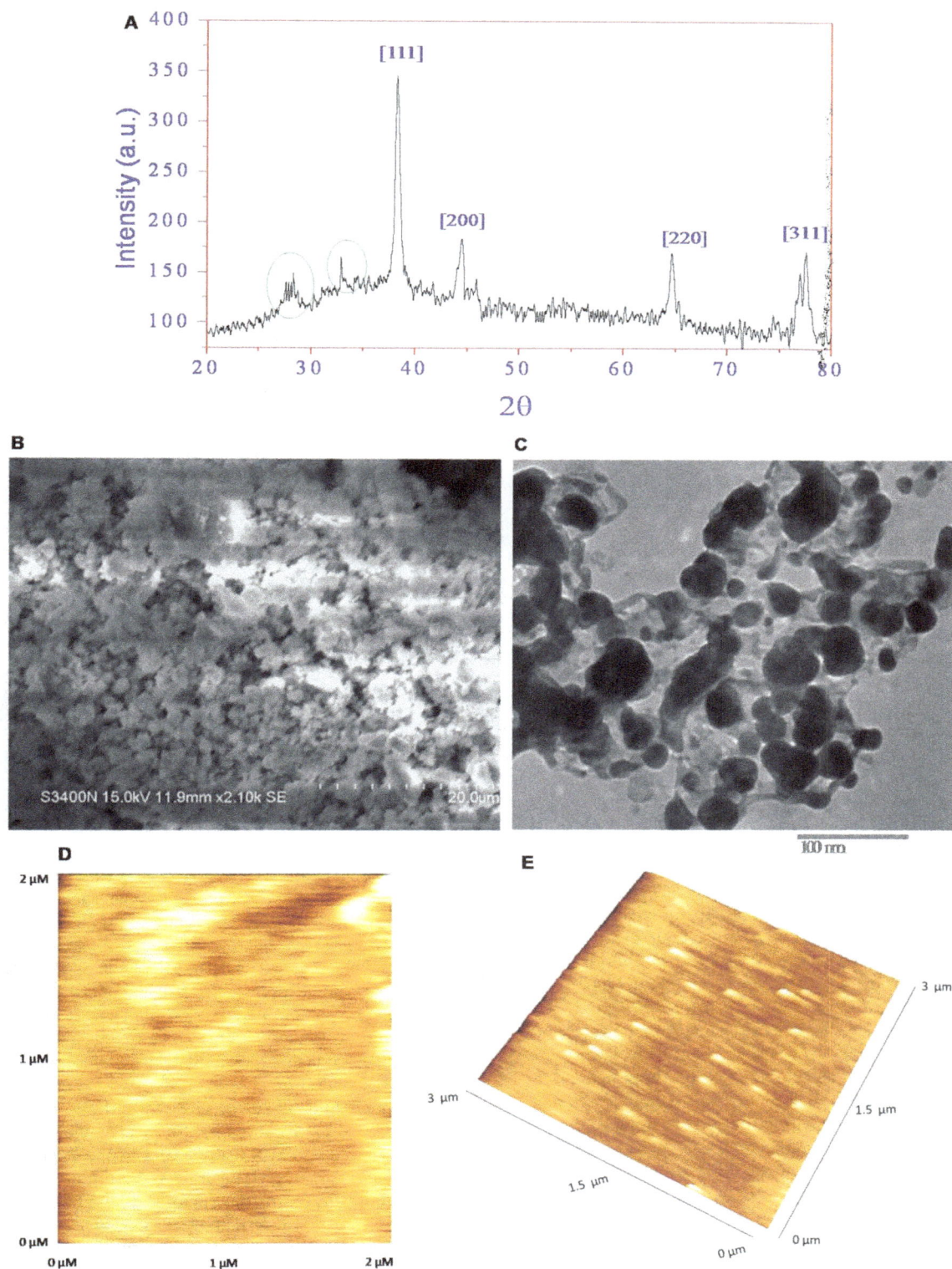

Figure 3. X-ray diffraction pattern (A), SEM (B) and TEM (C) micrographic images of bsAgNPs by *Serratia* sp. BHU-S4. AFM micrographs 2D (D) and 3D (E) illustrating the nanostructure of bsAgNPs.

of carbon and oxygen peaks revealed the presence of organic moieties on the surface of bsAgNPs.

The surface functionalization and stabilization of the bsAgNPs by the biological macromolecules present in the *Serratia* sp. BHU-S4 culture supernatant was confirmed by FTIR spectroscopy. The

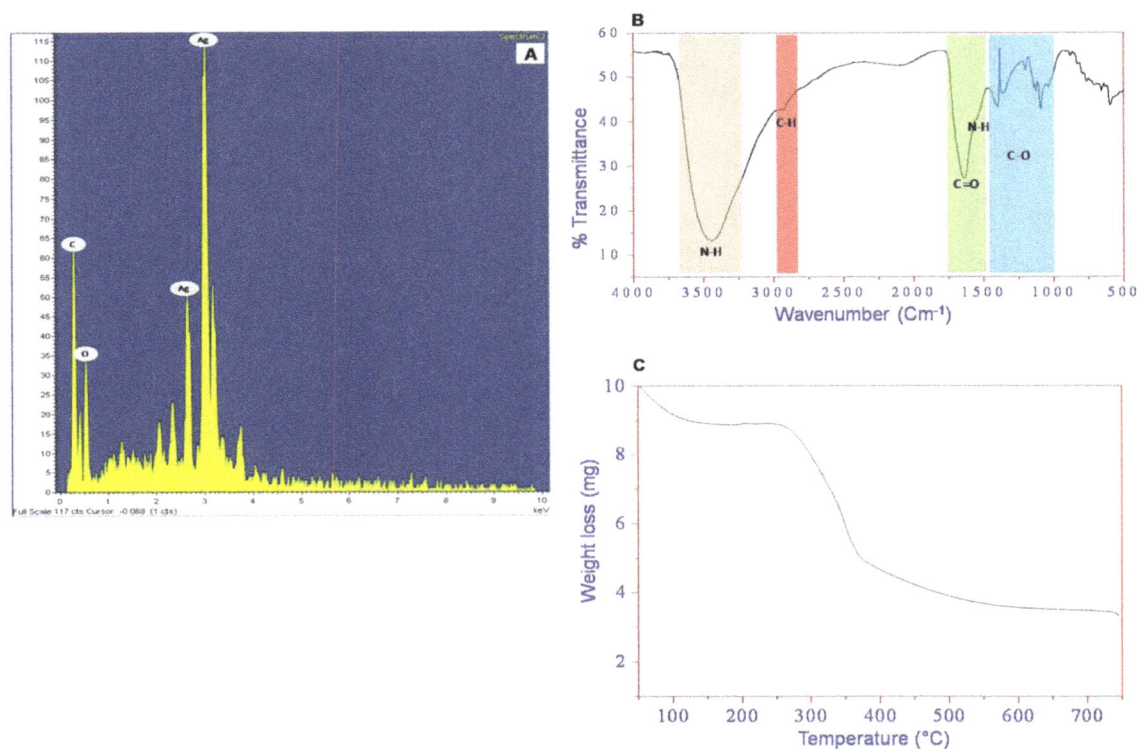

Figure 4. EDAX spectrum (A), FTIR spectrum (B) and TGA graph (C) of bsAgNPs by *Serratia* sp. BHU-S4.

FTIR spectrum recorded from the dried powder of bsAgNPs is shown in Figure 4B. The result suggests the release of extracellular proteins in the *Serratia* sp. BHU-S4 strain culture supernatant which are primarily responsible for the effective synthesis and stabilization of the AgNPs. The amide linkage between consecutive amino acid residues in proteins provides a well-known signature in infrared region of the electromagnetic spectrum. The vibration bands position at 3436.52 cm^{-1} and 2942.73 cm^{-1} were assigned to the stretching vibrations of primary and secondary amines, respectively. The corresponding bending vibrations were observed at 1641.93 cm^{-1} assigned to chelated –C = O or –OH group from –COO group in the extracellular proteins. Besides, the two bands observed at 1387.92 cm^{-1} and 1087.45 cm^{-1} can be assigned to the C–N stretching vibrations of aromatic and aliphatic amines, respectively. The FTIR data suggested the presence of amine and carbonyl groups in proteins present in the *Serratia* sp. BHU-S4 strain culture supernatant which act as reducing and stabilizing agents through the capping of bsAgNPs [44,45]. Thus, the higher stability of the bsAgNPs could be attributed to the complex nature of the *Serratia* sp. BHU-S4 strain culture supernatant [27,46–48].

The functionalization and stabilization of bsAgNPs by the biological macromolecules present in the *Serratia* sp. BHU-S4 strain culture supernatant was further confirmed by TGA. This analysis was performed under inert N$_2$ atmosphere and obtained thermogram is shown in Figure 4C. The thermogram revealed the two steps of weight loss in the temperature range of 50°C to 700°C. The first step weight loss (12%) was observed at ~100 to 200°C due to evaporation of adsorbed water molecules present in the bsAgNPs sample. The second step weight loss (51%) observed in the temperature range of ~300 to 450°C was a consequence of desorption of biological macromolecules like proteins and enzymes present on the surface of the bsAgNPs. It is evident from the above data that bsAgNPs were successfully synthesized using the culture supernatant of the *Serratia* sp. BHU-S4.

Biocontrol Potential of Biosynthesized AgNPs against *B. sorokiniana*

The bsAgNPs by *Serratia* sp. BHU-S4 were also examined for their antifungal activity against *B. sorokiniana*, the spot blotch pathogen of wheat under *in vitro* and *in vivo* conditions. *B. sorokiniana* causes serious foliar spot blotch disease in wheat leading to major yield loss [49]. The presence of thick-walled, elliptical conidia is the characteristic feature of this pathogen. The process of infection starts with conidial germination on the leaf surface followed by appresorium formation after which hyphae colonizes inter and intracellular leaf tissues [50]. Hence, the inhibitory effect of bsAgNPs on germination of *B. sorokiniana* conidia was tested under *in vitro* condition. Our results revealed that 2, 4 and 10 µg/ml concentrations of bsAgNPs accounted for total inhibition of conidial germination whereas in the absence of bsAgNPs, 100% conidial germination was recorded (Figure 5A). Furthermore, the effectiveness of bsAgNPs in obstructing the process of infection by *B. sorokiniana* was ascertained by detached leaf assay. Wheat leaves inoculated with conidial suspension of *B. sorokiniana* alone showed the characteristic disease symptom i.e. formation of spot blotch whereas on exposure to bsAgNPs and *B. sorokiniana* conidial suspension in combination failed to form spot blotch on wheat leaves which appeared to be as healthy as control leaves. This result was further confirmed after observing microscopic images which showed prominent conidial germination on wheat leaves inoculated with *B. sorokiniana* conidial suspension alone, while it was totally inhibited on treatment with bsAgNPs (Figure 5B).

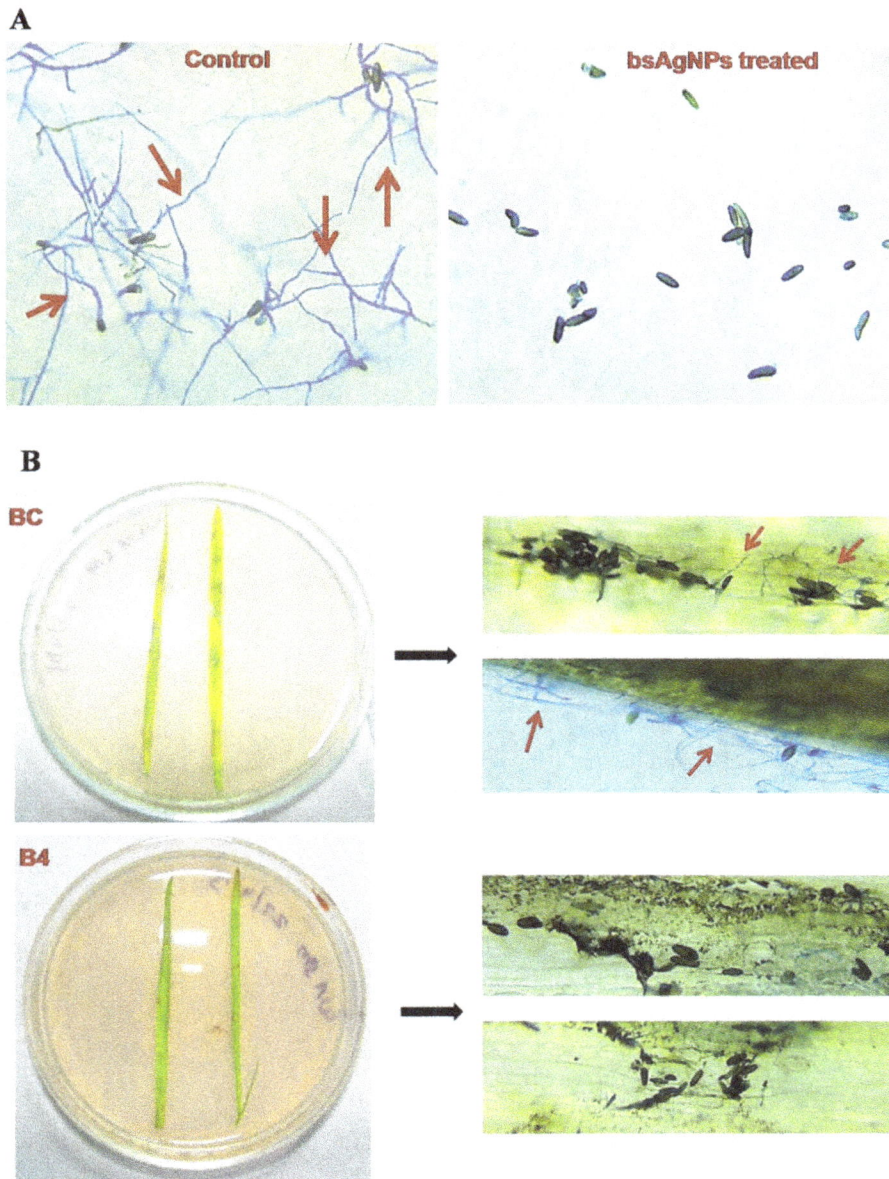

Figure 5. Inhibitory effects of bsAgNPs by *Serratia* sp. BHU-S4 on conidial germination of *B. sorokiniana* after 24 h as determined by cavity slide experiment. Conidial germination in control set (without bsAgNPs) was prominent as indicated by arrow (A). Detached leaf assay showing conidial germination over leaf surface in pathogen treated leaf (BC), while in presence of bsAgNPs (B4) conidial germination was totally inhibited (B).

These findings validate the biocontrol potential of bsAgNPs against *B. sorokiniana* under *in vitro* conditions. However, for further validation, greenhouse experiment was also conducted where application of bsAgNPs strongly controlled *B. sorokiniana* infection in wheat (Figure 6 A, B). Due to *B. sorokiniana* infection there was 46.73, 28.93, 68.47, 29.03% significant decrease in root length, shoot length, root dry wt. and shoot dry wt., respectively as compared to healthy control plants. It is interesting to note that decrease in plant growth after pathogen challenge was overcome when bsAgNPs treatment was given to the wheat plants that resulted into 11.55, 30.67, 60.83% increase in root length, shoot length and shoot dry wt. respectively, as compared to pathogen challenged plants (Figure 6D). Furthermore, to obtain an idea on the effect of different treatments on the wheat leaf proteome, we

examined SDS-PAGE profile that clearly showed major visible variations in protein banding pattern among different treatments (Figure 6C). The reduced intensity of protein bands justified the protein damage caused by *B. sorokiniana* infection. However, no protein damage occurred on bsAgNPs treatment as it showed banding pattern similar to control plants. Moreover, histochemical analysis of wheat stems of different treatments gave a clue about the defense strategy adopted by the plants against *B. sorokiniana* infection. Histochemical staining showed the pattern of lignification which is the main disease resistance factor as reported by Mandal and Mitra [51]. Notable difference in lignin deposition among different treatments was observed. Maximum lignification of vascular bundles was observed in control (C) and pathogen challenged treated with bsAgNPs plants (B4) while minimum

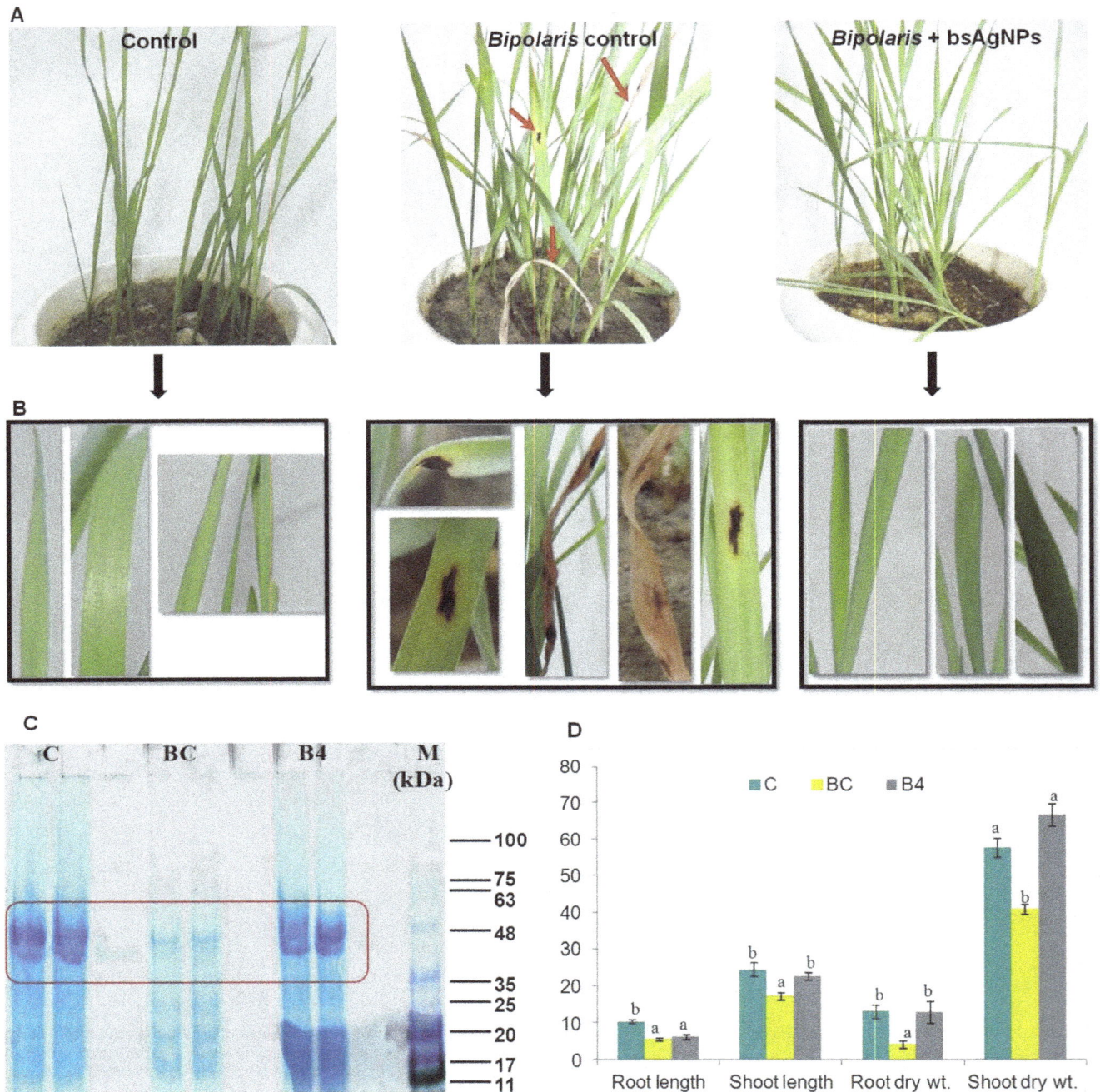

Figure 6. Efficacy of bsAgNPs in controlling pathogen attack under greenhouse conditions (A). Arrow indicates lesion developed as a result of *B. sorokiniana* infection. (B) Symptoms developed on leaves of *B. sorokiniana* infected plants (*Bipolaris* control). (C) SDS-PAGE profile of wheat leaf protein in different treatments. (D) Effect of different treatments on plant parameters are shown by bar graph. Control (C), *Bipolaris* control (BC); *Bipolaris* + bsAgNPs by *Serratia* sp. BHU-S4 (B4). Root length, shoot length (cm); root dry wt., shoot dry wt. (mg).

lignin deposition was found in pathogen challenged plants (BC) (Figure 7). It is evident from previous reports that lignin deposition plays a crucial role in plant development as well as disease resistance by forming a physical barrier against pathogen attack [52,53]. In addition, the content and composition of lignin is also reported to vary when plants are exposed to various stresses. Consequently, we hypothesize that the treatment of bsAgNPs enhanced lignification which could have worked as a hindrance against pathogen attack in plants treated with pathogen and

bsAgNPs in combination (B4), while least lignin deposition in pathogen challenged plants (BC) favored pathogen attack. This finding helps us in understanding the role of bsAgNPs in controlling *B. sorokiniana* infection in wheat plants. However, a more detailed study is required to ascertain this role of bsAgNPs using culture supernatant of *Serratia* sp. BHU-S4.

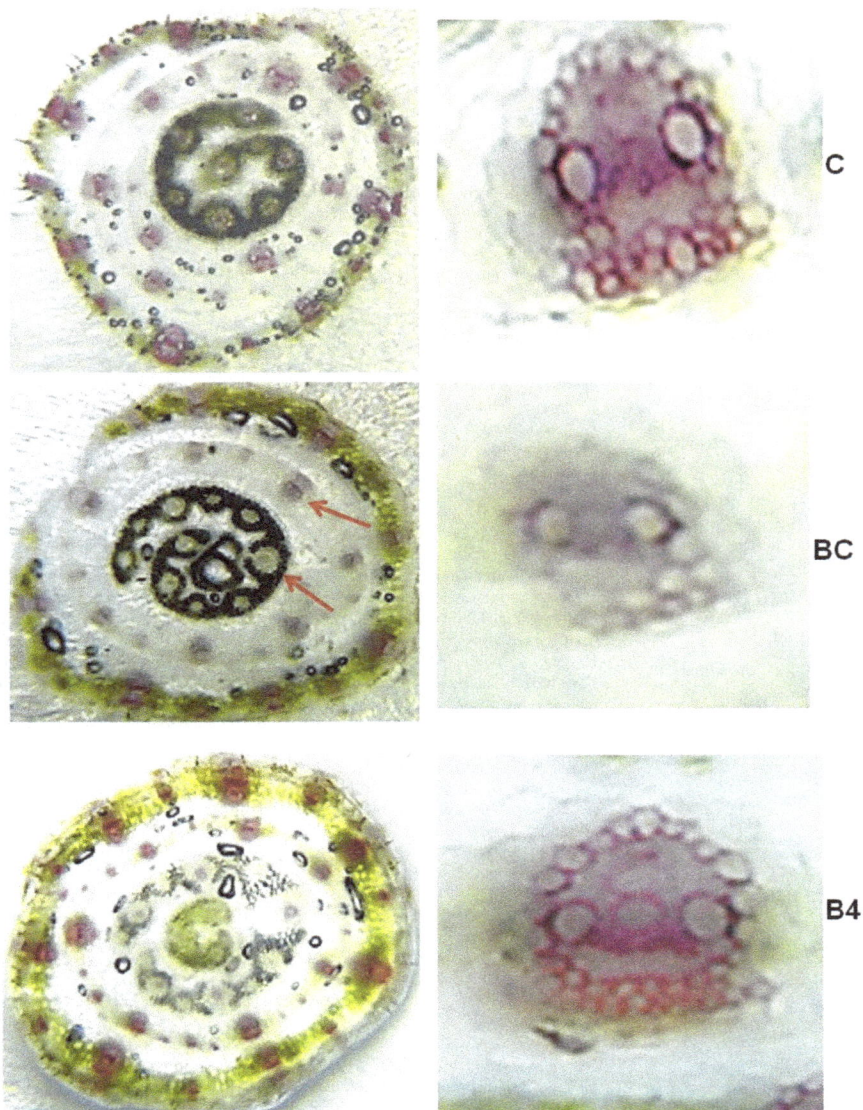

Figure 7. Effect of different treatments on lignification in wheat stem by histochemical staining. Control (C), *Bipolaris* control (BC); *Bipolaris* + bsAgNPs by *Serratia* sp. BHU-S4 (B4).

Conclusions

Present study encompasses the pivotal role of AgNPs synthesized by an agriculturally important bacterium, *Serratia* sp. BHU-S4 for plant disease control. The bsAgNPs exhibited strong antifungal activity against *Bipolaris sorokiniana* and successfully controlled its infection in wheat plants. The biocontrol potential of bsAgNPs was found to be promising under both *in vitro* and *in vivo* conditions as depicted in the graphical abstract (Figure S1). The foregoing data clearly reveals the robust application of bsAgNPs in agriculture, particularly for plant disease management. Till date many studies have reported antifungal activity of AgNPs under *in vitro* conditions but little is known about its activity under *in vivo* condition. In this regard, our findings certainly authenticate the application of bsAgNPs in controlling plant diseases, thereby pointing to the exciting possibilities of nanofungicide. However, further research is required to verify the effect of bsAgNPs on different phytopathogens causing serious crop losses under field conditions. Moreover, future studies should also be employed for developing efficient delivery systems for large scale application of AgNPs.

Supporting Information

Figure S1   Graphical abstract of the study.

Acknowledgments

Authors thank Prof. R. Chand, Banaras Hindu University, Varanasi, for generously providing the culture of *Bipolaris sorokiniana*. We also gratefully acknowledge the help from Centre of Excellence in Materials Science (Nanomaterials), Department of Applied Physics, Z.H. College of Engineering and Technology, Aligarh Muslim University, Aligarh 202002, India for providing the nanomaterials characterization facility.

Author Contributions

Conceived and designed the experiments: HBS. Performed the experiments: SM BRS AS CK. Analyzed the data: HBS SM BRS AHN. Wrote the paper: HBS SM BRS AHN.

References

1. Nair R, Varghese SH, Nair BG, Maekawa T, Yoshida Y, et al. (2010). Nanoparticulate material delivery to plants. Plant Sci 179: 154–163.
2. Ghormade V, Deshpande MV, Paknikar KM (2011) Perspectives for nano-biotechnology enabled protection and nutrition of plants. Biotechnol Adv 29: 792–803.
3. Bakshi M, Singh HB, Abhilash PC (2014) The unseen impact of nanoparticles: More or less? Curr Sci 106: 350–352.
4. Navrotsky A (2000) Technology and applications Nanomaterials in the environment, agriculture, and technology (NEAT). J Nanopart Res 2: 321–323.
5. Hu C, Lan YQ, Qu JH, Hu XX, Wang AM (2006) Ag/AgBr/TiO2 visible light photocatalyst for destruction of azodyes and bacteria. J Physical Chem B 110: 4066–4072.
6. Moonjung C, Kyoung-Hwan S, Jyongsik J (2010) Plasmonic photocatalytic system using silver chloride/silver nanostructures under visible light. J Colloid Interface Sci 341: 83–87.
7. Morones JR, Elechiguerra JL, Camacho A, Holt K, Kouri JB, et al. (2005) The bactericidal effect of silver nanoparticles. Nanotechnology 16: 2346–2354.
8. Tian J, Wong KK, Ho CM, Lok CN, Yu WY, et al. (2007) Topical delivery of silver nanoparticles promotes wound healing. Chem Med Chem 2: 129–136.
9. Prakasha P, Gnanaprakasama P, Emmanuel R, Arokiyaraj S, Saravananc M (2013) Green synthesis of silver nanoparticles from leaf extract of Mimusops elengi, Linn. for enhanced antibacterial activity against multi drug resistant clinical isolates. Colloids Surf B Biointerfaces 108: 255–259.
10. Oves M, Khan MS, Zaidi A, Ahmed AS, Ahmed F, et al. (2013) Antibacterial and cytotoxic efficacy of extracellular silver nanoparticles biofabricated from chromium reducing novel OS4 strain of Stenotrophomonas maltophilia. PLoS ONE 8(3): e59140.
11. Kim SW, Kim KS, Lamsal K, Kim YJ, Kim SB, et al. (2009) An in vitro study of the antifungal effect of silver nanoparticles on oak wilt pathogen Raffaelea sp. J Microbiol Biotechnol 19: 760–764.
12. He L, Liu Y, Mustapha A, Lin M (2011) Antifungal activity of zinc oxide nanoparticles against Botrytis cinerea and Penicillium expansum. Microbiol Res 166: 207–215.
13. Krishnaraj C, Ramachandran R, Mohan K, Kalaichelvan PT (2012) Optimization for rapid synthesis of silver nanoparticles and its effect on phytopathogenic fungi. Spectrochim Acta Part A Mol Biomol Spectrosc 93: 95–99.
14. Gopinath V, Velusamy P (2013) Extracellular biosynthesis of silver nanoparticles using Bacillus sp. GP-23 and evaluation of their antifungal activity towards Fusarium oxysporum. Spectrochim Acta Part A Mol Biomol Spectrosc 106: 170–174.
15. Lee KJ, Park SH, Govarthanan M, Hwang PH, Seo YS, et al. (2013) Synthesis of silver nanoparticles using cow milk and their antifungal activity against phytopathogens. Mater Lett 105: 128–131.
16. Saari EE (1998). Leaf blight diseases and associated soilborne fungal pathogens of wheat in south and southeast Asia. In: Duveiller E, Dubin HJ, Reeves J, McNab A, editors. Helminthosporium Blights of Wheat: Spot Blotch and Tan Spot. Mexico, D.F., Mexico: CIMMYT, 37–51.
17. Singh RV, Singh AK, Singh SP (1997) Distribution of pathogens causing foliar blight of wheat in India and neighbouring countries. In: Duveiller E, Dubin HJ, Reeves J, McNab A, editors. Helminthosporium blight of wheat: spot blotch and tan spot. CIMMYT El Batan, Mexico, DF, 59–62.
18. Mehta YR (1994) Manejo Integrado de Enfermedadas de Trigo-santa Cruz. Bolivia: CIAT/IAPAR. 314.
19. Kumar U, Joshi AK, Kumar S, Chand R, Roder MS (2009) Mapping of resistance to spot blotch disease caused by Bipolaris sorokiniana in spring wheat. Theor Appl Genet 118: 783–792.
20. Holt JG, Krieg NR, Sneath PHA, Staley JT, Williams ST (1994) Bergey's Manual of Determinative Bacteriology. Williams and Wilkins Press, Baltimore, 544–551.
21. Mishra S, Nautiyal CS (2012) Reducing the allelopathic effect of Parthenium hysterophorus L. on wheat (Triticum aestivum) by Pseudomonas putida. Plant Growth Regul 66: 155–165.
22. Singh BR, Singh BN, Singh HB, Khan MW, Naqvi AH (2012) ROS-mediated apoptotic cell death in prostate cancer LNCaP cells induced by biosurfactant stabilized CdS quantum dots. Biomaterials 33: 5753–5767.
23. Mehta S, Nautiyal CS (2001) An efficient method for qualitative screening of phosphate solubilizing bacteria. Curr Microbiol 43: 51–56.
24. Brick M, Bostock RM, Silverstone SE (1991) Rapid in situ assay for indole acetic acid production by bacteria immobilized on nitrocellulose membrane. Appl Environ Microbiol 57: 535–538.
25. Dwivedi SR, Singh BR, Al-Khedhairy AA, Musarrat J (2011) Biodegradation of isoproturon using a novel Pseudomonas aeruginosa strain JS-11 as a multi-functional bioinoculant of environmental significance. J Hazardous Materials 185: 938–944.
26. Jeyaraja M, Varadan S, Anthony KJP, Murugan M, Raja A, et al. (2013) Antimicrobial and anticoagulation activity of silver nanoparticles synthesized from the culture supernatant of Pseudomonas aeruginosa. J Ind Eng Chem 19: 1299–1303.
27. Musarrat J, Dwivedi S, Singh BR, Al-Khedhairy AA, Azam A, et al. (2010) Production of antimicrobial silver nanoparticles in water extracts of the fungus Amylomyces rouxii strain KSU-09. Bioresour Technol 101: 8772–8776.
28. Shoeb M, Khan JA, Singh BR, Singh BN, Singh HB, et al. (2013) ROS-mediated anticandidal activity of zinc oxide nanoparticles synthesized by using egg albumen as a biotemplate. Adv Nat Sci Nanosci Nanotechnol 4: 035015.
29. Singh S, Singh BR, Khan W, Naqvi AH (2014) Synthesis and characterization of carbon nanotubes/titanium molybdate nanocomposite and assessment of its photocatalytic activity. J Mol Struct 1056: 194–201.
30. Khan JA, Qasim M, Singh BR, Shoeb M, Singh S, et al. (2013) Synthesis and characterization of structural, optical, thermal and dielectric properties of polyaniline/CoFe2O4 nanocomposites with special reference to photocatalytic activity. Spectrochim Acta Part A Mol Biomol Spectrosc 109: 313–321.
31. Khan JA, Qasim M, Singh BR, Shoeb M, Khan W, et al. (2014) Polyaniline/CoFe2O4 nanocomposite inhibits the growth of Candida albicans 077 by ROS production. Comptes Rendus Chimie 17: 91–102.
32. Siddique YH, Fatima A, Jyoti S, Naz F, Rahul, et al. (2013) Evaluation of the toxic potential of graphene copper nanocomposite (GCNC) in the third instar larvae of transgenic Drosophila melanogaster (hsp70-lacZ)Bg9. PLoS ONE 8: e80944.
33. Rao RAK, Singh S, Singh BR, Khan W, Naqvi AH (2014) Synthesis and characterization of surface modified graphene-zirconium oxide nanocomposite and its possible use for the removal of chlorophenol from aqueous solution. J Environ Chem Eng 2: 199–210.
34. Sarkar A, Rakwal R, Agarwal SB, Shibato J, Ogawa Y, et al. (2010) Investigating the impact of elevated levels of ozone on tropical wheat using integrated phenotypical, physiological, biochemical, and proteomics approaches. J Prot Res 9: 4565–4584.
35. Bradford M (1976) A rapid and sensitive method for the quantitation of microgram quantities of protein utilizing the principle of protein-dye binding. Anal Biochem 72: 248–254.
36. Jensen WA (1962) Botanical histochemistry: principles and practices. London: WH Freeman and Co.
37. Benhamou N, Gagne S, Quere DL, Dehbi L (2000) Bacteria-mediated induced resistance in cucumber. Beneficial effect of the endophytic bacterium Serratia plymuthica on the protection against infection by Pythium ultimum. Biochem Cell Biol 90: 45–56.
38. Kamensky M, Ovadis M, Chet I, Chemin L (2003) Soil-borne strain IC14 of Serratia plymuthica with multiple mechanisms of antifungal activity provides biocontrol of Botrytis cinerea and Sclerotinia sclerotiorum diseases. Soil Biol Biochem 79: 584–589.
39. So-Yeon K, Cho KS (2009) Isolation and characterization of a plant growth promoting rhizobacterium, Serratia sp. SY5. J Microbiol Biotechnol 19: 1431–1438.
40. Kleemann W (1993) Random-field induced antiferromagnetic, ferroelectric and structural domain states. Int J Mod Phys B 7: 2469–2507.
41. Kanchana A, Agarwal I, Sunkar S, Nellore J, Namasivayam K (2011) Biogenic silver nanoparticles from Spinacia oleracea and Lactuca sativa and their potential antimicrobial activity. Dig J Nanomat Bios 6: 1741–1750.
42. Kumar CG, Mamidyala SK (2011) Extracellular synthesis of silver nanoparticles using culture supernatant of Pseudomonas aeruginosa. Coll Surf B Biointerfaces 84: 462–466.
43. Shivaji S, Madhu S, Singh S (2011) Extracellular synthesis of antibacterial silver nanoparticles using psychrophilic bacteria. Process Biochem 46: 1800–1807.
44. Balaji D, Basavaraja S, Deshpande S, Bedre R, Mahesh D, et al. (2009) Extracellular biosynthesis of functionalized silver nanoparticles by strains of Cladosporium cladosporioides fungus. Colloids Surf. B 68: 88–92.
45. Kasthuri J, Veerapandian S, Rajendiran N (2009) Biological synthesis of silver and gold nanoparticles using apiin as reducing agent. Colloids Surf B 68: 55–60.
46. Quester K, Avalos-Borja M, Vilchis-Nestor AR, Camacho-Lopez MA, Castro-Longoria E (2013) SERS properties of different sized and shaped gold nanoparticles biosynthesized under different environmental conditions by Neurospora crassa. PLoS ONE 8: e77486.
47. Viet Long N, Ohtaki M, Yuasa M, Yoshida S, Kuragaki T, et al. (2013). Synthesis and self assembly of gold nanoparticles by chemically modified polyol methods under experimental control. J Nanomaterial Article ID 793125: 8 pages.
48. Malhotra A, Dolma K, Kaur N, Rathore YS, Ashish, et al. (2013) Biosynthesis of gold and silver nanoparticles using a novel marine strain of Stenotrophomonas. Bioresour Technol 142: 727–731.

49. Kumar J, Schäfer P, Hückelhoven R, Langen G, Baltruschat H, et al. (2002) *Bipolaris sorokiniana*, a cereal pathogen of global concern: cytological and molecular approaches towards better control. Mol Plant Pathol 3: 185–195.

50. Domiciano GP, Rodrigues FA, Guerra AMN, Vale FXR (2013) Infection process of *Bipolaris sorokiniana* on wheat leaves is affected by silicon. Trop Plant Pathol 38: 258–263.

51. Mandal S, Mitra A (2007) Reinforcement of cell wall in roots of *Lycopersicon esculentum* through induction of phenolic compounds and lignin by elicitors. Physiol Mol Plant Pathol 71: 201–209.

52. Moershbacher BM, Noll U, Gorrichon L, Reisener HJ (1990) Specific inhibition of lignification breaks hypersensitive resistance of wheat to stem rust. Plant Physiol 93: 465–470.

53. Shi H, Liu Z, Zhu L, Zhang C, Chen Y, et al. (2012) Overexpression of cotton (*Gossypium hirsutum*) *dirigent 1* gene enhances lignification that blocks the spread of *Verticillium dahliae*. Acta Biochim Biophys Sin 44: 555–564.

15

High Diversity and Low Specificity of Chaetothyrialean Fungi in Carton Galleries in a Neotropical Ant–Plant Association

Maximilian Nepel[1], Hermann Voglmayr[2,3], Jürg Schönenberger[1], Veronika E. Mayer[1]*

1 Division of Structural and Functional Botany, Department of Botany and Biodiversity Research, University of Vienna, Vienna, Austria, **2** Division of Systematic and Evolutionary Botany, Department of Botany and Biodiversity Research, University of Vienna, Vienna, Austria, **3** Institute of Forest Entomology, Forest Pathology and Forest Protection, Department of Forest and Soil Sciences, BOKU-University of Natural Resources and Life Sciences, Vienna, Austria

Abstract

New associations have recently been discovered between arboreal ants that live on myrmecophytic plants, and different groups of fungi. Most of the – usually undescribed – fungi cultured by the ants belong to the order Chaetothyriales (Ascomycetes). Chaetothyriales occur in the nesting spaces provided by the host plant, and form a major part of the cardboard-like material produced by the ants for constructing nests and runway galleries. Until now, the fungi have been considered specific to each ant species. We focus on the three-way association between the plant *Tetrathylacium macrophyllum* (Salicaceae), the ant *Azteca brevis* (Formicidae: Dolichoderinae) and various chaetothyrialean fungi. *Azteca brevis* builds extensive runway galleries along branches of *T. macrophyllum*. The carton of the gallery walls consists of masticated plant material densely pervaded by chaetothyrialean hyphae. In order to characterise the specificity of the ant–fungus association, fungi from the runway galleries of 19 ant colonies were grown as pure cultures and analyzed using partial SSU, complete ITS, 5.8S and partial LSU rDNA sequences. This gave 128 different fungal genotypes, 78% of which were clustered into three monophyletic groups. The most common fungus (either genotype or approximate species-level OTU) was found in the runway galleries of 63% of the investigated ant colonies. This indicates that there can be a dominant fungus but, in general, a wider guild of chaetothyrialean fungi share the same ant mutualist in *Azteca brevis*.

Editor: Petr Karlovsky, Georg-August-University Göttingen, Germany

Funding: This work was supported by a University of Vienna, Austria, KWA fellowship to MN; https://international.univie.ac.at/graduate-students/kurzfristige-auslandsstipendien-kwa. The funders had no role in study design, data collection and analysis, decision to publish, or preparation of the manuscript.

Competing Interests: The authors have declared that no competing interests exist.

* Email: veronika.mayer@univie.ac.at

Introduction

It is now clear that microorganisms are major partners in obligate interactions between ants and plants. Ant–fungus associations have been recognised since the mid-19th century (e.g. [1]), and the best-studied examples are the fungal gardens of the leaf-cutter ants in the tribe Attini. Leaf-cutter ants grow monocultures of basidiomycetes on shredded leaf material and feed on the nutrient-rich tips of the fungal hyphae [2,3]. Other examples of ant–plant–fungus interactions have also been found recently in different groups of non-attine ants, where ascomycete fungi are cultivated in domatia (nesting spaces provided by host plants) or on a cardboard-like construction material (named "carton" in ant-plant literature) [4–7]. Such ant–plant–fungus associations have been described from Africa, America and Asia and involve a wide range of plant lineages associated with an equally wide range of ant groups [7].

There is evidence that the fungi cultivated within the domatia are used as a food source [8], whereas those in the carton-like material do not appear to be consumed. Rather, they seem to serve to stabilise the carton mechanically. Carton structures with fungi were first documented in nest walls of the European ant *Lasius fuliginosus* inside hollow tree-trunks [1,9,10]. They have since been found in the walls of free-hanging canopy ant nests in the Palaeotropics [11,12] and in the Neotropics, where ants use fungus-infused, carton-like material to construct tunnel systems called "runway galleries" along branches of their host trees [6,13,14] (Figure 1A–D).

In the tripartite ant–plant–fungus interactions involving non-attine ants studied so far, the vast majority of the fungi have belonged to the ascomycete order Chaetothyriales, the so-called "black yeasts" [7]. These are usually dark, melanised, slow-growing fungi that often colonise extreme environments [15–18], but little is known about the order's ecology and diversity.

A recent survey based on molecular phylogenetics showed that ant-associated chaetothyrialean fungi belong to four clades within the order: a domatia-symbiont clade, two clades with carton fungi, and a mixed clade containing both domatia symbionts and carton fungi [7]. Only a few isolates were placed outside these four clades, and ant-fungi cultivated in the domatia seemed to be specific to each ant species [7]. Carton structures have been less well investigated, and studies to date have produced disparate results: in the *Hirtella* (Chrysobalanaceae)/*Allomerus* (Formicidae) asso-

Figure 1. Carton runway galleries built by *Azteca brevis* ants, consisting of mycelia and various organic particles. (A, B) Galleries on the lower side of branches of *Tetrathylacium macrophyllum*: note the scattered circular openings in the gallery walls. (B) Alarmed workers wait with open mandibles below the holes for prey or intruders. (C, D) Scanning electron microscope images of the gallery walls infused with different types of hyphae. (E, F) Light-microscope images of both hyphal types: (E) thin-walled hyaline hyphae typical for carton clades 2 and 3; (F) pigmented thick-walled hyphae typical for carton clade 1 (see Figure 2–3). Bars: (A, B) 1 cm; (C, D) 100 μm; (E, F) 20 μm.

ciation, it was reported that a specific fungus is cultivated by the ants in the wall material of their galleries [14]; in contrast, a wider guild of fungi seems to be involved in the structurally analogous galleries of the *Tetrathylacium/Azteca* association [6]. The aim of the present investigation was (1) to unravel the diversity and geographical pattern of the carton fungi found in the carton galleries, and (2) to investigate the hyphal morphology of the relevant fungal strains with respect to the proposed function of the galleries.

Material and Methods

Species and study site

Azteca brevis Forel, 1899 (Formicidae, Dolichoderinae) is a reddish-brown ant, c. 4 mm long, known from wet forests of the southern Pacific lowlands of Costa Rica [19]. Colonies have been found on *Tetrathylacium macrophyllum* (Salicaceae), *Licania* sp. (Chrysobalanaceae), *Grias* sp. (Lecythidaceae), *Myriocarpa* sp. (Urticaceae), *Ocotea nicaraguensis* (Lauraceae) [19] and *Lonchocarpus* sp. (Fabaceae). The nesting chambers inside the stems are connected externally by runway galleries dotted with small, circular holes (Figure 1A, B).

The most common host plant for *Azteca brevis* is *Tetrathylacium macrophyllum* Poepp. (Salicaceae), a small tree (c. 8 m) that grows

on the Pacific slopes of Central and South America in areas characterised by high annual rainfall (>5000 mm). It is found chiefly on steep slopes near rivers and streams in primary forest [20]. About 30% of *T. macrophyllum* trees are occupied by *Azteca brevis* [21,22]. The ants start by colonising hollow chambers in the branches that the plant forms through pith degeneration. As the colony grows, the ants excavate the remaining pith between adjacent naturally formed chambers, and build large nest sites inside the branches.

We sampled within a 5-km circle around the Tropical Research Station La Gamba, Costa Rica (8° 42′ 03″ N, 83° 12′ 06″ W) along the Waterfall Trail, Bird Trail, Río Gamba, Río Bolsa and Río Sardinal. Carton samples, ants and plant parts for herbarium specimens were collected from 18 *T. macrophyllum* trees and one *Lonchocarpus* tree colonised by *Azteca brevis*, under permission from SINAC – Sistema Nacional de Areas de Conservación de Costa Rica of the Ministry of Environment and Energy (MINAE) to M.N. and V.E.M. (No. 182-2010-SINAC). In recent years, trees colonised by *Azteca brevis* have become inexplicably rare at the study site (VEM, pers. obs.), limiting the sample size to 19 trees colonised by *Azteca brevis*. At least three carton pieces per tree and colony, each c. 1 cm long, were taken from runway galleries and stored in 1.5-mL reaction tubes sealed with air-permeable cotton wool. The reaction tubes were kept in a sealed plastic bag with

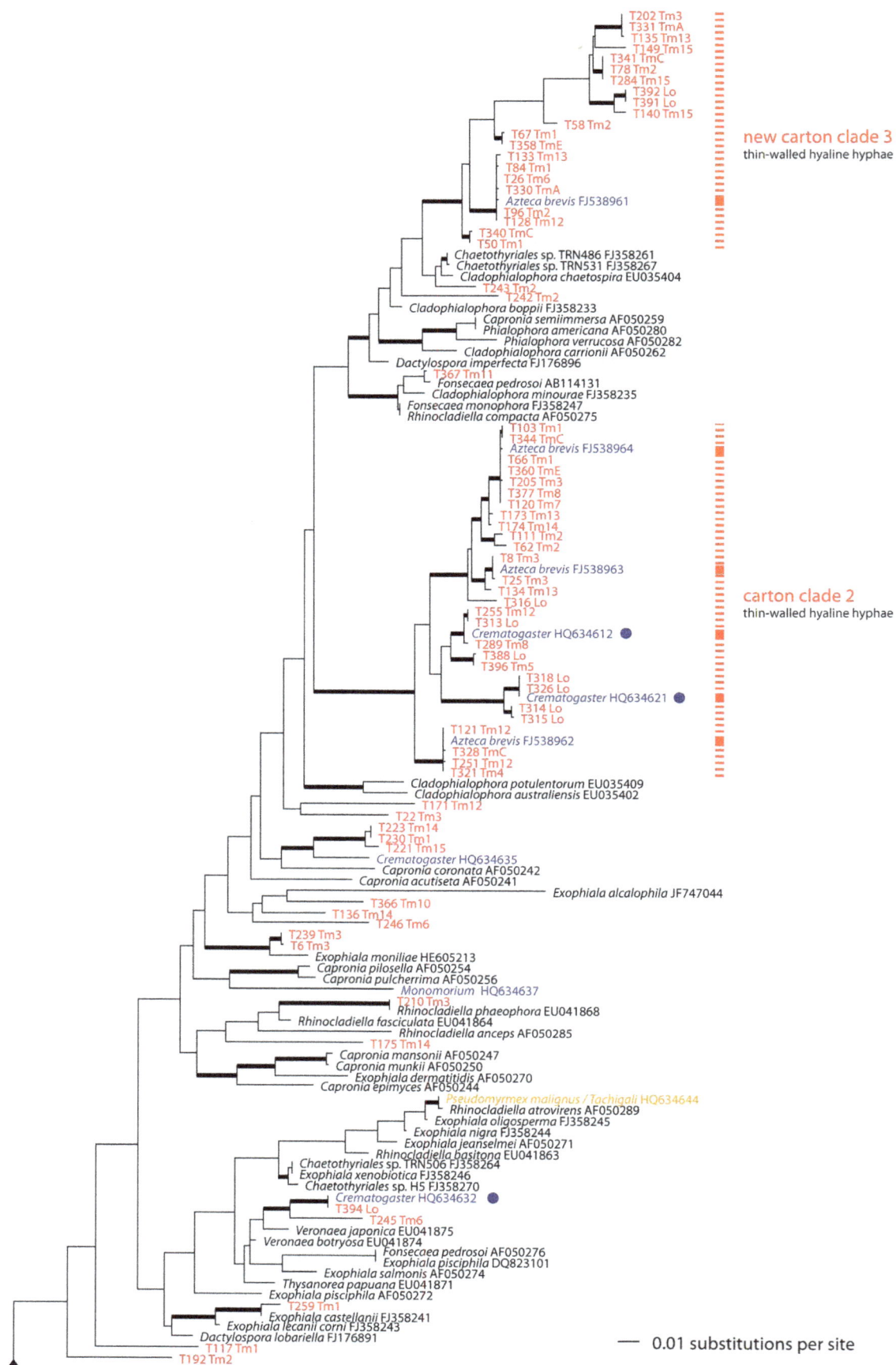

T202 Tm3
T331 TmA
T135 Tm13
T149 Tm15
T341 TmC
T78 Tm2
T284 Tm15
T392 Lo
T391 Lo
T140 Tm15
T58 Tm2
T67 Tm1
T358 TmE
T133 Tm13
T84 Tm1
T26 Tm6
T330 TmA
Azteca brevis FJ538961
T96 Tm2
T128 Tm12
T340 TmC
T50 Tm1

new carton clade 3
thin-walled hyaline hyphae

Chaetothyriales sp. TRN486 FJ358261
Chaetothyriales sp. TRN531 FJ358267
Cladophialophora chaetospira EU035404
T243 Tm2
T242 Tm2
Cladophialophora boppii FJ358233
Capronia semiimmersa AF050259
Phialophora americana AF050280
Phialophora verrucosa AF050282
Cladophialophora carrionii AF050262
Dactylospora imperfecta FJ176896
T367 Tm11
Fonsecaea pedrosoi AB114131
Cladophialophora minourae FJ358235
Fonsecaea monophora FJ358247
Rhinocladiella compacta AF050275

T103 Tm1
T344 TmC
Azteca brevis FJ538964
T66 Tm1
T360 TmE
T205 Tm3
T377 Tm8
T120 Tm7
T173 Tm13
T174 Tm14
T111 Tm2
T62 Tm2
T8 Tm3
Azteca brevis FJ538963
T25 Tm3
T134 Tm13
T316 Lo
T255 Tm12
T313 Lo
Crematogaster HQ634612 ●
T289 Tm8
T388 Lo
T396 Tm5
T318 Lo
T326 Lo
Crematogaster HQ634621 ●
T314 Lo
T315 Lo
T121 Tm12
Azteca brevis FJ538962
T328 TmC
T251 Tm12
T321 Tm4

carton clade 2
thin-walled hyaline hyphae

Cladophialophora potulentorum EU035409
Cladophialophora australiensis EU035402
T171 Tm12
T22 Tm3
T223 Tm14
T230 Tm1
T221 Tm15
Crematogaster HQ634635
Capronia coronata AF050242
Capronia acutiseta AF050241
Exophiala alcalophila JF747044
T366 Tm10
T136 Tm14
T246 Tm6
T239 Tm3
T6 Tm3
Exophiala moniliae HE605213
Capronia pilosella AF050254
Capronia pulcherrima AF050256
Monomorium HQ634637
T210 Tm3
Rhinocladiella phaeophora EU041868
Rhinocladiella fasciculata EU041864
Rhinocladiella anceps AF050285
T175 Tm14
Capronia mansonii AF050247
Capronia munkii AF050250
Exophiala dermatitidis AF050270
Capronia epimyces AF050244
Pseudomyrmex malignus / Tachigali HQ634644
Rhinocladiella atrovirens AF050289
Exophiala oligosperma FJ358245
Exophiala nigra FJ358244
Exophiala jeanselmei AF050271
Rhinocladiella basitona EU041863
Chaetothyriales sp. TRN506 FJ358264
Exophiala xenobiotica FJ358246
Chaetothyriales sp. H5 FJ358270
Crematogaster HQ634632 ●
T394 Lo
T245 Tm6
Veronaea japonica EU041875
Veronaea botryosa EU041874
Fonsecaea pedrosoi AF050276
Exophiala pisciphila DQ823101
Exophiala salmonis AF050274
Thysanorea papuana EU041871
Exophiala pisciphila AF050272
T259 Tm1
Exophiala castellanii FJ358241
Exophiala lecanii corni FJ358243
Dactylospora lobariella FJ176891
T117 Tm1
T192 Tm2

—— 0.01 substitutions per site

Figure 2. Phylogram of Chaetothyriales, top part. The maximum-likelihood tree is shown, based on partial SSU, complete ITS and 5.8S, and partial LSU rDNA regions. Bold branches are supported in all three analyses: BA probabilities higher than 0.9, ML and MP bootstrap support above 70%. Red labels denote fungal genotypes isolated in this study from ant-built carton structures on *Tetrathylacium macrophyllum* and *Lonchocarpus* sp. trees; orange and blue mark domatia fungi and carton fungi, respectively, from Voglmayr *et al.* [7]; GenBank accession numbers follow taxon names; solid red, violet and orange vertical lines indicate clade definitions and captions from Voglmayr *et al.* [7]; dotted lines mark clade extensions from this study. Blue dots point out three sequences from other continents (2× Cameroon; 1× Thailand) differing by only three mutations from our Costa Rican genotypes. Note the high diversity of isolated genotypes (the large clade extensions compared to Voglmayr *et al.* [7] are due to a greater number of samples and the new monophyletic carton-fungi cluster (new carton clade 3).The tree is continued in Figure 3.

silica gel for three weeks before culturing the fungi; the bags were transported and stored at room temperature.

Fungal cultures and DNA-extraction

At the University of Vienna, a c. 10-mm^2 piece of carton was placed into a droplet (c. 20 μL) of sterile water and fragmented with sterile forceps to make a mycelial suspension. An aliquot of mycelial suspension was then diluted with 1 mL sterile water and spread over each of two 2% malt extract agar plates (MEA) containing 0.5% penicillin and 0.5% streptomycin. This was carried out on average for three samples per colonised tree. The plates were stored at room temperature and visually checked under a dissecting microscope at least once a day for 11 days. Fast-growing "weeds" (*Aspergillus*, *Cladosporium*, *Fusarium*) were excised to prevent overgrowth of the slower-growing carton fungi. The thick, darkly pigmented hyphae that are typical for the carton usually started to grow after 2–4 days and were then transferred to new 2% malt extract agar plates. Several carton samples from each ant colony and tree were processed in this way to minimise any cultivation bias.

Sections of approximately 25 mm^2 were cut out from mycelia on pure-culture agar plates and stored in 2-mL reaction tubes at −20°C. The frozen samples were subsequently freeze-dried overnight and ground with five glass beads (3 mm diameter) for 10 min at 30 Hz in an MM 400 mixer mill (Retsch, Germany), after which DNA was extracted using NucleoSpin 96 Plant II kits (Macherey-Nagel, Düren, Germany).

PCR and cleanup

A 1.5–3.5-kb nuclear ribosomal DNA (rDNA) fragment comprising partial small subunit (SSU), complete ITS1–5.8S–ITS2 (ITS) and partial long subunit (LSU) sequences was amplified with the fungal primers V9G [23] and LR5 [24] using Thermo Scientific 2.0× ReddyMix Extensor PCR Master Mix and 1.1× ReddyMix PCR Master Mix (ABgene, Epsom, UK) (for primer sequences and detailed PCR protocol, see Table S1 in Appendix S1). The PCR products were purified with 6 U exonuclease I and 0.6 U FastAP thermosensitive alkaline phosphatase (Fermentas, St. Leon-Rot, Germany) [25]; the PCR product was then incubated for 30 min at 37°C, followed by enzyme deactivation for 15 min at 85°C.

Sequencing

DNA was cycle-sequenced with ABI PRISM BigDye Terminator Cycle Sequencing Ready Reaction v. 3.1 (Applied Biosystems, Warrington, UK) using the PCR primers and primers LR3 [24] and ITS4 [26]. For sequences with large indels, the additional primers LR2R-A, LR2-A [27], F5.8Sr, F5.8Sf [28] and LR3-CH (5′-GGT ATA GGG GCG AAA GAC TAA TC-3′) were necessary to obtain full-length sequences (see Appendix S1 for detailed sequencing protocol). Sequencing was performed on an ABI 3730xl Genetic Analyzer automated DNA sequencer (Applied Biosystems). One sequence of each genotype was deposited in

GenBank. The complete list of accession numbers for the SSU–ITS–LSU locus can be found in Table S2 in Appendix S1.

Analysis of sequence data

After a BLAST search (Basic Local Alignment Search Tool in GenBank, http://blast.ncbi.nlm.nih.gov/Blast.cgi) of the nuITS1–5.8S–ITS2–LSUr DNA sequences obtained from the ant carton (423 in total), 381 sequences were identified as Chaetothyriales and used for further analyses. For phylogenetic analyses, identical sequences from carton samples were reduced to a single sequence per genotype. Cases where two sequences differed only in homopolymer regions were also merged to a single genotype.

For alignment, the chaetothyrialean sequences used in Voglmayr *et al.* [7] and the closest sequences to our isolates from GenBank were added. *Verrucaria denudata*, *V. csernaensis* and *V. andesiatica* (Verrucariales) were included as outgroups. Ambiguously aligned regions in ITS1/ITS2 and leading gap regions were excluded. The matrix of 258 sequences contained 7767 alignment positions, with the longest sequence comprising 3425 nucleotides. Alignments were produced with Muscle 3.8.31 [29] and revised in BioEdit 7.1.3.0 [30]. (GenBank accession numbers of the sequences included in the phylogenetic analyses are listed in Table S2 in Appendix S1.)

For Bayesian analyses, MrBayes 3.2.1 [31] was run through the Bioportal web service of the University of Oslo [32]. The six-parameter general time-reversible substitution model was used, with a proportion of invariant sites and a gamma distribution for the remaining sites (GTR + I + G), as determined by Modeltest 3.7 [33]. Three parallel runs of four chains were performed over 30 million generations, sampling 30 000 trees in each run. For 90% majority-rule consensus trees, the first 2000 trees of each run were discarded as burn-in.

Maximum-parsimony (MP) bootstrap analyses were performed with PAUP* 4.0b10 [34], using 1000 replicates of a heuristic search with 10 rounds of random sequence addition during each bootstrap replicate and a limit of 100 000 rearrangements per replicate. TBR branch swapping was used, allowing multitrees, and steepest descent was set to 'no'. Gaps were treated as missing data, and no weighting of nucleotides was applied.

The maximum-likelihood (ML) analyses used RAxML [35], as implemented in the programme raxmlGUI 0.95 [36]. The GTRGAMMAI nucleotide substitution model was applied for the ML heuristic search and the ML rapid bootstrap analysis.

Fungal distribution at the genotype and species levels

To evaluate the specificity of the fungi to *Azteca brevis*, the frequency of occurrence on the sampled trees was analyzed. For each tree, a matrix containing genotypes and sampled trees was compiled (Table S3 in Appendix S1) and Bray–Curtis similarity indices among sampled trees were plotted by non-metric multidimensional scaling (NMDS) with 1000 restarts. In addition, an ANOVA of similarity (ANOSIM) was conducted with up to 999 permutations, based on different geographical study sites

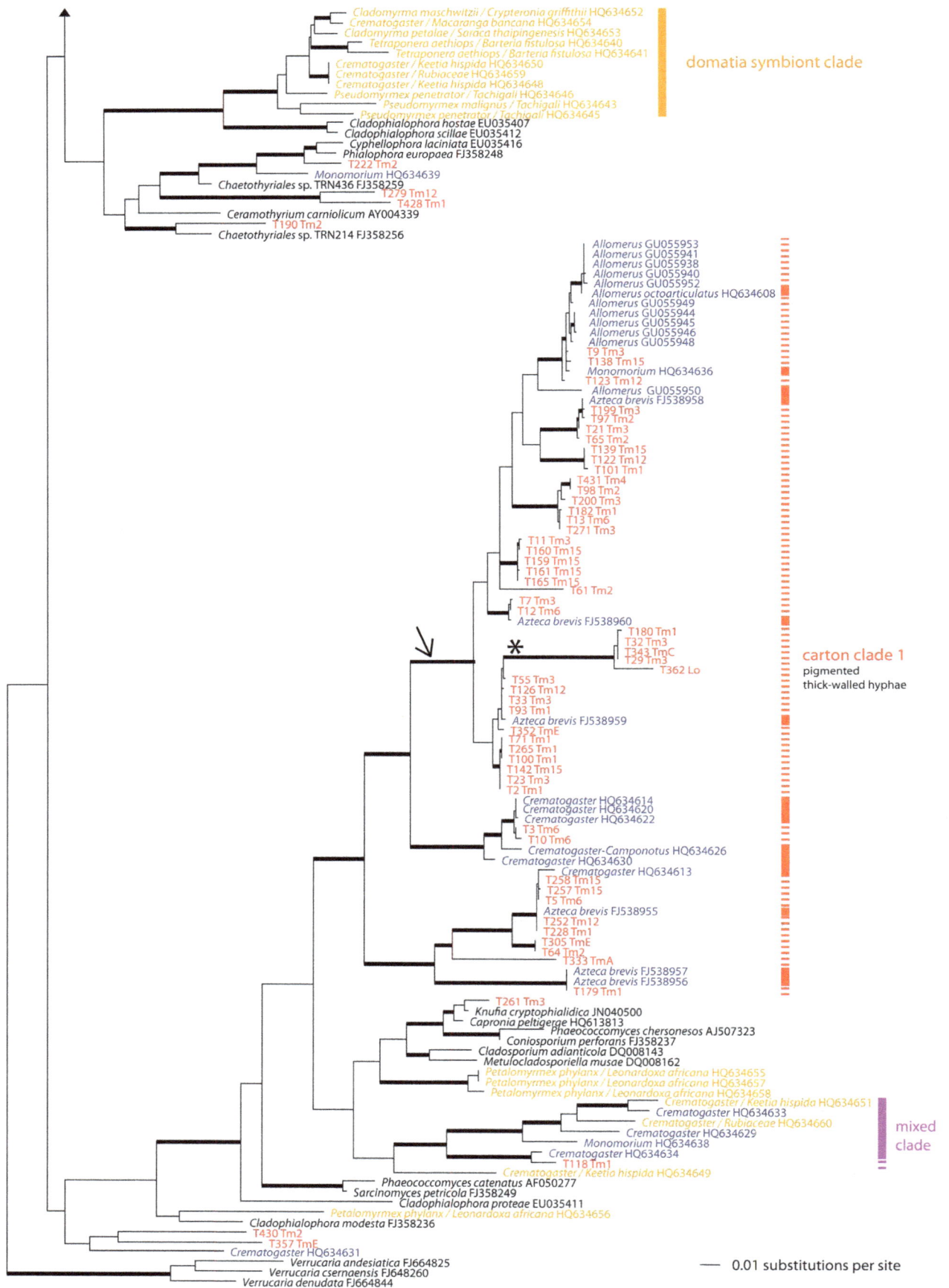

Cladomyrma maschwitzii / Crypteronia griffithii HQ634652
Crematogaster / Macaranga bancana HQ634654
Cladomyrma petalae / Saraca thaipingenesis HQ634653
Tetraponera aethiops / Barteria fistulosa HQ634640
Tetraponera aethiops / Barteria fistulosa HQ634641
Crematogaster / Keetia hispida HQ634650
Crematogaster / Rubiaceae HQ634659
Crematogaster / Keetia hispida HQ634648
Pseudomyrmex penetrator / Tachigali HQ634646
Pseudomyrmex malignus / Tachigali HQ634643
Pseudomyrmex penetrator / Tachigali HQ634645

domatia symbiont clade

Cladophialophora hostae EU035407
Cladophialophora scillae EU035412
Cyphellophora laciniata EU035416
Phialophora europaea FJ358248
T222 Tm2
Monomorium HQ634639
Chaetothyriales sp. TRN436 FJ358259
T279 Tm12
T428 Tm1
Ceramothyrium carniolicum AY004339
T190 Tm2
Chaetothyriales sp. TRN214 FJ358256

Allomerus GU055953
Allomerus GU055941
Allomerus GU055938
Allomerus GU055940
Allomerus GU055952
Allomerus octoarticulatus HQ634608
Allomerus GU055949
Allomerus GU055944
Allomerus GU055945
Allomerus GU055946
Allomerus GU055948
T9 Tm3
T138 Tm15
Monomorium HQ634636
T123 Tm12
Allomerus GU055950
Azteca brevis FJ538958
T199 Tm3
T97 Tm2
T21 Tm3
T65 Tm2
T139 Tm15
T122 Tm12
T101 Tm1
T431 Tm4
T98 Tm2
T200 Tm3
T182 Tm1
T13 Tm6
T271 Tm3
T11 Tm3
T160 Tm15
T159 Tm15
T161 Tm15
T165 Tm15
T61 Tm2
T7 Tm3
T12 Tm6
Azteca brevis FJ538960
T180 Tm1
T32 Tm3
T343 TmC
T29 Tm3
T362 Lo
T55 Tm3
T126 Tm12
T33 Tm3
T93 Tm1
Azteca brevis FJ538959
T352 TmE
T71 Tm1
T265 Tm1
T100 Tm1
T142 Tm15
T23 Tm3
T2 Tm1
Crematogaster HQ634614
Crematogaster HQ634620
Crematogaster HQ634622
T3 Tm6
T10 Tm6
Crematogaster-Camponotus HQ634626
Crematogaster HQ634630
Crematogaster HQ634613
T258 Tm15
T257 Tm15
T5 Tm6
Azteca brevis FJ538955
T252 Tm12
T228 Tm1
T305 TmE
T64 Tm2
T333 TmA
Azteca brevis FJ538957
Azteca brevis FJ538956
T179 Tm1

carton clade 1

pigmented
thick-walled hyphae

T261 Tm3
Knufia cryptophialidica JN040500
Capronia peltigerae HQ613813
Phaeococcomyces chersonesos AJ507323
Coniosporium perforans FJ358237
Cladosporium adianticola DQ008143
Metulocladosporiella musae DQ008162
Petalomyrmex phylanx / Leonardoxa africana HQ634655
Petalomyrmex phylanx / Leonardoxa africana HQ634657
Petalomyrmex phylanx / Leonardoxa africana HQ634658
Crematogaster / Keetia hispida HQ634651
Crematogaster HQ634633
Crematogaster / Rubiaceae HQ634660
Crematogaster HQ634629
Monomorium HQ634638
Crematogaster HQ634634
T118 Tm1
Crematogaster / Keetia hispida HQ634649

**mixed
clade**

Phaeococcomyces catenatus AF050277
Sarcinomyces petricola FJ358249
Cladophialophora proteae EU035411
Cladophialophora modesta FJ358236
T430 Tm2
T357 TmE
Crematogaster HQ634631
Verrucaria andesiatica FJ664825
Verrucaria csernaensis FJ648260
Verrucaria denudata FJ664844

— 0.01 substitutions per site

Figure 3. Phylogram of Chaetothyriales, bottom part. Continuation of Figure 2 with arrowheads indicating the connection. For label and colour descriptions see legend to Figure 2. The clade labelled with an asterisk (*) is placed more basally (arrow) in the MP analysis than in ML and BA analyses. Note the domatia-symbiont clade, which remains distinct from carton fungi.

(Table S2 in Appendix S1). All analyses were carried out with Primer 5 5.2.9 (PRIMER-E, 2002).

Slightly different genotypes can, however, represent the same species: the maximum number of mutations between two individuals of the same species (mutation limit) varies with the DNA region and type of organism. For fungi, the ITS region is more variable than SSU or LSU [37]. Because ITS is the main DNA fragment in the present study, the mutation limit was determined by multiplying the maximum intraspecific variation of ITS (0.58%) [37] by the average sequence length. We define an OTU as a species with a maximum genotype variation of 12.02 mutations. The abundance of OTUs was also analyzed using a modified presence–absence matrix (Table S4 in Appendix S1).

Light microscopy and SEM analysis of carton material and fungal hyphae

Pieces of carton from 16 trees and aerial hyphae of the pure cultures were investigated using a Zeiss AxioImager A1 compound microscope with a Zeiss AxioCam ICc3 digital camera. SEM investigations were made with a Jeol JSM-T 300 scanning electron microscope (SEM) at 10 kV.

Results

Cultures

Pure cultures of carton fungi were obtained from carton material of host trees colonised by *Azteca brevis*. Because different species could not be distinguished morphologically, all the pigmented hyphae that germinated were transferred to MEA plates to obtain pure cultures. In total, 423 pure cultures were sequenced, resulting in 128 different genotypes after identical sequences were removed.

Molecular phylogenetic analyses

After adding relevant sequences from GenBank, the final alignment, including outgroups, consisted of 258 sequences and 7767 alignment positions. In the best tree from the ML search (shown as a phylogram in Figure 2–3), backbone support is mostly low or absent, but most of the subclades are well-supported. Consensus trees across the three analyses differed topologically only in one point: one clade of five sequences containing a large indel of c. 1000 bp within the LSU region is located more basally in the topology resulting from the MP analysis than in the topologies from BA and ML analyses (marked with an asterisk and an arrow in Figure 3). Bootstrap support for this node in the MP analysis was only 73%, so this difference was not considered further.

The phylogenetic reconstruction revealed three main clades containing 73% (100 out of 128) of the isolated genotypes (Figure 2–3). Carton clades 1 and 2 were described by Voglmayr *et al.* [7], but carton clade 3, which was represented by a single sequence in previous analyses, is new. The remaining 28 genotypes are distributed across the phylogenetic tree of Chaetothyriales, except for the "domatia-symbiont clade". Three sequences from carton material from Cameroon and Thailand are nearly identical to some fungi sequenced in this study. Each of those three fungal genotypes (marked by blue dots in Figure 2) differs by only three mutations from fungi grown from carton structures collected in Costa Rica.

Fungal distribution at the genotype and species levels

The 128 different genotypes isolated from carton material of 19 *Azteca*-inhabited trees were analyzed with a presence–absence matrix (Table S3 in Appendix S1). The matrix showed that no genotype was found on all trees, with the most common one isolated from nine out of 19 trees. On average, 10.5 different genotypes occurred on each tree, of which 46% were unique to single trees. In the carton sample of one *Lonchocarpus* tree inhabited by *Azteca brevis*, 11 out of 12 genotypes (92%) were unique.

Bray–Curtis similarity indices between sampled trees were calculated to investigate the correlation between genotype composition and collection site. The non-metric multidimensional scaling (NMDS) plot showed no clustering of trees from the same collection site (Figure 4), and the analysis of similarity (ANOSIM) showed a significance level of only $P = 0.32$. A correlation between genotype composition and collection site can therefore be ruled out. One sampled tree (Tm10) had to be excluded because only a single, unique fungus could be isolated, and the Bray–Curtis distance to the other sampled trees was too great for Tm10 to be displayed without clustering all remaining trees too tightly together.

The results were similar at the approximated species level. The 128 genotypes were reduced to 62 OTUs, and the most common OTU in the modified presence–absence matrix, represented by nine genotypes, was found on 12 out of 19 trees (63%) (Table S4; Figure S1 in Appendix S1). Three other OTUs were found on a total of 9 out of 19 trees (47%). The correlation of fungal community and collection site at the species level is weak (NMDS plot, Figure S2 in Appendix S1) and not significant (ANOSIM: $P = 0.16$).

Light microscopy of carton material

Dark, melanised moniliform hyphae with thick cell walls (Figure 1F) appeared to be dominant in each sample. The cell width of these thick-walled hyphae ranged from about 6 to 9 μm. Hyaline hyphae with a cell width less than 5 μm were also present, but appeared less abundant (Figure 1E). Aerial hyphae of pure cultures representing the three carton clades were also examined and, surprisingly, darkly pigmented, thick-walled hyphae are largely restricted to carton clade 1, whereas thin-walled hyaline hyphae are found in carton clades 2 and 3 (Figure 2–3).

Discussion

There is growing evidence that multicellular organisms are shaped by symbioses with smaller partners – often microbial – that contribute to their host's nutrition, protection and even to their normal development [38]. In obligate interactions between ants and plants, for example, it has only recently become apparent that micro-organisms are major partners in interactions that go far beyond the relationship between the ant and the plant. In ant–plant symbioses from Africa, America and Asia, ascomycete fungi are cultivated in domatia and on ant-built carton structures, involving a wide range of distantly related ants and plant families [5,7]. This study is, however, the first dealing with fungi in ant-built cardboard-like carton material in which pure cultures of several samples were made per colony. This resulted in 381 pure cultures and 128 chaetothyrialean genotypes, the highest number

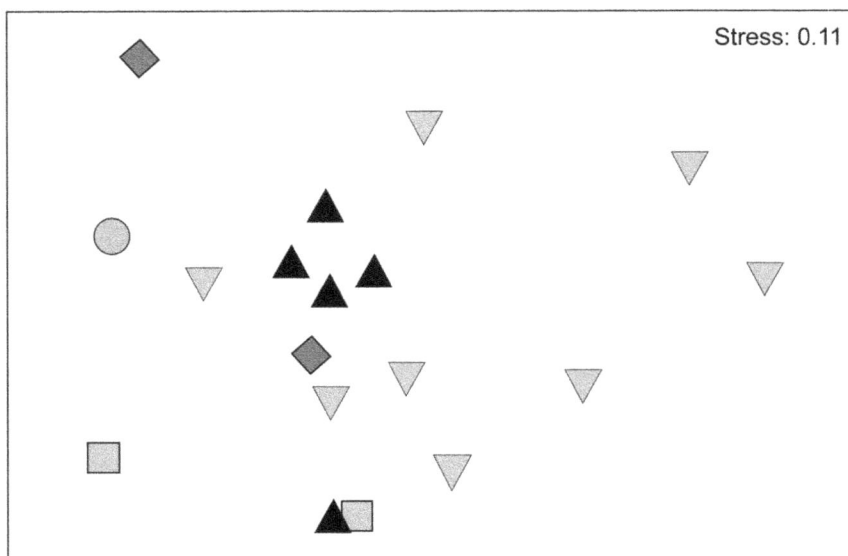

Figure 4. Correlation analysis of genotype sets and collection sites. Non-metric multidimensional scaling (NMDS) plot based on Bray–Curtis similarities between fungal genotype sets occurring on carton material of 18 sampled *Tetrathylacium macrophyllum* trees, and one *Lonchocarpus* sp., colonised by *Azteca brevis*. Different symbols represent trees from different collection sites (squares: Waterfall Trail; triangles: Bird Trail; circle: Río Gamba; diamonds: Río Bolsa; inverted triangle: Río Sardinal).

of carton-associated chaetothyrialean symbionts ever found associated with a single ant species.

This high number of fungus genotypes of the ascomycete order Chaetothyriales ("black yeasts") is astonishing, because these fungi generally seem to have weak competitive abilities. They are slow-growing and often extremophilic. They occur on nutrient-poor substrates, such as leaf or rock surfaces [15,39], or in toxic environments [17,18], and quickly disappear under less extreme conditions [7]. It is not yet known why "black yeasts" are so dominant on the carton material of ants. The frequent germination of fast-growing "weeds" on isolation plates indicates that the spores of moulds probably occur on the carton surface, but that their growth is inhibited. It may be the gallery substrate, the weeding and grooming behaviour of *Azteca brevis*, some ant-specific compounds or antifungal substances released from the fungi themselves that cause this inhibition. One indication that the construction material also shapes the fungal community on the galleries is the fact that 11 of 12 genotypes (92%) isolated from the samples collected from carton on a *Lonchocarpus* tree were unique and not present on the *Tetrathylacium* trees. *Azteca brevis* workers use particles of bark, excavated pith tissue and epiphylls from the host trees as materials for gallery construction, and plant secondary compounds may disfavour fungi other than Chae-tothyriales. Furthermore, *Azteca brevis* was observed to groom the carton galleries constantly, and even to nourish them (M. Nepel & V. Mayer, unpubl.). Antibacterial and antifungal compounds produced by ants' exocrine glands, such as the metapleural gland, may play an important role in preventing other moulds from growing [40,41]. In contrast, Chaetothyriales are able to tolerate and even to metabolise aromatic hydrocarbons [18] and can therefore cope and may even use ant-produced antifungal compounds metabolically. Finally, Chaetothyriales themselves might produce bioactive substances against competing fungi [42]. The combination of these factors may account for the relationship between ants and those fungi.

Molecular phylogenetic analyses and fungal diversity

There was no ubiquitous fungus (genotype or OTU) among the 128 genotypes found in this study; 78% of the sequenced carton fungi clustered into three clades. Two of those clades were already established [7], and we have discovered a third (Figure 2). The 28 genotypes that were not assigned to any of those clades were scattered across the whole phylogenetic tree (Figure 2–3), but none of the 128 fungal genotypes isolated from the carton samples arose in the "domatia-symbiont clade" [7]. The fungi belonging to this special clade are highly distinct from carton fungi in terms of their hyphal morphology and growth form (hyaline or less pigmented and tending to produce spores in domatia fungi) [7] probably due to their different functions. Domatia fungi are used as food for the larvae [8], whereas carton fungi probably improve the stability of carton walls in ant nests or runway galleries. The specificity and coevolutionary dynamics between domatia and carton symbionts may differ.

Fungus specificity at the genotype and species level

In most known insect–fungus symbioses (e.g. termites [43,44] or leaf-cutter ants [45]), the associated fungus is cultivated for food, whereas *Azteca brevis* is not likely to eat the fungi, but cultivates them for their nest architecture. In this association, the ants as a single host use a group of multiple fungus species for this purpose. The interpretation of the degree of specificity depends on the taxonomic level and differs between the fungi involved. At higher taxonomic levels, the interaction specificity between *Azteca brevis* and Chaetothyriales is high: chaetothyrialean fungi were found in every carton sample analyzed. At the genotype level, a much more modest degree of interaction specificity was seen (Table S3 in Appendix S1): the most common genotype (T121_Tm12) was associated with 9 out of 19 ant colonies (c. 47%). No correlation between fungal community and collection site was found, and the fungal community seems not to be habitat-specific (Figure 4). Merging genotypes into OTUs based on the maximum intraspe-cific variation of ITS [37] increased the occurrence of the most common fungus (represented by T66 in Table S4 in Appendix S1) to 63% of the ant colonies and the three next most common OTUs to 47% (Table S4 in Appendix S1). Surprisingly, a mean of eight different OTUs were found per carton sample, but no specialist fungal partner obligate to all *Azteca brevis/Tetrathylacium macrophyllum* associations was found. Environmental samples for the most common OTU could not be analyzed, because no primers could be developed which were specific enough for reliable separation of the OTUs, and only culturing was possible. It might be argued that culturing introduces a bias, in that a genotype could have been missed. Because we sampled the carton galleries of every colony and tree several times each, and also performed the isolation procedure on average three times per sample, it is unlikely that we missed any genotype. A heterogeneous spatial distribution of the genotypes along the branches can also be excluded, because we always took samples at positions from the branching point to the tip of a branch, and would therefore have included any heterogeneity.

Moreover, some fungal sequences from *Azteca brevis* carton are nearly identical at the SSU–ITS–LSU region with those of fungi from cartons of other ant species and from samples collected on other continents. Two sequences from *Crematogaster* carton nests from Cameroon (*Crematogaster* HQ634612 and *Crematogaster* HQ634621), and one found in Thailand (*Crematogaster* HQ634632) [7] differ by only three mutations from the Costa Rican sequences (blue dots in Figure 2). This indicates that at least some ant-associated chaetothyrialean fungi have a transcontinental distribution and are associated with more than one ant genus rather than being specific to *Azteca brevis* hosts.

Fungus selection and origin of the fungi

The carton runway galleries of *Azteca brevis* are used as a defence against intruders and as ambush traps for capturing prey (V. Mayer, pers. observ.), analogous to the carton galleries described for *Allomerus decemarticulatus* [13]. The fungal symbionts may play a particularly important functional role by increasing the stability of the carton produced by the ants [6,13]. The hypothesis that fungi are used for reinforcement was first raised by Lagerheim for the carton constructions of *Lasius fuliginosus* [10] and it has been accepted by many authors [6,11,13,46]. Trimming and grooming of the carton fungi has been observed and, due to the reinforcement demand, it may be expected that *Azteca brevis* would select fungal symbionts with a particular phenotype. Because we conducted a pure-culture method, we were able to examine the hyphae microscopically and correlate genotype or OTU with hyphal type. Two types of hyphae were regularly found: thin-walled, hyaline hyphae (Figure 1E), which represent two of our three major carton clades. The other type, thick-walled, melanised hyphae (Figure 1F), are less abundant in the species in our phylogenetic tree. When investigating pieces of carton, however, the thick-walled hyphae visibly dominated the biomass. This indicates that the second hyphae type is favoured on the carton. This may be either due to the substrate or due to the ants' preference. Fungi with thick-walled, melanised hyphae are likely to be better for the carton's stability than thin-walled hyphae.

Unfortunately, no alate queens were found, and interpretations of the transmission of the fungi must therefore be made without experimental evidence. The lack of any strong specificity indicates that the fungi growing on the *Azteca brevis* carton are either transmitted horizontally or environmentally acquired, although the degree of host–symbiont specificity is not always correlated with the transmission mode [47]. *De novo* acquisition of fungal symbionts from the environment in each ant generation would,

however, explain the high number of genotypes and OTUs in the carton samples. *Azteca brevis* workers may collect spores or hyphal fragments from the environment, as termites do [48], but detailed field observations of the worker ants' behaviour are needed to prove this. Also, the plant material used for construction (typically bark and epiphylls) may already be infected with chaetothyrialean spores. The branches and stems of trees inhabited by *Azteca brevis* appear to be completely cleaned of epiphylls (algae, bryophytes and lichens) and epiphytes. In fact, the newly described Chaetothyriales family Trichomeriaceae was found to grow on the surface of living leaves [49]. Investigation of the host plants' surface (stem, branches and leaves) and the epiphylls are needed to clarify whether Chaetothyriales are already present. The chaetothyrialean fungus community of the carton galleries might be a subset of the fungal community found on the host plant's surface.

Two analogous systems: *Allomerus* sp. and *Azteca brevis*

A plant–ant–fungus association with runway galleries that are structurally similar to those in the present study is seen in Amazonian *Allomerus* ants living on *Hirtella physophora* (Chrysobalanaceae) and *Cordia nodosa* (Boraginaceae) [14]. In contrast to the *Azteca brevis* carton, with its guild of numerous species of Chaetothyriales, the *Allomerus* ants are described as cultivating one specific fungal symbiont on the carton galleries (see upper part of carton clade 1; Figure 3). *Allomerus* foundress queens apparently store a pellet with the specific carton fungus represented by a monophyletic group of haplotypes on the domatium wall. While building their runway galleries, *Allomerus* workers were observed to glue pellets from scraped epidermis and mesophyll of the inner domatia walls onto the gallery frame built from trichomes of the host plant [14]. In *Azteca brevis* colonies, no inoculation pellet was found, an observation that supports the hypothesis that fungal symbionts are acquired *de novo* from the environment.

Although the sampling methods were different (only newly produced carton material was collected from the *Allomerus* galleries, whereas it was mainly "mature" black material that was sampled from the *Azteca brevis* galleries), the frequency of the most common fungus or OTU was more or less the same: on 57% (138 out of 240) of the *Allomerus* galleries, and on 63% (12 out of 19) of the *Azteca brevis* galleries. A fungal species found in a little over half of the analyzed carton samples should not, however, be regarded as specific.

Conclusions

We give an insight into the diversity of Chaetothyriales present in the carton galleries of *Azteca brevis*. Our results refute the initial hypothesis that *Azteca brevis* forms a symbiosis with a specific fungus. At the genotype level as well as that of approximated species (OTUs), the fungi we isolated appear to be a guild of different Chaetothyriales. An obligate mutualism with the fungi found in carton galleries of *Azteca brevis* is found for the host-ant with the ascomycete order Chaetothyriales; on the level of fungal species, no obligate mutualism is found. Moreover, *Azteca brevis* does not seem to strongly select for a particular morphological type, as both hyaline, thin-walled hyphae and pigmented, thick-walled hyphae are present in the carton. *Azteca brevis* cultivates and uses many different kinds of Chaetothyriales, and future research is needed to clarify the origins of these fungi. The reasons for the general preference of black yeasts in such ant–plant–fungus associations are still unclear. Knowledge of the diversity, coevolutionary processes and functional role of fungi in ant–plant symbioses is currently very fragmentary and further investigation is needed.

Acknowledgments

We would like to thank the staff of the Tropical Research Station La Gamba, Costa Rica. Research permission (No 182-2010-SINAC) was kindly granted by Javier Guevara of SINAC – Sistema Nacional de Areas de Conservación de Costa Rica of the Ministry of Environment and Energy (MINAE). Many thanks are due to Tamara Bernscherer, Florian Etl, Katharina Kneissl and Rafael Ramskogler for helping us collect samples. We thank Jack Longino for identifying *Azteca brevis*. Two anonymous reviewers are cordially thanked for their great input, which improved the paper substantially. Chris Dixon eliminated the Germanisms, and made the syntax clearer and less painful to read for the native English-speaking community.

Author Contributions

Conceived and designed the experiments: VEM. Performed the experiments: MN VEM. Analyzed the data: MN HV VEM. Contributed reagents/materials/analysis tools: VEM MN JS HV. Wrote the paper: MN VM JS. Other: Received research permission from SINAC – Sistema Nacional de Areas de Conservación de Costa Rica of the Ministry of Environment and Energy (MINAE) (No. 182-2010-SINAC): MN VEM.

References

1. Fresenius G (1852) Beiträge zur Mykologie. vol. 2, Heinrich Ludwig Bönner, Frankfurt am Main.
2. Mueller UG, Rehner SA, Schultz TR (1998) The evolution of agriculture in ants. Science 281: 2034–2038.
3. Mueller UG, Scott JJ, Ishak HD, Cooper M, Rodrigues A (2010) Monoculture of leafcutter ant gardens. PLoS One 5.
4. Schlick-Steiner BC, Steiner FM, Konrad H, Seifert B, Christian E, et al. (2008) Specificity and transmission mosaic of ant nest-wall fungi. Proc Natl Acad Sci 105: 940–943.
5. Defossez E, Selosse MA, Dubois MP, Mondolot L, Faccio A, et al. (2009) Ant-plants and fungi: a new threeway symbiosis. New Phytol 182: 942–949.
6. Mayer VE, Voglmayr H (2009) Mycelial carton galleries of *Azteca brevis* (Formicidae) as a multi-species network. Proc R Soc B 276: 3265–3273.
7. Voglmayr H, Mayer V, Maschwitz U, Moog J, Djieto-Lordon C, et al. (2011) The diversity of ant-associated black yeasts: insights into a newly discovered world of symbiotic interactions. Fungal Biol 115: 1077–1091.
8. Blatrix R, Djiéto-Lordon C, Mondolot L, La Fisca P, Voglmayr H, et al. (2012) Plant-ants use symbiotic fungi as a food source: new insight into the nutritional ecology of ant-plant interactions. Proc R Soc B 279: 3940–3947.
9. Elliott JSB (1915) Fungi in the nests of ants. Trans Br Mycol Soc 5: 138–142.
10. Lagerheim G (1900) Über *Lasius fuliginosus* (Latr.) und seine Pilzzucht. Entomol Tidskr 21: 17–29.
11. Weissflog A (2001) Freinestbau von Ameisen (Hymenoptera, Formicidae) in der Kronenregion feuchttropischer Wälder Südostasiens. Bestandsaufnahme und Phänologie, Ethoökologie und funktionelle Analyse des Nestbaus. PhD thesis, J W Goethe Univ Frankfurt am Main, Germany.
12. Kaufmann E, Maschwitz U (2006) Ant-gardens of tropical Asian rainforests. Naturwissenschaften 93: 216–227.
13. Dejean A, Solano PJ, Ayroles J, Corbara B, Orivel J (2005) Arboreal ants build traps to capture prey. Nature 434: 973.
14. Ruiz-González MX, Malé P-JG, Leroy C, Dejean A, Gryta H, et al. (2011) Specific, non-nutritional association between an ascomycete fungus and Allomerus plant-ants. Biol Lett 7: 475–479.
15. Selbmann L, Isola D, Zucconi L, Onofri S (2011) Resistance to UV-B induced DNA damage in extreme-tolerant cryptoendolithic Antarctic fungi: detection by PCR assays. Fungal Biol 115: 937–944.
16. Gueidan C, Villaseñor CR, de Hoog GS, Gorbushina AA, Untereiner WA, et al. (2008) A rock-inhabiting ancestor for mutualistic and pathogen-rich fungal lineages. Stud Mycol 61: 111–119.
17. Seyedmousavi S, Badali H, Chlebicki A, Zhao J, Prenafeta-Boldú FX, et al. (2011) Exophiala sideris, a novel black yeast isolated from environments polluted with toxic alkyl benzenes and arsenic. Fungal Biol 115: 1030–1037.
18. Zhao J, Zeng J, de Hoog GS, Attili-Angelis D, Prenafeta-Boldú FX (2010) Isolation and identification of black yeasts by enrichment on atmospheres of monoaromatic hydrocarbons. Microb Ecol 60: 149–156.
19. Longino JT (2007) A taxonomic review of the genus *Azteca* (Hymenoptera: Formicidae) in Costa Rica and a global revision of the *aurita* group. Zootaxa 1491: 1–63.
20. Janzen DH, editor (1983) Costa Rican natural history. Chicago, IL: The University of Chicago Press.
21. Tennant LE (1989) A new ant-plant, *Tetrathylacium costaricense*. Symposium: Interaction between Ants Plants. Oxford, Abstracts vol.: Linnean Society of London. p. 27.
22. Schmidt M (2001) Interactions between Tetrathylacium macrophyllum (Flacourtiaceae) and its live-stem inhabiting ants. Master thesis, Univ Vienna, Austria.
23. De Hoog GS, Gerrits van den Ende AHG (1998) Molecular diagnostics of clinical strains of filamentous Basidiomycetes. Mycoses 41: 183–189.
24. Vilgalys R, Hester M (1990) Rapid genetic identification and mapping of enzymatically amplified ribosomal DNA from several *Cryptococcus* species. J Bacteriol 172: 4238–4246.
25. Werle E, Schneider C, Renner M, Völker M, Fiehn W (1994) Convenient single-step, one tube purification of PCR products for direct sequencing. Nucleic Acids Res 22: 4354–4355.
26. White TJ, Bruns T, Lee S, Taylor J (1990) Amplification and direct sequencing of fungal ribosomal RNA genes for phylogenetics. PCR Protocols: a guide to methods and applications. San Diego, USA: Academic Press. pp. 315–322.
27. Voglmayr H, Rossman AY, Castlebury LA, Jaklitsch WM (2012) Multigene phylogeny and taxonomy of the genus *Melanconiella* (Diaporthales). Fungal Divers 57: 1–44.
28. Jaklitsch WM, Voglmayr H (2011) *Nectria eustromatica* sp. nov., an exceptional species with a hypocreaceous stroma. Mycologia 103: 209–218.
29. Edgar RC (2004) MUSCLE: multiple sequence alignment with high accuracy and high throughput. Nucleic Acids Res 32: 1792–1797.
30. Hall TA (1999) BioEdit: a user-friendly biological sequence alignment editor and analysis program for Windows 95/98/NT. Nucleic Acids Symp 41: 95–98.
31. Huelsenbeck JP, Ronquist F (2001) MRBAYES: Bayesian inference of phylogenetic trees. Bioinformatics 17: 754–755.
32. Kumar S, Skjaeveland A, Orr RJS, Enger P, Ruden T, et al. (2009) AIR: A batch-oriented web program package for construction of supermatrices ready for phylogenomic analyses. BMC Bioinformatics 10: 357.
33. Posada D, Crandall KA (1998) MODELTEST: testing the model of DNA substitution. Bioinformatics 14: 817–818.
34. Swofford DL (2003) PAUP*. Phylogenetic Analysis Using Parsimony (*and Other Methods). Sunderland, Massachusetts, USA: Version 4. Sinauer Associates.
35. Stamatakis A (2006) RAxML-VI-HPC: maximum likelihood-based phylogenetic analyses with thousands of taxa and mixed models. Bioinformatics 22: 2688–2690.
36. Silvestro D, Michalak I (2012) raxmlGUI: a graphical front-end for RAxML. Org Divers Evol 12: 335–337.
37. Schoch CL, Seifert K a, Huhndorf S, Robert V, Spouge JL, et al. (2012) Nuclear ribosomal internal transcribed spacer (ITS) region as a universal DNA barcode marker for fungi. Proc Natl Acad Sci 109: 6241–6246.
38. McFall-Ngai M, Hadfield MG, Bosch TCG, Carey HV, Domazet-Lošo T, et al. (2013) Animals in a bacterial world, a new imperative for the life sciences. Proc Natl Acad Sci U S A 110: 3229–3236. doi:10.1073/pnas.1218525110.
39. Cannon PF, Kirk PM (2007) Fungal families of the world. Wallingford, UK: CABI Publishing.
40. Schlüns H, Crozier RH (2009) Molecular and chemical immune defenses in ants (Hymenoptera: Formicidae). Myrmecological News 12: 237–249.
41. Yek SH, Nash DR, Jensen AB, Boomsma JJ (2012) Regulation and specificity of antifungal metapleural gland secretion in leaf-cutting ants. Proc R Soc B 279: 4215–4222.
42. El-Elimat T, Figueroa M, Raja HA, Graf TN, Adcock AF, et al. (2012) Benzoquinones and Terphenyl compounds as Phosphodiesterase- 4B inhibitors from a fungus of the order Chaetothyriales (MSX 47445). J Nat Prod 76: 382–387.
43. Aanen DK, Ros VID, de Fine Licht HH, Mitchell J, de Beer ZW, et al. (2007) Patterns of interaction specificity of fungus-growing termites and *Termitomyces* symbionts in South Africa. BMC Evol Biol 7: 115. doi:10.1186/1471-2148-7-115.
44. Nobre T, Koné NA, Konaté S, Linsenmair KE, Aanen DK (2011) Dating the fungus-growing termites' mutualism shows a mixture between ancient codiversification and recent symbiont dispersal across divergent hosts. Mol Ecol 20: 2619–2627. doi:10.1111/j.1365-294X.2011.05090.x.
45. Mikheyev a S, Mueller UG, Boomsma JJ (2007) Population genetic signatures of diffuse co-evolution between leaf-cutting ants and their cultivar fungi. Mol Ecol 16: 209–216. doi:10.1111/j.1365-294X.2006.03134.x.
46. Maschwitz U, Hölldobler B (1970) Der Kartonnestbau bei *Lasius fuliginosus* Latr. (Hym. Formicidae). Z Vgl Physiol 66: 176–189.

47. Fabina NS, Putnam HM, Franklin EC, Stat M, Gates RD (2012) Transmission mode predicts specificity and interaction patterns in coral-*Symbiodinium* networks. PLoS One 7: 1–9. doi:10.1371/journal.pone.0044970.

48. Korb J, Aanen DK (2003) The evolution of uniparental transmission in fungus-growing termites (Macrotermitinae). Behav Ecol Sociobiol 53: 65–71. doi:10.1007/sOQ265-002-0559-y.

49. Chomnunti P, Bhat DJ, Jones EBG, Chukeatirote E, Bahkali AH, et al. (2012) Trichomeriaceae, a new sooty mould family of Chaetothyriales. Fungal Divers 56: 63–76.

16

RNA-Seq Analysis of the *Sclerotinia homoeocarpa* – Creeping Bentgrass Pathosystem

Angela M. Orshinsky[1*◊], **Jinnan Hu**[1◊], **Stephen O. Opiyo**[2], **Venu Reddyvari-Channarayappa**[3], **Thomas K. Mitchell**[1], **Michael J. Boehm**[1]

1 Department of Plant Pathology, The Ohio State University, Columbus, Ohio, United States of America, 2 Molecular and Cellular Imaging Center-Columbus. The Ohio State University, Columbus, Ohio, United States of America, 3 Dale Bumpers National Rice Research Center, Stuttgart, Arizona, United States of America

Abstract

Sclerotinia homoeocarpa causes dollar spot disease, the predominate disease on highly-maintained turfgrass. Currently, there are major gaps in our understanding of the molecular interactions between *S. homoeocarpa* and creeping bentgrass. In this study, 454 sequencing technology was used in the *de novo* assembly of *S. homoeocarpa* and creeping bentgrass transcriptomes. Transcript sequence data obtained using Illumina's first generation sequencing-by-synthesis (SBS) were mapped to the transcriptome assemblies to estimate transcript representation in different SBS libraries. SBS libraries included a *S. homoeocarpa* culture control, a creeping bentgrass uninoculated control, and a library for creeping bentgrass inoculated with *S. homoeocarpa* and incubated for 96 h. A Fisher's exact test was performed to determine transcripts that were significantly different during creeping bentgrass infection with *S. homoeocarpa*. Fungal transcripts of interest included glycosyl hydrolases, proteases, and ABC transporters. Of particular interest were the large number of glycosyl hydrolase transcripts that target a wide range of plant cell wall compounds, corroborating the suggested wide host range and saprophytic abilities of *S. homoeocarpa*. Several of the multidrug resistance ABC transporters may be important for resistance to both fungicides and plant defense compounds. Creeping bentgrass transcripts of interest included germins, ubiquitin transcripts involved in proteasome degradation, and cinnamoyl reductase, which is involved in lignin production. This analysis provides an extensive overview of the *S. homoeocarpa*-turfgrass pathosystem and provides a starting point for the characterization of potential virulence factors and host defense responses. In particular, determination of important host defense responses may assist in the development of highly resistant creeping bentgrass varieties.

Editor: Ying Xu, University of Georgia, United States of America

Funding: This study was supported with state and federal dollars appropriated to The Ohio State University and the Ohio Agricultural Research and Development Center. The funders had no role in study design, data collection and analysis, decision to publish, or preparation of the manuscript.

Competing Interests: The authors have declared that no competing interests exist.

* E-mail: aorshinsky@gmail.com

◊ These authors contributed equally to this work.

Introduction

Sclerotinia homoeocarpa (Bennett) is an ascomycete fungus that causes dollar spot disease on turfgrass world-wide [1]. *S. homoeocarpa* can affect all species of turfgrass as well as some dicot plants [1]. Creeping bentgrass (*Agrostis stolonifera* L.) is a cool-season turfgrass that is common on golf course greens in the United States and Canada [2]. Many commonly used creeping bentgrass cultivars are highly susceptible to dollar spot disease and are subject to frequent, low mowing practices that promote disease outbreaks [1,3].

Symptoms of dollar spot include straw-colored, hourglass-shaped lesions with characteristic reddish-brown borders. Diseased areas grow to about 2.5 cm wide [1]. Under high disease pressure and favorable conditions the diseased areas will merge to form larger patches of diseased turf. Environmental conditions favoring disease development have been well documented [4,5,6]; however, prediction models designed to reduce fungicide inputs have been ineffective due to a lack of understanding of the *S. homoeocarpa* lifestyle, epidemiology, and disease etiology [1,4].

S. homoeocarpa is thought to overwinter in the thatch of turf swards as stroma. In the spring, *S. homoeocarpa* becomes active and infects the newly emerging leaf tissue through wounds, stomates, and directly with appressorium formation [7,8,9]. Only infertile apothecia have been recorded for North American isolates [10]; however, population studies suggest that genetic recombination in this fungus is possible [11]. Therefore, the possibility of ascospore production and the role of sexual or asexual spores as initial inoculum for dispersal of *S. homoeocarpa* cannot be discounted. Early studies noted root browning and cell death through the production of diffusible toxins [12]. Recent studies have identified several tetranorditerpenoid compounds that could be responsible for the root-browning and exhibit extremely phytotoxic properties; however, a correlation between the production of these compounds and disease symptoms was not established [13].

Cultural management strategies for dollar spot include maintaining adequate nitrogen balance, promotion of air flow to assist in dew removal, and using moderately resistant cultivars or species of turf [14,15]. On highly maintained areas such as golf courses, cultural practices are not sufficient for management of dollar spot, and fungicides are often applied biweekly to weekly. High amounts of fungicide use have resulted in resistance to several chemical classes commonly used on turf [1]. Other nonfungicide products that have been marketed for dollar spot control include plant defense activators that work by activating two different plant defense pathways: induced systemic resistance (ISR) and systemic acquired resistance (SAR). However, it is unclear which pathway would be most beneficial for preventing dollar spot epidemics.

A better understanding of the molecular interactions between *S. homoeocarpa* and creeping bentgrass will be essential for the development of more sustainable and practical management strategies, including the use of plant defense activators and the development of cultivars with increased resistance to *S. homoeocarpa*.

The introduction of next generation sequencing, also termed massively parallel sequencing (MPS), has enabled researchers to sequence the genomes and transcriptomes of organisms at a relatively low cost in return for a vast amount of data with quantitative properties [16,17,18,19]. In this paper, two MPS technologies were used to generate sequence data for RNA-Sequence (RNA-Seq) analysis: Illumina's sequencing-by-synthesis (SBS) and Roche's 454-pyrosequencing. The 454 reads were used for the *de novo* assembly of *S. homoeocarpa* and creeping bentgrass transcriptome libraries. SBS reads were mapped to the 454 assemblies to calculate transcript levels from *S. homoeocarpa* and creeping bentgrass during dollar spot disease development. The objective of this study was to identify transcripts that may be important for fungal virulence and creeping bentgrass defense. The results of the analysis will be used to form testable hypotheses for future studies on dollar spot etiology and turfgrass defense mechanisms.

Results

Transcriptomic Analysis

The *S. homoeocarpa* (SH) 454 sequencing data contained 600,760 reads total, and the *A. stolonifera* (AS) 454 sequence data contained 205,403 reads total (Table 1). The 454 read lengths ranged from 50 bp to >500 bp, with a majority of the reads between 400 and 500 bp in length. The transcriptome coverage for each of the 454 assemblies was $3.3\times$ coverage for AS and $17.2\times$ coverage for the SH assembly. These sequencing coverages were calculated by dividing the total number of sequence reads by the size of the respective, assembled transcriptome libraries.

The SBS data, which was used for calculating significant differences in transcript levels between libraries, resulted in 4.3–7.2 million reads (Table 2); however, these reads were only 16 bp long. This is due to the use of first generation Illumina sequencing protocols. The SBS reads provided ample coverage of the entire transcript assemblies: $14\times$ coverage for SH and $12\times$ for AS. The SBS library construction resulted in 9, 319 SH transcripts with lengths ranging from 400 bp to 3, 500 bp and a distribution peak

at 500 bp. Construction of the SBS transcript library for AS resulted in 20, 293 transcript sequences with lengths ranging from 200 bp to 8,500 bp and distribution peak at 480 bp. The full length of transcripts was not estimated since very few annotated coding sequences from either SH or AS were available at Genbank. The mapping percentage of SBS transcripts was quite low. Only 67.4% of SH PDA reads, 63.8% of SH PDB reads, and 33.3% of Interaction reads were mapped to the SH transcripts assembly, respectively (Table 2). Only 16.8% of AS only reads, and 5.7% of Interaction reads mapped to the AS transcripts library (Table 2). This is likely due to the short length of the SBS reads and stringent mapping parameters used in this experiment, which eliminated reads mapping to more than one location and did not allow for any base mismatches.

A Venn diagram was constructed to identify the distribution of the unique and common reads in both SBS and 454 SH, AS, and interaction libraries (Figure 1). There are discrepancies in the read distribution that appear between the SBS and the 454 libraries. The most obvious example is the increase in the number of reads that are unique to the SH-only library from 6.2% in the SBS to 58.9% in the 454 data. This is because there were more total SH reads in the 454 dataset, where four of the six libraries (69.8% of the reads) were constructed using RNA from *S. homoeocarpa* grown in pure culture. In the SBS dataset, only two out of four (31.7% of the reads) of the libraries were constructed from *S. homoeocarpa* grown in pure culture. The number of reads common to the interaction and control libraries was consistently low in both SBS and 454 datasets (Figure 1). The common reads are likely a result of similar sequences in both organisms, including those of transposable elements found in both organisms.

Table 2. Mapping characteristics of the *Sclerotinia homoeocarpa* (SH) and *Agrostis stolonifera* (AS) SBS reads to the SH and AS transcript assemblies.

SBS Library	SH PDA	SH PDB	AS 96h	Interaction
Number of Reads	6,101,988	4,350,510	6,885,250	7,182,868
Mapping Percentage to SH Assembly	67.4%	63.8%	0.9%	33.3%
Mapping Percentage to AS Assembly	2.6%	2.4%	16.8%	5.7%

Table 1. Characteristics of the 454 RNA-Seq data.

Transcript Assembly	Sequencing Libraries Included	Number of Reads	Number of Isotigs	Mean Isotig Length (bp)	Number of Singletons
Total SH	SH PDA 96h	600,760	10,101	1,172	51,502
	SH PDB 48h/96h/144h				
	Interaction 96h				
Total AS	AS 966h	205,403	5,017	898	58,446
	Interaction 96h				

SBS Reads

454 Reads

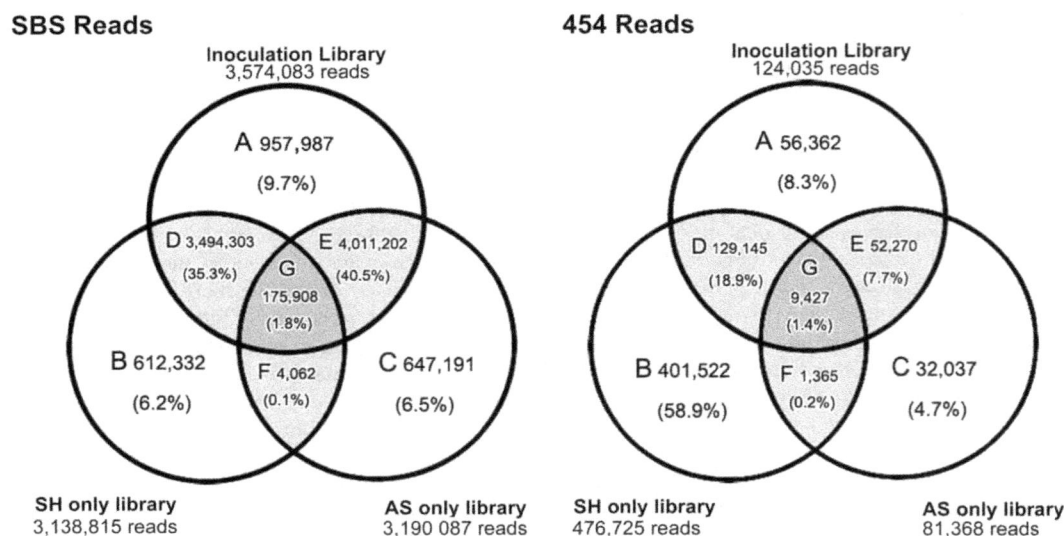

Figure 1. Venn diagram of RNA-Seq reads unique and common to the SH library, AS library, and Interaction library. A. Reads unique to the Inoculation library, B. Reads unique to the SH only library, C. Reads unique to the AS only library, D. Reads common to the Inoculation and SH library, E. Reads common to the Inoculation and AS library, F. Reads common to the SH and AS libraries, G. Reads common to the Inoculation, SH, and AS libraries.

Comparative analysis

The transcripts were blasted against the NCBI protein database for functional annotation using Blast2GO [20] blastx function. Fifty seven percent (5,257) of the 9,319 SH transcripts returned at least one annotation hit. Likewise, 57% (11, 551) of the 20, 293 AS transcripts returned at least one hit. From the blastn taxonomy report, top hits of the *S. homoeocarpa* sequences were primarily from *Sclerotinia sclerotiorum* (Lib.) de Bary (44. 1%) and *Botryotinia fuckeliana* (de Bary) Whetzel (34.9%) (Figure 2 a). This is not surprising considering their taxonomic similarity to *S. homoeocarpa*, and the abundance of *S. sclerotiorum* and *B. fuckeliana* sequence data available on GenBank. The 26 fungal species represented in the taxonomy report belong to various classes of the subphylum Pezizomycotina, with seven of the 26 species being pathogens of monocot plants. There were a small number of sequences related to monocots including rice (*Oryza sativa* Linnaeus, 1.6%), bamboo (*Phyllostachys edulis* (Carr.) Houz., 0.3%), and barley (*Hordeum vulgare* Linnaeus, 0.4%). We attributed this either to contamination of the fungal sequences by *A. stolonifera* sequences in the interaction library, to the presence of very highly conserved sequences, or to the presence of sequences occurring in both organisms, such as those of transposable elements. The top blast hit species for the *A. stolonifera* sequences included predominantly monocot grass species (Figure 2 b). Fungal sequences that were found included those of *Pyrenophora teres* Smed.-Pet. and *Phaeosphaeria nodorum* (E. Müll) Hedjar, pathogens of barley and wheat, respectively.

Comparison of RPKM ratios to Relative Transcription ratios

Reads Per Kilobase of exon model per Million mapped reads (RPKM) value was calculated of all SH and AS transcripts in different libraries and the Log_2 ratio was calculated to reflect the fold changes of interaction library compared to control libraries (SH PDA and AS only). Relative transcription data from real-time RT-PCR assays was obtained for selected transcripts with significantly different transcription levels during infection. The relative transcription data verified that the RPKM calculating and

statistical analysis of transcriptome data was accurate (Figure 3). The real time RT-PCR data confirms the upregulation of two fungal xylanases (SH_5411 and SH_5726), a fungal laccase (SH_7961), and a creeping bentgrass germin protein (AS_608) (Figure 2). Downregulated transcripts SH_6925 (scyatalone dehydrogenase) and SH_8369 (β-mannosidase) were also verified (Figure 2).

Transcripts Expression Analysis

To identify transcripts that were up-regulated in the interaction condition, transcripts with statistically significant changes of RPKM values during infection were examined to determine if there were patterns in the types of genes with increased or decreased transcription level. Table 3 summarizes some of the types of SH transcripts of interest during infection including a long list of glycosyl hydrolase transcripts and various proteases. Table 4 summarizes transcripts of interest in the AS library including a long list of retrotransposons, defense-related proteins including 13 different germin-like proteins, and various enzymes involved in plant hormone synthesis and regulation. Interestingly, all jasmonate-induced protein transcripts were significantly down regulated at 96 hpi. A full list of SH and AS transcripts that are significantly over or under expressed during infection are available as supplementary data (Tables S1 and S2, respectively).

Gene Ontology Analysis

The list of significantly upregulated SH and AS transcripts were sorted into gene ontology (GO) term categories for molecular function, biological processes, and cellular component using the Blast2GO [20] program (Figures 4 and 5). The molecular function of GO terms associated with upregulated SH transcripts include a large percentage of plant cell-wall degrading enzymes categories (Figure 4 a) including polygalacturonase activity (4.0%), cellulase activity (5.1%), endopeptidase activity (4.0%), α-arabinofuranosidase activity (2.9%), and serine-type peptidase activity (5.1%). Other upregulated molecular function of GO terms included those for xenobiotic-transporting activity (5.1%) and oxidoreductase

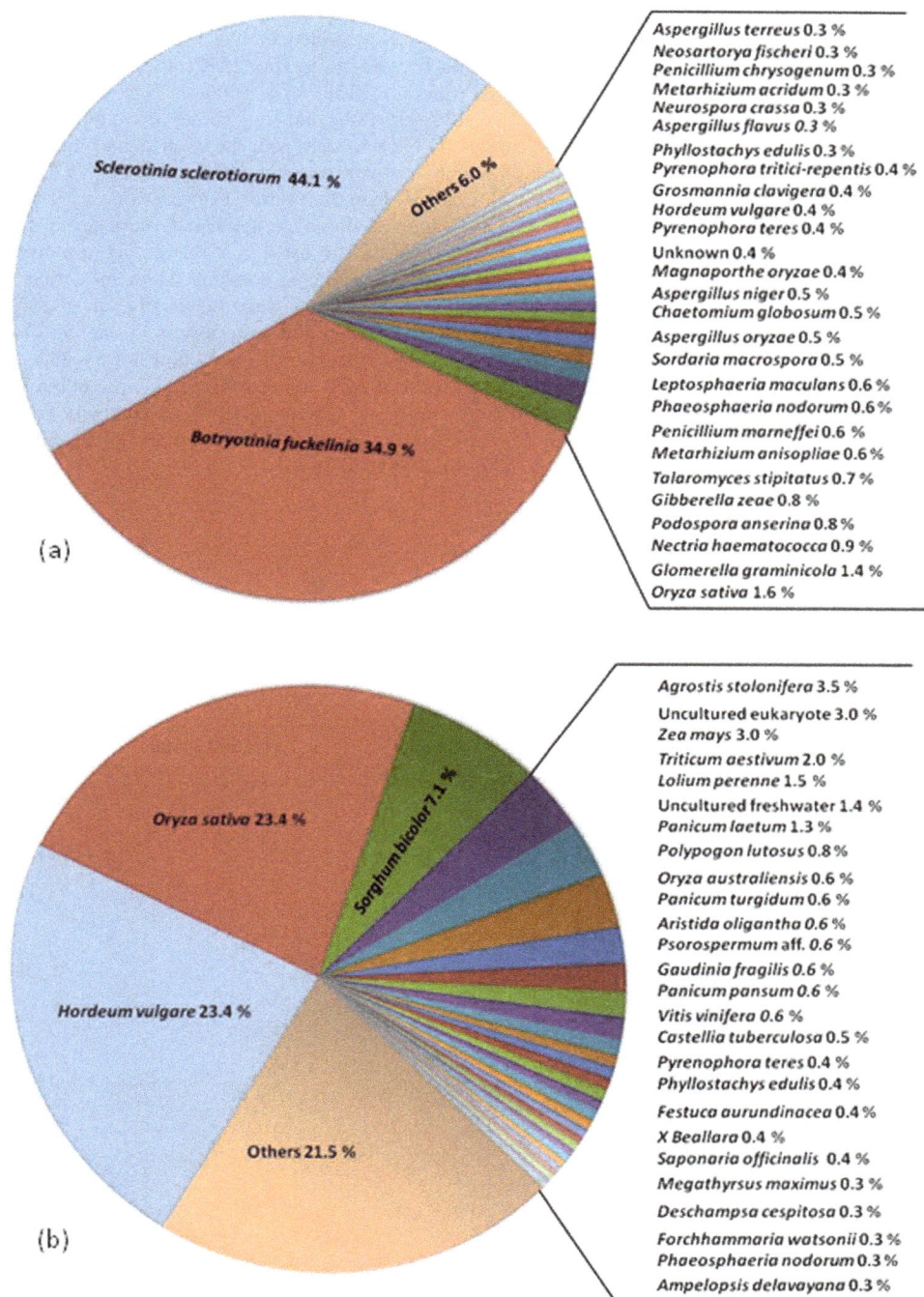

Figure 2. Top Species Blast hits: (a) SH transcript library (b) AS transcript library.

activity (7.4%). Biological process of upregulated SH transcript GO terms included auxin biosynthesis (11.4%), transmembrane transport (18.9%), and oxidation reduction (25.1%) (Figure 4 b). A majority of the upregulated SH cellular component GO terms were for those integral to the membrane (37.2%) and those for extracellular regions (34.0%) (Figure 4 c), perhaps reflecting the secretion of cell wall degrading enzymes and other secondary metabolites important for pathogenicity and virulence of *S. homoeocarpa*.

The molecular function of GO terms associated with the upregulated AS transcripts included those associated with hydrolase activity (19.2%), DNA binding activity (10.9%), and nucleotide binding (24.3%) (Figure 5 a). The biological process GO terms associated with the upregulated AS transcript library included those related to translation (10.8%), transport (14.6%), and DNA metabolic processes (11.1%) (Figure 5 b). Finally, an overwhelming number of cellular process GO terms from the AS upregulated library were associated with the mitochondrion (30.8%) and plastids (31.8%) (Figure 5 c).

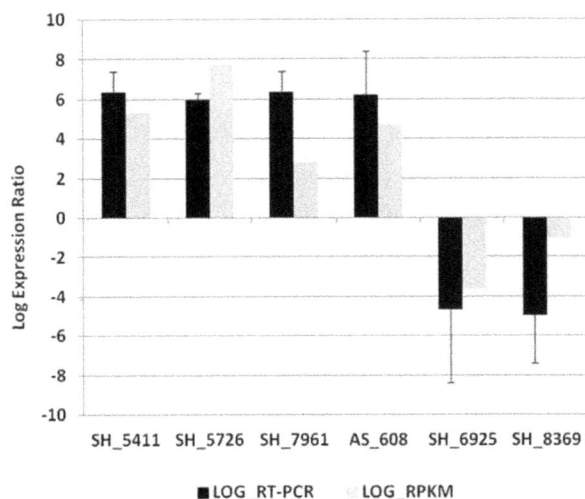

Figure 3. Validation of RPKM data for selected SH and AS transcripts using relative expression real time PCR.

Conserved Domain Analysis

Protein domains were identified in the *S. homoeocarpa* and *A. stolonifera* transcripts libraries using HMMER software (v3.0) [21]. Also among SH transcripts, 1588 were predicted to contain signal peptide by signalP (v4.0) [22], and thus could be candidates for secreted proteins. Three proteases with secretion signals included two subtilisin/sedolisin proteases with log-fold increases of 6.4 and 6.2 as well as a cuticle degrading serine protease with a log-fold increase of 4.0. Conserved domains significantly enriched in the interaction library ($P\leq0.01$) are listed in Table 5 (SH conserved domains) and Table 6 (AS conserved domains). Domains enriched in the SH interaction library include a variety of glycosyl hydrolases, proteases domains, and transporters (Table 5). Domains of interest in the enriched AS interaction library include cytochrome P450, various ABC transporters, and cupin domains, which include germin-class enzymes (Table 6). A complete list of SH and AS conserved domains can be found as supplementary data in tables S3 and S4, respectively.

Table 3. Summary of select *Sclerotinia homoeocarpa* transcript types that are significantly increased at 96 h post inoculation on creeping bentgrass.

Transcript Types	Description	Transcripts	Log fold change
Glycosyl hydrolases	ferruloyl esterase	2	4.8–11.5
	dihydroceramidase	1	3.1
	amylase	1	3.0
	arabinofuranosidase	7	2.4–9.9
	rhamnosidase	1	2.1
	endoglucanase	4	2.9–5.7
	glucosidase	4	3.1–4.9
	xylosidase	5	2.6–5.3
	cellobiohydrolase	4	4.3–6.9
	cellobiose dehydrogenase	1	4.1
	cellulase	1	3.6
	cutinase	1	3.0
	mannosidase	2	3.1–4.6
	xylanase	3	4.6–8.5
	glucanase	8	2.2–7.3
	polygalacturonase	3	3.0–8.1
	rhamnogalacturonase	2	2.9–8.1
	lysozyme	1	5.7
	pectinesterase	1	2.0
	Total	**52**	**2.0–11.5**
Proteases	serine protease	4	2.9–9.4
	aspergillopepsin-2	1	2.3
	peptidases	8	2.4–6.5
	neutral protease	1	5.8
	Total	**14**	**2.3–9.4**
Transporters	mdr-like abc transporter	7	4.1–6.6
	mfs multidrug	2	2.7–2.9
	Total	**9**	**2.7–6.6**

Table 4. Summary of select *Agrostis stolonifera* transcript types that are significantly increased at 96 h post inoculation with *Sclerotinia homoeocarpa.*

Transcript Types	Product	Transcripts	Log fold change
Transposons	athila retroelement	1	5.8
	copia-type retroelement	2	2.6–4.3
	gag-pol polyprotein	10	2.4–4.9
	mutator-like transposase	1	4.6
	retroelement pol poly	2	2.2–2.7
	retrotransposon line	1	3.0
	retrotransposon ty1-copia	6	3.4–5.8
	retrotransposon ty3-gypsy	13	2.5–8.6
	transposon en spm	15	2.0–5.2
	unidentified transposon	2	2.3–3.7
	retrotransposon unclassified	18	2.4–7.6
	Total	**72**	**2.0–8.6**
Defense	anthranilate synthase	1	4.5
	calcineurin	1	3.5
	cinnamoyl reductase	1	3.5
	cytochrome p450	5	2.0–5.5
	disease resistance nbs-lrr	3	2.3–4.0
	e3 ubiquitin-protein ligase	2	3.7–4.0
	fusarium resistance, i2c-5-like	1	2.9
	germin a	13	3.4–6.2
	mdr abc transporter	6	5.9–7.7
	pathogenesis associated pep2	1	9.1
	rust resistance kinase lr10	1	4.5
	terpene synthase	1	2.9
	ubiquitin	2	2.1–4.8
	ubiquitin-conjugating enzyme	1	2.1
	ubiquitin-specific protease	1	3.8
	zingiberene synthase	1	2.8
	Total	**47**	**2.0–9.1**

Discussion

Previous analyses using RNA-Seq have evaluated the reproducibility and accuracy of SBS [18] and 454 data for RNA transcription studies [23]. Both technologies have demonstrated higher-quality results than microarray and EST data [18,19]. However, SBS technologies were found to be superior due to the greater coverage and depth of sequences for a much lower cost. In this study, the mapping percentage of the SBS data to the *S. homoeocarpa* assembly ranged from 33.3–67.4% and *A. stolonifera* mapping percentages ranged from 5.7–16.8%. The mapping percentages were low due to the very stringent parameters that eliminated all reads with even a single mismatch and reads that mapped to more than a single location, as well as possible transcriptome misassembly. The use of these parameters and their resulting influence on the statistical analysis of the RPKM data was verified with real-time relative expression data. Mapping uncertainty in RNA-Seq analyses is one of the pitfalls of the SBS sequencing method. Short sequence lengths, as was the case in this study, make it difficult to account for paralogous genes, alternatively spliced isoforms, and low complexity sequences, all

of which result in multireads that are eliminated in the mapping process [24]. Low mapping percentage may also be due to the lower coverage of 454 transcriptome assemblies compared to the high SBS coverage.

RNA-Seq analysis in this study has made it possible to identify potential pathogenicity factors transcribed by *S. homoeocarpa* during infection and colonization of *A. stolonifera*. The most striking discovery was the number of enriched *S. homoeocarpa* transcripts within the interaction library that encoded glycosyl hydrolase enzymes. In fact, 52 of the upregulated SH transcripts encoded glycosyl hydrolase genes and 22.3% of molecular function GO terms were associated with glycosyl hydrolase activity. Within the total SH transcript library, a keyword search using search terms such as glycosyl hydrolase, cellulase and pectinase revealed over 100 transcripts annotated as glycosyl hydrolases in the SH library, representing 85 different families of the total 125 glycosyl hydrolase families [25,26]. The numerous transcripts representing such a varied group of glycosyl hydrolase families corroborate the ability of this fungus to infect and cause disease on a wide range of plant hosts. In this study, *S. homoeocarpa* glycosyl hydrolase genes showing upregulation were predominantly xylanases and arabi-

SH Molecular Function

- ■ Phosphate transmembrane-transporting ATPase activity, 5.1 %
- ■ Heme binding, 4.0 %
- ■ FAD binding, 2.9 %
- ■ Transferase activity (phosphorus-containing groups), 4.0 %
- ■ Polygalacturonase activity, 4.0 %
- ■ Xenobiotic-transporting ATPase activity, 5.1 %
- ■ ATP binding, 13.1 %
- ■ Cellulase activity, 5.1 %
- ■ Structural molecule activity, 2.9 %
- ■ Cellulose binding, 10.3 %
- ■ DNA binding, 6.3 %
- ■ Transcription regulator activity, 2.9 %
- ■ Oxidoreductase activity, 7.4 %
- ■ Zinc ion binding, 7.4 %
- ■ Endopeptidase activity, 4.0 %
- ■ Carboxylesterase activity, 2.9 %
- ■ Protein binding, 4.6 %
- ■ Alpha-N-arabinofuranosidase activity, 2.9 %
- ■ Serine-type peptidase activity, 5.1 %

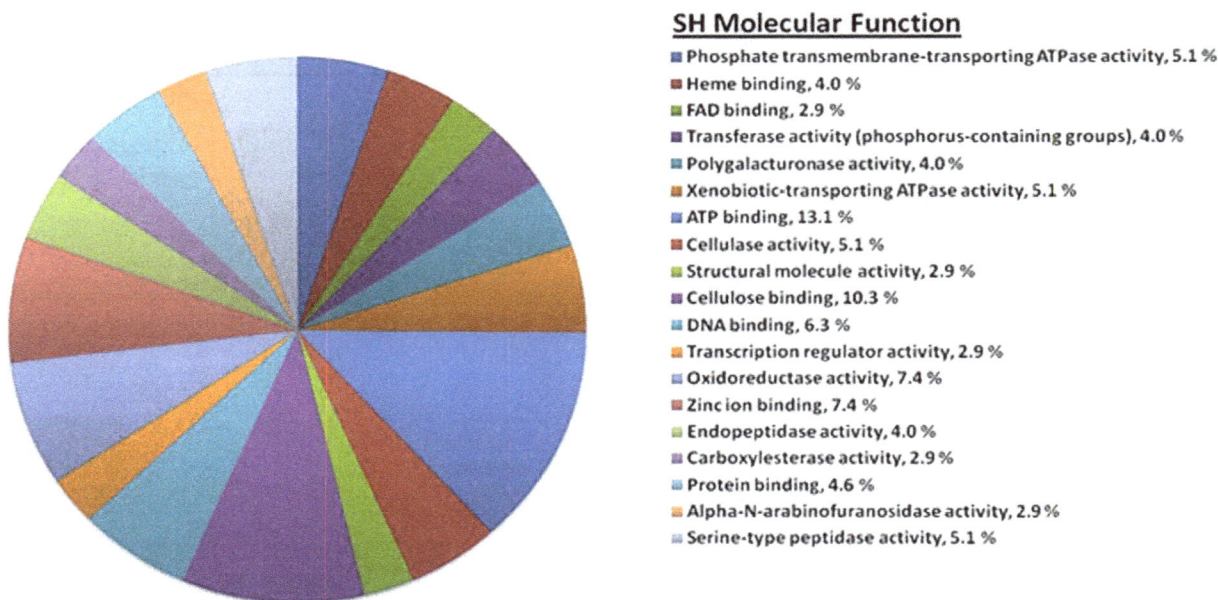

Figure 4. GO terms associated with upregulated SH transcripts. (a) Molecular Function, (b) Biological Process, (c) Cellular Component.

nases. This is not surprising since grass cell walls consist of approximately 40% xylans as well as glucuronoarabinoxylans (GAX) that make up a majority of monocot hemicelluloses [27]. Enzymes such as pectinases, xylanases, and cellulases are often virulence and pathogenicity factors for plant pathogenic fungi and bacteria. For example, *Botrytis cinerea* Pers. and *Septoria nodorum* (Berk.) Berk., demonstrate reduced virulence when a single gene encoding a cell wall degrading enzyme is knocked-out [28,29,30]. However, the large number of upregulated and functionally redundant cell wall degrading enzymes expressed by *S. homoeocarpa* interacting with creeping bentgrass will make verifying the role of any one enzyme extremely difficult.

Numerous proteinases, mostly serine proteases, were also enriched in the fungal interaction library. Secretion signals were detected on three of the proteases of interest, including two subtilisin/sedolisin proteases and one identified as a cuticle degrading serine protease. These three secreted proteases were upregulated in the interactoin library by 4.0–6.4 log-fold. Of the molecular function GO terms, 5.1% were associated specifically with serine-protease activity. Serine proteases are pathogenicity determinants for a variety of pathogenic fungi including nematode-parasitizing fungi, entomopathogens, and plant pathogenic fungi [31]. For example, protease deficient mutants of the hemibiotroph *Pyrenopeziza brassicae* B. Sutton & Rawl. were either avirulent or had dramatically reduced virulence [32]. Proteases in other plant pathogenic fungi have been characterized, but their role in pathogenicity has not necessarily been confirmed in each case [33,34].

In general, the presence of a variety of hydrolytic enzymes including glycosyl hydrolases and proteinases supports the broad host range reported for *S. homoeocarpa* [1]. The diversity and number of enzymes also support saprophytic ability of *S. homoeocarpa* [35]. For example, fungi with superior saprotrophic abilities, such as *Aspergillus* spp. were found to produce a more diverse collection of hydrolytic enzymes under a variety of conditions, while phytopathogens with inferior saprophytic abilities such as *Verticillium albo-atrum* Reinke & Berthold and *V. lecanii* (Zimm.) Viégas, produced a more targeted array of enzymes suited

to their particular host range [35]. The large number of hydrolytic enzymes expressed by *S. homoeocarpa* supports the hypothesis of superior saprophytic abilities of the dollar spot pathogen, which may play a role in the overwintering of the fungus within the thatch during the winter months.

Multi-drug resistance (MDR) ATP binding cassette (ABC) transporter proteins were also upregulated in the *S. homoeocarpa* interaction library, with seven of the MDR-ABC transporters being upregulated while only one MDR-ABC transporter transcript was downregulated. Members of the MDR-ABC transporter family have important roles in fungicide resistance, resistance to plant defense compounds and efflux of endogeneous toxins [36,37]. In *B. cinerea*, both ABC and MFS transporters have been implicated in resistance of the pathogen to a variety of fungicides including phenylpyrroles, azoles, anilopyrimidines, dicarboximides, and strobilurins [38,39,40]. The ABC transporter BcatrB of *B. cinerea* is strongly induced in the presence of fludioxonil, cyprodinil, cyproconazole, tebuconazole, and trifloxystrobin suggesting a role for this single transporter in resistance to multiple fungicide types [41]. Enrichment of a variety of transport proteins with similar function in the *S. homoeocarpa* interaction library may provide an explanation for the development of isolates that are resistant to more than one fungicide class. For example, isolates are known that are resistant to demethylation inhibitors as well as benzimidazoles and dicarboximides [1,42,43]. Aside from fungicide resistance, ABC transporters are also implicated in resistance to plant defense compounds and secretion of fungal toxins. For example, BcatrB not only provides resistance to fungicides but also to the plant defense compounds resveratrol and camalexin [39,40]. Characterization of these transporter proteins could lead to a better understanding of fungicide resistance mechanisms and *S. homoeocarpa*-creeping bentgrass interactions, resulting in more effective fungicide application programs and fungicide resistance management strategies.

Integrated fungicide resistance management recommendations often include using cultivars of grass that have increased pathogen resistance. Several cultivars of creeping bentgrass have decreased susceptibility to dollar spot disease [44,45]; however, the molecular

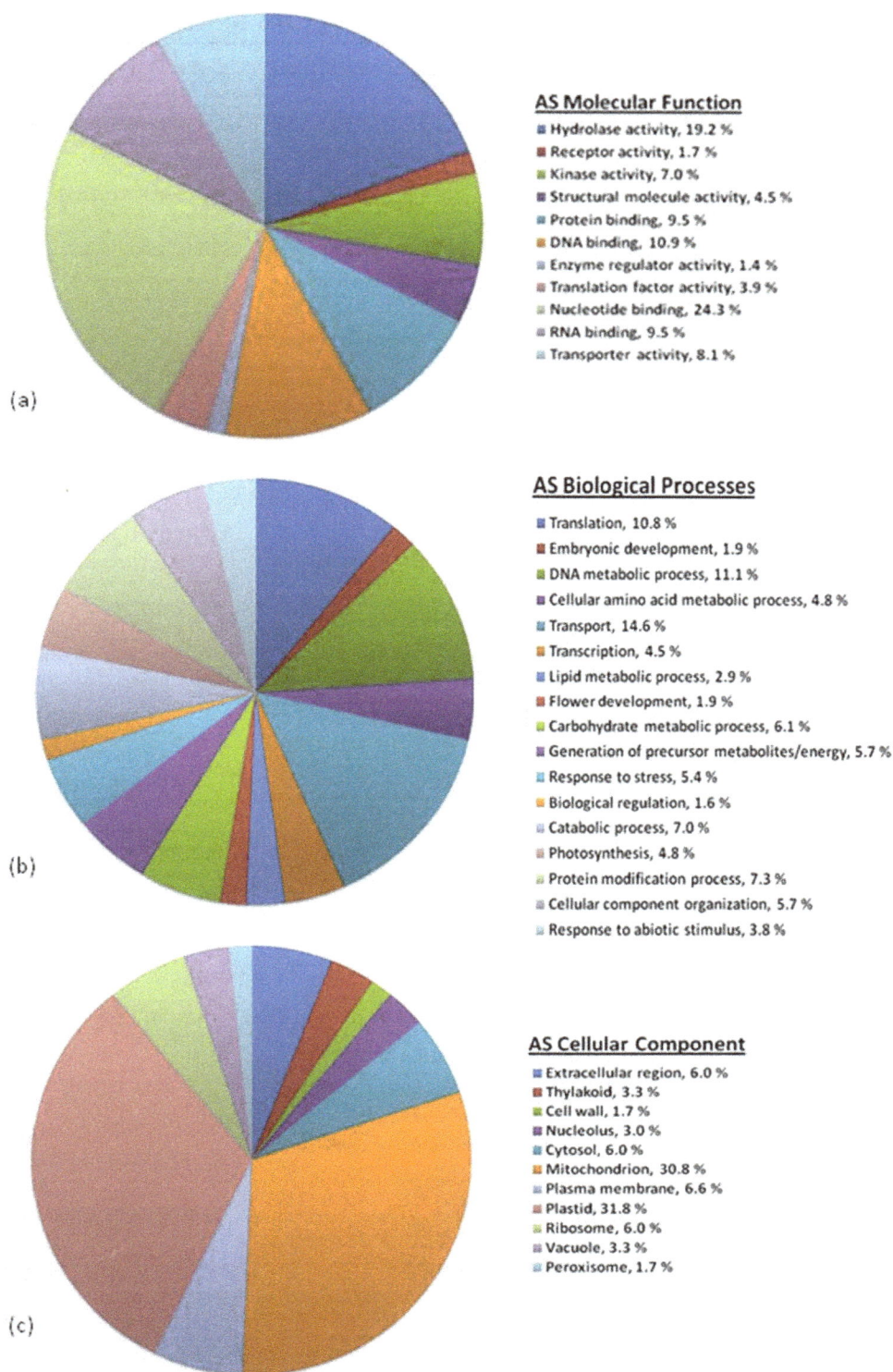

(a)

AS Molecular Function
- Hydrolase activity, 19.2 %
- Receptor activity, 1.7 %
- Kinase activity, 7.0 %
- Structural molecule activity, 4.5 %
- Protein binding, 9.5 %
- DNA binding, 10.9 %
- Enzyme regulator activity, 1.4 %
- Translation factor activity, 3.9 %
- Nucleotide binding, 24.3 %
- RNA binding, 9.5 %
- Transporter activity, 8.1 %

(b)

AS Biological Processes
- Translation, 10.8 %
- Embryonic development, 1.9 %
- DNA metabolic process, 11.1 %
- Cellular amino acid metabolic process, 4.8 %
- Transport, 14.6 %
- Transcription, 4.5 %
- Lipid metabolic process, 2.9 %
- Flower development, 1.9 %
- Carbohydrate metabolic process, 6.1 %
- Generation of precursor metabolites/energy, 5.7 %
- Response to stress, 5.4 %
- Biological regulation, 1.6 %
- Catabolic process, 7.0 %
- Photosynthesis, 4.8 %
- Protein modification process, 7.3 %
- Cellular component organization, 5.7 %
- Response to abiotic stimulus, 3.8 %

(c)

AS Cellular Component
- Extracellular region, 6.0 %
- Thylakoid, 3.3 %
- Cell wall, 1.7 %
- Nucleolus, 3.0 %
- Cytosol, 6.0 %
- Mitochondrion, 30.8 %
- Plasma membrane, 6.6 %
- Plastid, 31.8 %
- Ribosome, 6.0 %
- Vacuole, 3.3 %
- Peroxisome, 1.7 %

Figure 5. GO terms associated with upregulated AS transcripts. (a) Molecular Function, (b) Biological Process, (c) Cellular Component.

defense mechanisms responsible for the decrease in susceptibility are unknown. In this study, some of the most prominent, upregulated defense-related transcripts were the germin and germin-like proteins. Germins belong to the cupin superfamily and include the important defense enzyme, oxalate oxidase [46].

Germin-type oxalate oxidases are enzymes that are unique to Gramineae plants [46]. In creeping bentgrass cultivars, oxalate oxidase has been shown to be more active in the moderately resistant creeping bentgrass cultivar, L-93, compared to Crenshaw, a highly susceptible cultivar [47]. Oxalic acid is a pathoge-

Table 5. Top enriched *Sclerotinia homoeocarpa* domains at 96 h post inoculation on creeping bentgrass.

Domain name	Enrichment	P-value	Description
Glyco_hydro_45	17.0	0.0030	Endoglucanase
Glyco_hydro_62	17.0	0.0030	Alpha -L-arabinofuranosidase
Glyco_hydro_61	15.3	<0.0001	Endoglucanase
Peptidase_S28	13.6	0.0010	Serine carboxypeptidase S28
CBM_1	13.4	<0.0001	Fungal cellulose binding domain
E1_DerP2_DerF2	12.8	0.0003	Lipid recognition, recognition of pathogen products
Pectinesterase	12.8	0.0003	Pectinesterase
Eno-Rase_NADH_b	12.2	0.0003	NAD(P)H binding domain, trans-2-enoyl-CoA reductase
Glyco_hydro_43	11.4	<0.0001	Arabinanase
COX3	11.4	0.0016	Cytochrome c oxidase subunit III
Flavodoxin_2	11.4	0.0016	Flavodoxin-like fold
Glyco_hydro_11	11.4	0.0016	Xylanases
Glyco_hydro_12	11.4	0.0016	Endoglucanase and xyloglucan hydrolase
Nucleoplasmin	11.4	0.0016	Chromatin decondensation proteins
PLA2_B	11.4	0.0016	Lysophospholipase catalytic domain
Glyco_hydro_28	11.0	<0.0001	Polygalacturonase, Rhamnogalacturonase A
ABC_ATPase	10.8	<0.0001	Predicted ATPase of the ABC class
Syja_N	10.2	0.0006	SacI homology domain
Cellulase	9.7	0.0023	Cellulase
CFEM	9.7	0.0023	Cysteine rich, putative role in fungal pathogenesis
Flavodoxin_5	8.5	0.0030	Flavodoxin
FMN_red	8.5	0.0030	NADPH-dependent FMN reductase
Glyco_hydro_7	8.5	0.0030	Endoglucanase; cellobiohydrolase
Glyco_hydro_92	8.5	0.0030	Alpha-1,2-mannosidases
Mannosyl_trans3	8.5	0.0030	Mannosyltransferase putative
Yos1	8.5	0.0030	Transport between ER and the Golgi complex
7tm_1	8.5	0.0089	7 transmembrane receptor
Chorismate_synt	8.5	0.0089	Chorismate synthesis
COX6B	8.5	0.0089	Cytochrome oxidase c subunit VIb
Cu-oxidase_2	8.5	0.0089	Multicopper oxidase
Cupin_5	8.5	0.0089	Cupin superfamily
DHDPS	8.5	0.0089	Dihydrodipicolinate synthetase family
Fig1	8.5	0.0089	Ca2+ regulator and membrane fusion protein Fig1
Metallophos_2	8.5	0.0089	Calcineurin-like phosphoesterase
Pex2_Pex12	8.5	0.0089	N terminal of a number of known and predicted peroxins
Ribosomal_L35Ae	8.5	0.0089	Ribosomal protein L35Ae
SAPS	8.5	0.0089	SIT4 phosphatase-associated protein
UbiA	8.5	0.0089	UbiA prenyltransferase family
ABC_membrane	8.2	<0.0001	ABC transporter transmembrane region
Fungal_lectin	6.8	0.0061	Fungal fucose-specific lectin
OPT	6.2	0.0080	OPT oligopeptide transporter protein
Peptidase_S8	6.1	<0.0001	Subtilase family
ABC_tran	5.8	<0.0001	ABC transporter

nicity factor for *Sclerotinia sclerotiorum* (Lib.) de Bary [48], and oxalic acid is produced in large quantities by *S. homoeocarpa* as well [49]. Cloning oxalate oxidase into dicot hosts such as sunflower, soybean, canola, and tomato renders them resistant to *S. sclerotiorum* [50,51,52]. It is possible that increased production of oxalate oxidases by some cultivars of creeping bentgrass is responsible for their decreased susceptibility to *S. homoeocarpa*. Furthermore, the product of oxalate oxidase activity is H_2O_2, a plant defense signal that is capable of passing through cell membranes, is directly toxic to invading microorganisms, cross-

Table 6. Top enriched *Agrostis stolonifera* domains at 96 h post inoculation with *Sclerotinia homoeocarpa*.

Domain name	Enrichment	P-value	Description
eIF-5_eIF-2B	20.2	<0.0001	Zinc binding C4 finger.
ATS3	20.2	0.002	Embryo-specific protein 3
BDS_I_II	20.2	0.002	Antihypertensive protein BDS-I/II
CSD	20.2	0.002	'Cold-shock' DNA-binding domain
Folate_rec	20.2	0.002	Folate receptor family
MaoC_dehydratas	20.2	0.002	Synthesis of monoamine oxidase
SH3_1	20.2	0.002	Signal transduction related to cytoskeletal organisation
TAFII28	20.2	0.002	Assembly of the transcription preinitiation complex.
ABC2_membrane_3	15.1	<0.0001	ABC-2 type transporter
Fer4	13.5	<0.0001	Iron sulfur binding domain
Fer4_2	13.5	<0.0001	Iron sulfur binding domain
Fer4_6	13.5	<0.0001	Iron sulfur binding domain
Ribosomal_S2	13.5	<0.0001	Ribosomal protein S2
CALCOCO1	13.5	0.0077	Calcium binding and coiled-coil domain (CALCOCO1)
iPGM_N	13.5	0.0077	independent phosphoglycerate mutase
tRNA-synt_1c	12.6	<0.0001	tRNA synthetases class I (E and Q), catalytic domain
TMF_DNA_bd	12.1	0.0004	TATA element modulatory factor 1 DNA binding
UDPGT	12.1	0.0004	UDP-glucoronosyl and UDP-glucosyl transferase
Cupin_1	11.9	<0.0001	Cupins, including germins
Cupin_2	11.1	<0.0001	Cupin domain
ABC2_membrane	10.9	<0.0001	ABC-2 type transporter
GrpE	10.1	0.0011	GrpE nucleotide exchange factor
EFG_C	9.0	0.0002	Elongation factor G C-terminus
GTP_EFTU_D3	8.8	<0.0001	Elongation factor Tu C-terminal domain
ABC_ATPase	8.7	<0.0001	Predicted ATPase of the ABC class
IF-2B	8.7	0.0025	Initiation factor 2 subunit family
RNA_pol_Rpb1_5	8.7	0.0025	RNA polymerase Rpb1, domain 5
Cytochrom_B_C	7.6	0.0049	Cytochrome b/b6, C-terminal
SMC_N	7.3	<0.0001	structural maintenance of chromosomes protein
ABC_membrane	6.7	<0.0001	ABC transporter transmembrane region
ATP-synt_C	6.7	0.0086	ATP synthase subunit C
ATP-synt_ab_C	6.7	0.0012	ATP synthase alpha/beta chain
DUF1602	6.1	<0.0001	Protein of unknown function
GTP_EFTU	5.8	<0.0001	Elongation factor Tu GTP binding domain
ABC_tran	5.5	<0.0001	ABC transporters
Terpene_synth_C	5.4	0.0049	Terpene synthase family, metal binding domain
p450	5.0	<0.0001	Oxidative degradation of compounds
GTP_EFTU_D2	5.0	0.0005	Elongation factor Tu domain 2

links glycoproteins and phenolic molecules to form defense structures, and can induce systemic acquired resistance [53,54]. Clearly, more research on germin-like proteins and oxalate oxidase activity in creeping bentgrass is warranted.

In addition to the plant defense-related transcripts, other notable creeping bentgrass transcripts were identified as transposons and retrotransposons including pong, em, mutator, gypsy, and copia subclasses. This is not surprising since it has been documented that transposons make up approximately 80% of the DNA in cereal crops [55]. In a recent EST analysis of creeping bentgrass and colonial bentgrass, 1.4% and 0.18% of the total

ESTs were retrotransposon related, and retrotransposon transcripts were eight times higher in creeping bentgrass under disease stress than in healthy colonial bentgrass plants that were not under disease stress [56]. In the RNA-Seq data analysed here, 878 of 20,493 (4%) of the total AS transcripts were identified as transposons and 731 (3.5%) were identified as retrotransposons. Seventy-two of the 1017 transcripts upregulated in the interaction library were identified as retrotransposons: that is 7% of all upregulated AS transcripts. This is in clear support of the previous study concluding that retrotransposons in creeping bentgrass are likely activated by disease stress [56]. In the shared library (reads present in both AS and SH), only 0.1% of the reads mapped to transcripts identified as transposons. This means that it is not likely that identical transposon sequences are shared between SH and AS. However, several of the classes of retrotransposons identified in the study, such as the *Ty* and *copia* classes, are known to exist in both fungi and plants [57,58]. The upregulation of 72 retrotransposons in this study provides direct support for activation of transposons in creeping bentgrass in response to disease stress as demonstrated previously [56].

It has been reported that retrotransposons are transcriptionally activated in plants that are undergoing stress such as during protoplast formation, wounding, and in the presence of salicylic acid and jasmonate [59]. The influence of retrotransposons, if any, on plant defense pathways has not yet been determined; however, their increased transcription during infection should be investigated. For example, it would be interesting to determine if any of these retroelements are located within or near genes involved in plant defense responses, where they could affect the expression of these genes.

The RNA-Seq analysis presented in this study demonstrates the immense value of MPS technology for developing a comprehensive understanding of disease pathosystems. Advances in SBS technology since the start of this study would now provide adequate read length and depth for the *de novo* transcriptome assemblies, making it even more powerful and cost efficient. The resulting data analysis will be used to support future research efforts to characterize *S. homoeocarpa* virulence factors and creeping bentgrass defense-related transcripts.

Materials and Methods

Preparation of RNA for SBS Libraries

Sclerotinia homoeocarpa isolate MB01 was isolated from creeping bentgrass at The Ohio State University Turfgrass Research and Education Facility in Columbus, Ohio. The fungus was cultured on potato dextrose agar (PDA, Difco, Franklin Lakes, NJ) and 5 mm plugs of actively growing mycelium were transferred to potato dextrose broth (PDB, Difco) medium and incubated at 26°C with shaking at 160 rpm for 96 h. Cultures grown on PDA overlayed with cellophane membranes were also used for SBS library construction. At 96 h, mycelia were filtered from PDB or scraped from the cellophane membrane on PDA, and total RNA was extracted using TriReagent (Sigma, St. Louis, MO) according to manufacturer's instructions. For interaction and creeping bentgrass libraries, creeping bentgrass cv. Crenshaw was grown in a growth chamber at 26°C day and 22°C night temperatures with a 12 h photoperiod and at 70% relative humidity. Three-week old seedlings were challenged with millet seeds colonized by *S. homoeocarpa* MB01. These plants were incubated in clear bags to create humidity for 48 h after which the plants were uncovered during the day and bagged again for the night. After 96 h, leaves from inoculated and noninoculated plants were harvested, homogenized in liquid nitrogen, and RNA was extracted using TriReagent (Sigma) according to the manufacturer's instructions.

The SBS library templates were prepared using the Illumina Duplex-Specific thermostable nuclease (DSN) normalization kit and were analyzed using the Illumina Genome Analyzer I (GAI) at the Molecular and Cellular Imaging Center of The Ohio State University in Wooster, Ohio. The resulting libraries included a 96 h PDB culture control, a 96 h PDA culture control, a 96 h creeping bentgrass control, and a 96 h *S. homoeocarpa* – creeping bentgrass interaction library.

Preparation of RNA for 454 Libraries

S. homoeocarpa isolate MB01 was cultured on PDA and 5 mm plugs of actively growing mycelium were transferred to PDB medium and incubated at 26°C with shaking at 160 rpm for 48, 96, or 144 h. At each time point, mycelia were filtered and total RNA was extracted using TriReagent (Sigma) according to manufacturer's instructions. For interaction and creeping bentgrass libraries, creeping bentgrass cv. Crenshaw was grown in a growth chamber as described previously. Three-week old seedlings were inoculated in one of two ways: with a 5 mm plug of actively growing mycelium from a PDA culture or misting of a mycelia homogenate, in water, from a 96 h PDB culture of *S. homoeocarpa* MB01. These plants were incubated as described for SBS library preparation. After 96 h, leaves from inoculated and noninoculated plants were harvested, homogenized in liquid nitrogen, and RNA was extracted using TriReagent according to the manufacturer's instructions. Total RNA was sent to the Core Genomics Facility at Purdue University, Indiana for preparation and sequencing using Roche GS-FLX (454) sequencer with Titanium chemistry. Seven RNA-Seq libraries were prepared that included *S. homoeocarpa* PDB cultures at 48, 96, and 144 h after inoculation, *S. homoeocarpa* grown on PDA at 96 h, creeping bentgrass noninoculated controls, and for creeping bentgrass incubated for 96 h with *S. homoeocarpa* applied as a PDA culture plug or a mycelia homogenate in water.

Sequencing, Assembly, and Library Construction

The 454 raw sequence reads were assembled using GS Data Analysis Software (v2.5, Roche, Indianapolis, IN) after removal of adaptor sequences. *S. homoeocarpa* (SH) reads from the PDA, PDB, and 96 hpi Interaction libraries, and the *A. stolonifera* (AS) reads from the 96 h control and 96 h Interaction libraries were assembled using GS *de novo* assembler (v2.5, Roche) with parameter setting as 90% identity and a minimum 40 bp overlap. To eliminate SH reads in AS assembly library, reads from the *S. homoeocarpa* (SH) Interaction libraries that matched creeping bentgrass (*Agrostis stolonifera*) GenBank EST data were removed from the SH transcript assembly. Then reads from the *A. stolonifera* (AS) Interaction library matching SH assembly were also removed. Both *S. homoeocarpa* and *A. stolonifera* transcriptome assembly as used in this study was submitted to GenBank TSA database (Accession: PRJNA84359).

Full length SBS reads were directly mapped to the SH and AS transcriptome libraries using the Maq (v 0.7) [60] alignment. Mapping parameters allowed for only unique mapped reads with no mismatches. The number of mapped SBS reads for each transcript was counted by "soap.coverage" in SOAPAligner package.

Transcript Expression Analysis

RPKM values were calculated for each transcript by dividing number of mapped reads by length of the transcripts and number of total sequenced reads in this library. Fisher's Exact Test, a test used for analyzing unreplicated transcript data [61] was used to determine statistical significance of transcript RPKM value change. The test was done using a pairwise comparison of Interaction RPKM values versus RPKM from fungus grown on PDA, or for Interaction RPKM values versus uninoculated control grass RPKM values. The fungal data contained 9,319 unique annotated transcripts and the grass data contained 20,293 unique annotated transcripts. Therefore, the statistical significance of the Fisher's Test was evaluated against a Bonferroni corrected *P*-value of 1.86×10^{-5} and 2.43×10^{-6} for fungal and grass transcripts, respectively. Furthermore, only transcripts with at least 2-log fold change in transcript abundance were selected as upregulated. These conservative criteria were applied to avoid false positives.

The log-fold RPKM values of selected transcripts were validated using real time reverse transcription polymerase chain (RT-PCR) reaction on an iQ5 real time thermocycler (Biorad, Hercules, CA). RNA was extracted from fungus grown on PDA, noninoculated grass, and infected grass according to the previously described methods. RNA was extracted using Trizol (Invitrogen, Carlsbad, CA) and RNA extracts were treated with RQ1 DNAse (Promega, St. Louis Obispo, CA) according to the manufacturer's directions, except that RNase Inhibitor (Promega) was added to each 100 µl of RNA extract. Complementary DNA was created from 1 µg of total RNA using iScript (BioRad, Hercules, CA). PCR was conducted using the SYBR supermix (Biorad) as per manufacturer's directions. The cycle conditions were 95°C for 3 minutes followed by 40 cycles of 95°C for 30 s, 58.5°C for 30 s, and 72°C for 30 s. The melting temperature profiles and gel electrophoresis was used to evaluate the specificity of the reactions and the absence of primer dimers. Real time RT-PCR was conducted with three biological replicates for each of grass, interaction, and fungal cDNA samples. Three technical replicates of each cDNA sample were used in each experiment.

Reaction efficiency and relative expression data were analyzed using the relative expression software tool (REST) program (Qiagen; [62]). Log base two values of relative expression ratios calculated in the REST program and the corresponding log base two RPKM expression ratios were compared graphically.

Library annotation

Taxonomic and functional annotation of the SH and AS transcripts were conducted by using Blast2GO [20] software to run blastx and blastn algorithms against non-redundant nucleotide/protein database from the National Center for Biotechnology Information (National Institutes of Health). Within the SH library, any transcripts that resulted in top blastn hit species of *Hordeum*, *Oryza*, *Sorghum*, *Agrostis*, or *Vitis* were removed. Similarly, any transcripts in the AS library that had top blastn hit species of *Sclerotinia*, *Botryotinia*, *Glomerella*, *Ajellomyces*, or *Nectria* were removed. Combined graphs for GO terms associated with statistically upregulated transcripts are presented in the results section.

Conserved Domain Analysis

Transcript sequences were translated to proteins in all 6 possible frames, and conserved protein domains were identified using HMMER (v3.0) [21] to screen the Pfam-A database. All the predicted domains with E-value <0.01 were reported. There were a total of 14,161 domains identified in the SH transcripts assembly and 18,591 domains identified in the AS transcripts assembly. The enrichment ratio and significant value was calculated based on a Fisher's Exact Test. Descriptions for enriched domains were found on the Pfam website.

Supporting Information

Table S1 A complete list of *Sclerotinia homoeocarpa* transcripts that are over or under represented at 96 h post inoculation on creeping bentgrass.

Table S2 A complete list of *Agrostis stolonifera* transcripts that are over or under represented at 96 h post inoculation with *Sclerotinia homoeocapra*.

Table S3 A complete list of *Sclerotinia homoeocarpa* predicted protein conserved domains.

Table S4 A complete list of *Agrostis stolonifera* predicted protein conserved domains.

Acknowledgments

The authors of this study would like to acknowledge the Molecular and Cellular Imaging Center located at the Ohio Agricultural Research and Development Center in Wooster, Ohio for conducting SBS sequencing using Illumina sequencer. We would also like to acknowledge the Genomics Facility at Purdue University for conducting the 454 pyrosequencing and for subsequent troubleshooting advice.

Author Contributions

Conceived and designed the experiments: MJB TKM VRC. Performed the experiments: VRC AMO JH. Analyzed the data: AMO JH SO. Contributed reagents/materials/analysis tools: MJB TKM SO. Wrote the paper: AMO JH.

References

1. Walsh B, Ikeda S, Boland GJ (1999) Biology and management of dollar spot (*Sclerotinia homoeocarpa*); an important disease of turfgrass. HortScience 34: 13–21.
2. Chakraborty N, Chang T, Casler MD, Jung G (2006) Response of bentgrass cultivars to isolates representing 10 vegetative compatibility groups. Crop Science 46: 1237–1244.
3. Pigati RL, Dernoeden PH, Grybauskas AP, Momen B (2010) Simulated rainfall and mowing impact fungicide performance when targeting dollar spot in creeping bentgrass. Plant Disease 94: 596–603.
4. Burpee LL, Goulty LG (1986) Evaluation of two dollarspot forecasting systems for creeping bentgrass. Canadian Journal of Plant Science 5: 345–351.
5. Couch HB, Bloom JR (1960) Influence of environment on disease of turfgrasses. II. Effect of nutrition, pH, and soil moisture on Sclerotinia Dollar Spot. Phytopathology 257: 761–765.
6. Hall R (1984) Relationship between weather factors and dollar spot of creeping bentgrass. Canadian Journal of Plant Science 64: 167–174.
7. Orshinsky AM (2010) Characterization of the growth and host colonization of virulent, asymptomatic, and hypovirulent isolates of *Sclerotinia homoeocarpa*. Guelph: University of Guelph. 203 p.
8. Endo RM (1966) Control of dollar spot of turfgrass by nitrogen and its probable basis. Phytopathology 56: 877.
9. Monteith J, Dahl AS (1932) Turfgrass diseases and their control. Bulletin of the United States Golf Association 12: 85.
10. Orshinsky AM, Boland GJ (2011) *Ophiostoma mitovirus 3a*, ascorbic acid, glutathione, and photoperiod affect the development of stromata and apothecia by *Sclerotinia homoeocarpa*. Canadian Journal of Microbiology 57: 398–407.
11. Hsiang T, Mahuku GS (1999) Genetic variation within and between southern Ontario populations of *Sclerotinia homoeocarpa*. Plant Pathology 48: 83–94.
12. Endo RM (1963) Influence of temperature on rate of growth of five fungus pathogens of turfgrass and on rate of disease spread. Phytopathology 53: 857–877.
13. Bandara-Herath HMT, Herath WHMW, Carvalho P, Khan SI, Tekwani BL, et al. (2009) Biologically active tetranorditerpenoids from the fungus *Sclerotinia homoeocarpa* causal agent of dollar spot in turfgrass. Journal of Natural Products 72: 2091–2097.
14. Bonos SA, Buckley RJ, Clarke BB (2007) An integrated approach to dollar spot disease in turfgrasses. New Brunswick, NJ: Rutgers Cooperative Extension.
15. Latin R (2009) Turfgrass disease profiles: Dollar Spot. West Lafayette: Purdue Extension.
16. Huse SM, Huber JA, Morrison HG, Sogin ML, Welch D (2007) Accuracy and quality of massively parallel DNA pyrosequencing. Genome Biology 8: R143.
17. Birch P, Kamoun S (2000) Studying interaction transcriptomes: coordinated analyses of gene expression during plant-microorganism interactions. New technologies for life sciences: A Trends Guide 0: 1471–1931.
18. Marioni JC, Mason CE, Mane SM, Stephens M, Gilad Y (2008) RNA-seq: An assessment of technical reproducibility and comparison with gene expression arrays. Genome Research 18: 1509–1517.
19. Wang Z, Gerstein M, Snyder M (2009) RNA-Seq: a revolutionary tool for transcriptomics. Nature Reviews: Genetics 10: 57–63.

20. Conesa A, Gotz S, Garcia-Gomez JM, Terol J, Talon M, et al. (2005) Blast2GO: a universal tool for annotation, visualization and analysis in functional genomics research. Bioinformatics 21: 3674–3676.

21. Finn RD, Clements J, Eddy SR (2011) HMMER web server: interactive sequence similarity searching. Nucleic Acids Research 39: W29–W37.

22. Petersen TN, Brunak S, von Heijne G, Nielsen H (2011) SignalP 4.0: discriminating signal peptides from transmembrane regions. Nature Methods 8: 785–786.

23. Vera JC, Wheat CW, Fescemyer HW, Frilander MJ, Crawford DL, et al. (2008) Rapid transcriptome characterization for a nonmodel organism using 454 pyrosequencing. Molecular Ecology 17: 1636–1647.

24. Li B, Ruotti V, Stewart RM, Thomson JA, Dewey CN (2009) RNA-Seq gene expression estimation with read mapping uncertainty. Bioinformatics 26: 493–500.

25. Henrissat B (1991) A classification of glycosyl hydrolases based on amino acid sequence similarities. Biochemical Journal 280: 309–316.

26. Henrissat B, Bairocht A (1993) New families in the classification of glycosyl hydrolases based on amino acid sequence similarities. Biochemical Journal 293: 781–788.

27. Vogel J (2008) Unique aspects of the grass cell wall. Current Opinion in Plant Biology 11: 301–307.

28. Kars I, van Kan JAL (2007) Extracellular enzymes and metabolites involved in pathogenesis of Botrytis. Elad Y, Williamson, B., Tudzynski, P., and Delen, N., editor. Norwell, MA, USA.: Kluwer Academic Publishers.

29. Rowe HC, Kliebenstein DJ (2007) Elevated genetic variation within virulence-associated Botrytis cinerea polygalacturonase loci. Molecular Plant-Microbe Interactions 20: 1126–1137.

30. Lehtinen U (1993) Plant cell-wall degrading enzymes of Septoria nodorum. Physiological and Molecular Plant Pathology 43: 121–134.

31. Li J, Yu L, Yang J, Dong L, Tian B, et al. (2010) New insights into the evolution of subtilisin-like serine protease genes in Pezizomycotina. BMC Evolutionary Biology 10: 68.

32. Ball AM, Ashby AM, Daniels MJ, Ingram DS, Johnstone K (1991) Evidence for the requirement of extracellular protease the pathogenic interaction of Pyrenopeziza brassicae with oilseed rape. Physiological and Molecular Plant Pathology 38: 147–161.

33. Dobinson KF, Lecomte N, Lazarovits G (1997) Production of an extracellular trypsin-like protease by the fungal plant pathogen Verticillium dahliae. Canadian Journal of Microbiology 43: 228–233.

34. Sreedhar L, Kobayashi DY, Bunting TE, Hillman BI, Belanger FC (1999) Fungal proteinase expression in the interaction of the plant pathogen Magnaporthe poae with its host. Gene 235: 121–129.

35. St. Leger RJ, Roberts DW (1997) Adaptation of proteases and carbohydrases of saprophytic, phytopathogenic and entomopathogenic fungi to the requirements of their ecological niches. Microbiology 143: 1983–1992.

36. de Waard MA, Andrade AC, Hayashi K, Schoonbeek H-j, Stergiopoulos I, et al. (2006) Impact of fungal drug transporters on fungicide sensitivity, multidrug resistance and virulence. Pest Management Science 62: 195–207.

37. Del Sorbo G, Schoonbeek H-j, De Waard MA (2000) Fungal transporters involved in efflux of natural toxic compounds and fungicides. Fungal Genetics and Biology 30: 1–15.

38. Kretschmer M, Leroch M, Mosbach A, Walker A–S, Fillinger S, et al. (2009) Fungicide-driven evolution and molecular basis of multidrug resistance in field populations of the grey mould fungus Botrytis cinerea. PLoS One 5.

39. Schoonbeek H, Sorbo GD, Waard MAD (2001) The ABC transporter BcatrB affects the sensitivity of Botrytis cinerea to the phytoalexin resveratrol and the fungicide fenpiclonil. Molecular Plant-Microbe Interactions 14: 562–571.

40. Stefanato FL, Abou-Mansour E, Buchala A, Kretschmer M, Mosbach A, et al. (2009) The ABC transporter BcatrB from Botrytis cinerea exports camalexin and is a virulence factor on Arabidopsis thaliana. The Plant Journal 58: 499–510.

41. Vermeulen T, Schoonbeek H, Waard MAD (2001) The ABC transporter BcatrB from Botrytis cinerea is a determinant of the activity of the phenylpyrrole fungicide fludioxonil. Pest Management Science 57: 393–402.

42. Detweiler AR, Vargas JM Jr, Danneberger TK (1983) Reistance of Sclerotinia homoeocarpa to iprodione and benomyl. Plant Disease 67: 627–630.

43. Golembiewski RC, Vargas JM Jr, Jones AL, Detweiler AR (1995) Detection of demethylation inhibitor (DMI) resistance in Sclerotinia homoeocarpa populations. Plant Disease 79: 491–493.

44. Bonos SA (2006) Heritability of Dollar Spot resistance in creeping bentgrass. Phytopathology 96: 808–812.

45. Guo Z, Bonos S, Meyer WA, Day PR, Belanger FC (2003) Transgenic creeping bentgrass with delayed dollar spot symptoms. Molecular Breeding 11: 95–101.

46. Davidson RM, Reeves PA, Manosalva PM, Leach JE (2009) Germins: A diverse protein family important for crop improvement. Plant Science 177: 499–510.

47. DaRoche AB, Hammerschmidt R (2004) Increase in oxalate oxidase activity in bentgrass response to Sclerotinia homoeocarpa and chemical treatment. Phytopathology 94 S: 157.

48. Chipps TJ, Gilmore B, Myers JR, Stotz HU (2005) Relationship between oxalate, oxalate oxidase activity, oxalate sensitivity, and white mold susceptibility in Phaseolus coccineus. Phytopathology 95: 292–299.

49. Venu RC, Beaulieu RA, Graham TL, Medina AM, Boehm MJ (2009) Dollar spot fungus Sclerotinia homoeocarpa produces oxalic acid. International Turfgrass Journal 11: 263–270.

50. Lu G (2003) Engineering Sclerotinia sclerotiorum resistance in oilseed crops. African Journal of Biotechnology 2: 509–516.

51. Hu X (2003) Overexpression of a gene encoding hydrogen peroxide-generating oxalate oxidase evokes defense responses in sunflower. Plant Physiology 133: 170–181.

52. Walz A, Zingen-Sell I, Loeffler M, Sauer M (2008) Expression of an oxalate oxidase gene in tomato and severity of disease caused by Botrytis cinerea and Sclerotinia sclerotiorum. Plant Pathology 57.

53. Wojtaszek P (1997) Oxidative burst : an early plant response to pathogen infection. Biochem J 322: 681–692.

54. Zhou T, Boland GJ (1999) Mycelial growth and production of oxalic acid by virulent and hypovirulent isolates of Sclerotinia sclerotiorum. Canadian Journal of Plant Pathology 21: 93–99.

55. Sabot F, Simon D, Bernard M (2004) Plant transposable elements, with an emphasis on grass species. Euphytica 139: 227–247.

56. Rotter D, Bharti AK, Li HM, Luo C, Bonos SA, et al. (2007) Analysis of EST sequences suggests recent origin of allotetraploid colonial and creeping bentgrasses. Molecular Genetics and Genomics 278: 197–209.

57. Kejnovsky E, Hawkins J, Feschotte C (2012) Plant Transposable Elements: Biology and Evolution. In: Wendel JF, Greilhuber J, Dolezel J, Leitch IJ, editors. Plant Genome Diversity. Wien: Springer-Verlag 18–19.

58. Daboussi M-J, Capy P (2003) Transposable elements in filamentous fungi. Annual Review of Microbiology 57: 275–299.

59. Grandbastion M-A (1998) Activation of plant retrotransposons under stress conditions. Trends in Plant Science 3: 181–187.

60. Li H, Ruan J, Durbin R (2008) Mapping short DNA sequencing reads and calling variants using mapping quality scores. Genome Res 18: 1851–1858.

61. Auer PL, Doerge RW (2010) Statistical design and analysis of RNA sequencing data. Genetics 185: 405–416.

62. Pfaffl MW, Horgan GW, Dempfle L (2002) Relative expression software tool for group-wise comparison and statistical analysis of relative expression results in real time PCR. Nucleic Acids Research 30: e36.

17

Association between Virulence and Triazole Tolerance in the Phytopathogenic Fungus *Mycosphaerella graminicola*

Lina Yang[1,2], **Fangluan Gao**[2], **Liping Shang**[2], **Jiasui Zhan**[1,2]*, **Bruce A. McDonald**[3]

1 Key Lab for Biopesticide and Chemical Biology, Ministry of Education, Fujian Agriculture and Forestry University, Fuzhou, Fujian, People's Republic of China, **2** Laboratory of Plant Virology of Fujian Province, Institute of Plant Virology, Fujian Agriculture and Forestry University, Fuzhou, Fujian, People's Republic of China, **3** Institute of Integrative Biology, ETH Zurich, Zürich, Switzerland

Abstract

Host resistance and synthetic antimicrobials such as fungicides are two of the main approaches used to control plant diseases in conventional agriculture. Although pathogens often evolve to overcome host resistance and antimicrobials, the majority of reports have involved qualitative host – pathogen interactions or antimicrobials targeting a single pathogen protein or metabolic pathway. Studies that consider jointly the evolution of virulence, defined as the degree of damage caused to a host by parasite infection, and antimicrobial resistance are rare. Here we compared virulence and fungicide tolerance in the fungal pathogen *Mycosphaerella graminicola* sampled from wheat fields across three continents and found a positive correlation between virulence and tolerance to a triazole fungicide. We also found that quantitative host resistance selected for higher pathogen virulence. The possible mechanisms responsible for these observations and their consequences for sustainable disease management are discussed.

Editor: Sung-Hwan, Soonchunhyang University, Republic of KoreaYun

Funding: The project was supported by Chinese National Science Foundation Grant no. 31071655 and Swiss Federal Institute of Technology Grant TH-49a/02-1. The funders had no role in study design, data collection and analysis, decision to publish, or preparation of the manuscript.

Competing Interests: The authors have declared that no competing interests exist.

* E-mail: Jiasui.zhan@fafu.edu.cn

Introduction

Knowledge of the evolutionary biology of plant pathogens is needed for sustainable disease management in agricultural systems [1]. The development of host resistance through plant breeding and applications of synthetic fungicides are two major approaches used to control fungal diseases. Plants have evolved an array of chemical, structural and enzymatic defenses to protect themselves against pathogens [2], [3], [4], [5], [6]. Chemical defenses include the production of secondary metabolites that are toxic to pathogens [7], [8]. Like the fungicidal secondary metabolites produced by plants, synthetic fungicides disrupt fungal metabolism, either inhibiting development and growth or killing the fungus outright.

The widespread use of host resistance and fungicides selects for pathogen individuals or populations that can overcome the host defense systems or that are resistant to the applied fungicides. For qualitative host – pathogen interactions following the gene-for-gene model and fungicides targeting a single fungal protein, the emergence of pathogenicity (here defined as the qualitative capacity of a parasite to infect and cause disease on a host, [9]) or fungicide resistance often results from single point mutations that occur at random in pathogen populations [10], [11], [12], [13]. Under selection, these mutations increase in frequency and can spread rapidly over large areas through natural or human-mediated gene flow. When resistance is quantitative or a fungicide targets several proteins or biochemical pathways, the emergence of

virulence (here defined as the degree of damage caused to a host by parasite infection, [14]) or fungicide resistance in pathogen populations is more complex and occurs more slowly, likely involving recurring cycles of mutation-selection-recombination. Natural selection increases the frequency of phenotypes with higher fitness. New mutations or recombination among the selected phenotypes will create new genetic variation for the next cycle of selection.

The majority of studies on the evolution of plant pathogens have involved qualitative host – pathogen interactions or antimicrobials targeting a single pathogen protein or metabolic pathway. Studies that jointly consider the evolution of virulence and antimicrobial resistance are limited. Yet this type of study is important to understand the emergence of infectious diseases and to devise sustainable disease management in agriculture and medicine. In this study, we used the wheat-*Mycosphaerella graminicola* system to address the interaction of the evolution of virulence and antimicrobial resistance in agricultural ecosystems. The objectives of this study were: 1) to determine whether there is an association between virulence and resistance to fungicides; and 2) to determine whether host resistance affects the evolution of virulence and fungicide resistance.

Mycosphaerella graminicola (Fückel) Schroeter (anamorph *Septoria tritici*) is the causal agent of septoria leaf blotch on wheat [15], [16]. The pathogen has a global distribution and can cause up to 40% yield loss in many areas of the world [17]. The life cycle of the pathogen involves both sexual and asexual reproduction [15],

[16]. Wind-dispersed ascospores produced by the teleomorph contribute significantly both to initiation and further development of disease epidemics [18] and are likely to be one of the main mechanisms contributing to long distance gene flow [19] and host adaptation [20]. Genetic variation in *M. graminicola* populations is high [21] as a result of frequent sexual recombination [18], [20], high gene flow [22] and large effective population size [22]. Results from experimental evolution and population genetic studies indicate that the genetic structure of the pathogen can change significantly over a single growing season in response to host selection [23], while local adaptation leads to significant population differentiation for virulence [24], fungicide resistance [25] and temperature sensitivity [26].

Though both quantitative and qualitative resistances have been identified in wheat hosts, the majority of resistant cultivars used in commercial production display quantitative resistance (QR) to the pathogen [27], [28]. QR is believed to be more durable because natural selection is thought to operate more slowly on quantitative traits. Unlike qualitative resistance (also called major gene resistance), QR is thought to be mediated by several genes each contributing small but additive effects to the overall host resistance [29]. It is thought that mechanisms underlying QR in plants involve preformed, constitutive, physical and chemical barriers, Pathogen-Associated Molecular Pattern (PAMP)-triggered responses [5] and pathogen life-history traits [30]. Interactions of these mechanisms hinder the growth, penetration, reproduction and transmission of a pathogen. QR in plants slows down but does not prevent epidemics, thus effective disease control may require supplementary applications of fungicides.

Triazoles represent a major category of fungicides used widely in agriculture and medicine. This group of fungicides inhibits cytochrome P450 sterol 14 alpha-demethylase, an enzyme required for the biosynthesis of ergosterol in many fungi [31]. Resistance to triazoles is thought to be polygenic [32] and mediated by several mechanisms including mutations in the target protein gene *CYP51* and increased active efflux by ABC transporters [33], [34], [35], [36]. Cyproconazole is a triazole fungicide that has been used for many years to control *M. graminicola* [32].

Materials and Methods

Ethics Statements

We confirm that no specific permits were required for the described field study and to collect samples from these locations. We further confirm that the locations were not privately-owned or protected in any way and the field study did not involve endangered or protected species.

Pathogen populations

Five *M. graminicola* populations sampled from four geographical locations, including one population each from Australia, Israel and Switzerland and two populations from Oregon, USA [21], [26], were used for this study. The Australian population (AUS) was collected near Wagga Wagga in 2001. The Israel population (ISR) was collected near Nahal Oz in 1992 and the Swiss population (SWI) was sampled near Winterthur, kanton Zurich in 1999. The two USA populations (ORER and ORES) were collected on the same day in 1990 from a field planted to the partially resistant cultivar Madsen and the highly susceptible cultivar Stephens, respectively. The fungal isolates were stored in silica gel at −80°C after they were isolated from infected leaves.

Isolates in each population were genotyped using restriction fragment length polymorphisms (RFLPs) and DNA fingerprints.

The genotype data were published earlier [21]. Only isolates with a distinct multi-locus RFLP haplotype and DNA fingerprint were chosen for virulence and fungicide resistance tests. A total of 141 genetically distinct isolates were included in the experiment. Each population was represented by 25–30 isolates.

Measurement of cyproconazole tolerance

M. graminicola isolates retrieved from silica gel long-term storage were grown on potato dextrose agar (PDA) amended with 50 mg/L kanamycin and placed at 18°C for seven days. Blastospores formed on these plates were transferred into 50 mL Falcon tubes containing 30 ml yeast sucrose broth (YSB) supplemented with 50 mg/L kanamycin. The tubes were placed at 18°C at 140 rpm for seven days. Spore concentrations for each isolate were determined on the day of inoculation using a haemocytometer and adjusted to 200 spores per mL. 500 µL of the calibrated spore suspension was inoculated onto a PDA plate containing 0.1 ppm cyproconazole while another 500 µL of the spore suspension was inoculated onto a PDA plate without cyproconazole. Our preliminary experiments showed that 0.1 ppm provided the best resolution with the least experimental error. Many isolates did not grow when we used higher concentrations while growth rates of many isolates did not change when we used lower concentrations. Five isolates, one from each of the five populations, were replicated ten times. All other isolates were replicated twice. The inoculated plates were kept at 18°C and colonies on the plates were recorded with a digital camera five days after inoculation. All inoculations and photographs were made by the same person during a single day.

Colony sizes were measured with the image analysis software Assess 2.0 [37]. Cyproconazole tolerance for each isolate was determined by calculating the relative colony size with and without the fungicide, as described previously [24], [25]. Colony sizes were calculated as the average value for the ~20–70 colonies formed on each plate. Only colonies that clearly developed from single spores were used for the analysis. Fused colonies originating from two or more spores were excluded from the analysis.

Measurement of virulence

Five 10 cm plastic pots were filled with Ricoter garden soil (Ricoter Erdaufbereitung AG, Switzerland) and sown with ten seeds each of either wheat cultivar Toronit or Greina. Cultivar Toronit was classified as moderately resistant to *M. graminicola* while cultivar Greina was classified as susceptible. The plastic pots were placed in a greenhouse for 21 days at 60% relative humidity and 20°C during daytimes and 40% relative humidity and 16°C during nighttimes. Seedlings were supplemented with 50 kLux florescent light to provide 16 h day-lengths.

M. graminicola isolates retrieved from long-term storage were placed on yeast maltose agar plates amended with 50 mg/L kanamycin and kept at 20°C for seven days. Blastospores formed on these plates were transferred into sterile flasks containing 50 ml YSB supplemented with 50 mg/L kanamycin. The inoculated flasks were placed at 20°C with continuous shaking for a week. Spore suspensions were calibrated to a concentration of 5×10^6 spores per ml on the day of inoculation using a haemocytometer.

Inoculations were made at 21 days after sowing, at approximately growth stage 11 [38]. Seedlings in each pot were thinned to the five most uniform ones and inoculated with 50 ml of the calibrated spore suspension. Leaves of both cultivars were inoculated until run-off with 50 ml of the spore suspension using a semi-automatic sprayer. The inoculated seedlings were placed at 100% relative humidity and 21°C for two days in greenhouse chambers. New plant leaves formed after the inoculation were removed at three day intervals. At 22 days after the inoculation an average of 8–10 leaves was collected from each isolate-cultivar combination. These leaves were mounted onto blue paper sheets and their images were digitized. Percentage Leaf Area Covered by Pycnidia (PLACP) and Percentage Leaf Area Covered by Lesions (PLACL) were measured with the image analysis software Assess 2.0 [37]. All inoculations and all virulence assessments were made during a single day to minimize environmental variance among treatments.

Data analysis

Frequencies of cyproconazole resistance (natural logarithm transformed) and both PLACL and PLACP (square root transformed) in the fungal isolates were grouped using a binning approach and each group was labelled with the mid-point value of the lower and upper boundaries of the corresponding bins. Analyses of variance for cyproconazole tolerance, PLACL and PLACP were performed using the general linear model procedure implemented in SAS [39]. The raw data were leftward skewed. After applying a natural logarithm transformation to cyproconazole tolerance and a square root transformation to the PLACL and PLACP datasets, the transformed data showed a more even distribution than the non-transformed data, so the transformed data were used in the ANOVA. Least significant differences [40] were used to compare cyproconazole tolerance, PLACL and PLACP among populations sampled from different regions and hosts.

The environmental variance of cyproconazole tolerance, PLACL and PLACP in each population was estimated using the among-replicate variance [41], [42]. In common garden experiments with asexually reproducing species, any variance among replicates can be attributed to environmental effects because individuals in different replicates have the same genotype (i.e. they are identical clones). Therefore, variance among replicates in this case is equivalent to the environmental variance of cyproconazole tolerance, PLACL and PLACP (for details see [24], [41], [42]). Our earlier analyses [24] indicated that the environmental variance estimated using the large number of isolates with two replicates was not significantly different from the variance estimated using the limited number of isolates with 10 replicates. Therefore, we included all isolates in the analysis of environmental variance. Genetic variance in each M. graminicola population was estimated by subtracting the environmental variance from the phenotypic variance in the corresponding population. The association between virulence and fungicide tolerance in the pathogen populations was evaluated by simple linear correlation using transformed data [39].

Results

Variation in PLACL, PLACP and cyproconazole tolerance in the Mycosphaerella graminicola populations

PLACLs ranged from 1–90% with an average (95% confidence interval) of 26% ($\pm 3\%$) on the moderately resistant cultivar Toronit and from 1–90% with an average (95% confidence interval) of 39% ($\pm 4\%$) on the susceptible cultivar Greina, while PLACPs ranged from 0–50% with an average (95% confidence interval) of 9% ($\pm 2\%$) on Toronit and from 0–71% with an average (95% confidence interval) of 24% ($\pm 3\%$) on Greina, respectively.

Frequency distributions of cyproconazole tolerance (natural logarithm transformed) and both PLACL and PLACP (square root transformed) were visualized by grouping isolates into 11, 10 and 13 bins differing by 0.56, 1.0 and 0.70 units, respectively. Our

analyses revealed that these bin allocations yielded the optimum distribution to display the trait frequency, with enough strains occupying each bin to calculate and display a frequency with equal spacing between bin means.

The square root transformed distributions of PLACL were unimodal and symmetric, peaking at the post-transformation level of 6 (Fig. 1A, 1B) for both cultivars. The square root transformation of PLACP displayed a bimodal distribution on both cultivars with one major peak and one minor peak (Fig. 1C, 1D). On Toronit, the majority of isolates displayed a lower level of virulence, forming a major peak at the post-transformation level of 1 and a minor peak at the post-transformation level of 4. On Greina, the major peak was shifted to the right at the post-transformation level of 5 while the minor peak was shifted to the left at the post-transformation level of 2.

20–70 colonies were formed on each plate and there was no systematic difference in the number of colonies formed between the plates supplemented and not supplemented with cyproconazole. Tolerance to cyproconazole ranged from 0.01–2.30 with an average of 0.34. Six out of 141 (4%) isolates grew better in the presence of cyproconazole than without the fungicide, resulting in values of fungicide tolerance larger than 1. Like PLACP, the natural logarithm of tolerance to cyproconazole within a pathogen population displayed a bimodal distribution with a major peak at the post-transformation level of −2.6 and a minor peak at the post-transformation level of −0.4 (Fig. 2).

Comparison of cyproconazole tolerance and virulence among M. graminicola populations

Isolate and origin of isolate contributed significantly ($p < 0.0001$) to cyproconazole tolerance, PLACL and PLACP. Cultivar also contributed significantly ($p < 0.0001$) to PLACL and PLACP. Least significant difference analyses showed that the population from Switzerland displayed the highest levels of cyproconazole tolerance and virulence while the population from Australia displayed the lowest levels of cyproconazole tolerance and virulence (Table 1). The population from the resistant host Madsen displayed a higher cyproconazole tolerance and virulence than the population from the susceptible host Stephens (Table 1). All six isolates growing better in the presence of cyproconazole had a Swiss origin.

Correlation between cyproconazole tolerance and virulence of M. graminicola isolates

There were significant correlations between cyproconazole tolerance and PLACP on both Toronit ($r_{91} = 0.21$, $p = 0.04$ Fig. 3A) and Greina ($r_{91} = 0.22$, $p = 0.03$, Fig. 3B). The correlation coefficient between cyproconazole tolerance and PLACL on Toronit was positive and significant ($r_{97} = 0.21$, $p = 0.04$, Fig. 3C). Though the correlation coefficient between cyproconazole tolerance and PLACL on Greina was also positive, it was not significant ($r_{98} = 0.13$, $p = 0.20$, Fig. 3D), Variances and population means of cyproconazole tolerance and virulence were also positively associated, but none of them were significant (Fig. 4 & 5).

Discussion

We assayed virulence and tolerance to a triazole fungicide in a large collection of M. graminicola isolates sampled across several host genotypes and geographic locations. We found positive correlations between virulence and fungicide tolerance (Fig. 3), suggesting an association between these two quantitative traits. In an earlier experiment conducted in Oregon, USA, Cowger and Mundt [43] also found that M. graminicola isolates from cultivars

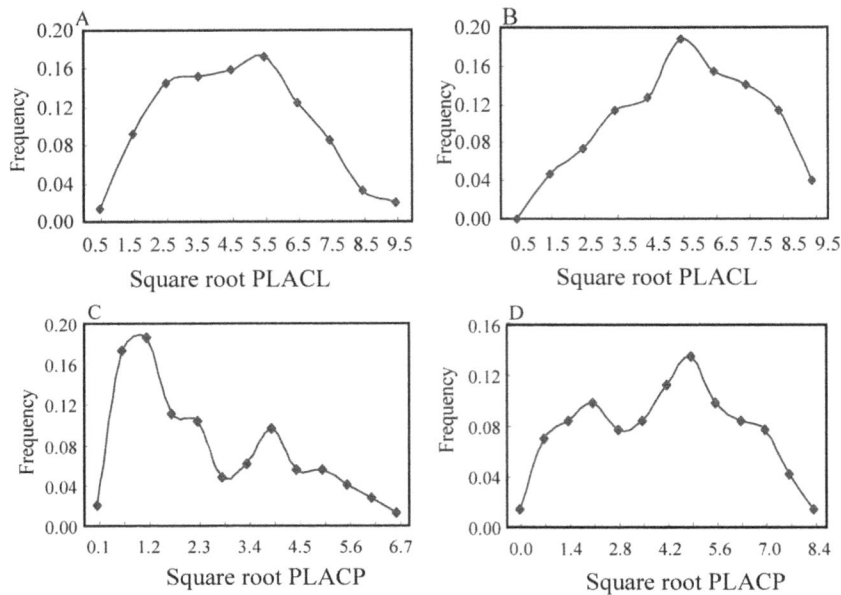

Figure 1. Frequency distribution of Percentage Leaf Area Covered by Lesions (PLACL) and Percentage Leaf Area Covered by Pycnidia (PLACP) in 141 isolates of *Mycosphaerella graminicola* **evaluated on two Swiss wheat cultivars.** Both PLACL and PLACP were square root transformed and labelled using the mid-point values of the corresponding bins: A) PLACL on Toronit; B) PLACL on Greina: C) PLACP on Toronit; and D) PLACP on Greina.

treated with the protectant fungicide chlorothalonil were more aggressive than isolates sampled from the same cultivars in nearby, untreated fields. It is not clear whether the positive correlation between virulence and fungicide tolerance observed in pathogens sampled from agricultural ecosystems will also be found in pathogens sampled from natural ecosystems. Additional studies with other agricultural pathogens and with pathogens collected from natural systems will be needed to determine the generality of these findings. The lack of significant correlations between variances and means in virulence and cyproconazole tolerance at the population level could be due to the small number of data points available for this comparison. Because only five populations originating from four geographic locations were included in this study, associations would need to be very high $(r>0.89)$ to detect a significant correlation with such a small number of data points.

Local adaptation and population differentiation can affect the estimate of association between ecological characters [44], [45]. Extensive utilization of fungicides and quantitative resistance in some regions may result in both high virulence and high fungicide tolerance. In *M. graminicola*, we found that the Australian population had the lowest overall virulence and cyproconazole tolerance while the Swiss population had the highest overall virulence and cyproconazole tolerance [25], consistent with significant local adaptation and a high level of population differentiation for the two characters. To eliminate the possible effect of this population structure on our conclusions, the association between fungicide tolerance and virulence was further evaluated using a randomisation procedure [46]. The fungicide and virulence datasets in the Switzerland and Australia populations were randomized and then added to the original dataset (without randomization) of the other three populations to calculate

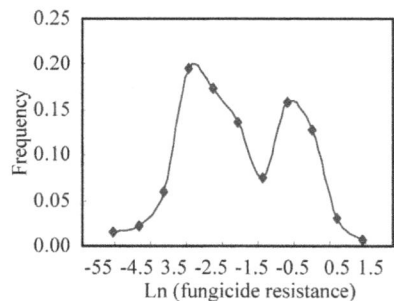

Figure 2. Frequency distribution of cyproconazole resistance in 141 isolates of *Mycosphaerella graminicola*. Cyproconazole resistance was determined by calculating the relative colony size of an isolate grown on Petri plates with and without the fungicide. Data were natural logarithm transformed and labelled using the mid-point values of the corresponding bins.

Table 1. LSD test for differences in cyproconazole resistance and virulence among the five *Mycosphaerella graminicola* populations sampled from Australia, Israel, Switzerland and USA.

Populations	Cyproconazole resistance	PLACL (%)[1]	PLACP (%)[2]
SWI	0.82 a[3]	37.8 a	20.7 a
ORE. R	0.29 b	35.1 a	17.3 a
ISR	0.26 bc	29.3 a	16.9 ab
ORE. S	0.16 c	33.3 a	13.2 bc
AUS	0.15 c	20.5 b	7.5 c

[1]Percentage Leaf Area Covered by Lesions.
[2]Percentage Leaf Area Covered by Pycnidia.
[3]Values followed by different letters are significantly different at P≤0.05.

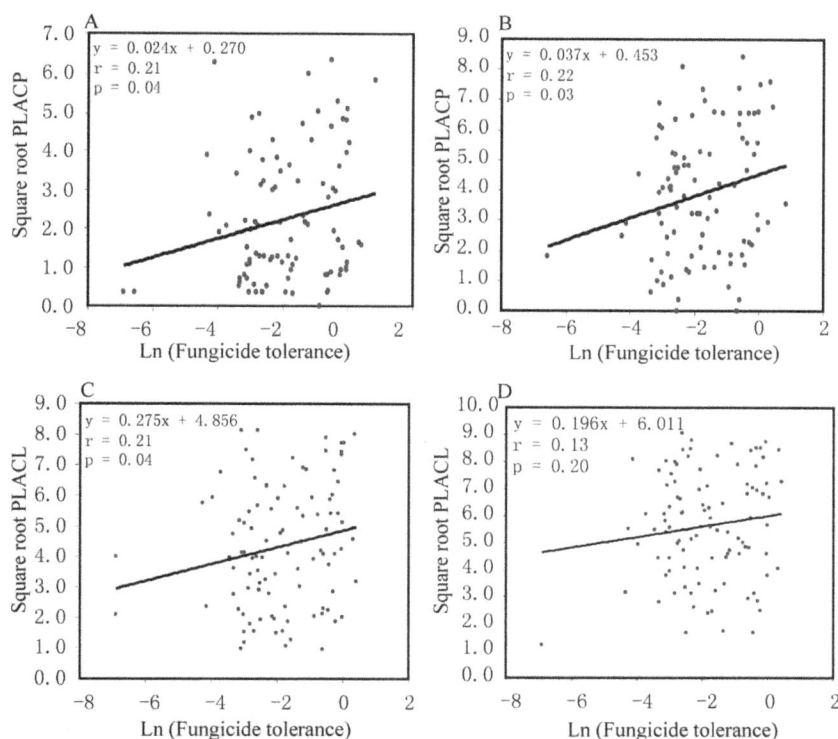

Figure 3. Correlations between cyproconazole resistance and two measures of virulence in 141 isolates of *Mycosphaerella graminicola* evaluated on two Swiss wheat cultivars. Cyproconazole resistance was determined by calculating the relative colony size of an isolate grown on Petri plates with and without the fungicide: A) Percentage Leaf Area Covered by Pycnidia (PLACP) on Toronit; B) Percentage Leaf Area Covered by Pycnidia (PLACP) on Greina; C) Percentage Leaf Area Covered by Lesions (PLACL) on Toronit; and D) Percentage Leaf Area Covered by Lesions (PLACL) on Greina.

correlation coefficients. The process was repeated 10000 times. Results from the randomization analysis revealed that the observation of a positive association between the two traits could not be attributed to local adaptation or population differentiation (data not shown).

We also found significantly higher PLACL, PLACP and cyproconazole tolerance in the pathogen population sampled from the resistant wheat cultivar Madsen than the susceptible cultivar Stephens (Table 1). These two pathogen populations were sampled from the same field at the same point in time and therefore most likely originated from the same source population. Because no triazole fungicides were applied to this field, and fungicide use was rare in this region, we do not believe that the difference in triazole tolerance between the *M. graminicola* populations from the resistant and susceptible hosts is due to selection for fungicide tolerance. This interpretation is supported by the lack of *CYP51* sequence variation among isolates from the two hosts [25]. Instead, we hypothesize that the resistant host selected for higher pathogen virulence, which in turn was linked to or had secondary functions related to triazole tolerance (see below for details). Resistant hosts selecting for higher pathogen virulence has already been predicted theoretically [47], [48] and reported from experiments [49], [50]. Because quantitative host resistance decreases pathogen growth rate, pathogens can compensate for lower growth rates by evolving towards an increasing competitive ability, which in turn can result in increased virulence [46].

Both linkage (i.e. hitch-hiking) and pleiotropic effects could lead to a positive association between virulence and fungicide tolerance in pathogens, but hitch-hiking is unlikely to be the cause in this

case. First, hitch-hiking refers to the process through which an allele increases in frequency because it is linked to an allele that is under positive selection [51], [52]. Cyproconazole tolerance, PLACL and PLACP are quantitative traits that display continuous variation within populations (Figs. 1–2). It is possible that each of these traits is affected by many minor genes, but it is unlikely that all or most of the genes contributing to the increase of cyproconazole tolerance are closely linked to the genes governing the increase of PLACL or PLACP. Second, recombination rate plays a key role in determining the degree of hitch-hiking [53]. Hitch-hiking effects are expected to be lower in populations with high recombination rates. *M. graminicola* populations display a high degree of sexual recombination both during and between growing seasons [18], [20] and the populations included in this study were at gametic equilibrium [21], [22]. Thus, even if there were close linkage between the genes encoding cyproconazole tolerance and virulence, the high recombination rate observed in populations of *M. graminicola* would lead to a rapid decay in disequilibrium.

We hypothesize that the observed correlation is due to pleiotropic effects of genes that affect both virulence and cyproconazole tolerance. Host defense systems usually involve the production of compounds that have lethal or inhibitory effects on the penetration, survival and reproduction of pathogens [54]. These defense-related compounds may share some structural or functional characteristics with synthetic antimicrobials. Pathogen strains having the ability to detoxify the compounds produced by resistant hosts may also have the ability to detoxify synthetic antimicrobial compounds, leading to a simultaneous increase in virulence and antimicrobial resistance. This detoxification process

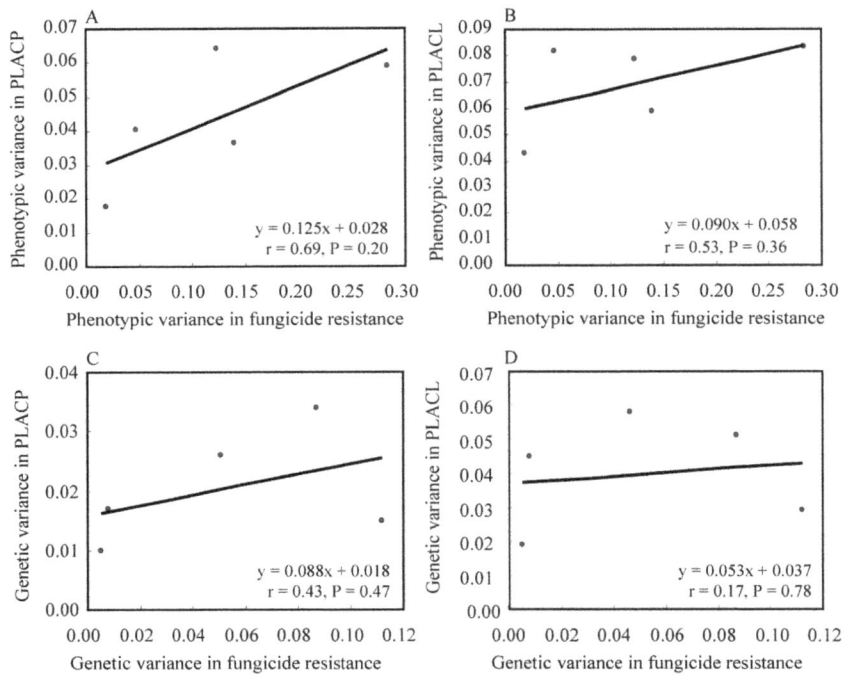

Figure 4. Correlation between variation in cyproconazole resistance and variation in two measures of virulence for five populations of *Mycosphaerella graminicola*. Cyproconazole resistance was determined by calculating the relative colony size of an isolate grown on Petri plates with and without the fungicide. Correlation was estimated at the population level: A) phenotypic variation in Percentage Leaf Area Covered by Pycnidia (PLACP); B) phenotypic variation in Percentage Leaf Area Covered by Lesions (PLACL); C) genetic variation in Percentage Leaf Area Covered by Pycnidia (PLACP); and D) genetic variation in Percentage Leaf Area Covered by Lesions (PLACL).

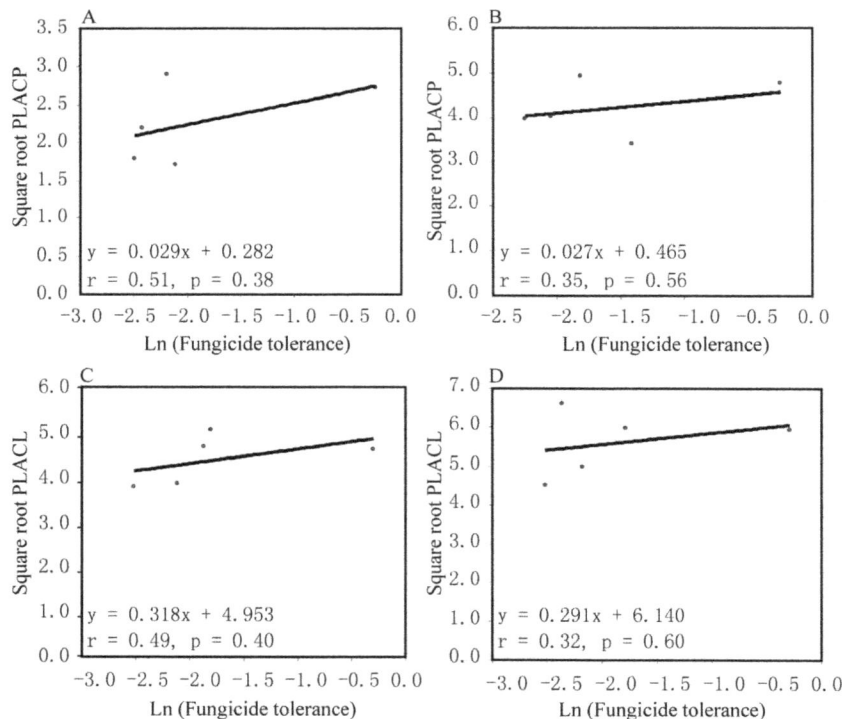

Figure 5. Correlation between population mean in cyproconazole resistance and two measures of virulence in *Mycosphaerella graminicola*. Cyproconazole resistance was determined by calculating the relative colony size of an isolate grown on Petri plates with and without the fungicide: A) Percentage Leaf Area Covered by Pycnidia (PLACP) on Toronit; B) Percentage Leaf Area Covered by Pycnidia (PLACP) on Greina; C) Percentage Leaf Area Covered by Lesions (PLACL) on Toronit; and D) Percentage Leaf Area Covered by Lesions (PLACL) on Greina.

could involve mechanisms such as reducing the entry of natural and synthetic compounds into pathogen cells through the action of efflux pumps located in the cytoplasmic membrane. It has been reported that some efflux pumps, such as ABC transporters and MgrA protein, have the ability to transport a broad range of structurally unrelated compounds during pathogen infection, therefore affecting both virulence and antimicrobial resistance, in many plant and human pathogens [55], [56], [57], [58], [59].

The positive correlation between virulence and cyproconazole tolerance could also be due to pathogen metabolites that can destroy or modify the structures and functions of both natural and synthetic antimicrobials. An example of such a defense metabolite is melanin. Melanin is composed of dark-brown or black pigments formed by the oxidative polymerization of phenolic compounds and can be produced by a broad array of plant and human pathogens [60], [61], [62]. It has been documented that melanin can increase antimicrobial resistance (see [63] for review) by reducing the susceptibility of melanized cells to antimicrobials [64], [65], [66] and increase virulence by interfering with numerous host defense mechanisms [67], [68], [69], [70], [71], [72] in many human pathogens. In *M. graminicola* strain IPO323, Mehrabi et al. [59] found that disruption of *MgSlt2* in *M. graminicola* led to a loss of melanization on potato dextrose agar, a loss of virulence and increased sensitivity to several fungicides including cyproconazole. Choi and Goodwin [60] also found that the velvet gene *MVE1* is involved in the synthesis of melanin in *M. graminicola*. *MVE1* mutants produced significantly less melanin. In *Fusarium graminearum*, deletion of the homologous velvet gene (*FgVEA*) reduced virulence and increased fungicide sensitivity [73].

The finding of a positive association between pathogen virulence and tolerance to synthetic antimicrobials coupled with the knowledge that resistant plant hosts can select for higher pathogen virulence has many implications for sustainable disease management in agroecosystems. It suggests that one unforeseen consequence of widespread deployment of quantitatively resistant cultivars or intensive application of synthetic antimicrobials might be selection for a higher basal level of antimicrobial resistance and enhanced virulence in pathogen populations, which would pose a greater threat to agricultural production. In this case, more dynamic disease management programs that incorporate more rapid spatial and temporal turnover of host resistance or synthetic antimicrobials may be important for sustainable disease control [74]. More rapid spatial and temporal turnover of host resistance or antimicrobials is expected to generate fluctuating selection against pathogens that could prevent the emergence of pathogen individuals and populations with higher virulence and antimicrobial resistance. However, the effectiveness of the proposed strategy for disease management depends largely on the fitness costs associated with virulence or antimicrobial resistance. If there are no fitness costs, then there may be no benefit derived from spatial and temporal deployments of host resistance or antimicrobials.

Author Contributions

Conceived and designed the experiments: JZ BAM. Performed the experiments: JZ. Analyzed the data: LY FG LS JZ. Wrote the paper: LY FG LS JZ BAM.

References

1. Zhan J (2009) Population Genetics of Plant Pathogens. In: ELS. John Wiley & Sons Ltd, Chichester. http://www.els.net [doi: 10.1002/9780470015902.a0021269].
2. Nicholson RL, Hammerschmidt R (1992) Phenolic compounds and their role in disease resistance. Annu Rev Phytopathol 30: 369–389.
3. Cordero MJ, Raventos D, San Segundo B (1994) Differential expression and induction of chitinases and β-1,3-glucanases in response to fungal infection during germination of maize seeds. Mol Plant Microbe Interact 7: 23–31.
4. Bhuiyan NH, Selvaraj G, Wei Y, Kong J (2009) Role of lignification in plant defense. Plant Signaling Behavior 4: 158–159.
5. Vergne E, Grand X, Ballini E, Chalvon V, Saindrenan P, et al. (2010) Preformed expression of defense is a hallmark of partial resistance to rice blast fungal pathogen Magnaporthe oryzae. BMC Plant Biol 10:206 doi: 1471-2229/10/206.
6. Amil-Ruiz FR, Blanco-Portales R, Muñoz-Blanco J, Caballero JL (2011) The strawberry plant defense mechanism: a molecular review. Plant Cell Physiol 52: 1873–1903.
7. Wink M (1988) Plant breeding: importance of plant secondary metabolites for protection against pathogens and herbivores. Theor Appl Genet 75: 225–233.
8. Morrissey JP, Osbourn AE (1999) Fungal resistance to plant antibiotics as a mechanism of pathogenesis. Microbiol Mol Biol Rev 63: 708–724.
9. Sacristán S, García-Arenal F (2008) The evolution of virulence and pathogenicity in plant pathogen populations. Mol Plant Pathol 9: 369–384.
10. Orbach MJ, Farrall A, Sweigard JA, Chumley FG, Valent B (2000) A telomeric avirulence gene determines efficacy for the rice blast resistance gene Pi-ta. Plant Cell 12: 2019–2032.
11. Janzac B, Montarry J, Palloix A, Navaud O, Moury B (2010) A point mutation in the polymerase of potato virus y confers virulence toward the pvr4 resistance of pepper and a high competitiveness cost in susceptible cultivar. Mol Plant-Microbe Interact 23:. 823–830.
12. Liu X, Yin Y, Wu J, Jiang J, Ma Z (2010) Identification and characterization of carbendazim-resistant isolates of Gibberella zeae. Plant Dis 94: 1137–1142.
13. Van der Wouw AP, Cozijnsen AJ, Hane JK, Brunner PC, McDonald BA, et al. (2010) Evolution of linked avirulence effectors in Leptosphaeria maculans is affected by genomic environment and exposure to resistance genes in host plants. PLoS Pathog 6: e1001180. doi:10.1371/journal.ppat.1001180.
14. Read A (1994) The evolution of virulence. Trends Microbiol 2: 73–76.
15. Palmer C, Skinner W (2002) Mycosphaerella graminicola: latent infection, crop devastation and genomics. Mol Plant Pathol 3: 63–70.
16. Orton ES, Deller S, Brown JKM (2011) Mycosphaerella graminicola: from genomics to disease control. Mol Plant Pathol 12: 413–424.
17. Eyal Z (1999) The Septoria tritici and Stagonospora nodorum blotch diseases of wheat. Eur J Plant Pathol 105: 629–641.
18. Zhan J, Mundt CC, McDonald BA (1998) Measuring immigration and sexual reproduction in field populations of Mycosphaerella graminicola. Phytopathology 88: 1330–1337.
19. Shaw MW, Royle DJ (1989) Airborne inoculum as a major source of Septoria tritici (Mycosphaerella graminicola) infections in winter wheat crops in the UK. Plant Pathol 38: 35–43.
20. Zhan J, Mundt CC, McDonald BA (2007) Sexual reproduction facilitates the adaptation of parasites to antagonistic host environment: evidence from field experiment with wheat-Mycosphaerella graminicola system. Intl J Parasitol 37: 861–870.
21. Zhan J, Pettway RE, McDonald BA (2003) The global genetic structure of the wheat pathogen Mycosphaerella graminicola is characterized by high nuclear diversity, low mitochondrial diversity, regular recombination, and gene flow. Fungal Genet Biol 38: 286–297.
22. Zhan J, McDonald BA (2004) The interaction among evolutionary forces in the pathogenic fungus Mycosphaerella graminicola. Fungal Genet Biol 41: 590–599.
23. Zhan J, Mundt CC, Hoffer MH, McDonald BA (2002) Local adaptation and effect of host genotype on the evolution of pathogen: an experimental test in a plant pathosystem. J Evol Biol 15: 634–647.
24. Zhan J, Linde CC, Juergens T, Merz U, Steinebrunner F, et al. (2005) Variation for neutral markers is correlated with variation for quantitative traits in the plant pathogenic fungus Mycosphaerella graminicola. Mol Ecol 14: 2683–2693.
25. Zhan J, Stefanato F, McDonald BA (2006) Selection for increased cyproconazole tolerance in Mycosphaerella graminicola through local adaptation and in response to host resistance. Mol Plant Pathol 7: 259–268.
26. Zhan J., McDonald BA (2011) Thermal adaptation in the fungal pathogen Mycosphaerella graminicola. Mol Ecol 20: 1689–1701.
27. Risser P, Ebmeyer E, Korzun V, Hartl L, Miedaner T (2011). Quantitative trait loci for adult-plant resistance to Mycosphaerella graminicola in two winter wheat populations. Phytopathology 101: 1209–1216.
28. Kelm C, Ghaffary SMT, Bruelheide H, Röder MS, Miersch S, et al. (2012). The genetic architecture of seedling resistance to Septoria tritici blotch in the winter wheat doubled-haploid population Solitär × Mazurka. Mol Breeding 29: 813–830.
29. Lannou C (2012) Variation and selection of quantitative traits in plant pathogens. Annu Rev Phytopathol 50: 319–338.
30. Kroener A, Hamelin G, Andrivon D, Val F (2011) Quantitative resistance of potato to Pectobacterium atrosepticum and Phytophthora infestans: integrating PAMP-triggered response and pathogen growth. PLoS One 6: e23331. doi:10.1371/journal.pone.0023331.

31. Dahl C, Biemann HP, Dahl J (1987) A protein kinase antigenically related to pp60v-src possibly involved in yeast cell cycle control: positive in vivo regulation by sterol. Proc Natl Acad Sci USA 84: 4012–4016.

32. Gisi U, Hermann D, Ohl L, Steden C (1997) Sensitivity profiles of *Mycosphaerella graminicola* and *Phytophthora infestans* populations to different classes of fungicides. Pestic Sci 51: 290–298.

33. Stergiopoulos I, van Nistelrooy JGM, Kema GHJ, De Waard MA (2003) Multiple mechanisms account for variation in base-line sensitivity to azole fungicides in field isolates of *Mycosphaerella graminicola*. Pest Manag Sci 59: 1333–1343.

34. Akins RA (2005) An update on antifungal targets and mechanisms of resistance in *Candida albicans*. Med Mycol 43: 285–318.

35. Ma Z, Michailides TZ (2005) Advances in understanding molecular mechanisms of fungicide resistance and molecular detection of resistant genotypes in phytopathogenic fungi. Crop Prot 24: 853–863.

36. Leroux P, Walker A (2011) Multiple mechanisms account for resistance to sterol 14α-demethylation inhibitors in field isolates of *Mycosphaerella graminicola*. Pest Manag Sci 67: 44–59.

37. Lamari L (2002) Assess: Iimage Analysis Software for Plant Disease Quantification. The American Phytopathological Society, St. Paul, USA.

38. Zadoks JC, Chang TT, Konzak CF (1974) Decimal code for growth stages of cereals. Weed Res 14: 415–421.

39. SAS Institute Inc. (1990) SAS User's Guide to Statistics, Version 6. 4th edn. SAS Institute Inc., Cary, NC.

40. Ott RL (1992) An Introduction to Statistical Methods and Data Analysis 4th edn. Duxbury Press, Belmont, USA.

41. Spitze K (1993) Population structure in *Daphnia obtusa*: quantitative genetic and allozymic variation. Genetics 135: 367–374.

42. Lynch M, Pfrender M, Spitze K, Lehman N, Hicks J, et al. (1999) The quantitative and molecular genetic architecture of a subdivided species. Evolution 53: 100–110.

43. Cowger C, Mundt CC (2002) Aggressiveness of *Mycosphaerella graminicola* isolates from susceptible and partially resistant wheat cultivars. Phytopathology 92: 624–630.

44. Goodnight CJ (1989). Population differentiation and the correlation among traits at the population level. Am Nat 133: 888–900.

45. Andersson S (1993) Population differentiation in *Crepis tectorum* (*Asteraceae*): patterns of correlation among characters. Biol J Linn Soc 49: 185–194.

46. Roff DA (2006) Introduction to computer-intensive methods of data analysis in biology, Cambridge University Press, Cambridge, UK.

47. Dwyer G, Levin SA, Buttel L (1990) A simulation model of the population dynamics and evolution of myxomatosis. Ecol Monogr 60: 423–447.

48. Gandon S, Michalakis Y (2000) Evolution of parasite virulence against qualitative or quantitative host resistance. Proc R Soc Lond B 267: 985–990.

49. Pink DAC, Lot H, Johnson R (1992) Novel pathotypes of lettuce mosaic virus – Breakdown of a durable resistance? Euphytica 63: 169–174.

50. Pariaud B, Robert C, Goyeau H, Lannou C (2009) Aggressiveness components and adaptation to a host cultivar in wheat leaf rust. Phytopathology 99: 869–878.

51. Kojima K, Schaffer HE (1967) Survival processes of linked mutant genes. Evolution 21: 518–531.

52. Maynard-Smith J, Haigh J (1974) The hitch-hiking effect of a favorable gene. Genet Res 23: 23–35.

53. Takuno S, Fujimoto R, Sugimura T, Sato K, Okamoto S, et al. (2007) Effects of recombination on hitchhiking diversity in the *Brassica* self-incompatibility locus complex. Genetics 177: 949–958.

54. Kliebenstein DJ (2012) Plant defense compounds: systems approaches to metabolic analysis. Annu Rev Phytopathol 50: 155–173.

55. Urban M, Bhargava T, Hamer JE (1999) An ATP-driven efflux pump is a novel pathogenicity factor in rice blast disease. EMBO J 18: 512–521.

56. Truong-Bolduc QC, Dunman PM, Strahilevitz J, Projan SJ, Hooper DC (2005) MgrA is a multiple regulator of two new efux pumps in *Staphylococcus aureus*. J Bacteriol 187: 2395–2405.

57. Chen PR, Bae T, Williams WA, Duguid EM, Rice PA, et al. (2006) An oxidation-sensing mechanism is used by the global regulator MgrA in *Staphylococcus aureus*. Nature Chem Biol 2: 591–595.

58. Sun CB, Suresh A, Deng YZ, Naqvi NI (2006) A multidrug resistance transporter in *Magnaporthe* is required for host penetration and for survival during oxidative stress. Plant Cell 18: 3686–3705.

59. Gupta A, Chattoo BB (2008) Functional analysis of a novel ABC transporter ABC4 from *Magnaporthe grisea*. FEMS Microbiol Lett 278: 22–28.

60. Cousin A, Mehrabi R, Guilleroux M, Dufresne M, Van der Lee M, et al. (2006) The MAP kinase-encoding gene *MgFus3* of the non-appressorium phytopathogen *Mycosphaerella graminicola* is required for penetration and in vitro pycnidia formation. Mol Plant Pathol 7: 269–278.

61. Mehrabi R, Van der Lee T, Waalwijk C, Kema GHJ (2006) *MgSlt2*, a cellular integrity MAP kinase gene of the fungal wheat pathogen *Mycosphaerella graminicola*, is dispensable for penetration but essential for invasive growth. Mol Plant-Microbe Interact 19: 389–398.

62. Choi YE, Goodwin SB (2011) *MVE1*, encoding the velvet gene product homolog in *Mycosphaerella graminicola*, is associated with aerial mycelium formation, melanin biosynthesis, hyphal swelling, and light signaling. Appl Environ Microbiol 77: 942–953.

63. Nosanchuk JD, Casadevall A (2006) Impact of melanin on microbial virulence and clinical resistance to antimicrobial compounds. Antimicrobial Agents Chemoth 50: 3519–3528.

64. Ikeda R, Sugita T, Jacobson ES, Shinoda T (2003) Effects of melanin upon susceptibility of *Cryptococcus* to antifungals. Microbiol Immunol 47: 271–277.

65. Taborda CP, da Silva MB, Nosanchuk JD, Travassos LR (2008) Melanin as a virulence factor of *Paracoccidioides brasiliensis* and other dimorphic pathogenic fungi: a mini-review. Mycopathologia 165: 331–339.

66. Liaw SJ, Lee YL, Hsueh PR (2010) Multidrug resistance in clinical isolates of *Stenotrophomonas maltophilia*: roles of integrons, efux pumps, phosphoglucomutase (SpgM), and melanin and biofilm formation. Intl J Antimicrobial Agents 35: 126–130.

67. Nosanchuk JD, Rosas AL, Lee SC, Casadevall A (2000) Melanization of *Cryptococcus neoformans* in human brain tissue. Lancet 355: 2049–2050.

68. Romero-Martinez R, Wheeler M, Guerrero-Plata A, Rico G, Torres-Guerrero H (2000) Biosynthesis and functions of melanin in *Sporothrix schenckii*. Infect Immun 68: 3696–3703.

69. Morris-Jones R, Youngchim S, Gómez BL, Aisen P, Hay RJ, et al. (2003) Synthesis of melanin-like pigment by *Sporothrix schenckii* in vitro and during mammalian infection. Infect Immun 71: 4026–4033.

70. Youngchim S, Morris-Jones R, Hay RJ, Hamilton AJ (2004) Production of melanin by *Aspergillus fumigatus*. J Med Microbiol 53: 175–181.

71. Mednick AJ, Nosanchuk JD, Casadevall A (2005) Melanization of *Cryptococcus neoformans* affects lung inammatory responses during cryptococcal infection. Infect Immun 73: 2012–2019.

72. Ngamskulrungroj P, Meyer W (2009) Melanin production at 37°C is linked to the high virulent *Cryptococcus gattii* Vancouver Island outbreak genotype *VGIIa*. Austr Mycol 28: 9–14.

73. Jiang J, Liu X, Yin Y, Ma Z. 2011. Involvement of a velvet protein *FgVeA* in the regulation of asexual development, lipid and secondary metabolisms and virulence in *Fusarium graminearum*. PLoS One 6: e28291. doi:10.1371/journal.pone.0028291.

74. Zhan J, Yang L, Zhu W, Shang L, Newton AC (2012) Pathogen populations evolve to greater race complexity in agricultural systems – evidence from analysis of *Rhynchosporium secalis* virulence data. PLoS One 7: e38611. doi:10.1371/journal.pone.0038611.

18

Effects of Picoxystrobin and 4-n-Nonylphenol on Soil Microbial Community Structure and Respiration Activity

Marianne Stenrød*, Sonja S. Klemsdal, Hans Ragnar Norli, Ole Martin Eklo

Norwegian Institute for Agricultural and Environmental Research (Bioforsk), Ås, Norway

Abstract

There is widespread use of chemical amendments to meet the demands for increased productivity in agriculture. Potentially toxic compounds, single or in mixtures, are added to the soil medium on a regular basis, while the ecotoxicological risk assessment procedures mainly follow a chemical by chemical approach. Picoxystrobin is a fungicide that has caused concern due to studies showing potentially detrimental effects to soil fauna (earthworms), while negative effects on soil microbial activities (nitrification, respiration) are shown to be transient. Potential mixture situations with nonylphenol, a chemical frequently occurring as a contaminant in sewage sludge used for land application, infer a need to explore whether these chemicals in mixture could alter the potential effects of picoxystrobin on the soil microflora. The main objective of this study was to assess the effects of picoxystrobin and nonylphenol, as single chemicals and mixtures, on soil microbial community structure and respiration activity in an agricultural sandy loam. Effects of the chemicals were assessed through measurements of soil microbial respiration activity and soil bacterial and fungal community structure fingerprints, together with a degradation study of the chemicals, through a 70 d incubation period. Picoxystrobin caused a decrease in the respiration activity, while 4-n-nonylphenol caused an increase in respiration activity concurring with a rapid degradation of the substance. Community structure fingerprints were also affected, but these results could not be directly interpreted in terms of positive or negative effects, and were indicated to be transient. Treatment with the chemicals in mixture caused less evident changes and indicated antagonistic effects between the chemicals in soil. In conclusion, the results imply that the application of the fungicide picoxystrobin and nonylphenol from sewage sludge application to agricultural soil in environmentally relevant concentrations, as single chemicals or in mixture, will not cause irreversible effects on soil microbial respiration and community structure.

Editor: Hauke Smidt, Wageningen University, The Netherlands

Funding: This work was performed as part of the strategic institute programme 'Bioavailability and biological effects of chemicals - Novel tools in risk assessment of mixtures in agricultural and contaminated soil' funded by the Norwegian Research Council, project number 186901/i30. The funders had no role in study design, data collection and analysis, decision to publish, or preparation of the manuscript.

Competing Interests: The authors have declared that no competing interests exist.

* E-mail: marianne.stenrod@bioforsk.no

Introduction

Various practices have been promoted to meet the demands for increased productivity of agricultural areas, including the use of mineral fertilizers and pesticides as well as the use of organic amendments. The potential risk of non-target effects of pesticides in soils is evident, and has been widely studied. Specific attention is now given to the use of organic amendments originating from sewage sludge. Sludge has been shown to contain high levels of many organic chemicals (e.g. nonylphenols, PAHs a.o.) that might exert ecotoxicological effects upon soil addition, e.g., [1,2]. Potentially toxic compounds, single or in mixtures, are added to the soil medium on a regular basis, while the ecotoxicological risk assessment procedures mainly follow a chemical by chemical approach focusing on establishing dose-response relationships for soil fauna. The use of sewage sludge on agricultural fields where pesticides are sprayed regularly as part of conventional farming practices calls for increased attention to the potential combined effects of pesticides and known contaminants in sewage sludge.

Sewage sludge, potentially used for application on agricultural soil, is known to contain considerable amounts of nonylphenol [2–4] – an industrial by-product and degradation product of

nonylphenol ethoxilate plasticizers. The occurrence, fate and toxicity of nonylphenol in the environment have been reviewed by Soares and co-workers [4], pointing at knowledge gaps as well as identified challenges with nonylphenol in soil including occurrence and possible accumulation in soil following sludge application, reduced degradation rates in soil due to sorption and reduced bioavailability, and potential toxic effects on soil microorganisms. Nonylphenol is also included on the list of priority substances of the Water Framework Directive [5] and, hence, require further attention to clarify its potential effects in soil. Picoxystrobin is a strobilurin fungicide [6] for spraying in cereals, with a maximum of one spraying per season in Norway. Picoxystrobin inhibits mitochondrial respiration by blocking electron transfer at the Qo centre of cytochrome bc1 [7]. Picoxystrobin is reported to have a high acute toxicity to earthworms (LC_{50} at 6.7 mg kg^{-1}) [8], and field assays of earthworm toxicity indicate acute toxic effects even at recommended doses for use [9], which might be caused by heavy rain shortly after spraying forcing the earthworm to migrate to the surface. Potentially negative effects on soil microbial nitrogen and carbon mineralization activity are shown to be transient (dose: 750 g ha^{-1}, duration: 28 days) [8]. Picoxystrobin require continued attention due to its demonstrated potentially negative effects to earthworms.

The importance of soil microbes and their activity in the functioning of soils, e.g., [10,11] justify their thorough investigation in risk assessments [12–14]. The development of genomic techniques over the last decades has made detailed studies of the soil microbial community possible, beyond the scope of broad-scale measures like substrate induced respiration. DNA extraction from soil followed by different molecular approaches to determine the genetic diversity and quantify the presence of single organisms or groups of related organisms in a soil sample, have been employed successfully in studies of species and functional diversity in agricultural soils [15–17]. According to OECD guidelines for the testing of chemicals carbon [18] and nitrogen [19] transformation tests (with cut-off criteria of 25% effect) are the recommended methods to assess effects concentrations of chemicals on the soil microbial community. In research, soil respiration is commonly used to assess effects of pesticides and other chemicals on soil microbes. These are measures linked to the activity level of the soil microbial community, but are crude measures that do not necessarily reveal all relevant effects. The low percentage of soil microorganisms we are able to culture ex situ stresses the need for employing molecular and genomic methods suitable for terrestrial ecotoxicological studies. T-RFLP-analysis is one much used technique showing good results when looking at effects of different environmental conditions and chemical stressors on soil microbial communities, e.g., [16,17,20]. But there are many important methodological aspects to be considered when interpreting the results, e.g., [21,22], including choice of primers and restriction enzymes, procedure for noise reduction and profile alignment, and statistical analysis using relative abundance or presence/absence data.

The main objective of this study was to assess the effects of picoxystrobin and nonylphenol, as single chemicals and mixtures, on soil microbial community structure and respiration activity in an agricultural sandy loam soil. The choice of compounds is based on their individual occurrence and effects in the environment, and expected potential for co-occurrence in the field, although the latter is scarcely documented. Further, they were chosen to represent an example of combined effects of agricultural and industrial contaminants with their expected independent effect mechanisms due to differences in mode of action. Effects of the chemicals were assessed through measurements of soil microbial respiration activity and soil bacterial and fungal community structure fingerprints, together with a degradation study of the chemicals, through a 70 d incubation period. The different measures showed corresponding results in support of a conclusion that a mixture situation with picoxystrobin and nonylphenol in soil will not increase the potential negative effects of picoxystrobin on the soil microbial community.

Materials and Methods

Soil

Bulk soil sampling in the top 10 cm of the plough-layer of an agricultural field at Norderås, Ås, South East Norway (59°41′14″ N, 10°46′22″ E), was done in middle of August 2008. After sieving (4 mm mesh) the soil was physically and chemically characterized (Table 1; all analyses performed in accordance with recognized laboratory standards at Analycen AS/Eurofins Norway) and stored moist at 4°C for about 3 weeks before use in laboratory experiments. No specific permits were required for the described field soil sampling and subsequent lab studies.

Table 1. Selected physical and chemical characteristics of the studied soil.

Parameter	Value	Unit
pH	6.3	
TOC – Organic C	1.1	%
Kjeldahl N	0.0986	%
C/N	11	
Tot C	2.0	%
Tot N	0.07	%
CEC	11.2	meq/100 g
Coarse (>2 mm)	13.7	%
Sand (2.000-0.060)	68.3	%
Silt (0.060-0.002)	20.7	%
Clay (<0.002)	11.0	%

Analysis performed by Analycen AS/Eurofins Norway.

The Water Holding Capacity (WHC) was estimated on sieved soil gently packed in plastic columns. After wetting with Milli-Q-water overnight, excess water was allowed to drip off during 1 h the next day, and the gravimetric moisture content was measured after drying overnight at 105°C.

Chemicals

Treatment solutions were prepared from analytical grade picoxystrobin (Riedel-de Häen) and 4-n-nonylphenol (Fluka) purchased from Sigma-Aldrich. The chemicals were dissolved in acetone (4 ml portions; Lab-Scan) and added to quartz sand (27 g portions; Sigma-Aldrich). After evaporation of the acetone (about 30 minutes) the quartz sand with chemical was mixed into batches of moist soil (2.7 kg dry weight equivalent (dw eq.), 60% WHC) giving 2 or 10 mg picoxystrobin kg^{-1} soil (dw eq.), or 0.5 or 10 mg 4-n-nonylphenol kg^{-1} soil (dw eq.) as single treatments as well as all mixture combinations (Table 2). A solvent control (acetone added to quartz sand, left to evaporate before addition to soil) and an untreated control (soil with no addition of chemical) were included in the experiment.

The lowest test concentration for picoxystrobin was set from reported levels for transient effects on carbon and nitrogen mineralization (750 g ha^{-1}) [8], considering that picoxystrobin has been shown to sorb to the top cm of soil (50–70% interception, sorption in top 1 cm of soil), although recommended doses in Norway are below these levels. The worst case concentration was set to 10 mg kg^{-1} surpassing reported acute toxicity levels for earthworm ($LC_{50} = 6.7$ mg kg^{-1}) [8]. The lowest test concentration for nonylphenol was set to the Predicted Initial Environmental Concentration (PIEC) estimated through a risk assessment made by the Norwegian Scientific Committee on Food Safety [2]; application of sludge containing mean expected amounts of nonylphenol (32 mg kg^{-1}) at rates in accordance with Norwegian regulations (40 tonnes ha^{-1}). From this a worst case concentration was set to 10 mg kg^{-1} considering a risk of reduced incorporation depth and possible initial effects.

Experimental Set-up

An incubation experiment was set up with all chemical treatments in five replications and the control treatments in four replications. Subsamples of about 100 g dw eq. of sieved, moist soil

Table 2. Overview of picoxystrobin and 4-n-nonylphenol treatments in the incubation experiment.

Treatment	Picoxystrobin (PI)		4-n-nonylphenol (NP)	
	2 mg kg^{-1} dry soil	10 mg kg^{-1} dry soil	0.5 mg kg^{-1} dry soil	10 mg kg^{-1} dry soil
PI low	x			
PI high		x		
NP low			x	
NP high				x
Mix low	x		x	
Mix NP high	x			x
Mix PI high		x	x	
Mix high		x		x
Control				

(60% WHC) in plastic (polypropylene) containers were placed in air-tight glass jars together with small beakers with NaOH-solution (10 mL, 1 M) to trap CO_2 evolved from the soil. All treatments were set up with five repeats to enable destructive sampling of five and four replicates, of chemical treatments and the control respectively, on each sampling occasion through the 70 d incubation period. A total of 244 samples were incubated at 20°C in the dark. Samples were taken on six time-points; before addition of chemicals (4 samples), and after 1, 7, 14, 28 and 70 days of incubation (48 samples on each occasion). Each sample was homogenized and subsamples taken for chemical residue analysis (25 g), and DNA extraction (30 g). In addition, the general microbial activity level in the soil was monitored by weekly replacement and analysis of the NaOH-traps, giving a total of 11 time-points with measurements.

Total Soil Microbial Activity

The total activity of soil microbial biomass was followed by respirometric measurements [23]. The CO_2 in the NaOH traps was determined by colourimetric continuous flow analysis (AutoAnalyzer3, Bran+Luebbe, Germany). Results are reported as mg CO_2-C kg^{-1} dry soil.

Soil Microbial Community Structure

DNA extraction procedure. Soil samples were homogenized (30 g moist soil +30 mL Milli-Q-water) in a mill (3 min forward +3 min reverse spin; Retsch PM 400) before extraction of DNA according to the Fast DNA Spin Kit for Soil (QBiogene) manual. In short, soil suspension (400 μL) was added to Lysing Matrix E Tubes before homogenization with the FastPrep® instrument (30 sec at speed 5.5; QBiogene, MP Biomedicals) and centrifugation (14000×g, 3 min). DNA was purified with guanidine thiocyanate (5.5 M, Sigma) twice (1 mL and 600 μL) after DNA binding to the Binding Matrix Suspension, before elution (100 μL sterile distilled water) from Spin^{TM}Filters. The eluate was further purified through Micro Bio-Spin Chromatography columns (Bio-Rad Laboratories Ltd.) packed with polyvinylpolypyrrolidone (PVPP) (Sigma) (4000×g, 5 min), and stored at −80°C.

PCR procedure. PCR reactions were run with fluorescently labelled PCR primer sets (Applied Biosystems), both forward and reverse primer labelled. After testing of 8 primer sets, chosen from previously published T-RFLP studies, PCR reactions were run for all samples with three selected primer sets and fluorescence labelling (Table 3).

All PCR reactions were run with 1 μL template and 1 μL of forward and reverse primer (10 pmol/μL) in TaqMan® Environ-

Table 3. Primer sets used in the experiments.

Primer set no.	Primer	Sequence	Reference
1	63F[1]	CAG GCC TAA CAC ATG CAA GTC	[24]
	1087R[2]	CTC GTT GCG GGA CTT ACC CC	[24]
2	EF4[1]	GGA AGG GRT GTA TTT ATT AG	[25]
	Fung5[2]	GTA AAA GTC CTG GTT CCC C	[25]
3	ITS1F[3]	CTT GGT CAT TTA GAG GAA GTA A	[26]
	ITS4[4]	TCC TCC GCT TAT TGA TAT GC	[27]

Fluorescence labeling: [1]6FAM, [2]VIC, [3]PET, [4]NED.

mental Master Mix 2.0 (12.5 μL; Applied Biosystems) made up to a total reaction volume of 25 μL with sterile distilled water, on GeneAmp® PCR system 9700 (Applied Biosystems) or PTC-200 (MJ Research) thermal cyclers (40 cycles). An annealing temperature of 55°C was used for all primer combinations. PCR-products were purified with GenElute^{TM}PCR clean-up kit (Sigma-Aldrich) according to the instructions from the supplier, before restriction digestion and T-RFLP.

T-RFLP procedure. After initial testing of a range of restriction enzymes (AluI, DdeI, HaeIII, HinfI, MboI, MspI, RsaI, TaqI, HhaI, MvnI, BstUI) on random samples, MspI and HaeIII, and MspI and HinfI were chosen for restriction of PCR products from bacteria and fungi, respectively. Restriction with the different enzymes were run in separate reactions and analysed separately. Analysis was performed on an ABI 3730 DNA Analyzer (Applied Biosystems, Foster City, USA) after mixing of 1 μl sample (template) with 8.75 μl Hi-Di formamid^{TM} (Applied Biosystems) and 0.25 μl GeneScan^{TM} 500 LIZ size standard (Applied Biosystems). The results reported for the T-RFLP analyses are based on relative comparisons of fingerprints, to show relative effects of different chemical treatments and comparison with a soil sample not treated with chemicals.

Degradation Study

Residues of picoxystrobin and 4-n-nonylphenol were extracted from 5 g subsamples of soil after mixing with 1.0 g of dehydrated $MgSO_4$ (purum, Fluka, Sigma-Aldrich GmbH) and 10 mL acetonitrile (Pestiscan, LAB-SCAN POCH SA, Gliwice, Poland) in 50 ml centrifugal tubes. One μg of 4-n-nonylphenol-2,3,5,6-d4-OD (99.4%, Chiron, Trondheim, Norway) was added as internal standard. After a short homogenization (10–15 sec. whirl mix) the samples were extracted by end-over-end shaking (1 h; Reax2, Heidolph). Extraction efficiencies derived from analysis of sterilised (autoclaved) soil at $117\pm2.0\%$ and $108\pm0.0\%$ for 4-n-nonylphenol and picoxystrobin, respectively. After centrifugation ($1800\times g$, 5 min), 1.5 ml of the supernatant was transferred to GC-vials for analysis on GC-MS. Calibration standards at 0.001, 0.01, 0.05, 0.2, 1.0 and 5 $\mu g\ ml^{-1}$ where prepared by diluting stock solutions of picoxystrobin (98%, Dr. Ehrenstorfer, Augsburg, Germany) and 4-n-nonylphenol (99.2%, Chiron, Trondheim, Norway) with acetonitrile. To balance the matrix in the samples a GC-vial was added 1 ml of a blank soil extract which was evaporated to dryness. One ml of each calibration standard was added to the GC-vial together with 0.1 μg 4-n-nonylphenol-2,3,5,6-d4-OD as internal standard. The measurements were performed on an Agilent 6890 gas chromatograph connected to an Agilent 5973 mass spectrometer using ChemStation Software version D.03.00. The gas chromatograph was equipped with a Gerstel (Mühlheim Ruhr, Germany) programmable temperature vaporizing (PTV) injector with a sintered liner. The separation was performed using a fused silica column (HP-5MSI 30 m width, 0.25 mm internal diameter, 0.25 μm film thickness, J&W Scientific) connected to a 2.5 m methyl deactivated pre column (Varian Inc., Lake Forest CA, USA) of same internal diameter as the analytical column. The temperature program was as follows; 65°C held for 1.5 min, 20°C min^{-1} to 120°C, held for 0 min, 20°C min^{-1} to 300°C, held for 0.5 min, total runtime 13.75 min. The PTV program was as follows: solvent vent temperature 60°C held for 1.40 min with a vent flow at 200 ml min^{-1}. After 1.42 min the split valve was closed and the injector temperature increased by 720°C min^{-1} to 250°C and kept for 2 min. Injection volume 10 μl. The mass spectrometer was operated in selected ion monitoring mode with target/qualifier ions as follows: 4-n-nonylphenol-D4: $m/z = 111/224$, 4-n-nonylphenol: $m/z = 107/220$ and picoxystrobin: $m/z = 145/335$. Transfer line temperature was set at 280°C, ion source temperature at 230°C and quadrupole temperature at 150°C.

Biological degradation of picoxystrobin and 4-n-nonylphenol was verified by spiking selected matrices (quartz sand, unsterile soil, sterilized soil) at 0.05 $\mu g\ g^{-1}$ and left in the dark at room temperature ($20\pm2°C$) for 24 h before extraction and analysis as described above. The soil was sterilised by autoclaving (121°C, 20 min, x2) (Matachana S1000).

Data Processing and Statistical Analysis

First-order rate constants for organic carbon mineralization were estimated by linear regression (SigmaPlot 11.0, SYSTAT) using mean cumulative values for CO_2-C evolved, from the equation $\ln S = \ln S_0 + kt$, where S is the amount of CO_2-C evolved at time t (mg kg^{-1} dry soil), S_0 is the initial amount of mineralization product, k is the rate constant (d^{-1}), and t is the time (d). Due to the non-normal distribution of our CO_2-measurements, non-parametric analysis was utilized to test for differences between treatments; i.e. Kruskall Wallis test for equality of medians for all treatments, and Mann-Whitney test for equality of medians between selected treatments (Minitab 15, Minitab Inc.).

Handling of T-RFLP data with binning/identification of alleles, normalization (within project; i.e. for all samples independent of plate) and (automatic) alignment of profiles was done with GeneMapper 4.0 (Applied Biosystems) based on a 4 basepair (bp) bin width (assumed to give the more stable number of bins based on a screening of bin widths from 0.5 to 10 bp in 0.5 bp increments) and a threshold of 50 (default). Obvious pull-up peaks were removed from the dataset before further analysis. No further trimming of the data was done (i.e. no small peaks removed) to avoid exclusion of potentially important peaks for the microbial diversity analysis. Size (bp), area and height data were exported to an excel spreadsheet (Excel 2007, Microsoft corp.) for calculation of average size of each identified peak and calculation of relative peak height and area for all peaks within each individual sample. All peak data for each sample was assembled before statistical analysis, keeping the data for each primer set separate. The terminal fragments labelled from the forward primer as well as the fragments labelled from the reverse primer from each of the restriction reactions, were included in a single data row for each sample before analysis. The data were found to be not normally distributed (i.e. skewed due to rarely occurring peaks). Relative peak heights were analysed with non-metric multidimensional scaling (nmMDS) in R-software (v. 2.9.1) [28] using the metaMDS routine in the vegan package [29] with k = 2 or 3 dimensions, the isoMDS routine in the MASS package [30] with k = 2 dimensions, and with principal coordinates analyses using the PCO routine in R. All analyses resulted in similar patterns and the results presented are from metaMDS (k = 3 dimensions) as these gave the best fit to the observed data (STRESS below 0.10) [31,32]. Results are reported as two and three-dimensional plots showing the clustering of treatments at the different time-points of analysis through the incubation period. The samples taken before addition of chemical and onset of the incubation period (original soil sample) were included in the analysis of all time-points as a reference. Further, analysis of similarity was performed with the ANOSIM routine in PAST-software [33] with Bray-Curtis distance measure utilized in all tests. Results are given as ANOSIM R of a value between 0 and 1, where values close to 1 indicate large differences between treatments.

The observed T-RFLP patterns for the untreated and solvent control were not statistically significant different as analysed by nmMDS, and the results presented focus on differences between chemical treatments and the solvent control. This was to avoid overestimating any effects of the studied chemicals. A similar clustering of T-RFLP fingerprints was observed for the results obtained from using primer sets 2 and 3 (i.e. soil fungal community fingerprints; Table 3). Hence, the presented results only include primer set 3.

Soil degradation half-lives (DT50) for picoxystrobin and 4-n-nonylphenol was estimated from results from the degradation study, and calculated according to first-order kinetics from the equation $DT50 = \log_{10}2/b$, where b is the regression coefficient estimated through a linear regression procedure. All measured values were utilized in the estimation of the DT50 values except for two extreme outliers being removed before statistical analysis of the data. Due to the five replicates for each treatment this could be done without compromising the results of the data analyses, as the analyses for the affected treatments were still based on four replicate samples. Due to some differences in the observed degradation pattern for the chemicals at low and high concentrations, the soil half-lives were estimated separately for these. The estimated half-lives are based on data from both the relevant single chemical and mixture treatment, as the observed degradation pattern did not differ significantly between these. As an example

DT50 for 2 mg picoxystrobin kg^{-1} dry soil was estimated using data from both the single chemical treatment and the two mixture treatments with this concentration of picoxystrobin. The estimation was done assuming bi-phasic degradation of both chemicals, with both phases sufficiently described by first-order degradation kinetics and, hence, possible to estimate through linear regression.

A significance level of 5% was used for hypothesis testing.

Results

Total Soil Microbial Activity

No statistically significant differences between the treatments could be detected from observations of cumulative CO_2-C development after 70 d and estimated first-order rate constants for organic C-mineralization (Table 4). However, cumulative respiration curves (Fig. 1) indicated a tendency of reduced respiration activity in treatments with picoxystrobin as a single chemical, and increased respiration activity in treatments with high concentrations of nonylphenol (single chemical and mixture). The untreated control showed a lower respiration rate in the beginning of the period and a higher respiration rate at the end of the period, as compared to the other treatments. This difference did not result in any statistically significant difference in cumulative CO_2-C respiration, but indicated an effect of the solvent (acetone) used in the experiments.

The indications from the latter were supported by analysis of the separate measurements of CO_2 development from the soil samples at 11 time-points through the 70 d incubation period. Kruskall Wallis test for equality of medians between the treatments indicated statistically significant differences (p<0.001) when looking at all the individual CO_2 measurements in the analysis separately (i.e. not cumulative values at 70 d). The individual ranking of the treatments (data not shown) gave the solvent control a rank close to the overall rank (i.e. mean), while treatments with picoxystrobin as a single chemical (2 and 10 mg kg^{-1} dry soil) were indicated to cause a reduction in respiration activity, through a rank statistically significant lower than the overall rank. High concentrations of 4-n-nonylphenol (10 mg kg^{-1} dry soil) were indicated to cause an increase in respiration activity, through a rank statistically significant higher than the overall rank. This was also the case for the untreated control.

These indications were further supported by results from Mann-Whitney tests for equality of treatment medians, comparing the treatments two by two (test criteria: p<0.05). High concentration of 4-n-nonylphenol (10 mg kg^{-1} dry soil) in mixture with low concentration of picoxystrobin (2 mg kg^{-1} dry soil) gave statistically significant higher CO_2 development than treatments with picoxystrobin as a single chemical (2 and 10 mg kg^{-1} dry soil), low concentration of 4-n-nonylphenol (0.5 mg kg^{-1} dry soil), low concentration mixture (2 and 0.5 mg kg^{-1} dry soil of picoxystrobin and 4-n-nonylphenol, respectively) as well as the solvent control. Further, treatments with picoxystrobin as a single chemical (2 and 10 mg kg^{-1} dry soil) gave statistically significant reduction in CO_2 development compared to all mixture treatments, high concentration of 4-n-nonylphenol (10 mg kg^{-1} dry soil) as a single chemical, and the untreated control.

Soil Microbial Community Structure

The fungicide picoxystrobin was found to have a statistically significant effect on the soil microbial community structure, as found from nmMDS and ANOSIM of relative peak height data from T-RFLP analyses (Table S1). Looking at the effects of single chemical treatments in bacteria (Fig. S1) the largest segregation between treatments was observed after 7 d, with an overall

ANOSIM R of 0.65 (p = 0.0001). Here there was a statistically significant segregation between the low and high concentration treatment (R = 0.42; p = 0.018), indicating a concentration effect, and also a significant effect compared to the solvent control at low levels of picoxystrobin in the soil (R = 0.44, p = 0.04). Large variability was found in the data for the effects of picoxystrobin on the soil fungal community structure, with the largest spread in the data as shown by nmMDS after 7 d (Fig. 2a, Fig. S2) (R = 0.47, p = 0.0001). However, an effect due to different treatment concentrations of picoxystrobin was evident until 28 d (R = 0.38, p = 0.017) (Fig. S2). There was still a statistically significant difference between the single chemical treatments and the solvent control at 70 d for both bacteria (Fig. S1) and fungi (Fig. 2b), with an overall ANOSIM R of 0.65 and 0.4 (p = 0.0001), respectively.

Less evident effects were observed for the single chemical treatments with 4-n-nonylphenol. Analysis of the bacterial community structure indicated an increasing change in the T-RFLP fingerprints with time, with statistically significant differences due to chemical treatment in nmMDS-analysis at 28 d (overall ANOSIM R = 0.56, p = 0.0001) and 70 d (overall ANOSIM R = 0.67, p = 0.0001) (Fig. 3, Fig. S3). A concentration effect was also evident at 28 d (R = 0.42, p = 0.035). The T-RFLP fingerprints of the fungal community upon treatment with 4-n-nonylphenol showed a similar pattern (Fig. S4), however, with larger variation between replicate samples giving lower ANOSIM R-values (range: 0.25–0.52).

The effects of the mixture treatments (Fig. S5 and S6) could only be separated from the solvent control after 28 and 70 d for the T-RFLP fingerprints of the soil bacterial community (R = 0.73, p = 0.007 and R = 0.44, p = 0.04, respectively) and after 70 d for the soil fungal community (R = 0.72, p = 0.007). When comparing the mixture treatments with the single treatments, the more prominent alteration in T-RFLP fingerprints originated from picoxystrobin as a single chemical (10 mg kg^{-1} dry soil), with the largest separation between treatment effects observed after 28 d both for the soil bacterial (Fig. 4a) and soil fungal (Fig. 4b) community structure (overall ANOSIM R = 0.61 for both analyses). At this time-point also the single chemical treatment with 4-n-nonylphenol (10 mg kg^{-1} dry soil) could be singled out, with statistically significant changes in T-RFLP fingerprint as compared to the solvent control and other chemical treatments.

The duration of the observed changes in T-RFLP fingerprints of the soil microbial communities under the different chemical treatments was examined through nmMDS analysis of the grouping of the time-points for T-RFLP analyses. In general, these results indicated that the observed changes were transient. There was a more rapid re-establishment of the original T-RFLP fingerprints in the soil bacterial (i.e. closer resemblance between time-points 0 and 70) as compared to the soil fungal community (Fig. 5).

Degradation of Chemicals

Soil chemical residue was analysed day 0, 1, 7, 14, 28 and 70 of the incubation period. A lag phase was observed before the degradation of picoxystrobin commenced (Fig. 6a), while 4-n-nonylphenol was rapidly degraded in the soil (Fig. 6b). Calculations of soil half-lives (Table 5) were based on data from both single chemicals and mixtures as the degradation patterns were similar for these. Due to very rapid initial degradation of 4-n-nonylphenol in the low concentration treatments (0.5 mg kg^{-1} dry soil), the soil half-lives were here estimated for the 0–14 days and 14–70 days separately. For the other treatments the estimates were made for 0–28 days and 28–70 days. Soil half-life for picoxystrobin for the 28–70 day period was not calculated due to a measured

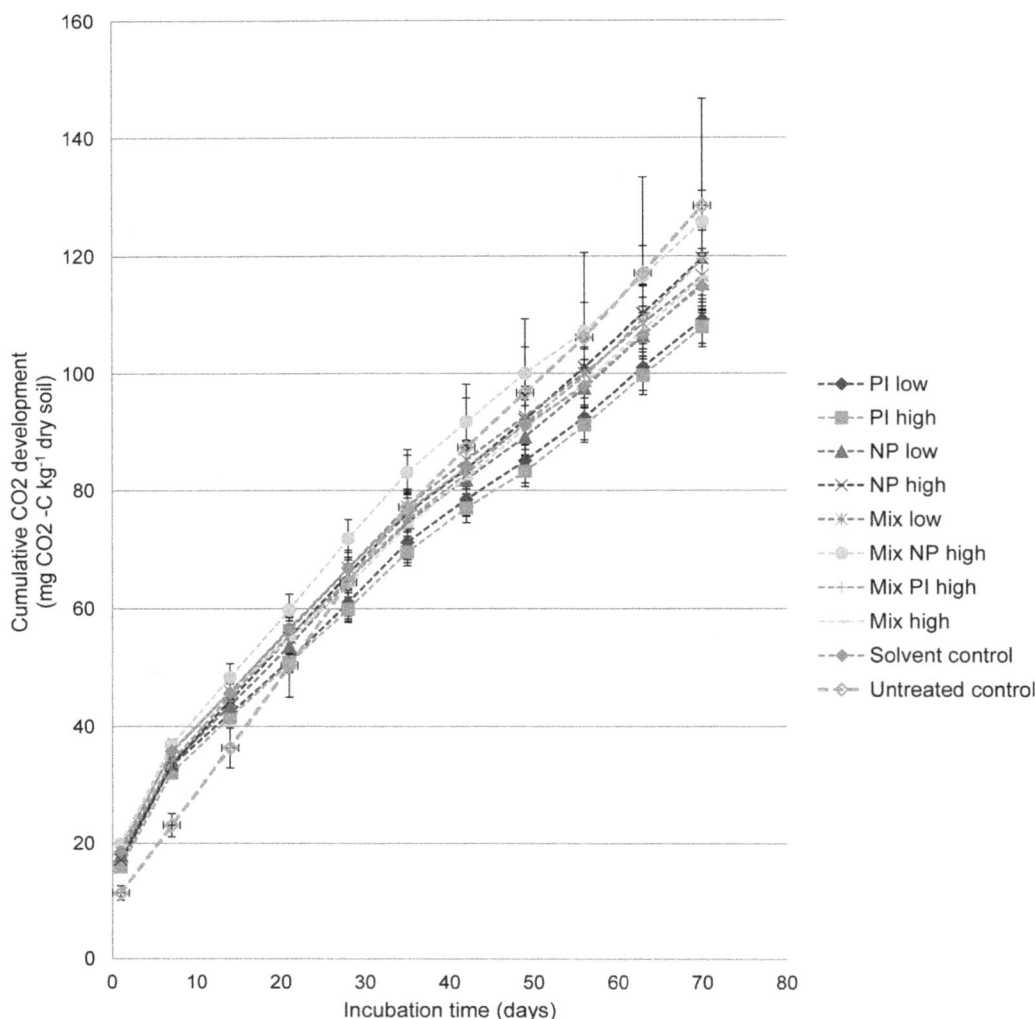

Figure 1. Effects of picoxystrobin and 4-n-nonylphenol on soil microbial respiration activity. Cumulative CO_2 development in a sandy loam soil during a 70 d incubation period after treatment with picoxystrobin and 4-n-nonylphenol as single chemicals and in mixture. Mean values ± SD. Legends: Treatments with picoxystrobin (PI) or 4-n-nonylphenol (NP) or both chemicals (Mix), in low (2 mg kg^{-1} PI, 0.5 mg kg^{-1} NP) or high (10 mg kg^{-1}) concentrations. Cf. Table 2 for full details.

increase in the soil residual concentrations at the end of the incubation period, hence, not allowing a reliable estimate of the long term degradation rate.

The rapid initial degradation of 4-n-nonylphenol was confirmed and shown to be microbially mediated (Fig. 7).

Discussion

Significance and Duration of Observed Effects

Total soil microbial activity. Our findings of a negative effect of picoxystrobin on individual measurements of CO_2 development, but not cumulative values, were in accordance with previous reports. Approval guidelines for plant protection products require documentation of effects on carbon and nitrogen mineralization, and only transient effects on nitrogen and carbon mineralization have been reported for picoxystrobin [8]. This fungicide inhibits mitochondrial respiration by blocking electron transfer at the Qo centre of cytochrome bc1 [7], and, hence, can be expected to affect the soil fungal community. We did not, however, expect an adverse effect in bacteria. The observed lag-

phase before on-set of degradation of picoxystrobin (Fig. 6a) could correspond with a negative initial effect on respiration activity and microbial degradation activity. This may, however, not be concluded from the results as bacterial growth on anthropogenic substrates in soil is generally preceded by a lag-phase to allow adaptation to a new substrate. The observed slowing of the degradation rates after 28 days (Table 5) indicates limitations to microbial activity (e.g. oxygen limitation, occurrence of inhibiting factors), reduced bioavailability of the compound (e.g. time-dependent sorption) and/or a negative effect of picoxystrobin on the soil microbial community.

Our results indicated stimulating effects of high levels of 4-n-nonylphenol on soil respiration that could, in part, be explained by an increase in soil microbial activity level. As an endocrine-disrupting agent, we did not expect adverse effects of 4-n-nonylphenol on the soil bacterial community. Our findings are in accordance with others reporting low risk of adverse effects of nonylphenol in environmentally relevant concentrations on soil fungi, but with a stated need for more in-depth studies [34]. This study shows stimulation of specific fungal strains during long-term

Table 4. Organic C mineralization in soil.

Treatment	Organic C mineralization		
	Cumulative CO_2-C	Rate constant	
	(mg CO_2-C/kg dry soil 70d)	(k_{totC}, 10^{-4} day^{-1})	r^2
PI low	109.2±4.1[a]	1.7±0.17[b]	0.82
PI high	108.0±3.5	1.6±.0.17	0.83
NP low	115.3±4.6	1.7±0.18	0.83
NP high	119.6±5.1	1.8±0.17	0.85
Mix low	116.6±4.5	1.8±0.19	0.81
Mix NP high	125.7±5.3	1.9±0.20	0.83
Mix PI high	119.5±6.8	1.8±0.18	0.84
Mix high	116.4±7.9	1.8±0.19	0.81
Solvent control	114.9±3.9	1.8±0.19	0.80
Untreated control	128.5±18.2	1.7±0.19	0.96
Total		1.8±0.05	0.83

[a]SD (standard deviation of measurements),
[b]SE (standard error of estimate).

exposure, as well as appreciable sorption of nonylphenol in soil. Others have reported toxic effects of nonylphenol on soil microbes [4]. However, the general picture is that of rapid and complete mineralization in a wide range of soils [35].

Soil microbial community structure and degradation of chemicals. The observed changes in the T-RFLP patterns of the soil bacterial and fungal communities during the lag-phase before onset of picoxystrobin degradation (Fig. 6a), are in agreement with the general assumption of adaptation of the microbial community to a new substrate. The strobilurin mode of action is, however, typically very fast acting [6], and it may not be ruled out that the observed rapid effects on T-RFLP patterns are, in part, caused by negative effects of the fungicide. We observed a bi-phasic degradation, with initial degradation rate (DT50) in correspondence with previous reports [8]. However, in our studies the levels of picoxystrobin stabilized after 28 days, possibly due to sorption and reduced bioavailability [8,36]. The estimated degradation half-lives (DT50) for picoxystrobin assume that the analysed fraction is the total residual fraction available for

degradation. The test for initial rapid degradation (Fig. 7, left) did not indicate a rapid sorption in moist soil, but specific sorption studies were not performed. Picoxystrobin is, however, classified as only slightly mobile [36], and non-bioavailable residues are expected to be above 20% after 100 days [8]. This would explain the lack of a concentration effect beyond 28 days. Alternatively this could be due to a lasting effect on the soil microbial community, as indicated by the lasting significant differences in T-RFLP-fingerprints in picoxystrobin treated soils compared to the solvent control. Similar percentage levels of degradation of picoxystrobin were observed for the two concentrations tested, meaning that actual concentration levels remaining in the treatments with high concentrations of picoxystrobin were much higher after 28 days than for the treatments with low concentrations. There was, however, no observable concentration effect at 70 days between the T-RFLP-profiles.

This observed levelling out of the residual soil concentration of picoxystrobin after 28 days together with the lack of a concentration effect at 70 days, indicated that the observed transience of the effects on the soil microbial community was coupled to decreasing bioavailability of the chemical. Time-dependent sorption might be the reason for reduced degradation and bioavailability of the pesticide after 70 days [37]. Observed major metabolites of picoxystrobin from degradation under aerobic conditions in soil are reported to have mean half-lives in the range of 14–29 days [8]. The chemical analyses performed in our study did not include metabolites, and we may, hence, not rule out any effects of these in our short-term laboratory experiment.

Although 4-n-nonylphenol was found to stimulate soil respiration, the effects on the microbial community structure were less evident. The observed changes in the T-RFLP fingerprints for the soil bacterial community developed more slowly after 4-n-nonylphenol treatment than for picoxystrobin. This indicated a change induced by microbial growth on the chemical, in accordance with the rapid disappearance of 4-n-nonylphenol from the soil through microbial degradation (Fig. 7, right). Residual amounts approached zero both when measured in percentage and in actual amounts. Nonylphenol originates from anthropogenic activity and accumulates and persists in environmental compartments characterized by high organic content, such as sewage sludge [3]. We observed rapid degradation in agricultural soil of comparatively low organic carbon content (quite on the average for large parts of the Norwegian agricultural area). This rapid degradation in an aerobic soil environment is in accordance with previous studies [35,38]. There are, however,

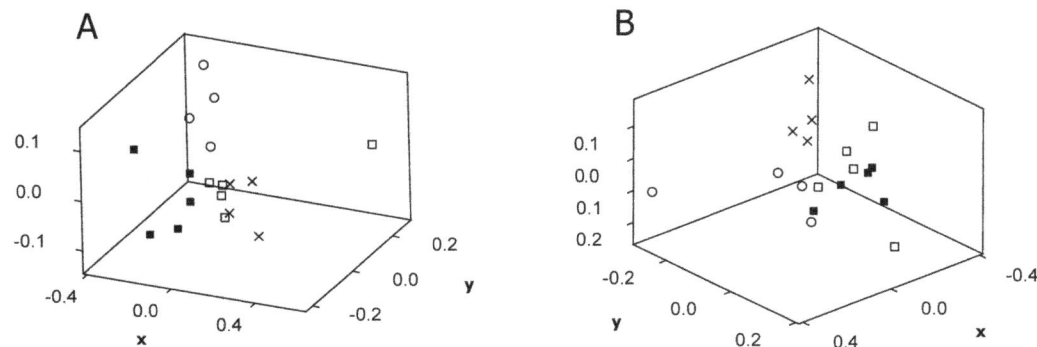

Figure 2. Effects of picoxystrobin on soil fungal community structure. Effects of picoxystrobin (PI) on soil fungal community structure, as shown from nmMDS of data from T-RFLP analyses after 7 (a) and 70 (b) days of incubation. Legends: Treatment PI low (□; 2 mg kg^{-1} dry soil) and PI high (■; 10 mg kg^{-1} dry soil), solvent control (×; no chemical added) and original soil sample (○).

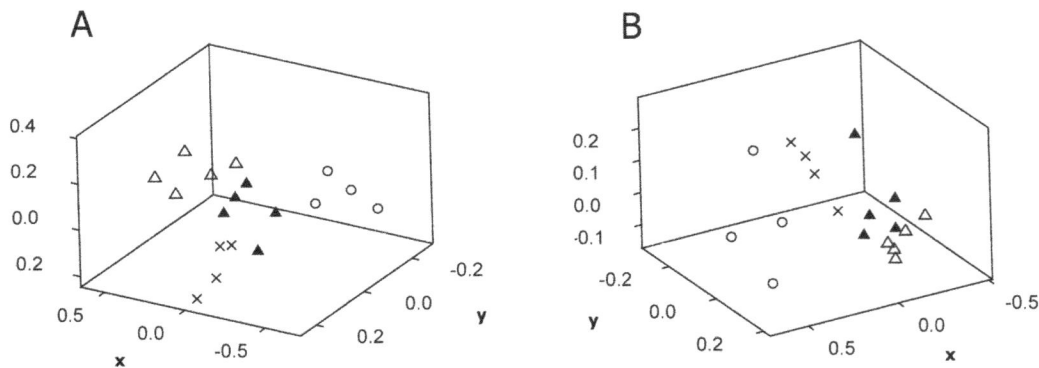

Figure 3. Effects of 4-n-nonylphenol on soil bacterial community structure. Effects of 4-n-nonylphenol (NP) on soil bacterial community structure, as shown from nmMDS of data from T-RFLP analyses after 28 (a) and 70 (b) days of incubation. Legends: Treatment NP low (Δ; 0.5 mg kg^{-1} dry soil) and NP high ($\blacktriangle$; 10 mg kg^{-1} dry soil), solvent control ($\times$; no chemical added) and original soil sample ($\bigcirc$).

reports of field studies with fast initial degradation, while some residues might persist [4]. The occurrence of metabolites of 4-n-nonylphenol was not measured in this study, but the observed stimulation of respiration activity by 4-n-nonylphenol addition as compared to the solvent control, was not large enough to account for the observed disappearance of the chemical through the incubation period. This indicated that the chemical was not completely mineralized. Results from a mineralization study with a range of soils show the conversion of around 40% of applied nonylphenol to CO_2 during a 40 day incubation period [39].

Our results indicated shorter duration of the effects of the chemicals in the soil bacterial as compared to the soil fungal community. This was expected due to shorter generation times enabling a faster re-establishment of the community structure. Overall, our results were in accordance with this, showing smaller differences in T-RFLP fingerprints between start and end of the experiment for the former (Fig. 5). The more statistically significant effects of the treatments were however shown for bacteria, while the changes in the soil fungal community were less consistent and varied a lot between replicate samples.

Implications of a Mixture Situation on the Effects of Picoxystrobin on the Soil Microbial Community

Analyses of the effects of mixture treatments on the soil microbial community structure indicated that the two chemicals might have antagonistic effects. Established concepts to estimate biological effects of chemical mixtures rely on data available for single chemicals [40,41]. Single chemicals vary in how they excerpt their effect and the effects of a mixture cannot be found directly from the different constituents' independent effects. One can observe a wide variety of effects due to synergism, antagonism or other forms of interactions, and the common concepts of concentration addition and independent action often come short. Despite the expected and indicated differences in modes of action of the two chemicals, we observed less evident changes in T-RFLP patterns in soils treated with the chemicals in mixture, as compared to the single chemical treatments. Our results showed that statistically significant different T-RFLP patterns, for both the soil bacterial and fungal communities, resulted from treatments with picoxystrobin, 4-n-nonylphenol, and a mixture of these. However, the changes in T-RFLP patterns evolved more slowly and were less pronounced for the mixture than for the single chemical treatments, as compared to the solvent control. Further, the negative effects of low levels of picoxystrobin on respiration were apparently remediated when mixing with high concentration

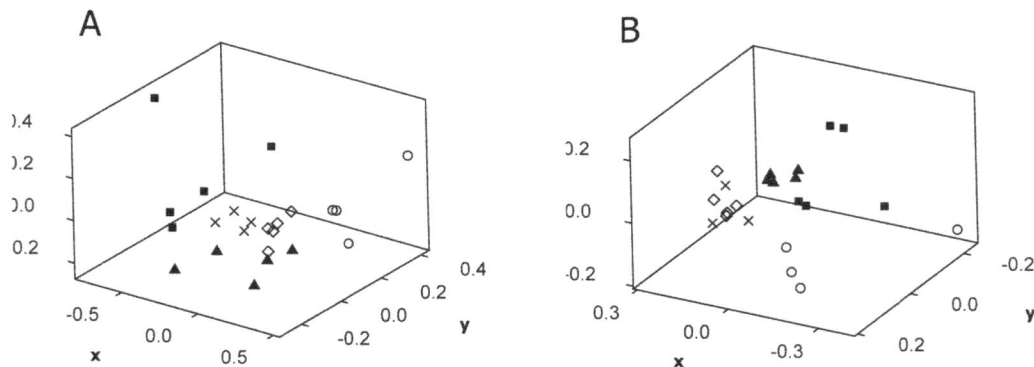

Figure 4. Effects of picoxystrobin and 4-n-nonylphenol on soil microbial community structure. Effects of picoxystrobin (PI) and 4-n-nonylphenol (NP) as single chemicals and in mixture on soil bacterial (a) and fungal (b) community structure, as shown from nmMDS of data from T-RFLP analyses after 28 days of incubation. Legends: Treatment PI high ($\blacksquare$; 10 mg kg^{-1} dry soil), NP high ($\blacktriangle$; 10 mg kg^{-1} dry soil), Mix high ($\diamondsuit$; 10 mg picoxystrobin and 10 mg 4-n-nonylphenol kg^{-1} dry soil), solvent control ($\times$; no chemical added) and original soil sample ($\bigcirc$).

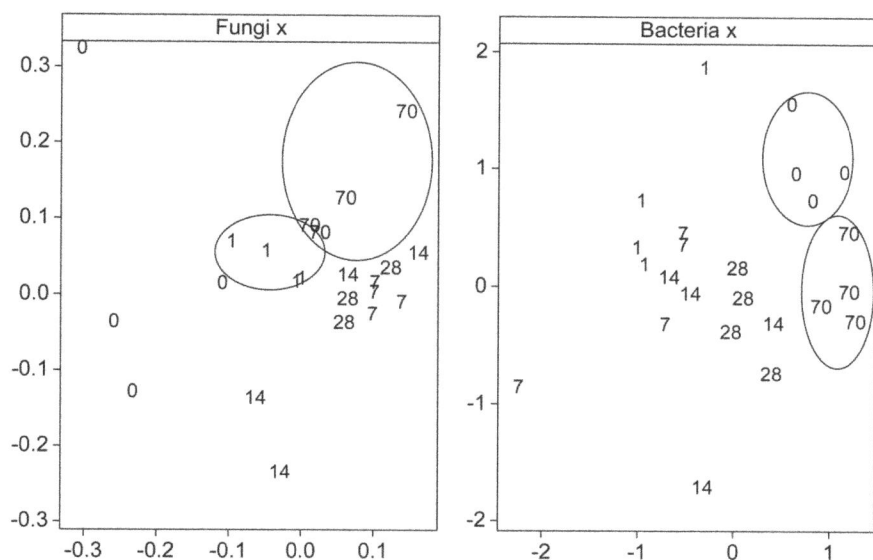

Figure 5. Duration of effects of chemicals on soil microbial community structure. Representative pattern (exemplified by the solvent control treatment) for observed clustering of T-RFLP fingerprints of soil microbial communities at day 0, 1, 7, 14, 28, and 70 after addition of chemicals, shown for fungi (left panel) and bacteria (right panel). Small distance between start (0) and end (70) of the incubation period indicates only minor differences in T-RFLP fingerprints, and imply transient effects of the chemicals.

of 4-n-nonylphenol (Mix NP high), while high levels of picoxystrobin nulled the positive effects of high levels of 4-n-nonylphenol (Mix high). In a practical mixture situation in the field, the expected environmental concentrations of nonylphenol can from this not be expected to increase the potentially adverse effects of picoxystrobin.

To assure the validity in extrapolation of the observed effects to a risk assessment situation in the field, the results need to be verified in field/semi-field/mesocosm trials. These lab studies do not take into account the additional stress the soil microbial community experience in nature from climatic conditions (drought, frost), predators, competition, or food shortage, nor the effects of alteration in the soil microbial community structure on other parts of the soil fauna. The studied mixture is relevant in an agricultural context and the results should be possible to extrapolate to a certain degree to other nonylphenols and strobilurin fungicides. However, care should be taken to consider the variety of physico-chemical and toxicological properties of the different strobilurins [6] as well as the differences in estrogenicity, sorption properties and degradation rates of different isomers of

linear and branched nonylphenols [42]. Further, due to the general importance of soil properties for the sorption of chemicals, extrapolation of results from this sandy loam soil with low organic carbon content should be done with care.

Methodological Considerations

Our results showed that both soil microbial respiration activity and community structure was affected by the chemicals. They also illustrate the need for the use of several measures to be able to assess effects of chemicals in soil with a minimum degree of certainty. We found statistically significant negative effects of picoxystrobin on activity levels as shown by CO_2 development, but no statistically significant differences on cumulative CO_2-values (Table 4) due to too large variability between replicate samples. This was despite observable trends in the data (Fig. 1). Further, the T-RFLP results indicated a need for frequent sampling shortly after addition of chemicals, and possibly the resolution of the sampling was too low to capture the effects on the soil fungal community structure as the largest spread in these data as shown by nmMDS was found after 7 days (Fig. 2). The community fingerprints arising from T-RFLP analyses depend on the PCR-primers and restriction enzymes utilized, and cannot easily be interpreted in terms of ecological relevance of the observed effects. This was ameliorated through correspondence with effects on soil respiration activity. Our studies showed that the T-RFLP technique could be used as a valuable tool in elucidating how rapid and to what extent effects (positive or negative) of the chemicals in the soil microbial community can be expected.

Our results for the fungicide picoxystrobin show that the mere measurement of changes in soil chemical concentrations for this moderately sorbing chemical could not be used as an indication of when to expect effects, while exposure assessment in ecological risk assessment is often restricted to external exposure like concentrations in water, soil or sediment. However it has been recognized that in natural ecosystems, in particular in soils and sediments, the amount of chemicals truly available for uptake into organisms is frequently only a fraction of the total amount present, due to a

Table 5. Estimated degradation half-lives (DT50) for picoxystrobin and 4-n-nonylphenol in a sandy loam soil.

Start concentrations	DT50 initial	r^2	DT50 2nd phase	r^2
2 mg picoxystrobin kg^{-1} dry soil[a]	20.0±2.5	0.87	300±170	0.74
10 mg picoxystrobin kg^{-1} dry soil[a]	26.0±4.0	0.74		
0.5 mg 4-n-nonylphenol kg^{-1} dry soil[b]	3.0±1.0	0.85	42±2	0.58
10 mg 4-n-nonylphenol kg^{-1} dry soil[a]	6.5±0.5	0.92	52±21	0.89

[a]Initial phase 0–28 days, 2nd phase 28–70 days,
[b]Initial phase 0–14 days, 2nd phase 14–70 days.
Calculations are based on results from both single chemical and mixture experiments. (The test statistic of $p<0.05$ is valid for all estimated DT50-values.).

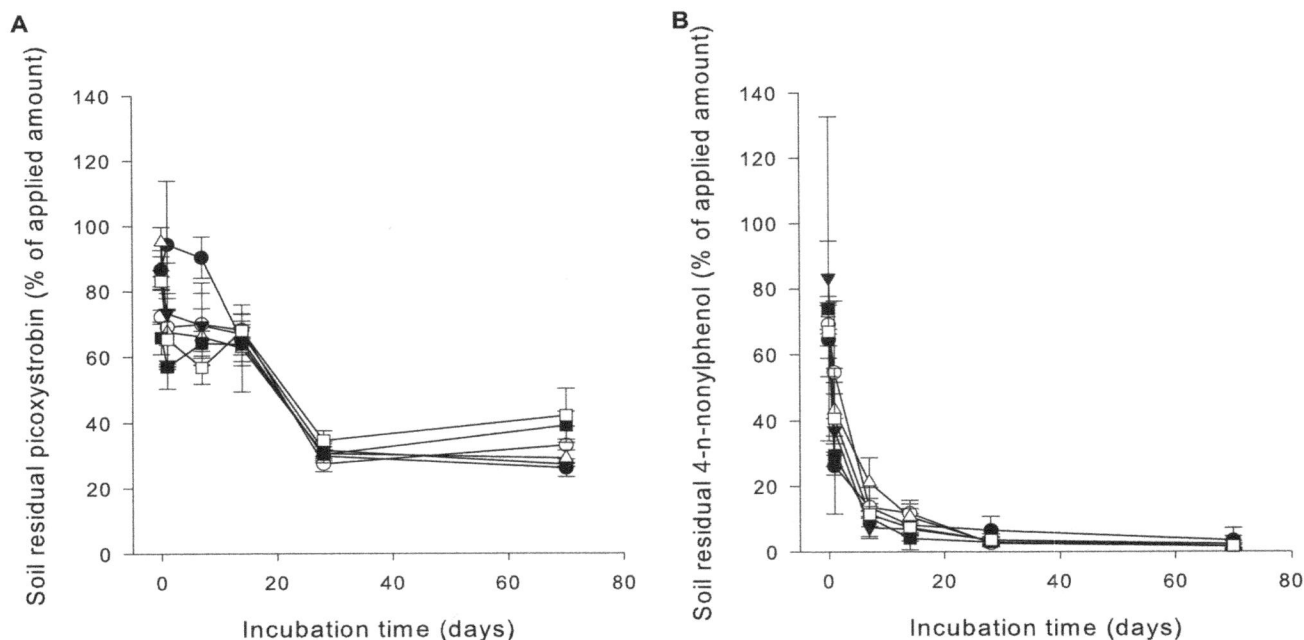

Figure 6. Degradation of picoxystrobin and 4-n-nonylphenol in soil. Degradation curve for picoxystrobin (a) and 4-n-nonylphenol (b) in soil, showing similar pattern and similar percentage degradation during the incubation period for all treatments with picoxystrobin and 4-n-nonylpyhenol, respectively, regardless of initial concentration or mixture situation. Mean values ± SD. Legends: Treatment PI low (a) and NP low (b) (●; PI = 2 mg and NP = 0.5 mg kg^{-1} dry soil), PI high (a) and NP high (b) (○; PI and NP = 10 mg kg^{-1} dry soil), Mix low (▼; PI = 2 mg and NP = 0.5 mg kg^{-1} dry soil), Mix NP high (△;PI = 2 mg and NP = 10 mg kg^{-1} dry soil), Mix PI high (■;PI = 10 mg and NP = 0.5 mg kg^{-1} dry soil), and Mix high (□; PI and NP = 10 mg kg^{-1} dry soil).

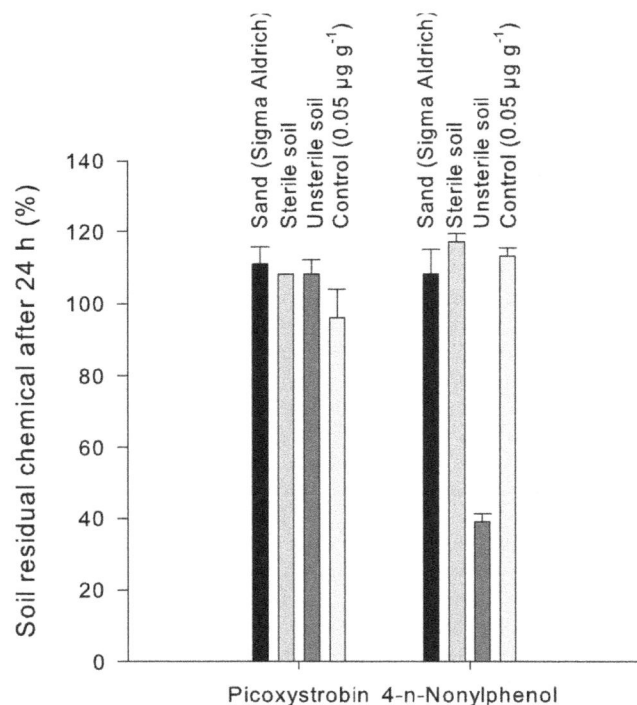

Figure 7. Microbial degradation of picoxystrobin and 4-n-nonylphenol in soil. Laboratory test results for degradation of picoxystrobin and 4-n-nonylphenol in selected media during 24 h, showing rapid degradation of 4-n-nonylphenol in unsterile soil.

complex environmental behaviour [43]. Measurements of residual chemical concentrations in our test system was, however, shown to be a valuable tool for explaining the observed changes in soil microbial community structure and respiration activity and elucidating the concept of a chemicals bioavailability, and should be included in effects assessments.

Conclusions

In summary, we have studied the degradation and effects of picoxystrobin and 4-n-nonylphenol on the soil microbial community structure and respiration activity in a sandy loam soil. The fungicide picoxystrobin was shown to decrease the soil microbial activity, while 4-n-nonylphenol caused an increase. These effects were accompanied by statistically significant changes in the T-RFLP fingerprints for the soil microbial community that were still detectable, but small, after 70 days. A mixture situation relevant for assessing the environmental risk of fungicide application in years with sewage sludge amendment of agricultural soil was tested. These results implied that a mixture of picoxystrobin and 4-n-nonylphenol will not have more adverse effects on soil respiration or the soil microbial community structure than the single chemicals. They indicated that the chemicals affected different parts of the soil microbial community, and resulted in a low net effect of the chemicals in mixture. The presented results do, however, not allow us to propose an explanation for the mechanisms causing this response.

In conclusion, our results imply that the application of picoxystrobin and nonylphenol to agricultural soil in environmentally relevant and worst-case concentrations will not cause irreversible effects on soil microbial respiration and community structure. Further, a mixture situation with the fungicide picoxystrobin and nonylphenol from sewage sludge application, will not increase any potentially adverse effects of picoxystrobin on the soil microbial community. In a wider perspective, these results illustrate that there is a need and possibility to refine today's risk assessment procedures to encompass the study of both soil bacterial and fungal communities, both broad-scale measures and genomics, and interpret the results in relation to residual soil chemical concentrations.

Supporting Information

Figure S1 Effects of picoxystrobin on soil bacterial community structure. Effects of picoxystrobin (PI) on soil bacterial community structure, as shown from nmMDS of data from T-RFLP analyses. Legends: Treatment PI low ($\square$; 2 mg kg^{-1} dry soil) and PI high ($\blacksquare$; 10 mg kg^{-1} dry soil), solvent control (X; no chemical added) and original sample ($\bigcirc$).

Figure S2 Effects of picoxystrobin on soil fungal community structure. Effects of picoxystrobin (PI) on soil fungal community structure, as shown from nmMDS of data from T-RFLP analyses. Legends: Treatment PI low ($\square$; 2 mg kg^{-1} dry soil) and PI high ($\blacksquare$; 10 mg kg^{-1} dry soil), solvent control (X; no chemical added) and original sample ($\bigcirc$).

Figure S3 Effects of 4-n-nonylphenol on soil bacterial community structure. Effects of 4-n-nonylphenol (NP) on soil bacterial community structure, as shown from nmMDS of data from T-RFLP analyses. Legends: Treatment NP low (Δ; 0.5 mg kg^{-1} dry soil) and NP high ($\blacktriangle$; 10 mg kg^{-1} dry soil), solvent control (X; no chemical added) and original sample ($\bigcirc$).

Figure S4 Effects of 4-n-nonylphenol on soil fungal community structure. Effects of 4-n-nonylphenol (NP) on soil fungal community structure, as shown from nmMDS of data from T-RFLP analyses. Legends: Treatment NP low (Δ; 0.5 mg kg^{-1} dry soil) and NP high ($\blacktriangle$; 10 mg kg^{-1} dry soil), solvent control (X; no chemical added) and original sample ($\bigcirc$).

References

1. JRC (2001). Organic contaminants in sewage sludge for agricultural use. Study coordinated by European Commission and Joint Research Centre and the Institute for Environment and Sustainable Soil and Waste Unit. Report prepared by UMEG Center for Environmental Measurements, Environmental Inventories and Product Safety, 18 October 2001. 73 p. Available: http://ec.europa.eu/environment/waste/sludge/pdf/organics_in_sludge.pdf. Accessed 21 May 2013.
2. Norwegian Scientific Committee for Food Safety (2009). Risk assessment of contaminants in sewage sludge applied on Norwegian soils – Opinion from the Panel on Contaminants in the Norwegian Scientific Committee for Food Safety. Oslo: VKM. 208 p. Available: http://www.vkm.no/dav/2ae7f1b4e3.pdf. Accessed 21 May 2013.
3. Hansen AB, Lassen P (2008). Screening of phenolic substances in the Nordic environments. TemaNord 2008:530. Copenhagen: Nordic Council of Ministers. 145 p.
4. Soares A, Guieysse B, Jefferson B, Cartmell E, Lester JN (2008). Nonylphenol in the environment: A critical review on occurrence, fate, toxicity and treatment in wastewaters. Environ Int 34, 1033–1049.
5. EC (2000). Directive 2000/60/EC of the European Parliament and of the Council of 23 October 2000 establishing a framework for Community action in the field of water policy.
6. Bartlett DW, Clough JM, Godwin JR, Hall AA, Hamer M, Parr-Dobrzanski B (2002). The strobilurin fungicides. Pest Manag Sci 58: 649–62.
7. BCPC (2011). ePesticide Manual V5.2 2011–2012.
8. EC (2003). Review report for the active substance picoxystrobin. SANCO/10196/2003-Final. Available: http://ec.europa.eu/food/plant/protection/evaluation/newactive/list1_picoxystrobin_en.pdf. Accessed 21 May 2013.
9. Norwegian Food Safety Autority (2005). Vurdering av plantevernmidlet Acanto Prima – pikoksystrobin+cyprodinil – vedrørende søknad om godkjenning. (In

Figure S5 Effects of picoxystrobin and 4-n-nonylpyhenol on soil bacterial community structure. Effects of picoxystrobin (PI) and 4-n-nonylphenol (NP) on soil bacterial community structure, as shown from nmMDS of data from T-RFLP analyses. Legends: Treatment PI high ($\blacksquare$; 10 mg kg^{-1} dry soil), NP high ($\blacktriangle$; 10 mg kg^{-1} dry soil), Mix high ($\diamond$; 10 mg picoxystrobin and 4-n-nonylphenol kg^{-1} dry soil), solvent control (X; no chemical added) and original sample ($\bigcirc$).

Figure S6 Effects of picoxystrobin and 4-n-nonylpyhenol on soil fungal community structure. Effects of picoxystrobin (PI) and 4-n-nonylphenol (NP) on soil fungal community structure, as shown from nmMDS of data from T-RFLP analyses. Legends: Treatment PI high ($\blacksquare$; 10 mg kg^{-1} dry soil), NP high ($\blacktriangle$; 10 mg kg^{-1} dry soil), Mix high ($\diamond$; 10 mg picoxystrobin and 4-n-nonylphenol kg^{-1} dry soil), solvent control (X; no chemical added) and original sample ($\bigcirc$).

Table S1 Summary of results from nmMDS (STRESS; k = 3 dimensions, Bray-Curtis distance measure) and ANOSIM (overall R) of T-RFLP fingerprints for the soil microbial community after different chemical treatments.

Acknowledgments

Monica Skogen, Hege Særvold Steen, Grete Lund and Elameen Abdelhameed, Norwegian Institute for Agricultural and Environmental Research, are acknowledged for technical assistance in the lab. Many thanks to Tim Daniell, James Hutton Institute, for advice on the experimental set-up, and analyses of T-RFLP-data. We thank the anonymous reviewers for a thorough review and helpful comments.

Author Contributions

Conceived and designed the experiments: MS SSK OME. Performed the experiments: MS HRN. Analyzed the data: MS HRN. Contributed reagents/materials/analysis tools: MS SSK HRN OME. Wrote the paper: MS SSK HRN OME.

Norwegian). Available: http://www.mattilsynet.no/mattilsynet/multimedia/archive/00014/Plantevernmidler__Ra_14615a.pdf. Accessed 21 May 2013.
10. Coleman DC (2008). From peds to paradoxes: Linkages between soil biota and their influences on ecological processes. Soil Biol Biochem 40: 271–289.
11. Wu L, Wang H, Zhang Z, Lin R, Zhang Z, et al. (2011). Comparative metaproteomic analysis on consequtively *Rehmannia glutinosa*-monocultured rhizosphere soil. PloS ONE 6(5): e20611. Doi:10.1371/journal.pone.0020611.
12. Winding A, Hund-Rinke K, Rutgers M (2005). The use of microorganisms in ecological soil classification and assessment concepts. Ecotox Environ Safe 62: 230–248.
13. EFSA (2010). EFSA Panel on Plant Protection Products and their Residues (PPR); Scientific Opinion on the development of specific protection goal options for environmental risk assessment of pesticides, in particular in relation to the revision of the Guidance Documents on Aquatic and Terrestrial Ecotoxicology (SANCO/3268/2001 and SANCO/10329/2002). EFSA Journal 8: 1821. 55 p. Available: http://www.efsa.europa.eu/en/efsajournal/doc/1821.pdf. Accessed 21 May 2013.
14. Imfeld G, Vuilleumier S (2012). Measuring the effects of pesticides on bacterial communities in soil: A critical review. Eur J Soil Biol 49: 22–30.
15. Øvreås L, Torsvik V (1998). Microbial diversity and community structure in two different agricultural soil communities. Microbial Ecol 36: 303–315.
16. Bending GD, Rodríges-Cruz MS, Lincoln SD (2007). Fungicide impacts on microbial communities in soils with contrasting management histories. Chemosphere 69: 82–88.
17. Griffiths RI, Thomson BC, James P, Bell T, Bailey M, et al. (2011). The bacterial biogeography of British soils. Environ Microbiol 13: 1642–1654.
18. OECD (2000). Test No. 217: Soil Microorganisms: Carbon Transformation Test, OECD Guidelines for the Testing of Chemicals, Section 2, OECD Publishing. doi: 10.1787/9789264070240-en.

19. OECD (2000). Test No. 216: Soil Microorganisms: Nitrogen Transformation Test, OECD Guidelines for the Testing of Chemicals, Section 2, OECD Publishing. doi: 10.1787/9789264070226-en.

20. MacDonald CA, Clark IM, Zhao F-J, Hirsch PR, Singh BK, et al. (2011). Long term impacts of zinc and copper enriched sewage sludge additions on bacterial, archeal and fungal communities in arable and grassland soils. Soil Biol Biochem 43: 932–941.

21. Schütte UME, Abdo Z, Bent SJ, Shyu C, Williams CJ, et al. (2008). Advances in the use of terminal restriction fragment length polymorphism (T-RFLP) analysis of 16S rRNA genes to characterize microbial communities. Appl Microbiol Biot 80: 365–380.

22. Aiken JT (2011). Terminal restriction fragment length polymorphism for soil microbial community profiling. Soil Sci Soc Am J 75: 102–111.

23. Chaussod R, Nicolardot B, Catroux G (1986). Mesure en routine de la biomasse microbienne des sols par la méthode de fumigation au chloroforme. Science du Sol - Bulletin AFES 2: 201–211.

24. Singh BK, Nunan N, Ridgeway KP, McNicol J, Peter W, et al. (2008). Relationship between assemblages of mycorrhizal fungi and bacteria on grass roots. Environ Microbiol 10: 534–541.

25. Anderson IC, Cairney JWG (2004). Diversity and ecology of soil fungal communities: increased understanding through the application of molecular techniques. Environ Microbiol 6: 769–779.

26. Gardes M, Bruns TD (1993). ITS primers with enhanced specificity for basidiomycetes: application to the identification of mycorrhizae and rusts. Mol Ecol 2: 113–118.

27. White TJ, Bruns T, Lee S, Taylor J (1990). Amplification and direct sequencing of fungal ribosomal RNA genes for phylogenetics. In: Innis MA, Gelfand DH, Sninsky JJ, White TJ, editors. PCR Protocols: a guide to methods and applications. New York: Academic Press. 315–322.

28. R Core Team (2012). R: A language and environment for statistical computing. R Foundation for Statistical Computing, Vienna, Austria. ISBN 3–900051–07–0, URL http://www.R-project.org/. Accessed 28 February 2013.

29. Oksanen J, Blanchet FG, Kindt R, Legendre P, Minchin PR, et al. (2013). vegan: Community Ecology Package. R package version 2.0–6. Available: http://CRAN.R-project.org/package = vegan. Accessed 28 February 2013.

30. Venables WN, Ripley BD (2002). Modern Applied Statistics with S. Fourth Edition. New York: Springer. ISBN 0-387-95457-0.

31. Kruskal JB (1964a), Multidimensional Scaling by Optimizing Goodness of Fit to a Nonmetric Hypothesis. Psychometrika 29: 1–27.

32. Kruskal JB (1964b). Nonmetric multidimensional scaling: a numerical method. Psychometrika 29: 115–129.

33. Hammer Ø, Harper DAT, Ryan PD (2001). PAST: Paleontological statistics software package for education and data analysis. Palaeontol Electron 4(1): art. 4. 9 p. Available: http://palaeo-electronica.org/2001_1/past/issue1_01.htm. Accessed 28 Februart 2013.

34. Kollmann A, Brault A, Touton I, Dubroca J, Chaplain V, et al. (2003). Effect of nonylphenol surfactants on fungi following the application of sewage sludge on agricultural soils. J Environ Qual 32: 1269–1276.

35. Corvini PFX, Chäffer A, Schlosser D (2006). Microbial degradation of nonylphenol and other alkylphenols – our evolving view. Appl Microbiol Biot 72: 223–243.

36. PPDB (2009). The Pesticide Properties Database (PPDB) developed by the Agriculture & Environment Research Unit (AERU), University of Hertfordshire, funded by UK national sources and the EU-funded FOOTPRINT project (FP6-SSP-022704). http://sitem.herts.ac.uk/aeru/footprint/en/index.htm. Accessed 21 May 2013.

37. Defra (2010). Development of guidance on the implementation of aged soil sorption studies into regulatory exposure assessments. Research report for DEFRA project PS 2235. The Food and Environment Research Agency and Alterra.

38. Hesselsøe M, Jensen D, Skals K, Olesen T, Moldrup P, et al. (2001). Degradation of 4-Nonylphenol in homogeneous and nonhomogeneous mixtures of soil and sewage sludge. Environ Sci Technol 35: 3695–3700.

39. Topp E, Starratt A (2000). Rapid mineralization of the endocrine-disrupting chemical 4-nonylphenol in soil. Environ Toxicol Chem 19: 313–318.

40. Loewe S (1927). Die Mischarznei. Versuch einer allgemeinen Pharmakologie der Arzneikombinationen. Klin Wochenschr 6: 1077–1085.

41. Bliss CI (1939). The toxicity of poisons applied jointly. Ann Appl Biol 26: 585–615.

42. Shan J, Jiang B, Yu B, Li C, Sun Y, et al. (2011). Isomer-Specific Degradation of Branched and Linear 4-Nonylphenol Isomers in an Oxic Soil. Environ Sci Technol 45: 8283–8289.

43. Alexander M (1995). How toxic are toxic chemicals in soil? Environ Sci Technol 29: 2713–2717.

Permissions

The contributors of this book come from diverse backgrounds, making this book a truly international effort. This book will bring forth new frontiers with its revolutionizing research information and detailed analysis of the nascent developments around the world.

We would like to thank all the contributing authors for lending their expertise to make the book truly unique. They have played a crucial role in the development of this book. Without their invaluable contributions this book wouldn't have been possible. They have made vital efforts to compile up to date information on the varied aspects of this subject to make this book a valuable addition to the collection of many professionals and students.

This book was conceptualized with the vision of imparting up-to-date information and advanced data in this field. To ensure the same, a matchless editorial board was set up. Every individual on the board went through rigorous rounds of assessment to prove their worth. After which they invested a large part of their time researching and compiling the most relevant data for our readers.

The editorial board has been involved in producing this book since its inception. They have spent rigorous hours researching and exploring the diverse topics which have resulted in the successful publishing of this book. They have passed on their knowledge of decades through this book. To expedite this challenging task, the publisher supported the team at every step. A small team of assistant editors was also appointed to further simplify the editing procedure and attain best results for the readers.

Apart from the editorial board, the designing team has also invested a significant amount of their time in understanding the subject and creating the most relevant covers. They scrutinized every image to scout for the most suitable representation of the subject and create an appropriate cover for the book.

The publishing team has been an ardent support to the editorial, designing and production team. Their endless efforts to recruit the best for this project, has resulted in the accomplishment of this book. They are a veteran in the field of academics and their pool of knowledge is as vast as their experience in printing. Their expertise and guidance has proved useful at every step. Their uncompromising quality standards have made this book an exceptional effort. Their encouragement from time to time has been an inspiration for everyone.

The publisher and the editorial board hope that this book will prove to be a valuable piece of knowledge for researchers, students, practitioners and scholars across the globe.

List of Contributors

Md. Hafizur Rahman, Latifur Rahman Shovan, Linda Gordon Hjeljord, Berit Bjugan Aam, Vincent G. H. Eijsink, Morten Sørlie and Arne Tronsmo
Department of Chemistry, Biotechnology and Food Science, Norwegian University of Life Sciences (NMBU), Ås, Norway

Sabine Fillinger and Pierre Leroux
INRA UR1290, BIOGER CPP, Thiverval-Grignon, France

Sakhr Ajouz, Philippe C. Nicot and Marc Bardin
INRA, UR407, Plant Pathology Unit, Montfavet, France

Olivier Lambert and Mélanie Piroux
LUNAM Université, Oniris, Ecole Nationale Vétérinaire, Agroalimentaire et de l'Alimentation Nantes-Atlantique, Plateforme Environnementale Vétérinaire, Centre Vété rinaire de la Faune Sauvage et des Ecosystémes des Pays de la Loire (CVFSE), Nantes, France
Clermont Université, Université Blaise Pascal, Laboratoire Microorganismes: Génome et Environnement, BP 10448, Clermont-Ferrand, France
CNRS, UMR 6023, LMGE, Aubiére, France

Frédéric Delbac
Clermont Université, Université Blaise Pascal, Laboratoire Microorganismes: Génome et Environnement, BP 10448, Clermont-Ferrand, France
CNRS, UMR 6023, LMGE, Aubiére, France

Sophie Puyo, Monique L'Hostis and Hervé Pouliquen
LUNAM Université, Oniris, Ecole Nationale Vétérinaire, Agroalimentaire et de l'Alimentation Nantes-Atlantique, Plateforme Environnementale Vétérinaire, Centre Vété rinaire de la Faune Sauvage et des Ecosystémes des Pays de la Loire (CVFSE), Nantes, France

Chantal Thorin
LUNAM Université, Oniris, Ecole Nationale Vétérinaire, Agroalimentaire et de l'Alimentation Nantes-Atlantique, Unité de Physiopathologie Animale et Pharmacologie Fonctionnelle, Nantes, France

Laure Wiest and Audrey Buleté
Université de Lyon, Institut des Sciences Analytiques, Département Service Central d'Analyse, UMR 5280 CNRS, Université de Lyon1, ENS-Lyon, Villeurbanne, France

Lei Chen, Xiaohong Lu, Zhili Pang, Meng Cai and Xili Liu
Department of Plant Pathology, College of Agriculture and Biotechnology, China Agricultural University, Beijing, China

Shusheng Zhu
Key Laboratory of Agro-Biodiversity and Pest Management of Education Ministry of China, Yunnan Agricultural University, Kunming, Yunnan, China

Noa Simon-Delso
Beekeeping Research and Information Centre, Louvain la Neuve, Belgium
Environmental Sciences, Copernicus Institute, Utrecht University, Utrecht, The Netherland

Gilles San Martin and Louis Hautier
Plant Protection and Ecotoxicology Unit, Life Sciences Department, Walloon Agricultural Research Centre, Gembloux, Belgium

Etienne Bruneau, Laure-Anne Minsart and Coralie Mouret
Beekeeping Research and Information Centre, Louvain la Neuve, Belgium

Jeffery S. Pettis
Bee Research Laboratory, USDA-ARS, Beltsville, Maryland, United States of America

Dennis van Engelsdorp, Jennie Stitzinger and Elinor M. Lichtenberg
Department of Entomology, University of Maryland, College Park, College Park, Maryland, United States of America

Michael Andree
Cooperative Extension Butte County, University of California, Oroville, California, United States of America

Robyn Rose
USDA-APHIS, Riverdale, Maryland, United States of America

Rasmus Bojsen
Department of Systems Biology, Technical University of Denmark, Kgs. Lyngby, Denmark

Rasmus Torbensen, Camilla Eggert Larsen and Birgitte Regenberg
Department of Biology, University of Copenhagen, Copenhagen, Denmark

Anders Folkesson
Department of Systems Biology, Technical University of Denmark, Kgs. Lyngby, Denmark
Section for Bacteriology, Pathology and Parasitology, National Veterinary Institute, Frederiksberg C, Denmark

Rumsaïs Blatrix, Sarah Debaud, Alex Salas-Lopez, Céline Born and Laure Benoit
Centre d'Ecologie Fonctionnelle et Evolutive (CEFE), CNRS/CIRAD-Bios/Université Montpellier 2, Montpellier, France

Doyle B. McKey
Centre d'Ecologie Fonctionnelle et Evolutive (CEFE), CNRS/CIRAD-Bios/Université Montpellier 2, Montpellier, France
Institut Universitaire de France, Montpellier, France

Christiane Attéké
Département de Biologie, Université des Sciences et Techniques de Masuku (USTM), Franceville, Gabon

Champlain Djié to-Lordon
Laboratory of Zoology, University of YaoundéI, Yaoundé, Cameroun
Elisa González-Domínguez and Josep Armengol
Instituto Agroforestal Mediterráneo, Universidad Politécnica de Valencia, Valencia, Spain

Vittorio Rossi
Istituto di Entomologia e Patologia vegetale, Universitá Cattolica del Sacro Cuore, Piacenza, Italy

Wei Sun, Noel Southall, Paul Shinn, John C. McKew and Wei Zheng
National Center for Advancing Translational Sciences, National Institutes of Health, Bethesda, Maryland, United States of America

Yoon-Dong Park, Janyce A. Sugui and Kyung J. Kwon-Chung
Laboratory of Clinical Infectious Diseases, National Institute of Allergy and Infectious Diseases, National Institutes of Health, Bethesda, Maryland, United States of America

Annette Fothergill
University of Texas Health Science Center, San Antonio, Texas, United States of America

Peter R. Williamson
Laboratory of Clinical Infectious Diseases, National Institute of Allergy and Infectious Diseases, National Institutes of Health, Bethesda, Maryland, United States of America
Section of Infectious Diseases, Immunology and International Medicine, University of Illinois College of
Medicine, Chicago, Illinois, United States of America

Swetha Tati, Woong Sik Jang, Rui Li, Rohitashw Kumar, Sumant Puri and Mira Edgerton
Department of Oral Biology, University at Buffalo, Buffalo, New York, United States of America

Michael C. R. Alavanja, Jonathan N. Hofmann, Stella Koutros, Kathryn H. Barry, Gabriella Andreotti, Jay H. Lubin, Aaron Blair and Laura E. Beane Freeman
Division of Cancer Epidemiology and Genetics, National Cancer Institute, Rockville, Maryland, United States of America

Charles F. Lynch
College of Public Health, University of Iowa, Iowa City, Iowa, United States of America

Cynthia J. Hines
National Institute for Occupational Safety and Health, Cincinnati, Ohio, United States of America

Joseph Barker and Dennis W. Buckman
IMS, Inc, Calverton, Maryland, United States of America

Kent Thomas
National Exposure Research Laboratory, U.S. Environmental Protection Agency, Research Triangle Park, North Carolina, United States of America

Dale P. Sandler and Jane A. Hoppin
Epidemiology Branch, National Institute for Environmental Health Sciences, Research Triangle Park, North Carolina, United States of America

Ida Karlsson and Paula Persson
Dept. of Crop Production Ecology, Swedish University of Agricultural Sciences (SLU), Uppsala, Sweden

Hanna Friberg
Dept. of Forest Mycology and Plant Pathology, SLU, Uppsala Sweden

Christian Steinberg
INRA, UMR 1347 Agroécologie, Pole IPM, Dijon, France

Xiao-yong Zhang, Gui-ling Tang, Xin-ya Xu, Xu-hua Nong and Shu-Hua Qi
Key Laboratory of Tropical Marine Bio-resources and Ecology/RNAM Center for Marine Microbiology/ Guangdong Key Laboratory of Marine Material Medical, South China sea Institute of Oceanology, Chinese academy of sciences, Guangzhou, China

Maximilian Nepel, Jürg Schönenberger and Veronika E. Mayer
Division of Structural and Functional Botany, Department of Botany and Biodiversity Research, University of Vienna, Vienna, Austria

Hermann Voglmayr
Division of Systematic and Evolutionary Botany, Department of Botany and Biodiversity Research, University of Vienna, Vienna, Austria
Institute of Forest Entomology, Forest Pathology and Forest Protection, Department of Forest and Soil Sciences, BOKU-University of Natural Resources and Life Sciences, Vienna, Austria

Angela M. Orshinsky, Jinnan Hu, Thomas K. Mitchell and Michael J. Boehm
Department of Plant Pathology, The Ohio State University, Columbus, Ohio, United States of America

Stephen O. O piyo
Molecular and Cellular Imaging Center-Columbus. The Ohio State University, Columbus, Ohio, United States of America

Venu Reddyvari-Channarayappa
Dale Bumpers National Rice Research Center, Stuttgart, Arizona, United States of America

Lina Yan and Jiasui Zhan
Key Lab for Biopesticide and Chemical Biology, Ministry of Education, Fujian Agriculture and Forestry University, Fuzhou, Fujian, People's Republic of China

Fangluan Gao and Liping Shang
Laboratory of Plant Virology of Fujian Province, Institute of Plant Virology, Fujian Agriculture and Forestry University, Fuzhou, Fujian, People's Republic of China

Bruce A. McDonald
Institute of Integrative Biology, ETH Zurich, Zürich, Switzerland

Marianne Stenrød, Sonja S. Klemsdal, Hans Ragnar Norli and Ole Martin Eklo
Norwegian Institute for Agricultural and Environmental Research (Bioforsk), Ås, Norway

Index

* 9 7 8 1 6 3 2 3 9 9 0 5 2 *